Artificial Intelligence

Lincoln Laboratory Series

James Ward

Perspectives on Defense Systems Analysis: The What, the Why, and the Who, but Mostly the How of Broad Systems Defense Analysis, William P. Delaney, 2015

Ultrawideband Phased Array Antenna Technology for Sensing and Communications Systems, Alan J. Fenn and Peter T. Hurst, 2015

Decision Making Under Uncertainty: Theory and Application, Mykel J. Kochenderfer, 2015

Applied State Estimation and Association, Chaw-Bing Chang and Keh-Ping Dunn, 2016

Perspectives in Space Surveillance, Ramaswamy Sridharan and Antonio F. Pensa, eds., 2017

Mathematics of Big Data: Spreadsheets, Databases, Matrices, and Graphs, Jeremy Kepner and Hayden Jananthan, 2018

Modern HF Signal Detection and Direction-Finding, Jay R. Sklar, 2018

Measurements-Based Radar Signature Modeling, Joseph T. Mayhan and John A. Tabaczynski, 2024

Artificial Intelligence: A Systems Approach from Architecture Principles to Deployment, David R. Martinez and Bruke M. Kifle, 2024

MIT Lincoln Laboratory is a federally funded research and development center that applies advanced technology to problems of national security. The books in the MIT Lincoln Laboratory Series cover a broad range of technology areas in which Lincoln Laboratory has made leading contributions. The books listed above and future volumes in the series renew the knowledge-sharing tradition established by the seminal MIT Radiation Laboratory Series published between 1947 and 1953.

Artificial Intelligence

A Systems Approach from Architecture
Principles to Deployment

David R. Martinez and Bruke M. Kifle

The MIT Press
Cambridge, Massachusetts
London, England

The MIT Press would like to thank the anonymous peer reviewers who provided comments on drafts of this book. The generous work of academic experts is essential for establishing the authority and quality of our publications. We acknowledge with gratitude the contributions of these otherwise uncredited readers.

This book was set in Adobe Garamond Pro and HelveticaNeue by Westchester Publishing Services. Printed and bound in the United States of America.

Library of Congress Cataloging-in-Publication Data

Names: Martinez, David R., author. | Kifle, Bruke, author.
Title: Artificial intelligence : a systems approach from architecture principles to deployment / David R. Martinez, Bruke Kifle.
Description: Cambridge : The MIT Press, 2024. | Series: Lincoln laboratory series | Includes bibliographical references and index.
Identifiers: LCCN 2023030187 (print) | LCCN 2023030188 (ebook) | ISBN 9780262048989 (hardcover) | ISBN 9780262378710 (epub) | ISBN 9780262378703 (pdf)
Subjects: LCSH: Artificial intelligence—Industrial applications. | Systems engineering.
Classification: LCC TA347.A78 M37 2024 (print) | LCC TA347.A78 (ebook) | DDC 006.3—dc23/eng/20231121
LC record available at https://lccn.loc.gov/2023030187
LC ebook record available at https://lccn.loc.gov/2023030188

10 9 8 7 6 5 4 3 2 1

Contents

Preface   xi

Acknowledgments   xiii

1   Overview   1

      1.1   AI Notable Events in the Past Decades   4
      1.2   AI Pipeline: A System Architecture Approach   9
      1.3   High-Level Description of AI System Architecture Building Blocks   11
      1.4   Effective AI Deployment   15
      1.5   AI Horizons: Content-Based Insights, Collaboration-Based Insights,
             and Context-Based Insights   17
      1.6   Chapter Road Map   19
      1.7   Main Takeaways   20
      1.8   Exercises   21
      1.9   References   22

PART I   AI SYSTEM ARCHITECTURE

2   Fundamentals of Systems Engineering   29

      2.1   Systems Engineering Common Definitions   31
      2.2   Characteristics Espoused within the Systems Engineering Discipline   33
      2.3   A Systems Engineering Approach Applied to Artificial Intelligence   39
      2.4   Architecture Framework: The What and the How   40
      2.5   Leadership: Systems Thinker   43
      2.6   Systems Engineering Challenges   48
      2.7   Main Takeaways   50
      2.8   Exercises   52
      2.9   References   53

3   Data Conditioning   57

      3.1   Exponential Data Growth   59
      3.2   Digital Transformation   63

3.3	Databases: Management and Evolution	66
3.4	Data Quality, Cleaning, and Preparation	70
3.5	Curated Data Set Examples and Attributes	74
3.6	Data Conditioning Challenges	79
3.7	Main Takeaways	82
3.8	Exercises	84
3.9	References	84

4	**Machine Learning**	**89**
4.1	Machine Learning Classes	91
4.2	Common Measures of Performance	100
4.3	Introduction to Deep Learning and Neural Nets	106
4.4	Training Neural Networks with Backpropagation	109
4.5	Designing a Neural Network	115
4.6	Introduction to Convolutional Neural Networks	120
4.7	Machine Learning Challenges	123
4.8	Main Takeaways	123
4.9	Exercises	125
4.10	References	127

5	**Modern Computing**	**131**
5.1	A Short History of Computing Technologies	134
5.2	Computing at the Enterprise versus Computing at the Edge	138
5.3	Neural Network: Key Computational Kernels	141
5.4	Arithmetic Precision	145
5.5	Confluence of ML Algorithm Improvements and Computing Technology	149
5.6	Domain-Specific Hardware and Software	152
5.7	Contemporary Computing Engines and Integrated Systems	155
5.8	Roofline as a Metric	160
5.9	Securing Modern Computing	163
5.10	Modern Computing Challenges	169
5.11	Main Takeaways	175
5.12	Exercises	178
5.13	References	180

6	**Human-Machine Teaming**	**191**
6.1	Augmenting Human Capabilities	193
6.2	AI as a Search-and-Discovery Tool	197

6.3   AI as a Teammate   199
6.4   Autonomy and Near-Term Barriers   201
6.5   Quantitative and Qualitative Performance Metrics   205
6.6   Human-Machine Teaming Challenges   210
6.7   Main Takeaways   214
6.8   Exercises   216
6.9   References   217

7   **Robust AI Systems**   **223**

7.1   Systems Perspective on AI Vulnerabilities   225
7.2   Classes of Adversarial Artificial Intelligence   229
7.3   Deepfakes and Examples   233
7.4   Explainable Artificial Intelligence   236
7.5   Mitigation Techniques   238
7.6   Methodology for Testing against Adversarial Attacks   242
7.7   Robust AI System Challenges   245
7.8   Main Takeaways   246
7.9   Exercises   249
7.10   References   251

8   **Responsible Artificial Intelligence**   **259**

8.1   AI and Society   261
8.2   Case Studies: Harms from Artificial Intelligence   263
8.3   Considerations for Sociotechnical Systems   267
8.4   Responsible AI Principles   270
8.5   RAI Considerations in the AI Development Life Cycle   273
8.6   RAI Challenges   276
8.7   Main Takeaways   278
8.8   Exercises   279
8.9   References   280

PART II   STRATEGIC PRINCIPLES

9   **AI Strategy and Road Map**   **287**

9.1   Introduction to Strategic Thinking   289
9.2   AI Strategic Development Model   294
9.3   Mission/Vision and Envisioned Future   299
9.4   Organization Core Values and Strategic Direction   302

9.5	AI Value Proposition	304
9.6	AI Strategic Road Map: A Blueprint	307
9.7	Strategy and Execution: A Complementary Duo	317
9.8	Main Takeaways	318
9.9	Exercises	321
9.10	References	323

10 AI Deployment Guidelines   327

10.1	Challenges in Deploying Artificial Intelligence	330
10.2	Ten Guidelines for Successfully Deploying AI Capabilities	332
10.3	A Process for Applying a Systems Engineering Discipline to AI Deployment	335
10.4	AI Adoption: Four Distinct Organizational Maturity Clusters	339
10.5	The AI Ecosystem	341
10.6	Gold Standard: Test Harness, Performance Metrics, and Benchmarks	348
10.7	AI Platform Characteristics and Benefits	351
10.8	Main Takeaways	353
10.9	Exercises	357
10.10	References	359

11 MLOps: Transitioning from Development to Deployment   363

11.1	Introduction to MLOps Fundamentals	365
11.2	AI System Architecture Implementation Using MLOps	367
11.3	MLOps Enabling Techniques and Contemporary Tools	375
11.4	MLOps Platforms, AutoML, and LCNC Application Development	382
11.5	AI Development and Deployment: Common Pitfalls	385
11.6	Main Takeaways	388
11.7	Exercises	392
11.8	References	393

12 Fostering an Innovative Team Environment   399

12.1	Organizational Culture	402
12.2	Organizational Structure and Innovation	405
12.3	AI Talent and the Future of Work	408
12.4	Preparing You for a Successful Career	411
12.5	AI Technical Depth and Breadth	414
12.6	Metrics for Measuring Progress and Results	417
12.7	AI Leadership and Resilience	422
12.8	Mentoring, Networking, and Recruiting AI Talent	427

12.9   Sustaining High-Performance Teams   431
12.10 Main Takeaways   433
12.11 Exercises   436
12.12 References   438

13 Communicating Effectively   443

13.1   VSN-C for Structuring Communications   446
13.2   Winston Star: Essentials for Being Remembered   448
13.3   Essentials of Outlining   451
13.4   Writing and Presentation Fundamentals   453
13.5   Main Takeaways   463
13.6   Exercises   464
13.7   References   465

PART III   HUMAN-MACHINE AUGMENTATION: USE CASES

14 Use-Case Example 1: Misty Companion Robot as Alzheimer's Application   469

14.1   Exercises   475
14.2   References   476

15 Use-Case Example 2: Bose AI-Powered Cycling Coach and Warning System   477

15.1   Exercises   483
15.2   Reference   483

16 Use-Case Example 3: Meal Evaluation and Attainment Logistics System (MEALS)   485

16.1   Exercises   492

17 Use-Case Example 4: Managing Energy for Smart Homes (MESH)   493

17.1   Exercises   499
17.2   References   499

18 Use-Case Example 5: AquaAI, an AI-Powered Modernized Marine Maintenance System 501

18.1 Exercises 506
18.2 Reference 507

Appendix 509

A.1 Representative AI Industries and Sample Applications 509
A.2 Setting up Your Interactive Development Environment (for Either PC or Mac OS Operating Systems) 510
A.3 ML Performance Metrics 512
A.4 Multilayer Perceptron Algorithm 517
A.5 CNN with MNIST Fashion Data Set 521
A.6 Raspberry Pi: Introduction and Setup 524

Abbreviations 525
Index 529

Preface

This book provides a comprehensive introduction to a systems approach to the architecture, design, development, and deployment of artificial intelligence (AI) capabilities. Readers will gain insights into seven areas:

- Understanding an end-to-end AI system architecture
- Learning the technical underpinnings of the AI pipeline building blocks
- Formulating a strategic vision and development road map focused on AI products or services
- Transitioning of AI developments into operations
- Fostering and leading innovative AI teams
- Communicating effectively the AI value proposition to stakeholders
- Receiving practical experience from use cases, exercises, and a large body of references

The book is intended for advanced undergraduates and graduate students, as well as working AI practitioners.

The book is divided into three parts. Part I focuses on the AI system architecture and its functional building blocks. Part II builds from the architecture principles to help with formulating a strategic development plan or blueprint. The blueprint employs a strategic development model used extensively by the authors in real-life applications. Part III concludes with a set of use cases stemming from courses taught by the authors at the Massachusetts Institute of Technology (MIT), both at the graduate level and for working professionals.

We decided to write this book because there are very limited resources that address a broad treatment of AI technologies, the processes necessary for successful AI implementations and deployments, and the importance of AI talent (meaning people) to these efforts. This triad of technologies, processes, and people is paramount to the successful design, development, and deployment of AI systems. Unfortunately, many businesses fail at transitioning AI concepts and initial prototypes into operations because they lack the tools and techniques to do so. This book provides a reference source that can address challenges in transitioning AI capabilities from the development stage to useful AI products or services.

Those readers with limited knowledge of the AI building blocks are encouraged to start with part I of the book. For those readers with an AI background who want to learn about practical tools and techniques in leading AI teams from architecture principles to deployment can start with part II. The use cases discussed in part III help cement the concepts discussed in parts I and II. The large number of figures, exercises, and references complement the discussion in each chapter.

Appendix A.1 presents representative AI industries and sample applications. This list can be used in the classroom to help students with choosing areas for developing a strategic road map as part of a class project. Appendixes A.2–A.5 provide guidelines for setting up the Anaconda environment to run a set of Jupyter notebooks, illustrating the implementation of a multilayer perceptron (MLP), a convolutional neural network (CNN), and using key machine learning (ML) performance metrics. The description in appendixes A.2–A.5 are adapted from an excellent book by Aurelien Geron, *Hands-On Machine Learning with SciKit-Learn, Keras and TensorFlow*, 2nd edition, O'Reilly (2019). The reader can find example code and quick start instructions at https://github.com/ageron/handson-ml2. In appendix A.6, we provide a short overview of the Raspberry Pi used in one of our MIT classes, which the students used to do a demonstration of ML neural networks to a panel of AI industry and academic experts.

The potential of AI is enormous, but it also poses significant risks if not managed properly. We hope the readers of this book adopt a systems approach—from architecture principles to deployment—to minimize the danger of AI falling outside the "guardrails."

As stated in the now famous Amara's Law:

> We tend to overestimate the effect of a technology in the short run and underestimate the effect in the long run.
> —Roy Amara, American researcher, scientist, and futurist

As AI continues to advance, we can expect it to increasingly augment human capabilities. However, we must proceed with caution as we architect, design, develop, and deploy these capabilities by establishing rigorous testing, verification, and validation methodologies with the stakeholders partaking in the process.

Acknowledgments

This book had been in progress for several years. The original work dates to when one of the authors, David Martinez, began to formulate tutorial material on artificial intelligence (AI) at MIT Lincoln Laboratory. This tutorial material served as the background and introduction to a comprehensive AI study performed at Lincoln Laboratory and led by David Martinez, Bill Streilein, and Nick Malyska. Brad Dillman contributed to the organization of the AI study report and all its graphics.

The book would not have been possible without the continued support by the leadership at Lincoln Laboratory. Eric Evans and Scott Anderson provided continuous encouragement and support from the beginning to the final book's completion. Jim Ward, chair of the Lincoln Laboratory Book Series (LLBS), was also instrumental in encouraging us to write the book as part of the LLBS. We also want to thank the leadership in the Cyber Security and Information Sciences Division (division 5) at Lincoln Laboratory for their continuing encouragement. Some of the material in part II of the book, on strategic planning and leadership, stems from years of applying these concepts in practice within division 5. Brad Dillman converted all the original lecture charts into the required book format. We also want to thank the division 5 administrative support staff, Kimberly Pitko and Renee Gylfphe. Pitko served as the administrative assistant during the book's proposal preparation and facilitated access to much of the referenced material.

We also want to thank Bob Hall in the Knowledge Services Department at MIT Lincoln Laboratory, who diligently searched a large body of references at the authors' request. Since this book covers a broad range of topics, narrowing the search to the most relevant material for each chapter in the book was a major effort.

It has been a great pleasure to work closely with Elizabeth Swayze and Matthew Valades of the MIT Press. They provided clear and concise guidance from the start of the book's proposal stage and throughout the feedback from anonymous reviewers. We also want to thank the anonymous reviewers for their input, which made the book much stronger in terms of quality, content, and fluidity.

The initial AI tutorial material served as the core set of charts for the lectures taught as part of a special subject course within the MIT School of Engineering. We want to give special thanks to the leaders and staff of the MIT Gordon Engineering Leadership (GEL) Program for their support during the teaching of this special subject course. We also want to thank Brad Dillman, who created the graphics used in the lecture material taught in this course, as well as Charlie Smith, who formulated a set of information technology (IT) instructions for use with the Raspberry Pi single-board computer, and all the graduate students who conceptualized and delivered innovative class projects, some of which we used in part III of the book.

Since the winter of 2021, David Martinez and Bruke Kifle have taught multiple courses, using the content of this book, for the MIT Professional Education certificate program on AI and machine learning (ML). The MIT Professional Education (MIT PE) was established by the School of Engineering in 2002, providing world-class educational opportunities for engineering and science professionals from around the globe. We are very grateful for the opportunities provided by the MIT PE center. More specifically, we want to extend our special thanks to Bhaskar Pant, Malgorzata Hedderick, Myriam Joseph, Lu Men, Lindsey Narron, and all the administrative support staff at the MIT PE office. We also want to extend our appreciation to all the students who helped strengthen the lecture material, and formulated class projects based on the teachings in parts I and II of the book. We used a selective set of these projects in part III.

Next, we acknowledge, individually, our closest supporters that have affected our book-writing journey. Undertaking the writing of this book required many dedicated hours by the authors, resulting in time away from their families, close friends, mentors, and colleagues.

David R. Martinez

I would like to deeply thank my wife, Denice, for her continuous encouragement. We took long walks that gave me an opportunity to tell her about the book's progress. She was a constant source of support. My daughter, Diane, and her husband, Sean, helped me by serving as a sounding board to explain some concepts in the book such that non-AI specialists could understand. My son, Robert, took time to help me with debugging the computer setup to make sure that the Jupyter notebooks would run properly, and with the earlier demonstration of ML code running in the Raspberry Pi.

I also want to thank a couple of mentors that have provided me with great opportunities. Professor Al Oppenheim (Massachusetts Institute of Technology/MIT) served as a role model academically, and by example, was someone who deeply cared about his graduate students. Ken Senne (MIT Lincoln Laboratory) influenced my passion for working on the most difficult problems using advanced technology.

It has been a joy working closely with Bruke Kifle during the writing of this book and while teaching our AI courses. He has a deep knowledge of the material presented in the book and also practices many of the approaches on transitioning AI capabilities from architecture principles to deployment.

Bruke Kifle

I would like to express my sincerest gratitude to my beloved family for their unwavering support throughout my academic and professional journey. In particular, I extend my heartfelt appreciation to my parents, Mesfin and Yeshimebet, who courageously emigrated from Ethiopia and made significant sacrifices to provide me and my siblings with opportunities in the US. Their selflessness and determination to overcome challenges serves as a constant source of inspiration, and their invaluable guidance, prayers, and encouragement have been instrumental in shaping my personal and professional growth and have played a crucial role in the writing and completion of this textbook. I am equally grateful to my three siblings, Tsegereda, Sentayehu, and Rediet, whose unwavering support has been a source of strength and motivation for me throughout my journey.

I also extend my appreciation to my close friends, colleagues, and mentors for their steadfast support and guidance, which have been vital in shaping my academic and professional endeavors. Without their input and support, the completion of this textbook would not have been possible.

Moreover, I would like to express my deep gratitude to the organizations that have contributed to my growth and development as a technologist and practitioner. My alma mater, MIT, has provided me with a world-class education and the opportunity to explore new fields of study alongside some of the brightest minds. The support, guidance, and mentorship provided by my professors and peers have pushed me to new heights, and I am deeply grateful for the opportunities, experiences, and friendships that MIT has afforded me. I am equally grateful for my time at Microsoft, where I have had the privilege to work with some of the most talented individuals in the industry in an environment that fosters innovation, creativity, and collaboration. The mentorship, support, and resources that Microsoft has provided me have allowed me to learn, grow, and contribute meaningfully. Both organizations have played a crucial role in my early career growth and development, and I am deeply grateful for their contributions to my journey and the support, guidance, and mentorship that they have offered along the way.

Lastly, I express my heartfelt gratitude to David Martinez for his unwavering belief in my abilities, for providing me with the opportunity to contribute to this book, and for being a constant source of encouragement and inspiration.

1

Overview

The greater danger for most of us lies not in setting our aim too high and falling short, but in setting our aim too low and achieving our mark.
—Michelangelo, c.1500

Many industries are transforming themselves to maintain a competitive edge and operate well in a fast-evolving digital environment. This environment, in any organization, includes those developing artificial intelligence (AI) capabilities, teams transitioning AI capabilities into operations, and those ultimately responsible for providing AI products or services to users and consumers.

However, in this digital transformation relevant to all AI stakeholders, there is a need to ascertain that AI will add value to the organizations, the respective industries, and humanity writ large. Thus, in this book, the focus is on a systems approach to AI from its architecture principles to deployment.

As said by Michelangelo, the famous Renaissance artist and architect, we need to aim high, even if we fall short, if we want to grow and prosper. This quote is as relevant today as it was when it was first said. In our context, we increase the odds of succeeding if we follow a rigorous systems approach.

The emphasis on *systems* implies a collection of subcomponents integrated to produce a useful AI capability. In this book, we use the phrases "AI systems approach" and "AI systems engineering approach" interchangeably. The latter emphasizes a rigorous delineation of engineering steps as we guide the readers from architecture principles through deployment.

It is obviously necessary but not sufficient to have good data, verified and validated algorithms, a computing environment to deploy the algorithms either in the cloud or at the edge (or both), and other subcomponents representative of enabling technologies.

But it is also necessary to attend to people and processes within the rubric of an AI ecosystem. This book introduces and explains architecture principles, development stages, and the integration of subsystem components into an end-to-end system, culminating in a successful deployment.

As a guide to the book's content, the reader can refer to figure 1.1, which illustrates the three parts of the book. In part I, we start with a detailed discussion of AI system architecture, concentrating primarily on the enabling technologies. In part II, we focus on strategic principles beginning with the critical steps that lead to a company's AI strategic blueprint (e.g., strategic vision, envisioned future, governance, culture, infrastructure, fostering an innovative team environment, attending to AI talent, and finally succeeding in the deployment and monitoring of AI capabilities). Again, in this book, we emphasize the importance of the people-process-technology triad to meet the business needs and goals, resulting in AI value that takes the form of products and services.

In part III, we bring parts I and II together for the reader by way of AI application use cases. All of our discussions center on *narrow AI* (see the simplified definition given in the "Narrow AI" box).

Thus, all the use cases represent AI capabilities augmenting human intelligence. We will elaborate on the importance of assessing these capabilities through a careful trade-off between "Confidence in the Machine Making the Decision" versus the "Consequence of Actions" (as they appear in the figure). Some narrow AI capabilities are best left to the machines. In other cases, the human is better at performing those tasks. However, as we will see later, many valuable AI tasks fall in between.

In the next section, we start by describing a set of important historical events. These events led to important successes, but we also point out some failures. Following the historical events, the AI system architecture is described. This architecture is foundational to understanding the development of AI as a system—thus, the book's title is specifically chosen to bring out the importance of treating AI as a complex system. We also highlight for the reader each of the building blocks, which are integrated into the end-to-end AI system architecture. In addition to the successful development of each of the building blocks (or subcomponents), the AI practitioners must pay equal attention to AI deployment.

Narrow AI

The theory and development of computer systems that perform tasks that augment human intelligence such as perceiving, learning, classifying, abstracting, reasoning, and acting.

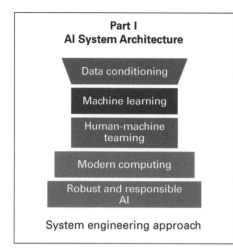

Part I
AI System Architecture

- Data conditioning
- Machine learning
- Human-machine teaming
- Modern computing
- Robust and responsible AI

System engineering approach

Part II
Strategic Principles

- Strategic vision
- Envisioned future
- Governance, culture, and infrastructure
- Innovative team environment
- Mentoring AI talent
- AI development and deployment (MLOps)

AI value proposition

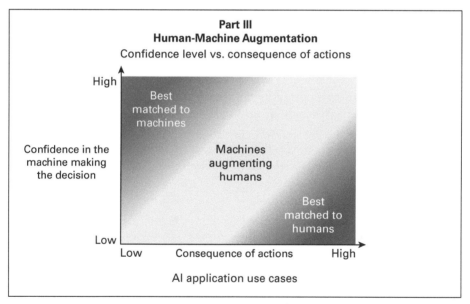

Part III
Human-Machine Augmentation
Confidence level vs. consequence of actions

Figure 1.1   Key parts of the book as a guide for the reader.

The journey in the deployment of an AI capability starts with a clear understanding of the business's needs. In this chapter, we introduce the reader to the importance of establishing an AI ecosystem that is put into practice by following ten guidelines for the successful deployment of AI capabilities. Later in the book, we elaborate on this topic in much more detail.

The last section in the chapter is devoted to the outlook for the future, divided into three AI horizons. Each horizon addresses AI capabilities that most organizations must strive for to deliver AI value to users and customers.

There is a subtle distinction between users and customers. Often users (or consumers) are (e.g., in the medical applications) clinicians or radiologists evaluating the output of an AI computer vision system; and customers can be those providing the financial resources needed to develop AI capabilities. In either case, both users/consumers and customers are invested in creating or using AI products or services.

1.1   AI Notable Events in the Past Decades

AI developments in the past several decades have enjoyed many successes, but also failures. Some of the early research in AI was in emulating very simple human brain architectures using analog circuitry. The quest to achieve human-level intelligence dates back to years before the seminal events of the 1950s. The initial models emulating how the brain connects neurons to synapses [1] led to scholars on a search for ways to replicate, in analog hardware, how humans achieve intelligence. Nils Nilsson [2] presents an excellent historical perspective of work leading to the seminal events in AI during the 1950s.

Figure 1.2 illustrates a number of notable AI events over the course of seven decades. In 1950, Alan Turing [3] wrote a seminal paper, "Computing Machinery and Intelligence," that led to what it is famously known as the *Turing test*. Turing's main discussion centered on the key question: Can machines think? The Turing test served to evaluate, blindly, the ability of an intelligent machine to respond to questions posed by judges, requiring them to determine if the answers originated from a machine or a human [4].

In 1956, a group of researchers and AI practitioners from industry convened at Dartmouth College to investigate approaches for machines to achieve human intelligence and to propose further research. This event is considered the dawn of AI. Professor John McCarthy, at the time a faculty member at Dartmouth College and the host of the event, coined the term "artificial intelligence" [5]. In addition to John McCarthy, other preeminent participants included Marvin Minsky, Claude Shannon (considered the father of digital communications), Oliver Selfridge of MIT Lincoln Laboratory, who worked on the initial demonstrations of machine perception, and other luminaries.

| 1950 | 1960 | 1970 | 1980 | 1990 | 2000 | 2010 | 2020 |

AI Winters 1974–1980 and 1987–1993

- 1950 - Computing Machinery and Intelligence "Turing Test" published by *MIND* vol. LIX

- 1955 - Western Joint Computer Conference Session on Learning Machines

- 1956 - Dartmouth Summer Research Project on AI J. McCarthy, M. Minsky, N. Rochester, O. Selfridge, C. Shannon, others

- 1957 - Frank Rosenblatt Neural Networks Perceiving and Recognizing Automation

- 1957 - Memory Test Computer, first computer to simulate the operation of neural networks

- 1958 - National Physical Laboratory in the UK Symposium on the Mechanization of Thought Processes

- 1959 - Arthur Samuel "Some Studies in Machine Learning Using the Game of Checkers" *IBM Journal of R&D*

- 1960 - Recognizing Hand-Written Characters, Robert Larson of SRI AI Center

- 1961 - James Slagle, Solving Freshman Calculus (Minsky Student) MIT

- 1979 - "An Assessment of AI from a Lincoln Laboratory Perspective" Internal MIT LL publication

James Forgie

- 1982 - Expert systems DENDRAL project

Ed Feigenbaum

- 1984 - Hidden Markov models

- 1986 to present - The return of neural networks

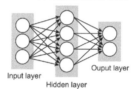

Input layer   Hidden layer   Ouput layer

- 1988 - Statistical Machine Translation

- 1989 - Convolutional Neural Networks

- 1994 - Human-level spontaneous speech recognition

- 1997 - IBM Deep Blue defeats reigning chess champion (Garry Kasparov)

- 2001 to present - The availability of very large data sets

- 2005 - Google's Arabic and Chinese to English translation

- 2007 - DARPA Grand Challenge ("Urban Challenge")

- 2011 - IBM Watson defeats former Jeopardy! champions (Brad Rutter and Ken Jennings)

- 2012 - Team from U. of Toronto (Geoff Hinton's lab) wins the ImageNet Large Scale Visual Recognition Challenge with deep-learning software

- 2014 - Google's GoogleNet Object classification at near human performance

- 2015 - DeepMind achieves human expert level of play on Atari games (using only raw pixels and scores)

- 2016 - DeepMind AlphaGo defeats top human Go player (Lee Sedol)

- 2017 - AlphaGo Zero wins against AlphaGo using reinforcement learning

Figure 1.2   Some notable AI events from 1950 to 2017.

Minsky continued to be an AI pioneer for many decades after the famous 1956 Dartmouth summer research project. He wrote a paper titled "Steps towards Artificial Intelligence," which established the vision leading to the establishment of the MIT Artificial Intelligence Laboratory, where Minsky served as its first director [6]. In 2003, the MIT Artificial Intelligence Laboratory merged with the MIT Laboratory for Computer Science to form the MIT Computer Science and Artificial Intelligence Laboratory (CSAIL), one of today's premier research centers in AI [7].

Another important research contribution that shaped the future of neural networks was work done by Frank Rosenblatt from Cornell University. Rosenblatt [8] formulated the architecture of a perceptron, which emulated the neurons and synapses of a human brain. The perceptron is a simple, one-stage predecessor to what we know now as a multilayer perceptron (e.g., a fully connected neural network). In chapter 4 and the appendix, we will do a deeper dive into an example of a neural network known as "multilayer perceptron (MLP)."

Marvin Minsky and Seymour Papert wrote a book titled *Perceptrons: An Introduction to Computational Geometry*, in which they elaborated on the mathematics of the perceptron architecture as one of the first examples of a machine that could be taught to perform simple tasks by using training data [9]. Perceptrons are considered an example of models within the rubric of "connectionists." However, Minsky and Papert felt that a perceptron could not be generalized to solve important AI problems. In contrast, other AI models showed impressive results, leveraging serial reasoning of symbolic expressions. These latter techniques belong under the rubric of "symbolists" [10]. Later in the book, we will compare and contrast the distinctions among these models.

Even though the early assessments by Minsky and Papert showed limitations with perceptrons, in the later edition of their book [11], they clarified that both connectionist learning (like perceptrons) and symbolist reasoning were important techniques within the scope of machine intelligence. In the 1980s, they predicted that connectionist approaches would flourish ("and we expect the future of network-based learning machines to be rich beyond imagining").

Another important AI milestone was the demonstration of an intelligent machine that could play the game of checkers against a human [12]. In 1959, Arthur Samuel, at the time working at IBM, showed that a machine could be programmed to play better than the human who programmed it. Samuel was one of the first AI researchers to introduce the term *machine learning* (ML). These were the early days of ML, mostly built on rule-based decision trees. Although simple in comparison to today's standards, this demonstration provided an initial indication that machines could be built with a capability to learn.

Despite all the initial successes achieved during the 1950s and 1960s, in the late 1970s and again in the late 1980s, as shown in figure 1.2, there were so-called AI winters, when research and development (R&D) funding in AI was very limited. These AI winters were driven by the hope of achieving artificial general intelligence (AGI). Although R&D funding came to almost a complete halt, there were significant accomplishments achieved during those periods, primarily based on expert systems [2].

An expert system encodes the knowledge of an expert into rules that help capture human expertise in a narrow field of specialty. For example, Ed Feigenbaum, a faculty member at Stanford University, demonstrated in 1982 a functional expert system, the DENDRAL project, that was applied to organic chemistry for the purpose of helping with the identification of organic molecules from their spectra. Although the project began in the mid-1960s, by the 1980s there were impressive results using an expert system based on a rule-based decision process. However, expert systems required significant effort in encoding human expertise and knowledge to be applicable to various classes of problems.

In 1997, there was a fundamental shift in recognizing what AI could do when IBM's Deep Blue defeated the reigning chess world champion Garry Kasparov [13]. The chess-playing program was written in the C programming language. It was capable of evaluating 200 million positions per second. In June 1997, Deep Blue was the 259th most powerful supercomputer in the world according to the well-known LINPACK benchmark used for evaluating supercomputers on the TOP500 list (delivering 11.38 billion floating-point operations per second). Kasparov points out that it was the ability of a computer to evaluate those 200 million positions per second that caused him to lose to IBM Deep Blue. In addition to the incredible demonstration of a machine defeating a human at as difficult a game as chess, which was revolutionary, it was very important for the field of AI to simultaneously use AI algorithms with a powerful computing platform, resulting in a major milestone in AI.

Since the 2000s, there have been significant AI milestones, as shown in figure 1.2. In addition to advances in AI algorithms, the availability of so-called big data and high-performance computing has led to many important AI accomplishments in a relatively short time. The 2007 Defense Advanced Research Projects Agency (DARPA) Grand Challenge demonstrated the ability of autonomous cars to navigate in an urban environment. In classical DARPA fashion, this successful Grand Challenge demonstration spun off a whole industry in autonomous systems, which we are witnessing and benefiting from today.

Another important AI milestone happened when IBM's Watson computer defeated former *Jeopardy!* champions Brad Rutter and Ken Jennings. This demonstration was impressive because in contrast to Deep Blue, which defeated the world chess champion by analyzing massive combinatorial chess moves, the challenge for Watson was to process natural language and search a massive database in real time to find the correct question to the answer in *Jeopardy!* terms.

In 2016, the company DeepMind Technologies Limited (which had been acquired by Google two years earlier) demonstrated the ability for a machine to defeat the top Go player, Lee Sedol from South Korea. This major AI accomplishment integrated advances in deep neural networks (DNNs) through reinforcement learning gleaned from many examples of human play. Reinforcement learning requires agents to set goals, policies, and then a neural network to take actions to try to achieve the goals constrained by the policies. The AI system was named AlphaGo [14] and as reported by Google, it used 1,202 central processing units (CPUs) and 176 graphics processing units (GPUs) in a distributed processing architecture. In 2017, DeepMind introduced a new AI system called AlphaGo Zero [15], which boasted the ability to defeat the previous system, AlphaGo, by self-play reinforcement learning. AlphaGo Zero was also remarkable because instead of depending on a large number of CPUs and GPUs in a distributed computing architecture, it used a single machine in the Google Cloud with four tensor processing units (TPUs) [16]. DeepMind continues to accomplish significant breakthroughs. For example, DeepMind's AlphaFold is a neural network that can predict protein structures with a high degree of accuracy. DeepMind's demonstration is a significant accomplishment in computational biology, which has vexed researchers for fifty years [17].

Several notable researchers are now working on bringing together the strengths of DNNs and symbolic AI to employ the benefits of both. In their Turing Award lecture, Yoshua Bengio, Yann LeCun, and Geoffrey Hinton [18] emphasized the need to consider both neural network approaches and rule-based approaches. Their discussion has generated great interest in the AI research community through an analogy to the main premise in the award-winning book by Daniel Kahneman, *Thinking, Fast and Slow*. A Nobel laureate, Kahneman defines system 1 as representative of a human's *thinking fast*. Deep learning networks encompass system 1–type reactive inference. Kahneman contrasts system 1 to system 2 representing a human's *thinking slow*. Symbolic AI is characterized as performing reasoning, or sequential reasoning, that is closely aligned with system 2. Kahneman also emphasizes that systems 1 and 2 can be thought of as "categories of mental processes" [19]. These analogies are very important since many researchers predict that in the next several decades, we will see AI systems integrating both approaches under the rubric of *neuro-symbolic AI* [20–24].

These notable AI accomplishments of the last several decades serve to illustrate the AI evolution leading to a potential future where rapid advancements will solve important classes of user problems. The major drivers behind this are the availability of a variety of sensors and sources of data, rapid innovation in ML models, and significant improvements in modern computing. This trend will continue as Internet of Things (IoT) devices become prolific, algorithms and simulation models continue to advance, and computing continues to accelerate. We elaborate on these enabling technologies in chapters 3–6.

In the next section, we introduce a system architecture approach for addressing AI capabilities through the lens of an end-to-end architecture, consisting of key building blocks. No AI capability is effectively deployed into operations by having only a well-performing set of ML algorithms. Highly accurate ML algorithms are necessary but not sufficient. There are several building blocks, as part of the AI architecture, that are also necessary and require effective development and integration to result in useful AI products and services.

1.2   AI Pipeline: A System Architecture Approach

As many researchers and practitioners predict, AI will continue to have a significant impact in many areas, including medicine, agriculture, energy, transportation, manufacturing, financial services, human resources, logistics, national security, and robotics process automation. These present and future AI applications are going to result in a change to the global economic landscape, education, workforce, and global competitiveness. The challenge to any organization, or more generally to any nation, will be in leveraging the commercial sector, academia, laboratories, and entrepreneurial companies to gather all the sources of AI innovation in a cohesive and coherent way.

There have been decades of technical developments guided by leveraging system architecture principles. Adopting a system architecture approach facilitates the formulation of AI capabilities (starting with business needs), design, implementation, prototyping, and deployment as stages of complex systems. In this book, we also employ a system architecture approach to address all the stages from AI architectural principles to deployment. In chapter 2, we elaborate in more detail on the fundamentals of systems engineering to more clearly expose the readers to using this technique in the development and deployment of AI capabilities. The formulation of a system architecture is a critical first step in the development of AI capabilities, as well as in following a rigorous systems engineering approach.

Our AI system architecture (shown in figure 1.3), which forms the bases of part I of the book, is also commonly referred to as a *functional architecture*. A functional architecture serves as the scaffolding structure to an engineered system. An *engineered system* can be defined as a combination of components that work in synergy to collectively perform

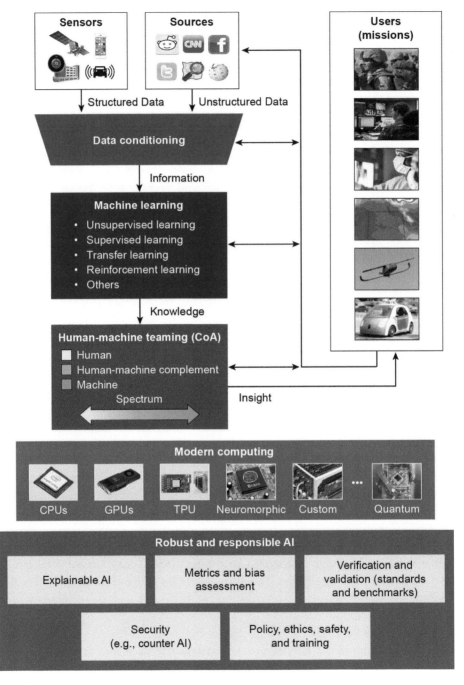

Figure 1.3 AI system architecture.

a useful function [25–27]. When developing AI systems, it is imperative that we think of this game-changing capability as an end-to-end system, not as a single-point solution, therefore requiring key subsystems to effectively be integrated together.

Andrew Moore, the former dean of the School of Computer Science at Carnegie Mellon University, envisioned and defined what he refers to as the "AI stack" [28]. Moore's AI stack shares many of the same key subsystem components of the AI system architecture shown in figure 1.3. As Moore points out [29], "AI isn't just one thing or a single piece of software; it is a massive collection of interrelated technology blocks called the AI stack." This perspective is very important for any organization investing in R&D and transitioning AI capabilities into operational systems. The AI stack, similar to our AI system architecture, can provide a framework for identifying and organizing all the technologies and capabilities required of an end-to-end AI system. The next section highlights each of the subcomponents shown in figure 1.3. In later chapters, we address each of these building blocks in greater detail.

1.3   High-Level Description of AI System Architecture Building Blocks

In this section, we describe at a high level the key subsystem components (also referred to as *building blocks*) of the AI system architecture shown in figure 1.3. In subsequent chapters, we elaborate in more detail on the role that each building block plays by working in synergy to perform a useful function—in this case, delivering useful AI capability. The format that we follow is to provide a short description, starting from input data and culminating with insight delivered to the stakeholders. An important subsystem component is what we referred to as *robust and responsible AI*. This subcomponent consists of multiple technologies and areas of emphasis to ascertain acceptance by AI consumers and users. If an end-to-end AI system fails to deliver trustworthy results with high confidence, while complying with transparency, privacy, ethics, reliability, security (to name only a few), the users will opt to revert to the prior approaches that they had used without the benefit of AI. This is because AI systems are held to a higher standard in avoiding errors compared to errors made by humans—until we get to be comfortable with the benefits provided by AI augmenting humans' cognitive tasks.

In the effective application of AI, data is one of three drivers contributing to the rapid progression of this field. As discussed earlier, the other two drivers are the ML algorithms and modern computing. The award-winning author Alexander Wissner-Gross stated very accurately: "Perhaps the most important news of our day is that datasets—not algorithms—might be the key limiting factor to development of human-level artificial intelligence" [30]. Although in this book, we are not on a quest to achieve human-level AI, as defined earlier in the discussion of AGI, the quote applies equally well to narrow AI, where the goal is to augment human intelligence.

We begin our high-level description of the subcomponents shown in the AI system architecture illustrated in figure 1.3 by providing a short description as follows.

Sensors and sources: This building block consists of data provided by physical sensors—like sensors mounted on or embedded in orbiting satellites, airplanes, driverless cars, smart phones, and IoT devices—and cybersources, such as different forms of social media, news articles, standard open websites, and the deep websites and dark websites; the latter two types of websites are commonly not accessible to the general public.

If metadata—which it is data about data—exists for these sensors, then the data is categorized as *structured data*. The raw digital data accompanies the metadata. *Metadata* describes the characteristics of the incoming data; for a radar system, it might be radar frequency, transmitter power, number of receiving channels, or sampling frequency of an analog-to-digital converter. In contrast, what we call *data sources* are accessible in the cyberdomain (or, for example, via handwritten notes), and these sources of data are categorized as *unstructured data*. There is, typically, no source of structure contained with those data. For example, reports from a medical doctor or a nurse, after seeing a patient, are in text form, void of any description of what is in the report until a human reads it. AI techniques in the field of natural-language processing (NLP) are advancing rapidly to determine what is contained in this type of report.

Data conditioning: Both structured and unstructured data need to go through a data conditioning stage before ML algorithms can be applied. Data conditioning, also referred to as *data munging* in the data science community, is quite involved. It is often cited that more than 90 percent of all data available in the world was generated in the past two years [31], and of this available data, more than 80 percent is unstructured. So data scientists and data engineers spend a considerable amount of time preparing the data prior to the next processing stage (i.e., ML techniques).

The main role of this subcomponent is to transform data into information. An example of information is a new sensor image (after data labeling) that we need to use to classify whether the object of interest is present in that image or not (like when performing image segmentation). Typical functions performed under this data conditioning subcomponent include standardization of data formats complying with data ontology, data labeling, highlights of missing or incomplete data, errors/biases in the data, and other functions discussed in chapter 3.

> **Machine learning:** During this processing stage, inputs are transformed from useful information into knowledge. Knowledge is more specific than information. For example, knowledge results from classifying what is in an image (e.g., a vehicle of specific make, model, and color).

There are many types of ML algorithms, including unsupervised learning, supervised learning, and reinforcement learning. One of the watershed moments in AI happened when the AI community experienced the confluence of labeled data and the demonstration of dramatic improvement in image classification. The labeled data set known as ImageNet was created by Fei-Fei Li (Stanford University) and her students [32]. They have created to date approximately 15 million labeled images. By 2010, they had labeled 1.5 million images by formulating a well-defined ontology and employing Amazon Mechanical Turk to have humans go through and label each image according to the ontology. In 2012, Geoffrey Hinton (presently at the University of Toronto) and colleagues used the labeled data from ImageNet and applied a deep CNN algorithm, using multiple GPUs for training [33]. They were able to demonstrate an error rate of 15.3 percent in image classification compared to about 26 percent error rate achieved in 2011. The demonstration was part of an annual competition called ImageNet Large-Scale Visual Recognition Challenge (ILSVRC). In 2016, the error rate was reduced to 3 percent, compared to humans, who typically have a 5 percent rate. It is important to point out that these error rates of 15.3 percent and 3 percent are from the Top-5 accuracy results, which means that the right answer is in the top five, with the highest probabilities at the output of the ML DNN algorithm. Humans are able to outperform a machine if the comparison is done as a single highest probable image (Top-1 accuracy).

These neural network demonstrations were impressive achievements leveraging the confluence of big data, ML algorithms, and modern computing.

> **Modern computing:** This subcomponent addresses classes of modern computing technologies suitable for ML training and ML inference stages. One of the leading processing engines is Google's TPU, which uses variable precision to gain in performance per watt. Similar techniques have been used for other applications where matrix–matrix and matrix–vector multiplies are essential for both training and application of the weights [34]. These linear algebra operations require not just high-speed computations, but also high-performance interconnects, as well as memory access, as discussed in chapter 5.

Human-machine teaming (HMT): HMT is the building block that represents the collaboration between humans and machines in the decision support stage of an AI system—augmented and collaborative intelligence. This collaboration requires operational speeds resulting in timely insights to users, plus a need for increasing scale and reducing the level of the consequence of actions, with respect to a recommended decision. Collaborative intelligence will permeate across many applications demanding different degrees of trust depending on the application domain [35–37].

Robust AI: There are many elements within the rubric of robust AI as shown in figure 1.3. For users to trust in and be confident with the results provided by AI, the overall system has to include the ability to explain how the ML algorithms arrived at their output (information to knowledge).

Similarly, there needs to be better metrics, not just for the ML algorithms, but across the end-to-end AI system architecture. Every system will need to undergo verification (i.e., ensuring that the software does what it was designed to do) and validation that the system performs as expected (e.g., meeting a set of system requirements). Security, both physical and in the cyberdomain, will be an important aspect of protecting the AI system. Finally, any AI capability needs to be designed to comply with a set of policies, meeting ethical standards, and ensure overall safety as discussed in more detail under responsible AI (RAI). In the context of augmenting human intelligence, there will need to be an active effort to train consumers and users. AI systems will be routinely used in operations, when we are able to demonstrate a high level of confidence versus an acceptable level in the consequence of actions. This will always involve a trade-off.

RAI: In this book, we address RAI as a separate chapter due to its importance. A successful AI deployment must include from the start all areas that fall under the umbrella of RAI. There are now many initiatives to establish regulatory arms both in the US, the European Union, and across many other countries, as well as several major AI commercial companies, to ascertain the responsible use of AI. The goal is to have some element of AI regulation to prevent either intentionally or unintentionally improper use of AI (e.g., under racial or gender-based biases). Government, industry, civil societies, and academic institutions must emphasize and enforce, in practice, that AI systems adhere to fairness, accountability, safety, transparency, ethics, privacy, and security. The Massachusetts Institute of Technology (MIT) established the Schwarzman College of Computing, and one of its major areas of emphasis is in research focusing on social and ethical responsibilities of computing (SERC). Finding the right balance between regulating AI and not suppressing innovation is a challenge, and approaches are still evolving. Many researchers and AI practitioners are wrestling with these issues [38–40].

Users: Although this stage in the overall AI system architecture is not a subcomponent per se, users (or AI customers and consumers) are ultimately the beneficiaries of an end-to-end AI system. Users must be able to provide feedback to the system designers, developers, and integrators in order to enable increased refinements and improvements. There are many types of users, ranging from being in an enterprise environment, at the edge, as well as autonomous systems. Autonomous systems will be able to operate without the aid of a human if and only if the confidence in the decisions made by the machine is high and the consequences of the action are low. For most other cases, the humans are ultimately the ones making the final decisions, at least as envisioned in the near future.

In the following section, we begin a discussion on the importance of taking into account the elements needed to achieve an effective AI deployment. More organizations fail in bringing AI pilots or prototypes into operations because of the lack of a supporting culture and a lack of a rigorous development approach—resulting in what is euphemistically called the "valley of death." AI systems must be tested, validated, and assessed in terms of desired performance from the beginning of the development process, including users in the loop.

1.4   Effective AI Deployment

There is great excitement about the value that AI brings, or potentially brings, to the business. For many companies, delivering AI products or services, growth is forecasted to increase year over year. However, there is a significant effort required in transitioning AI pilots or prototypes from experimental demonstrations—and in laboratory settings—into operations. Thus, there is an emerging new field referred to as *machine learning operations* (MLOps), which adopts several of the principles espoused by the *development operations* (DevOps) *community*. The DevOps community has established processes and guidelines to more effectively implement software in complex systems, as well as by employing rigorous software engineering practices.

AI businesses realize the potential value provided by integrating AI capabilities into operations. However, these organizations also realize the potential cost. Andrew Ng, a renowned researcher, innovator, and AI practitioner, has outlined areas to which organizations need to pay attention in order to successfully transition from development into operations within the context of MLOps [41]. Likewise, respected consulting

Table 1.1   Ten guidelines to successfully deploy AI capabilities

1.	Develop a clear strategic vision and project road map for the AI system
2.	Understand the stakeholder's AI needs
3.	Strengthen the AI team by fostering internal and external relationships
4.	Build a multidisciplinary and diverse team with complementary skills
5.	Provide measurable objectives while mentoring AI talent
6.	Continue to expand AI team skills as the future of work evolves
7.	Demonstrate an initial AI capability, and then iterate
8.	Verify individual subcomponents and validate the end-to-end AI system
9.	Secure the AI system both physically and against cyberthreats
10.	Attend to RAI

companies, such as McKinsey & Company, Boston Consulting Group in cooperation with the *MIT Sloan Management Review*, and others have performed an in-depth analysis of AI companies, looking at how successful they are at bringing AI capabilities into operations, including recommendations for achieving AI scalability [42–44].

Why is doing this so hard? There are several reasons; in this book, we address the development-to-operations barriers and provide recommendations to overcoming them. Table 1.1 enumerates ten guidelines to successfully deploy AI capabilities. At the core of these ten capabilities, there is a need to:

- Understand clearly the business needs and goals expected from AI; this is an element of guidelines 1 and 2
- Build a culture of innovation and foster AI teams that can work well within and across the organization, providing measurable objectives and team mentorship (guidelines 3–6)
- Implement an MLOps framework that facilitates rapid prototyping (piloting AI capabilities), rigorous AI system verification and validation (V&V), as well as preventing malicious attacks (guidelines 7–9)
- Attend to RAI principles (guideline 10)

Guidelines 7–9 will be addressed in chapter 10. We dedicate a full chapter to MLOps (chapter 11) because it is one of the most significant barriers to AI deployment. As discussed by staff at Google, there is a "hidden technical debt" that all organizations must recognize and properly address through the use of a well-operating MLOps platform (on-premise, in the enterprise cloud, or both) [45]. As we emphasized in earlier sections, successful deployment of ML techniques requires more than just good working ML code.

As we approach the end of this chapter, we conclude with a look at the future through the lens of three AI horizons. In chapters 3–8, we use this three-horizons framework to emphasize areas of R&D. We also recommend that the reader consult an excellent yearly report called the *Artificial Intelligence Index Report* [46]. This report is an independent initiative by Stanford University's Human-Centered Artificial Intelligence Institute (HAI), sponsored by several industry partners. As stated by its authors, the report is extremely useful for policy makers, researchers, executives, journalists, and the general public for tracking progress in the field of AI.

1.5   AI Horizons: Content-Based Insights, Collaboration-Based Insights, and Context-Based Insights

To effectively characterize advances and areas of R&D emphasis, we formulated the conceptual framework shown in figure 1.4, which is broken into three horizons: near-term, midterm, and far-term. The vertical axis is notional, representing AI capability impact. The horizontal axis represents a notional development time. A good reference, for respective times, is to think of near-term spanning over one to two years; midterm as three to four years out; and far-term as approximately five or more years to expect these capabilities to exist in day-to-day operations.

For example, for horizon 1, we foresee that AI advances should take place now, so the capabilities are operational in one to two years. Similarly, for horizon 2, we also envision starting investing now, so AI capabilities are available in operations in three to four years. For horizon 3, if we invest now, at the R&D level, we would look at those capabilities to be available in five or more years. Let us highlight each of the respective horizons. In later chapters, we provide more specifics relevant to each subsystem of interest.

Horizon 1 targets achieving robust and responsible content-based insight. Recall that the AI system architecture, illustrated in figure 1.3, represents a functional architecture for transforming data into insights that users can then use to augment their capabilities to make timely decisions. More specifically, these near-term advancements would be focused on applying AI to gain insight into interesting content that exists on disparate types of data—both structured and unstructured. The main benefits of these developments are the following:

- Reduced user workload
- Improved confidence in AI

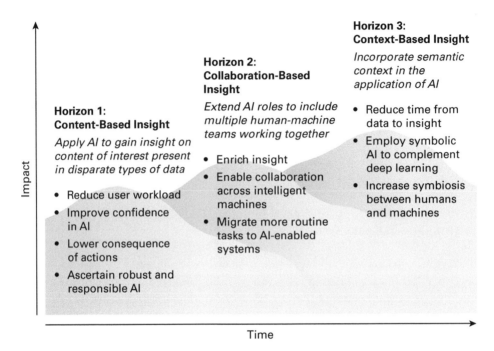

Figure 1.4 Areas of AI emphasis: near-term, midterm, and far-term horizons.

- Lower consequence of actions
- Improved robust AI and RAI [47]

Horizon 2 focuses on collaboration-based insights. This requires that we extend AI roles to include multiple human-machine teams working together. It has been demonstrated that a group of humans aided by a machine (or multiple machines) can outperform an expert in difficult cognitive tasks. Humans are much better than a machine at making subjective judgments, disambiguating options, and understanding context [35, 36]. The main benefits in achieving collaboration-based insights are the following:

- Enriched insight
- Collaboration across intelligent machines
- The ability to migrate more routine tasks to AI enabled systems

Horizon 3 addresses context-based insights. Unlike content-based insights, employing contextual information to perform intelligent tasks is an area where AI systems currently struggle. An AI system, incorporating context, should be able to refine its recommendations with a high degree of confidence by incorporating relevant knowledge from other related input. For example, in the application of terrorist countermeasures, one can envision an AI system that can classify a suspected vehicle but also refine the answer by knowing that the vehicle is parked in a compound identified as housing a terrorist cell. The main benefits of these investments are the following:

- Reduced time from data to reaching insight with high confidence
- Use symbolic AI to complement deep learning
- Increased symbiosis between humans and machines

The next section provides the readers with a road map outlining how each chapter follows the structure in figure 1.1.

1.6   Chapter Road Map

The purpose of this book is to give the readers an understanding of AI as an enabling technology for formulating, designing, building, and integrating complex systems to deliver business value. We employ a systems engineering approach, discussed in more detail in chapter 2. The AI systems engineering approach described in this book consists of three foundational pieces: the AI system architecture, strategic principles, and experiential and project-based learning.

Chapter 2 explains the systems engineering fundamentals by following a methodical and rigorous approach to ascertaining that the AI capabilities meet stakeholder needs. Systems engineering requires paying careful attention to the ultimate stakeholders (e.g., customers, users, and consumers). Chapters 3–8 elaborate, in much more detail, on each of the subcomponents that are necessary to design, build, and integrate the end-to-end AI system architecture shown in figure 1.3. We also include exercises and links to practical hands-on demonstrations. These first eight chapters form the content of part I, as shown in figure 1.1.

After the reader has a working knowledge of the key architecture subcomponents, we then elaborate on AI strategies, deployment techniques, and organizational structures to create an AI value proposition (and AI value capture) by following a set of strategic principles. In part II of the book, chapters 9–13 undertake a deeper dive into these topics.

Chapters 14–18 put parts I and II into practice. Several use cases are described to illustrate applications where AI is the enabling technology. Readers are exposed to the systems engineering methodology, starting with a mission and vision for a candidate AI product or service, formulating an envisioned future, and developing a value proposition centered on the people-process-technology triad. These use cases all demonstrate the development of a strategic road map (i.e., an AI blueprint). An important complementary part of the book is the appendix, where we help the readers work through representative ML examples employing a set of Project Jupyter notebooks.

In parts I and II, we conclude each chapter with a set of the main takeaways, exercises, and extensive references.

1.7   Main Takeaways

The intended audience for this book is AI developers and operational users. The book can also serve as a classroom text for advanced undergraduate- and graduate-level students. The text stems from classes taught by the authors at MIT and is written at a level where executives, technology leaders, and managers can get a broad knowledge in understanding the value proposition that AI can bring to their businesses. Readers do not need a computer science or AI background to understand the comprehensive knowledge taught in the book. After reading the book, readers will have a working knowledge to lead teams, develop AI capabilities, and deploy complex AI systems.

The book has three main parts. Part I presents the AI system architecture and its associated building blocks. Part II delves into the strategic principles to successfully deploy AI capabilities. Part III brings it all together, as the experiential learning component of the book, via use cases.

In chapter 1, we addressed:

- A short history of several notable AI events over the last several decades.
- A definition of narrow AI.
- An AI system architecture introduction, including highlights of the respective building blocks.
- Important challenges in the successful deployment of AI capabilities.
- A look into the future through the lens of three horizons. The near-term horizon addresses content-based insights; the midterm horizon focuses on collaboration-based insights; and the far-term horizon stresses the importance of context-based insights. This horizon framework is used in subsequent chapters to discuss areas of emphasis for each of the AI-pipeline building blocks.

In chapter 2, we describe the fundamentals of AI systems engineering. The field of systems engineering continues to evolve as technological innovations form the foundation for unique products and services. We apply an AI systems engineering approach from architecture principles through deployment. In the context of a system, it is necessary to attend to people and processes, in addition to technology, within the structure of an AI ecosystem. This book explains and introduces the readers to architecture principles, development stages, and integration of subsystem components into an end-to-end system, culminating in successful AI deployment.

1.8   Exercises

1. What does the term *systems* mean in the context of AI?
2. Elaborate on the reasons that led to the two AI winters.
3. What are the characteristics of a one-stage (i.e., single-layer) perceptron?
4. Which of the following is the most accurate definition of narrow AI?
   a. Replacement of all human cognitive capabilities
   b. The performance of tasks that augment human intelligence, such as perceiving, learning, classifying, abstracting, reasoning, and acting
   c. Both a and b
5. Is the confluence of large amounts of data, algorithms, and computing the cause of major advances in AI?
   a. Yes
   b. No
6. What was the primary ML technique responsible for AlphaGo Zero defeating AlphaGo?
   a. Unsupervised learning
   b. Supervised learning
   c. Reinforcement learning
7. Describe the differences between symbolic AI and a class of ML algorithms known as *connectionists*.
8. Outline the AI system architecture building blocks (i.e., architecture subcomponents), and include a short description for each.
9. Enumerate the ten guidelines for a successful deployment of AI capabilities.
10. What are the main benefits of developing AI capabilities as described in horizon 1?
11. What are the main benefits of developing AI capabilities as described in horizon 2?
12. What are the main benefits of developing AI capabilities as described in horizon 3?

13. In a recent question-and-answer session (performed by one of the book's authors) with a version of ChatGPT (developed by OpenAI) used in the search engine Microsoft Bing, the answer to the prompt: *Describe the benefits of leveraging an AI system architecture* resulted in what is shown in quotes. Discuss if the answer is accurate, or if not, why not.

    "An AI system architecture is a way of designing and implementing AI systems that considers various aspects such as data sources, algorithms, models, services, APIs, platforms, etc. Some benefits of leveraging an AI system architecture are:

    - It can help achieve better performance, scalability, reliability, security, and maintainability of AI systems.
    - It can help integrate AI systems with other technologies such as NLP, automation, and machine learning (ML) to enable digital transformation across enterprises and business sectors.
    - It can help create innovative and efficient solutions for various domains such as smart homes, parametric design, conversational tools, etc."

14. Formulate an AI control-based architecture and contrast it to the functional AI system architecture shown in figure 1.3.

15. Would quantum computers, when practical, be useful for other building blocks shown in figure 1.3 besides implementation of ML algorithms?

1.9   References

1. McCulloch, Warren S. and W. Pitts, A logical calculus of the ideas immanent in nervous activity. *Bulletin of Mathematical Biophysics*, 1943. 5(4): 115–133.

2. Nilsson, N. J., *The quest for artificial intelligence*. 2010, Cambridge University Press.

3. Turing, A. M., I., Computing machinery and intelligence. *MIND*, 1950. LIX (236): 433–460. https://doi.org/10.1093/mind/LIX.236.433.

4. Toronto, U., *The Turing test*. n.d. http://www.psych.utoronto.ca/users/reingold /courses/ai/turing.html

5. McCarthy, J., M. L. Minsky, N. Rochester, and C. Shannon, A proposal for the Dartmouth summer research project on artificial intelligence, August 31, 1955. *AI Magazine*, 2006. 27(4): 12.

6. Minsky, M., Steps toward artificial intelligence. *Proceedings of the IRE*, 1961. 49(1): 8–30.

7. *Establishment of MIT CSAIL, July 2003.* https://www.csail.mit.edu/about/mission-history.

8. Rosenblatt, F., The perceptron: A probabilistic model for information storage and organization in the brain. *Psychological Review*, 1958. 65(6): 386–408. https://doi.org/10.1037/h0042519.

9. Minsky, M., and S. Papert, *Perceptrons: An introduction to computational geometry.* 1969, MIT Press.

10. Domingos, P., *The master algorithm: How the quest for the ultimate learning machine will remake our world.* 2015, Basic Books.

11. Minsky, M. and S. A. Papert, *Perceptrons: An introduction to computational geometry.* 2017, Reissue of the 1988 Expanded Edition, MIT Press.

12. Samuel, A. L., Some studies in machine learning using the game of checkers. *IBM Journal of Research and Development*, 1959. 3(3): 210–229.

13. Kasparov, G., *Deep thinking: Where machine intelligence ends and human creativity begins.* 2017, PublicAffairs.

14. Silver, D., A. Huang, C. J. Maddison, et al., Mastering the game of Go with deep neural networks and tree search. *Nature (London)*, 2016. 529(7587): 484–489.

15. Silver, D., J. Schrittweiser, K. Simonyan, et al., Mastering the game of Go without human knowledge. *Nature*, 2017. 550(7676): 354–359. https://www.nature.com/articles/nature24270?sf123103138=1.

16. Holcomb, S., W. K. Porter, S. V. Ault, et al., Overview on DeepMind and its AlphaGo zero AI, in *Proceedings of the 2018 International Conference on Big Data and Education,* Honolulu. 2018, 67–71.

17. Jumper, J., R. Evans, A. Prinzel, et al., Highly accurate protein structure prediction with AlphaFold. *Nature*, 2021. 596(7873): 583–589. https://www.nature.com/articles/s41586-021-03819-2.

18. Bengio, Y., Y. Lecun, and G. Hinton, Deep learning for AI. *Communications of the ACM*, 2021. 64(7): 58–65.

19. Kahneman, D., *Thinking, fast and slow.* 2011, Farrar, Straus and Giroux.

20. *MIT-IBM Watson AI Lab,* https://mitibmwatsonailab.mit.edu/category/neuro-symbolic-ai/.

21. Cox, D., Invited speaker on neurosymbolic AI, in *AAAI 2020 Thirty-Fourth Conference on Artificial Intelligence,* New York City. 2020. AAAI.

22. Mao, J., C. Gan, P. Kohli, et al., *The neuro-symbolic concept learner: Interpreting scenes, words, and sentences from natural supervision.* arXiv preprint arXiv:1904.12584, 2019.

23. Sarker, M. K., L. Zhou, A. Eberhart, and P. Hitzler, *Neuro-symbolic artificial intelligence: Current trends.* arXiv preprint arXiv:2105.05330, 2021.

24. Nye, M., M. H. Tessler, J. B. Tennenbaum, and B. M. Lake, Improving coherence and consistency in neural sequence models with dual-system, neuro-symbolic reasoning. *Advances in Neural Information Processing Systems*, 2021. 34.

25. Martinez, D., AI canonical architecture and cybersecurity examples, in *Artificial Intelligence Conference Hosted by O'Reilly, September 4–7,* San Francisco, 2018. https://www.oreilly.com/videos/the-artificial-intelligence/9781492025832 /9781492025832-video322581/.

26. Martinez, D. MasterClass in artificial intelligence, in *EmTech 2020 leading with innovation.* 2020.

27. Martinez, D., N. Malyska, and W. Streilein, *Artificial intelligence: Short history, present developments, and future outlook.* 2019, MIT Lincoln Laboratory. 135.

28. Moore, A. W., M. Hebert, and S. Shaneman, The AI stack: A blueprint for developing and deploying artificial intelligence, in *Ground/Air multisensor interoperability, integration, and networking for persistent ISR IX.* 2018, International Society for Optics and Photonics. https://doi.org/10.1117/12.2309483.

29. Moore, A. W., Personal communication and discussions. 2018.

30. Alexander, Wissner-Gross, *Datasets over algorithms.* 2016. https://www.edge.org /response-detail/26587.

31. Jacobson, R., *IBM consumer products industry blog.* 2013. https://www.ibm.com /blogs/insights-on-business/consumer-products/2-5-quintillion-bytes-of-data -created-every-day-how-does-cpg-retail-manage-it/.

32. Deng, J., W. Dong, R. Socher, et al., Imagenet: A large-scale hierarchical image database, in *IEEE Conference on Computer Vision and Pattern Recognition, 2009.* IEEE.

33. Krizhevsky, A., I. Sutskever, and G. E. Hinton, Imagenet classification with deep convolutional neural networks, in *Advances in neural information processing systems.* 2012.

34. Martinez, D. R., M. M. Vai, and R. A. Bond, *High-performance embedded computing handbook: A systems perspective.* 2008, CRC Press.

35. National Research Council, *Frontiers in massive data analysis.* 2013, National Academies Press.

36. Quinn, A. J., and B. B. Bederson, Human computation: A survey and taxonomy of a growing field, in *Proceedings of the SIGCHI Conference on Human Factors in Computing Systems,* Toronto. 2011, ACM.

37. Wilson, H. J., and P. R. Daugherty, Collaborative intelligence: Humans and AI are joining forces. *Harvard Business Review*, 2018. 96(4): 114–123.

38. Acemoglu, D., *Redesigning AI*. 2021, MIT Press.

39. Coeckelbergh, M., *AI ethics*. 2020, MIT Press.

40. Shneiderman, B., Responsible AI: Bridging from ethics to practice. *Communications of the ACM*, 2021. 64(8): 32–35.

41. Ng, A., AI doesn't have to be too complicated or expensive for your business. *Harvard Business Review*, 2021. https://hbr.org/2021/07/ai-doesnt-have-to-be -too-complicated-or-expensive-for-your-business.

42. Davenport, T. H., J. Loucks, and D. Schatsky, Bullish on the business value of cognitive: Leaders in cognitive and AI weigh in on what's working and what's next. *2017 Deloitte State of Cognitive Survey*, 2017, 1–25.

43. McKinsey. *Your guide to deploying AI at scale*. 2021. https://www.mckinsey.com /featured-insights/themes/your-guide-to-deploying-ai-at-scale.

44. Ransbotham, S., D. Kiron, P. Gerbert, and M. Reeves, Reshaping business with artificial intelligence: Closing the gap between ambition and action. *MIT Sloan Management Review*, 2017. 59(1).

45. Sculley, D., G. Holt, D. Golovin, et al., Hidden technical debt in machine learning systems. *Advances in Neural Information Processing Systems*, 2015. 28: 2503–2511. https://proceedings.neurips.cc/paper/2015/file/86df7dcfd896fcaf2674f757a2463 eba-Paper.pdf.

46. Zhang, D., et al., *The AI index 2021 annual report*, AI Index and S. Committee, Editors. 2021, Stanford, CA: Stanford University.

47. Dietterich, T. G., Steps toward robust artificial intelligence. *AI Magazine*, 2017. 38(3): 3–24.

I

AI System Architecture

2

Fundamentals of Systems Engineering

> If America wants to put a man on the moon, which is really a tough engineering job, they just gather enough thousands of scientists, pour in the money, and the man will get there. He may even get back.
> —Vannevar Bush, former dean, professor, and presidential advisor

During the twentieth century, there were many revolutionary projects that were accomplished through the great scientific and engineering contributions of academic scientists and engineers. There were also visionaries, such as Vannevar Bush, a former dean and professor at the Massachusetts Institute of Technology (MIT) as well as an adviser to US presidents Franklin D. Roosevelt and Harry S. Truman, who emphasized the importance of scientific research enabling discoveries that were then transformed into actual systems [1]. Bush was a pioneer in the early development of analog versions of present-day technologies with relevance to artificial intelligence (AI). For example, he articulated the need for a personal assistant supplementing a human's memory, which he called "Memex." He envisioned Memex as follows: "A device in which an individual stores all his books, records, and communications, and which is mechanized so that it may be consulted with exceeding speed and flexibility" [2]. In addition to being the voice behind the creation of the National Science Foundation in the US, Bush was the chair of the National Advisory Committee for Aeronautics, which became the National Aeronautics and Space Administration (NASA). We chronicle this history, as well as the importance of having scholars with great vision such as Bush, because scientific investments are crucial as a precursor to bringing complex AI capabilities into the market while adhering to the systems engineering discipline.

NASA has a long history of demonstrating the use of rigorous systems engineering principles, as embodied in its comprehensive *NASA Systems Engineering Handbook* [3].

29

NASA has achieved unprecedented success and continues to expand the limits of the possible. There have also been catastrophic disasters, such as the *Challenger* explosion during its launch in 1986, and the space shuttle *Columbia* breakup during its return to Earth in 2003. After careful investigation of the cause of the *Columbia* disaster, it was concluded that there was a failure in leadership in not accepting input from the engineering ranks on the status of the foam attached to the structure of the external tank. It is important to recount this example since the failure was not with the systems engineering methodology. Instead, the grave mistake, which cost astronauts' lives, was leadership not accepting feedback. An important lesson learned, mapped to the topic of this book, is that leadership plays just as important a role, in the successful deployment of AI products and services, as the technology used to implement AI capabilities.

There is also an ethical component to systems engineering that, if not attended to properly, can result in significant costs to organizations and society as a whole. For example, Volkswagen failed to properly follow device under test procedures under normal operating conditions. Its testing approach allowed its diesel vehicles to pass emissions tests using a defeat device, violating the Environmental Protection Agency (EPA) regulations [4]. Another more recent example is when an Uber self-driving car killed a pedestrian in Tempe, Arizona [5–7]. The postaccident analysis indicated that the machine learning (ML; computer vision) algorithm did not have enough time to react to avoid the pedestrian. These are just a few examples where a more diligent approach to developing complex systems, championed and supported by the leadership and management, would have helped to avoid costly errors and fatal accidents.

As AI continues to advance the use of sophisticated real-time sensors and ML, such as laser detection and ranging (LADAR) and radar sensors mounted on driverless cars, sensors performing medical diagnosis, and detecting the fraudulent use of AI (e.g., deepfakes), there needs to be a more rigorous process of verification and validation (V&V) of AI capabilities at the full end-to-end system level. Human lives are at stake when AI is not performing as expected. Therefore, systems engineering is of paramount importance, employing decades of very well defined approaches that attend to all aspects of complex system architecture design, development, integration, verification/validation, deployment, and life-cycle monitoring.

In this chapter, we introduce the reader to the fundamentals of systems engineering, and then map its key characteristics to AI capability development and deployment [8]. In the next section, we begin with common systems engineering definitions. As we will elaborate shortly, the concept has several definitions, but they all have a common theme: an interdisciplinary approach incorporating technology enablers coupled to engineering management of people and processes forming part of integrated elements (subsystems), with the objective of achieving a useful function.

After setting the definition and how we employ a set of systems engineering guidelines, in section 2.2, we discuss key characteristics within the systems engineering discipline. In section 2.3, we delve deeper and apply the key characteristics to the development and deployment of AI products and services. In section 2.4, we map systems engineering principles to our AI system architecture and the complementary architectural framework for implementation.

As already stated in the introduction to this chapter, the people aspect of systems engineering is key to the successful development and deployment of AI capabilities. Therefore, we dedicate section 2.5 to leadership principles as highlighted in the context of systems engineering. With the fundamentals discussed in sections 2.1–2.5, the reader will have the background necessary to understand how systems engineering is a discipline that reduces the risk or likelihood of an AI system improperly functioning in operations.

We conclude this chapter, as we will do in chapters 3–8, with a discussion of the challenges ahead, the main takeaways, and a set of exercises. At that point, the reader will have a working knowledge of the important tools, tips, and techniques available to AI leaders, practitioners, and technologists when employing systems engineering guidelines.

2.1   Systems Engineering Common Definitions

It is important to establish up front a common definition for systems engineering since the field has evolved significantly from its earlier days. However, before we present common definitions of the term and summarize the definition used throughout this book, we begin with a short history on the evolution of systems engineering, with representative examples.

One of the most formal forums that conform with several other standards in systems engineering is the International Council on Systems Engineering (INCOSE) [9]. The *INCOSE Systems Engineering Handbook* (fourth edition) makes sure that the guidelines are also consistent with other standards, such as those presented in the *Guide to the Systems Engineering Body of Knowledge* (SEBook, 2014) [10], and the ISO/IEC/IEEE 15288 international standard [11]. Similarly, the National Institute of Standards and Technology (NIST) has created *Systems Security Engineering: An Integrated Approach to Building Trustworthy Resilient Systems*, conforming with INCOSE and the ISO/IEC/IEEE 15288 international standard [12]. All these publications emphasize the need to maintain the appropriate rigor, implementation discipline, and monitoring throughout the life cycle of a complex system. We reference these handbooks and standards because, as we adapt their guidelines to the application of systems engineering to AI, we need to be consistent with well-established international forums.

In the INCOSE handbook, the editors present important dates of the origins of systems engineering as a discipline (see table 2.1 and page 12 in [9]). For example, the Semi-Automatic Ground Environment (SAGE) was an air defense system developed at the MIT Lincoln Laboratory starting with input data from distributed radar sensors culminating with real-time insights provided to human operators. Its development in the 1950s exemplified a one-of-a-kind demonstration of a full systems engineering marvel, including many innovations in technologies and processes [13]. The architecture, design, development, and implementation of SAGE were led by Jay Forrester, a pioneering American computer engineer and systems scientist who invented the random access magnetic-core memory. As featured in the important dates in the origin of SE given by INCOSE, Jay Forrester also developed a system dynamics modeling language and is considered the founder of the system dynamics field as a result of his contributions [14–16]. In the celebration of the sixtieth anniversary of the system dynamics field, John Sterman's quote is as true then as it is today when seen through the lens of AI-engineered systems: "Today, more than ever, humanity faces grave threats arising from our mismanagement of increasingly complex systems" [16]. Other notable events and dates highlighted in the INCOSE handbook are the formation of the National Council on Systems Engineering in 1990, which then evolved in 1995 into INCOSE to incorporate a wider inclusion of an international community.

With this historical perspective as background, we can now present common systems engineering definitions. Here, we repeat a set of definitions to give the reader a view of what is commonly emphasized in the developing and deployment of systems. NASA defines "systems engineering" as follows [3]:

> A methodical, multi-disciplinary approach for the design, realization, technical management, operations, and retirement of a system. A "system" is the combination of elements that function together to produce the capability required to meet a need. The elements include all hardware, software, equipment, facilities, personnel, processes, and procedures needed for this purpose; that is, all things required to produce system-level results.

Similarly, INCOSE's definition of the term emphasized subsystems coming together as an individual system inclusive of support elements [9]—and that is the definition that we adapt for this book:

> An integrated set of elements, subsystems, or assemblies that accomplish a defined objective. These elements include products (hardware, software,

firmware), processes, people, information, techniques, facilities, services, and other support elements.

Another useful perspective is a systems engineering definition from the famous American engineer, entrepreneur, and author Simon Ramo, who started the successful company TRW (originally Thompson-Ramo-Wooldridge) [17]:

> A discipline that concentrates on the design and application of the whole system as distinct from the parts. It involves looking at a problem in its entirety, taking into account all the facets and all the variables, and relating the social to the technical aspect.

Howard Eisner, in his book *Systems Engineering: Building Successful Systems*, does an excellent job summarizing several additional common definitions as well [18]. We highlight these definitions because, although they do not explicitly mention AI, the definitions are inclusive of the triad of people, process, and technology, as we will discuss later in this book.

All these definitions have several key descriptors in common. They all bring out the importance of approaching the development of a system from a holistic viewpoint. In the context of technology, they point out that a system needs to address, systematically, all the underlying subsystems working in unison. From a process perspective, as we will discuss later in this book, stakeholder requirements, integration, verification, and validation, among others, must be addressed at each of the individual subcomponents, and as a full-system leveraging performance metrics. Last but certainly not least, people working on the system must have a systems engineering mindset and act as an integrated team to adhere to sound engineering principles, including having high ethics standards. In the next section, we dive deeper into the important characteristics most relevant to the systems engineering approach addressed here.

2.2 Characteristics Espoused within the Systems Engineering Discipline

The systems engineering discipline was officially endorsed internationally in a more formal way after the establishment of forums and standards discussed earlier. As mentioned previously, INCOSE evolved from the National Council on Systems Engineering in 1995. During the same decade, in a seminal paper, K. Forsberg and H. Mooz (from the Center for Systems Management) introduced the now well known "Vee-model" [19],

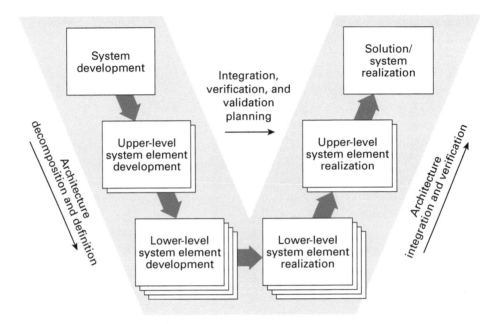

Figure 2.1   Modified Vee-model from the original depiction of system life-cycle development [19], starting with stakeholders' specifications through system realization [9]. © 2015 John Wiley & Sons, Inc.

shown in figure 2.1. Since then, several other organizations and authors have emphasized the importance of adopting the Vee-model representation when developing and deploying complex systems [9, 18, 20].

The Vee-model representation is well matched to our systems engineering approach as applied to AI. The relevance is a holistic view of the development of complex systems. Specifically, in their original paper, K. Forsberg and H. Mooz [19] emphasized the importance of starting at the highest level—stakeholders' system requirements—and continuing with a more detailed architecture decomposition, as shown on the left side of figure 2.1. Upon defining the key subcomponents (lower-level system elements), one then proceeds with the architecture integration, verification, and validation shown on the right side of figure 2.1, and culminates with a higher-level system capability. A very important function is to ascertain that the stakeholder requirements formulated at the start of the system development are met at the last stage of the solution/system realization. In sections 2.3 and 2.4, we capture the main stages of the

Vee-model process, but in the context of the AI system architecture illustrated in figure 1.3 in chapter 1.

As mentioned earlier—and encapsulated in the INCOSE systems engineering definition—the successful development and deployment (including implementation, verification, and validation) of a complex system demands that we follow a rigorous set of principles. Table 2.1 describes a set of key characteristics of systems engineering. These characteristics are formulated across the triad of people-process-technology, but they incorporate each of the building blocks (subsystems) illustrated in the AI system architecture shown in figure 1.3.

Table 2.1 outlines the key characteristics most relevant to AI within the context of using a systems engineering approach. There are several terms highlighted in the table that deserve further clarification. More specifically, we provide definitions for several terms:

- *Threshold requirements:* The minimum set of system requirements that an AI product or service must meet.
- *Target requirements:* The desired stakeholder's goals. However, there is always a need to perform a trade-off between threshold and target requirements by performing a risk analysis.
- *Risk analysis:* System trade-offs analysis at all stages of the Vee-model, from architecture decomposition and subcomponents definition through architecture integration-verification-validation. A very useful tool commonly used by systems engineers is the likelihood of occurrence versus consequence. This risk management trade-off will be discussed further in later chapters.

Before we conclude this section, it is important to refine our systems engineering definition based on the key systems engineering characteristics shown in table 2.1. Our AI systems engineering definition, which is most relevant to AI development and deployments, is adapted from the INCOSE definition (adapted to our AI context from their fourth edition handbook) discussed in section 2.1 [9]:

Our definition: "An integrated set of AI architecture elements, subsystems, or assemblies that accomplish a defined objective. These elements include *technologies* as enablers of AI products (hardware, software, firmware) or AI services, adhering to a set of *processes*, and using *people*, information, techniques, infrastructure, and other support elements. The integrated AI system must meet a minimum set of threshold requirements, while attending to FASTEPS principles."

The systems engineering key characteristics shown in table 2.1 map well to the systems approach discussed in the book. Chapters 3–8 are focused on technologies as AI

Table 2.1 Systems engineering key characteristics

Key Characteristics	Description	Risk Analysis
Technology		
System analysis trade-offs	Trade-offs must be performed incorporating a stakeholder's business goals and the AI capabilities provided. Clear articulation of the user/consumer/customer (stakeholder) AI application occurs at this early step. Perform trade-offs between threshold and target requirements.	Can the AI capability provided be scaled? Are the stakeholder's requirements realistic? Risks in achieving target requirements versus threshold requirements.
Functional architecture	System architecture building blocks.	End-to-end measures of performance.
Sensors and sources	Identify the data needed (structured or unstructured).	Are the required data available?
Data conditioning, ML, modern computing, human–machine teaming (HMT)	Formulate the data preparation needed; select the ML algorithms; identify the computing infrastructure. (enterprise and edge computing); address HMT levels of collaboration to achieve augmented intelligence.	Technology management and technology levels of maturity.
Robust AI	Vulnerability to adversarial AI.	What is the likelihood versus the consequence of unintentional or intentional adversaries?
Responsible AI (RAI)	Use tools and techniques to incorporate FASTEPS (fairness, accountability, safety, transparency, ethics, privacy, and security) principles.	To what degree are the harms mitigated and FASTEPS principles implemented?
Process		
Stakeholder needs (business goals)	Business goals drive the AI capabilities requirements.	Are the needs and goals realistic?
Integration-verification-validation (I-V&V)	I-V&V work in concert at each stage of the Vee-model. Verification is most often referred to as "building things right." Validation is about testing the full end-to-end system. Is the AI system enabling the right capability?	I-V&V must be exercised at all levels of the Vee-model, from system definition to system deployment, including a deployment readiness review.

Development, deployment, and monitoring	The characteristics given here are part of development, deployment, and monitoring, plus the required risk analysis
Operational excellence	Operational excellence is the ability to deliver business value (value capture) based on the AI value proposition, and to continually improve supporting processes and procedures through system monitoring. It cuts across all subcomponents identified in this table under "Technology." It must be assessed for each of the building blocks.
Cost optimization	Cost optimization starts at the stage of system development definition (see Vee-model), but it continues throughout the life cycle of the system. Minimize costs while still achieving the stakeholder's needs and goals.
Reliability and security	Each of the building blocks in the AI system architecture must be assessed in terms of their ability to meet the threshold and/or target requirements in a secured way. What is the likelihood of a failure and its associated consequence?
Operational metrics	These metrics include measures of performance (MoPs) for each of the subcomponents and the overall AI architecture. MoPs lead to the measure of effectiveness (MoE). Metrics should be quantitative based on well-established system benchmarks (i.e., gold standards). Including accuracy, precision, recall, F-scores, and other metrics, these are assessed relative to model performance and impact on the overall system.
Risk management	Includes technical risks (e.g., performance risk), management risks (e.g., cost and schedule risks), and organizational risks (e.g., societal risks). Must address the broad range of risks in the context of likelihood versus consequence (impact).
Quality assurance	A broad term that asserts that a product or service meets a set of specifications. Specs can include size, weight, power, shock, vibration, and humidity. In addition, ISO 9000 addresses manufacturing standards. Should be addressed from the design trade-offs through system deployment and monitoring.

(continued)

Table 2.1 (continued)

Key Characteristics	Description	Risk Analysis
Cross-cutting systems engineering	In this category, there are tools and approaches to assess performance of an AI system prior to deployment, such as modeling and simulation, emulation, and tabletop exercises simulating realistic environments.	Employing these tools and techniques should also include the use of operational metrics.
People		
Leadership and management	Leaders define and drive the AI strategic direction and road map. Management must be focused on execution.	Leaders are responsible for assessing the strategic risks. Managers are responsible for addressing execution risks. For example, is the necessary talent available to execute a task?
Systems thinker and systems engineer	Every AI team must include a person with responsibility for and authority over the overall AI system meeting a set of system requirements.	In concert with the AI leadership and management, the systems engineer must assess overall AI system risks.
Staff performance	Management must set clear staff performance goals, and assess progress several times during the AI project.	There are tools to measure what matters discussed later in the book.
Multicultural environment	AI projects require a multidisciplinary team including such members as systems engineers, data scientists, ML experts, computing technologists, and human-machine social engineers.	Assess the organization's mentorship and coaching approaches.
Effective communications	All members of an AI team must have the ability to communicate effectively at the subcomponent level up to the system level. Leadership and management also have the obligation to communicate effectively to their system stakeholders	Communication is paramount to address the problems elsewhere in this text. This attribute includes ethics in systems engineering.

system enablers. Chapters 9–11 elaborate in more detail on important processes to ascertain successful development and deployment of AI capabilities. People topics are covered in much greater detail in chapters 12 and 13. We then follow with a set of use cases in chapters 14–18.

In the next section, we elaborate further about the systems engineering discipline as it applies to AI. We also highlight recent work and discuss the relevance of this prior work to our methodology.

2.3   A Systems Engineering Approach Applied to Artificial Intelligence

Systems engineering is a discipline that, if employed properly, can minimize failures with AI-enabled systems. Ipek Ozkaya discusses what is different about engineering AI-enabled systems [21]. For example, Ozkaya points out the importance of integrating the data both as an enabler and a constraint, since the ML models rely on proper data as the input to an AI system. Verification is also challenging compared to more deterministic systems because AI-enabled systems are probabilistic by their very nature. Another prior study, which complements Ozkaya's work, is by Bosch and colleagues., where they address problems and challenges in transitioning ML-models into production [22]. They surveyed sixteen companies to identify a set of common difficulties among a broad range of industries. They categorized the difficulties with data quality, design methods, performance of models, and compliance. This prior research reinforces the need for a more rigorous systems engineering approach.

There are many lessons learned that one can use in the engineering of complex systems. For example, Martinez et al. employed a systems engineering approach in the development of a complex, multiprocessor system consistent with the Vee-model shown in figure 2.1, starting from the system architecture decomposition and definition through a rigorous integration, subcomponents verification, and system validation [23]. Martinez and Sobol also discussed tools appropriate for performing system analysis in the implementation of an AI expert system when applied to a simplified business use case drawn from the oil and gas exploration industry [24]. System analysis is the first stage of the set of key characteristics shown in table 2.1. These case examples serve to illustrate that the systems engineering approach, as a discipline, is proved to achieve success with the deployment of complex systems.

Most recently, Llinas presented a review of systems engineering for AI-based systems emphasizing the need for a rigorous approach to system development and deployment [25]. Several researchers have emphasized the importance of recognizing that the ML code, although it is difficult to make sure that it works properly for the AI application, is only a small part of the overall holistic view of an AI system. Sculley et al. articulated a very clear description of the "hidden technical debt," as a metaphor, of the potential large cost that can be incurred if other system components and infrastructure are not integrated together [26].

Ultimately, the business goal of any organization developing and delivering AI products and services is to provide value to users/consumers and customers. However, it is imperative that as the AI systems increase in complexity, we perform in-depth risk assessment and management of all stages shown in the Vee-model, exercised under different failure modes, even including a graceful degradation mode. Many of the companies considered AI trailblazers were not even in existence as recent as over a decade ago [27]. However, these organizations will not be able to maintain their history of AI successes, as their offerings increase in complexity, if they do not employ and practice a more rigorous systems engineering approach. In the next section, we pivot to the coupling of the AI system architecture, illustrated in figure 1.3 of chapter 1, and the employment of ML into operations but seen through the systems engineering lens.

2.4   Architecture Framework: The What and the How

Sections 2.1–2.3 set the stage for the systems engineering approach to AI, with discussions of the standards, techniques, and key characteristics of systems engineering (see table 2.1). In this section, we discuss how our AI system architecture shown in chapter 1, together with our strategic development model, enables the AI leaders and developers to formalize a rigorous methodology to successfully transition from early-stage prototyping (proof of concept) to production. Andrew Ng, the chief executive officer (CEO) and cofounder of Landing AI, has said: "Those of us in machine learning are really good at doing well on a test set, but unfortunately deploying a system takes more than doing well on a test set. All of AI [presently] . . . has a proof-of-concept-to-production gap" [28].

The AI system architecture, illustrated in figure 1.3, presents a functional architecture decomposition and definition of each of the building blocks from data inputs through insights created and delivered to the users/consumers or customers. We referred to this AI framework as the "what" the AI system consists of, which results in an AI product or service. This architectural framework forms the left side of the Vee-model from the system development stage through lower-system element developments.

Equally important is the "how" of an end-to-end system implementation—meaning the right side of the Vee-model from lower-level system realizations to end-to-end solution/system realization. For the implementation of a system, we are inspired by the guidelines described in the Amazon Web Services (AWS) well-architected framework [29]. This implementation framework is very well matched to a machine learning operations (MLOps) methodology. Let us provide the reader with a brief discussion on how the AWS framework relates to an MLOps methodology. We do a deeper dive into this topic in chapter 11.

MLOps is a methodology that has evolved from the software development community drawing from the principles of development, security, and operations (DevSecOps) throughout the full-system life cycle [30]. Our adaptation of an MLOps end-to-end life cycle, at the high level, starts from the business goals (business understanding), followed by the building blocks of the AI system architecture (from data conditioning through robust and responsible AI), but with the added stages of deployment, monitoring, and stakeholder assessment. Recall that as per our definition, a stakeholder can be a user/consumer or a customer.

Drawn from the AWS ML framework [31], note that in our "how" framework, we have adapted and overlaid additional attributes beyond the five attributes from the AWS framework, as shown in table 2.2, to include risk management and quality assurance. These additions are consistent with the description of the key characteristics described in

Table 2.2   Well-architected framework attributes through the AWS ML lens [29, 31]

Attribute Name	Description
Operational excellence	The ability to run and monitor systems to deliver business value and to continually improve supporting processes and procedures
Security	The ability to protect information, systems, and assets while delivering business value through risk assessments and mitigation strategies
Reliability	The ability of a system to recover from infrastructure or service disruptions, dynamically acquire computing resources to meet demand, and mitigate disruptions such as misconfigurations or transient network issues
Performance efficiency (or metrics)	The ability to use computing resources efficiently to meet system requirements, and to maintain that efficiency as demand changes and technologies evolve at an acceptable level of performance
Cost optimization	The ability to avoid or eliminate unneeded costs or suboptimal resources

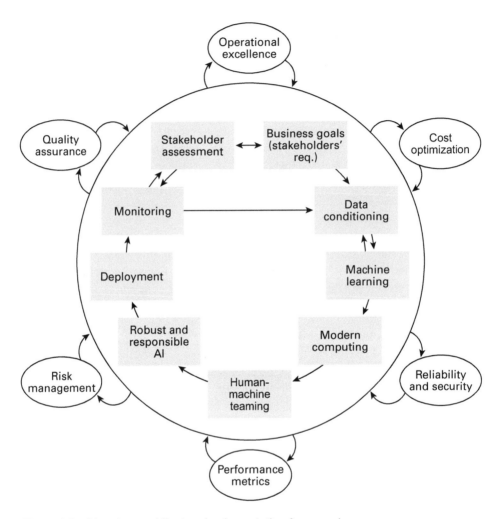

Figure 2.2   AI system architecture implementation framework.

table 2.1 under "Development, deployment, and monitoring." The "how" construct of our AI system architecture implementation framework is illustrated in figure 2.2.

Several of the inner-circle stages are the building blocks from our AI system's functional architecture (the "what"). We complement these building blocks (or stages of development) with business goals (or stakeholders' requirements), deployment, monitoring, and stakeholder assessment. In this last stage, the stakeholder (whether user/consumer or customer) must assess whether the business goals are being met. These stages are not strictly sequential, but rather iterative. During the AI system architecture implementation (the "how"), each of the inner-circle stages must be integrated-verified-validated against operational excellence, cost optimization, reliability and security, performance metrics, risk management, and quality assurance.

Thus far, we have discussed a systematic approach to the development and deployment of AI capabilities. With a more appropriate systems engineering definition applied to AI, we enumerated a set of systems engineering key characteristics. The key characteristics of systems engineering were used as a bridge between our AI system architecture at the functional level (the "what" in figure 1.3) to our AI system architecture implementation framework (the "how" in figure 2.2). All of this discussion, so far, has centered on technology and process. In the next section, we begin to incorporate another important piece of the puzzle—people and the associated systems engineering leadership.

2.5 Leadership: Systems Thinker

Systems engineering leadership is core to the successful development and deployment of AI capabilities. The field of AI evolves so rapidly that some of the important responsibilities of those in leadership roles are to accept change, to formulate a clear mission and vision for the AI team, to work closely with the technical team to identify the value proposition that uses the AI system architecture, and to ascertain a well-defined, strategic road map. The rapid advancements experienced with AI, as discussed in chapter 1, have been driven by the confluence of data availability, algorithms, and computing. Before we discuss systems engineering leadership roles, we present a very good proxy to past advancements in AI—compute usage in training AI systems—which have shown exponential growth.

OpenAI tracks closely the ML training demands in algorithm computation, as shown in figure 2.3 [32]. Compute usage in training AI systems is a very good proxy to show the exponential growth that AI has experienced in the past several decades. The vertical axis is the number of peta–floating-point operations per second–days (peta-FLOPS-days). Peta-FLOPS-days represents 10^{15} FLOPS executed over 24 hours, or equivalent to 10^{20} floating-point operations. The horizontal axis represents years. As we see, from the days of the perceptron (c. 1960) to about 2012, the algorithm computations have tracked closely with Moore's Law ($2x$ increase every 2 years). Starting at about 2012, the algorithm computation has increased two times every 3.4 months.

As per OpenAI, the AI algorithm era since 2012 has experienced exponential growth in the computing needs of about 128^5 every 10 years versus 2^5 for Moore's Law over the same 10-year period, or equivalently an increase of about 10^9 in exponential growth as

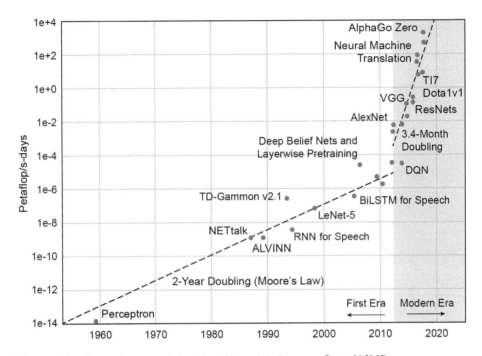

Figure 2.3   Compute usage in training AI models (source: OpenAI [32]).

a metric versus $2x$ every 2 years. This exponential growth (as an AI growth proxy) means that the opportunities for continuing to find new AI applications will continue to be astonishing. However, modern computing must enable such advances in AI algorithms; including computation in a low carbon footprint, which is quite challenging, as we will discuss in chapter 5. Also, the question that we have to address is how leadership in technology-based organization will adapt and manage this exponential growth opportunity.

In chapter 12, "Fostering an Innovative Team Environment," we elaborate in more detail the roles and responsibilities in leading AI teams. However, because of the importance of leadership in systems engineering, we begin by highlighting, as articulated by the famous scholar Warren Bennis [33], the four qualities of great leaders, which are also very relevant to our discussion in this section:

1. To create shared meaning (i.e., to have a vision)
2. To have a distinctive voice (i.e., to be authentic)

3. To have integrity (i.e., to be honest and to have ethics)
4. To have capacity (i.e., to be responsible for managing change)

These four leadership qualities are important as they apply to our AI systems engineering context, but for now, we want to look further at qualities 1 and 4. Leaders responsible for employing the systems engineering discipline can minimize risk and increase the likelihood of successfully developing and deploying AI capabilities by following the methodology shown in figure 2.4. In chapter 9, we describe each of the steps of this AI strategic development model (AISDM). Leaders must have a well-articulated, long-term mission and vision (leadership quality 1) for their AI products and services. They must also formulate a strategic direction consistent with the rapid changes occurring in AI (leadership quality 4).

The National Academy of Engineering (NAE) released *The Engineer of 2020: Visions of Engineering in the New Century*, a comprehensive study of the likely transformational changes ahead (envisioned future) and their impact on the desired engineer's competences [34]. In their report, the study members converge on a number of aspirational goals for engineers in 2020, but one aspiration worth repeating here, which is all-encompassing of our discussion in this chapter, is the following:

> We aspire to engineers in 2020 who will remain well grounded in the basics of mathematics and science, and who will expand their vision of design through a solid grounding in the humanities, social sciences, and economics. Emphasis on the creative process will allow more effective leadership in the development and application of next-generation technologies to problems of the future.

Engineers, and more generally all the talented staff working on advancing general purpose technologies such as AI, must be grounded in more than technological prowess, in order to include the softer sciences, such as leadership. The NAE study concludes with a set of key attributes that engineers—and again, more generally in our context, all AI workers—must strive to:

1. To be creative
2. To be good communicators (i.e., communicating convincingly to their stakeholders, as discussed further in chapter 13)
3. To have business and management acumen (i.e., the ability to execute)
4. To understand and practice leadership principles
5. To hold to the outmost high ethical standards and professionalism

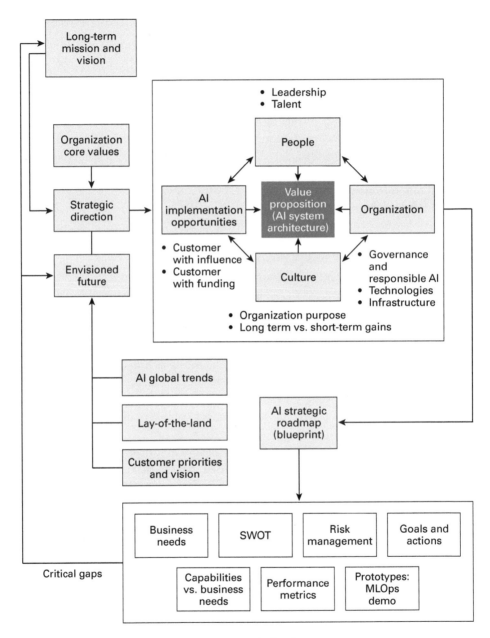

Figure 2.4   AI strategic development model (AISDM).

6. To demonstrate dynamism, agility, resilience, and flexibility (driven by technological and social-political-economic worldwide influences)
7. To be lifelong learners

These broad NAE attributes complement well the more specific INCOSE key competencies of systems engineers—namely, being systems thinkers [9]. Table 2.3 highlights the set of systems thinkers' skills formulated by INCOSE. An integral part of
our strategic development model is for AI leaders to be able to think strategically—
this means to be systems thinkers.

The methodology illustrated in figure 2.4, and discussed in more detail in chapter 9, facilitates a step-by-step process for AI leaders, working closely with systems
engineers and their AI teams to produce a strategic road map (a blueprint) for their AI
products or services. This strategic development model enables the practicing of the
NAE key attributes and the systems thinking skills enumerated in this chapter.

To recap, systems engineering must address the triad encompassed by people-process-
technology. In the previous sections, we have elaborated on all three of these foundational elements, viewed through the lens of the fundamentals of systems engineering.
We conclude this chapter by looking ahead to systems engineering challenges with
respect to developing and operationalizing AI capabilities.

Table 2.3   Systems thinkers' skills

Essential Skills of a Systems Thinker
Seeks to understand the big picture
Observes how elements within the system change over time, generating patterns and trends
Recognizes that a system's structure (elements and their interactions) generates behavior
Identifies the circular nature of complex cause-and-effect relationships
Surfaces and tests assumptions
Changes perspective to increase understanding
Considers an issue fully and resists the urge to come to a quick conclusion
Considers how mental models affect current reality and the future
Uses understanding of system structure to identify possible leverage actions
Considers both short- and long-term consequences of actions
Finds where unintended consequences emerge
Recognizes the impact of time delays when exploring cause-and-effect relationships
Checks results and changes actions if needed: "successive approximation"

2.6   Systems Engineering Challenges

Systems engineering as a discipline, as discussed in this chapter, adheres to a set of fundamentals that have stood the test of time. But in addition, there are several important challenges that must be addressed as we apply systems engineering to AI. In this section, we present those challenges focused on the development and deployment of AI products and services.

Similar to earlier sections, we structure the challenges based on the triad of people-process-technology. This structure helps us delineate challenges over the full systems engineering spectrum—AI is not just about the best technology. We must give equal attention to processes and people, as delineated in table 2.1. We identify respective challenges under these three categories:

Technology Challenges

1. Clearly articulating the stakeholder's requirements (both threshold and target requirements)
2. Functional architecture:
   - Design of a realistic architecture that can be implemented within the constraints of the available resources, schedule, and budget, while still meeting the stakeholder's requirements.
   - Availability of respective data (labeled "data" if using a supervised learning algorithm). Emphasis should be on good data (data that meet the performance desired of the ML algorithms) versus just big data.
   - Rapid iteration of candidate ML algorithms, leveraging an agile development process, for example (more on this topic in chapter 10).
   - The determination of whether the computing should be performed at the enterprise level, at the edge, or a combination of both.
   - Identification of the right operating point between machines performing a task versus humans being complemented by the machine (i.e., the trade-off between confidence of the machine providing the desired insight relative to the consequence of actions).
   - Susceptibility to adversarial (intentional) attacks or unintentional use of AI systems.
   - Balancing all the issues addressed under the rubric of FASTEPS.

Process Challenges

1. Standing up an AI ecosystem (discussed in chapter 10) that addresses the full life cycle of the system, from development through operations.
2. Methodically traversing the Vee-model (shown in figure 2.1) while performing the tasks of integration-verification-validation. Another common terminology for these tasks is test, evaluation, verification, and validation (TEV&V).
3. Risk management from development, deployment, and monitoring of the AI system, including model performance.
4. Incorporating various degrees of system fidelity that work in real, emulated, and simulated scenarios.

People Challenges

1. Competent systems thinkers at the leadership and management ranks with the ability to set the strategic plan and to execute the plan, respectively—plan the work and work the plan
2. Staff availability to execute, formulating clear and achievable staff performance goals and maintaining a high level of coaching and mentoring across the organization
3. Communicating effectively across all elements of the AI system architecture including low levels of system developments and realizations (as illustrated in the Vee-model construct)

The systems engineering discipline, following the prescribed step-by-step methodology discussed in this chapter, provides a set of tools, tips, and techniques to AI leaders, developers, system integrators, and to those responsible for deployment of a system. The systems engineering guidance helps to reduce risk and increase the likelihood of successful AI development and deployment. As emphasized by Luke et al. there is a critical need for better alignment between the real-world business applications and processes to provide business value—the use of new engineering practices facilitates such organizational goals [35]. All organizations across industry and research and development (R&D) laboratories must attend to AI from development through deployment [30, 36, 37]. Amershi et al. performed an in-depth interview within Microsoft to identify what makes the AI domain fundamentally different to other more traditional software application domains [38]. They highlighted a set of challenges that need to be addressed and are essential to the practicing of AI by software teams. Similarly, academia must teach not only the theoretical underpinnings of AI, but also practical approaches to realizing the great promise that AI provides across many industries.

The following chapters address in much greater detail a systems approach to AI through the lens of the fundamentals of systems engineering presented in this chapter. In the next section, we summarize this chapter's main conclusions. Then the exercises that follow help the reader cement the lessons of the chapter.

2.7   Main Takeaways

In this chapter, we have set the stage for the following chapters, which will emphasize the systems approach to AI, employing the fundamentals of systems engineering as a discipline. We adapt the systems engineering principles to the development process and to the deployment of AI capabilities throughout the book.

We began the chapter by recounting a history of the systems engineering discipline as described in the INCOSE handbook [9]. The INCOSE guidelines help us to stay consistent with well-established standards that have withstood the test of time. Because the field has evolved for many decades, we presented a set of definitions to clarify how we employ systems engineering. All these definitions have a common theme: an interdisciplinary approach incorporating technology enablers coupled to engineering management of people and processes that form part of integrated elements (subsystems), with the objective of achieving a useful function.

The definition of "AI systems engineering," used throughout our book and informed by the INCOSE's definition, is as follows:

> *Our definition:* An integrated set of AI architecture elements, subsystems, or assemblies that accomplish a defined objective. These elements include *technologies* as enablers of AI products (hardware, software, firmware) or AI services, adhering to a set of *processes*, and using *people*, information, techniques, infrastructure, and other support elements. The integrated AI system must meet a minimum set of threshold requirements, while attending to FASTEPS issues.

Within the systems engineering guidelines, practitioners can employ the Vee-model as a step-by-step methodology to reduce risks and improve the likelihood of a successful development and transition of a complex system into operations. The Vee-model representation is well matched to our systems engineering approach applied to AI. Its relevance to our systems engineering approach is the holistic view of developing and integrating a complex system.

We showed our AI system architecture in figure 1.3 of chapter 1. We referred to this functional architecture as "the what"—meaning what building blocks must be developed. In this chapter, we integrated these functional building blocks into an implementation architecture framework—"the how"—which focused on the implementation of the functional architecture following the Vee-model process. This framework was inspired by the AWS ML framework [31]. We incorporated additional implementation stages, shown in figure 2.2, including a stakeholder's assessment step.

We also enumerated a set of systems engineering key characteristics. This set of characteristics emphasized the importance of not just thinking about AI as a set of advanced technologies. We use the label of "people-process-technology" to describe this set of key characteristics. Advances in technology are necessary but not sufficient. More specifically, AI leadership and talent (people), as well as a rigorous process, are paramount to successfully bringing AI capabilities into operations.

Leadership must set the vision and develop the strategic plan, and management must execute the plan. We discussed leadership attributes, as well as the need to have a systems thinker mindset. The NAE recommended a set of engineer's attributes in its very comprehensive study from the perspective of a vision of engineering in the twenty-first century [34]. We outlined the NAE engineer's attributes, in addition to a more detailed set of systems thinker skills, as formulated in the INCOSE handbook [9]. In this book, under the label "engineer," we include all those people responsible for the successful engineering of an AI system, not just engineers by education.

We introduced an AI strategic development model that enables AI leaders, management, practitioners, and technologists to work together in the development of an AI strategic blueprint. Each of the elements of the strategic development model will be described in more detail in chapter 9. We introduced the model in this chapter to emphasize a process that AI leadership and management, together with their AI teams, can follow to reduce risks and improve the chances of a successful transition of a complex AI system into operations. This AI strategic development model enables AI practitioners to incorporate the NAE key attributes and the systems thinking skills enumerated in this chapter.

Although many decades of employing the systems engineering discipline have led to a very robust set of principles, additional systems engineering challenges must be attended to in the context of AI applications. We discussed these challenges in section 2.6.

AI leaders, developers, system integrators, educators, and more generally AI practitioners, armed with this set of systems engineering fundamentals, would be in an excellent position to succeed in defining, developing, and deploying valuable AI capabilities. With this systems approach, our ultimate goal is to provide tools, tips, and techniques to reduce risk and increase the likelihood of successful AI implementation.

2.8   Exercises

1. What are the attributes of the systems engineering definition, as we have adapted it to AI applications, compared to the standard INCOSE definition shown earlier in the chapter?

2. The left side of the Vee-model addresses which of the following?
   a. The implementation of lower-level realizations of the AI system architecture
   b. The definition of the AI system architecture
   c. The decomposition of the AI system architecture
   d. a and b
   e. a and c

3. The right side of the Vee-model addresses which of the following?
   a. Different levels of the AI system architecture realizations
   b. The upper level of the AI system architecture definition
   c. The decomposition of the AI system architecture
   d. a and b
   e. a and c

4. From the systems engineering key characteristics shown in table 2.1, pick what you consider the four most important characteristics and provide a short explanation for each of your choices.

5. Pick one AI industry application example of interest to you from the list in appendix A.1. Describe the business goals (i.e., the stakeholder's requirements) of that application.

6. What differentiates the first era versus the modern era with respect to the compute usage in training AI systems shown in figure 2.3.

7. The AI strategic development model shown in figure 2.4 can serve as a tool for which of the following?
   a. Only AI engineers
   b. AI leaders
   c. AI management
   d. Subcontractors
   e. AI leaders, management, systems engineers, and the AI team
   f. None of the above

8. AI leaders are primarily responsible for executing the plan, while management is primarily responsible for developing the strategic plan.
   a. True
   b. False
9. An example of a technology challenge is to determine if the computing should be performed at the enterprise level or at the edge.
   a. True
   b. False
10. An example of a process challenge is in incorporating various degrees of system fidelity leveraging real, emulated, and/or simulated scenarios.
    a. True
    b. False
11. An example of a people challenge is in communicating effectively only the low levels of system element developments and realizations shown in figure 2.1.
    a. True
    b. False
12. Based on your experience, describe how a systems engineering approach would have been valuable for achieving success. The example can be from your educational or work experience.
13. Select four skills from the essential skills of a systems thinker (as shown in table 2.3) that you consider the most critical to a successful implementation of an AI system and explain why.

2.9 References

1. Zachary, G. P., *Endless frontier: Vannevar Bush, engineer of the American century.* 2018, New York: Free Press.
2. Bush, V., As we may think. *Atlantic Monthly*, 1945. 176(1): 101–108.
3. *NASA systems engineering handbook.* 2016, National Aeronautics and Space Administration (NASA).
4. Hotten, R., Volkswagen: The scandal explained. *BBC News*, 2015. 10: 12.
5. Calo, R., Is the law ready for driverless cars? *Communications of the ACM*, 2018. 61(5): 34–36.

6.   Kohli, P. and A. Chadha. Enabling pedestrian safety using computer vision techniques: A case study of the 2018 uber inc. Self-driving car crash, in *Future of Information and Communication Conference*. 2019, San Francisco, Springer.

7.   Rice, D., The driverless car and the legal system: Hopes and fears as the courts, regulatory agencies, Waymo, Tesla, and Uber deal with this exciting and terrifying new technology. *Journal of Strategic Innovation and Sustainability*, 2019. 14(1): 134–146.

8.   Martinez, D. R., *AI: Why a systems engineering approach is essential*. 2020, MIT Lincoln Laboratory.

9.   *INCOSE: Systems engineering handbook: A guide for system life cycle processes and activities*. 4th ed., ed. D. D. Walden, et al. 2015, John Wiley & Sons.

10.   Board., B. E., *Guide to the systems engineering body of knowledge*. 2014.

11.   *ISO/IEC/IEEE international standard—Systems and software engineering—System life cycle processes*. ISO/IEC/IEEE 15288 First edition 2015-05-15, 2015.

12.   Zemrowski, K. M., NIST bases flagship security engineering publication on ISO/IEC/IEEE 15288: 2015. *Computer*, 2016. 49(12): 86–88.

13.   Delaney, W. P. and W. W. Ward, Radar development at Lincoln Laboratory: An overview of the first fifty years. *Lincoln Laboratory Journal*, 2000. 12(2): 147–166.

14.   Forrester, J. W., Lessons from system dynamics modeling. *System Dynamics Review*, 1987. 3(2): 136–149.

15.   Forrester, J. W., System dynamics-the next fifty years. *System Dynamics Review*, 2007. 23(2–3): 359–370.

16.   Sterman, J., System dynamics at sixty: The path forward. *System Dynamics Review*, 2018. 34(1–2): 5–47.

17.   Selvy, B., Systems engineering. *Aerospace America*, 2010. 48(11): 34.

18.   Eisner, H., Systems engineering: Building successful systems. In *Synthesis Lectures on Engineering, Science, and Technology*. Vol. 14, 1–139, 2011, Morgan & Claypool Publishers.

19.   Forsberg, K., and H. Mooz. The relationship of system engineering to the project cycle. In *INCOSE International Symposium*. 1991, Wiley Online Library.

20.   Rebovich, G., MITRE systems engineering guide. MITRE Corporation. http://www.mitre.org/publications/systems-engineering-guide/systemsengineering-guide. Accessed 2014. 2.

21.   Ozkaya, I., What is really different in engineering AI-enabled systems? *IEEE Software*, 2020. 37(4): 3–6.

22. Bosch, J., H. H. Olsson, and I. Crnkovic, Engineering AI systems: A research agenda. In *Artificial intelligence paradigms for smart cyber-physical systems,* 1–19. 2021, IGI Global.

23. Martinez, D. R., M. M. Vai, and R. A. Bond, *High-performance embedded computing handbook: A systems perspective.* 2008, CRC Press.

24. Martinez, D. R. and M. G. Sobol, Systems analysis techniques for the implementation of expert systems. *Information and Software Technology,* 1988. 30(2): 81–88.

25. Lawless, W. F., R. Mittu, D. A. Sofge, T. Shortell, and T. A. McDermott, eds., *Systems engineering and artificial intelligence.* 2021, Springer.

26. Sculley, D., G. Holt, D. Golovin, E. Davydov, and T. Phillips, Hidden technical debt in machine learning systems. *Advances in Neural Information Processing Systems,* 2015. 28: 2503–2511.

27. Diamandis, P. H. and S. Kotler, *The future is faster than you think: How converging technologies are transforming business, industries, and our lives.* 2020, Simon & Schuster.

28. Strickland, E., et al., *The great AI reckoning.* https://ieeexplore.ieee.org/abstract/document/9564272.

29. AWS, *AWS Well-Architected Tool documentation.* 2017, AWS. https://docs.aws.amazon.com/wellarchitected.

30. Rajapakse, R. N., M. Zahedi, M. A. Babar, et al., Challenges and solutions when adopting DevSecOps: A systematic review. *Information and Software Technology,* 2022. 141: 106700. https://www.sciencedirect.com/science/article/pii/S0950058 4921001543.

31. Amazon. *Machine learning lens—AWS well-architected framework.* December 20, 2021. https://docs.aws.amazon.com/wellarchitected/latest/machine-learning-lens/wellarchitected-machine-learning-lens.pdf#machine-learning-lens.

32. OpenAI. *AI and compute.* November 7, 2019. https://openai.com/blog/ai-and-compute/.

33. Bennis, W. G., *On becoming a leader.* 20th anniversary, rev. and updated ed. 2009, New York: Basic Books.

34. The engineer of 2020: Visions of engineering in the new century. *Research Technology Management,* 2004. 47(6): 119.

35. Luke, J., D. Porter, and P. Santhanam, *Beyond algorithms: Delivering AI for business.*

36. Baier, L., F. Jöhren, and S. Seebacher, *Challenges in the deployment and operation of machine learning in practice*. 2019.

37. Fischer, L., L. Ehrlinger, V. Geist, et al., AI system engineering—Key challenges and lessons learned. *Machine Learning and Knowledge Extraction*, 2021. 3(1): 56–83.

38. Amershi, S., A. Begel, C. Bird, et al., Software engineering for machine learning: A case study. In *2019 IEEE/ACM 41st International Conference on Software Engineering: Software Engineering in Practice (ICSE-SEIP)*. 2019, IEEE. https://www.microsoft.com/en-us/research/uploads/prod/2019/03/amershi-icse-2019_Software_Engineering_for_Machine_Learning.pdf..

3

Data Conditioning

Data is like garbage. You better know what you will do with it before you collect it.
—Mark Twain, American writer, humorist, entrepreneur, publisher, and lecturer

Among all the artificial intelligence (AI) researchers, practitioners, and operational users who are actively working on AI, the dependency on quality data is regarded as one of the biggest challenges to leveraging AI. Even though the quote from Mark Twain that starts this chapter dates back to the late 1800s, it is very appropriate to today's fielding of AI products and services. However, the operative words are "good and clean data," not just massive amounts of data. So-called big data, driving large volume, velocity, and variety, can still result in very poor results if the data is noisy and not representative of real-world operational environments.

It is also well accepted that data scientists and data engineers spend approximately 80 percent of their time on "data munging," where they must analyze, clean, and prepare the data prior to moving on to the machine learning (ML) stage, shown in figure 1.3 in chapter 1. In this data-conditioning stage, data scientists and data engineers collect data from external sensors and sources and transform raw data into information. Information means that the data contains in its samples the knowledge that we want ML algorithms to extract. For example, in medical scenarios, if the doctor is evaluating a patient to assess if a tumor is present in the lungs, many samples of X-rays are collected from sensors. In the data-conditioning stage, this data is carefully analyzed, cleansed, and prepared to best highlight the information needed by ML computer vision algorithms to determine if a tumor is present.

Ultimately, the primary goal of the data-conditioning stage is to reduce the noise in the data and augment and/or refine its content such that the ML algorithms can work best at improving the signal-to-noise ratio at the output of the ML building block shown in figure 1.3. As we discuss in this chapter, many diligent steps need to be undertaken to successfully transform data from external sensors and sources into information. However, it is important to emphasize that the data-conditioning stage and the ML stage must be carried out iteratively.

For many years, AI researchers and practitioners would focus primarily on optimizing the ML algorithms by iterating over and over on the model training and development. This approach has been commonly referred to as "model-centric AI." Most recently, the focus has pivoted to data-centric AI, where the data conditioning and the ML stages are performed iteratively until an acceptable level of performance is reached at the output (i.e., knowledge is attained). Furthermore, as we will discuss later in this chapter, there is also a need to continue monitoring the data conditioning and ML stages after deployment, as illustrated in figure 2.2 in chapter 2.

The logical flow adapted in this chapter is first to highlight the advances that have occurred with the exponential growth in data called "big data." This exponential growth has led to organizations having to become data-driven organizations to stay competitive. A requirement of data-driven organizations is to advance their internal operations through a digital transformation. A digitally transformed business is much better positioned to benefit from data-centric AI.

Therefore, we begin with an overview of the big data drivers. In this section, we elaborate in more detail on the data types and the associated data conditioning that must be performed to prepare the data prior to the ML stage, and we also ascertain that the data adheres to the FASTEPS principles discussed in chapter 8.

Undergoing a digital transformation—meaning to evolve an organization into being data-driven—permits more effective leveraging of its data. In section 3.2, we present the typical digital transformation steps found in many organizations. One important benefit of evolving into a digitally transformed enterprise is to break up the organizational silos that hinder the fluidity between business units. However, for an organization to be data-centric, regardless of its size, it must also pay attention to rapid access for performing data exploitation and insight discovery. In section 3.3, we elaborate on database management and formats. In addition to exponential growth in data over the past decades, there has been a need to improve the ability to perform online transactional processing (OLTP) and online analytical processing (OLAP), plus streaming data processing. These different types of processing have led to a set of database categories and required formats.

After discussions at the macro level regarding digital transformation, database management, and database formats, in section 3.4 we take a deeper dive into the data-conditioning techniques to perform data cleaning and preparation. Section 3.5 presents a snapshot in time of existing curated open-source data available to the AI community.

Finally, section 3.6 addresses data-conditioning challenges. Section 3.7 highlights the chapter's main takeaways, and in section 3.8, we conclude the chapter with a set of exercises.

3.1 Exponential Data Growth

Data continues to grow exponentially. At first glance, this growth is good and desirable for training ML algorithms. However, as Andrew Ng, a Stanford professor and the chief executive officer (CEO) and founder of Landing AI and DeepLearning AI, articulated during the NeurIPS Data-Centric AI workshop [1]: "It is not about having lots of data, we need good data." Ng pointed out that the AI community will need to focus more on iterating on good data for improving ML algorithms. Ng defines data-centric AI as the discipline of systematically engineering the data needed to successfully build an AI system. This emphasis demands that data scientists and data engineers use the exponential data growth that the AI community is experiencing while paying careful attention to the data-conditioning stage.

Recently, there has been an emphasis on understanding the roles of data engineers and software engineers. More and more, there needs to be close coupling between data scientists, data engineers, and software implementers. Hulten discusses ML engineering, from defining objectives through orchestration, to help in successful development and deployment [2].

Later in the chapter, we will look in more detail at data types. One can categorize data into two distinct classes: structured and unstructured (or semistructured) data. Structured data is typically stored in row-column format—for example, medical patient or inventory records. Unstructured data is typically raw data, such as documents, videos, tweets, images, and speech, and it can be stored in different formats. MongoDB, one of the premier organizations in database management, points out an annual growth in unstructured data of about 27 percent, and in structured data of about 20 percent. They estimated >120 exabytes (1 exabyte is equivalent to 10^{18} bytes) of stored unstructured digital data (c.2020).

This exponential growth in data, especially unstructured data, will continue to drive the total amount of data available worldwide. As shown in figure 3.1, courtesy of Domo [3], data from many popular social media platforms dominated data creation. Nearly 57 percent of the approximately 8 billion global population in 2021 were active social media users (about 4.55 billion users) [4]. This dramatic growth in data reinforces the need for a data-centric approach to AI, as discussed earlier.

Within the end-to-end AI system architecture, at the first stage of data conditioning shown in figure 1.3, data scientists and data engineers must address the characteristics of

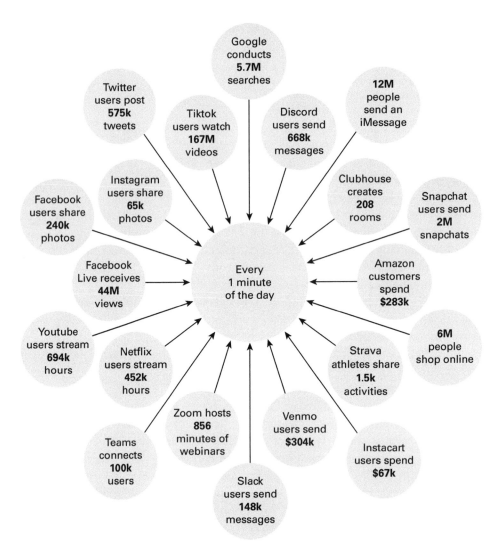

Figure 3.1   Data created every minute among the most popular social media in 2021 [3].

this big data. Commonly, big data is parsed into 4 "Vs": variety, velocity, volume, and veracity. Each of these characteristics is illustrated in figure 3.2 and defined next. As an example, the constellations shown in each of the 4 "Vs" illustrate proxy logs found in web traffic, representing very diverse and difficult data occupying all four quadrants, starting with the top-right quadrant and progressing clockwise:

- **Variety**: A data representation with different levels of dimensionality (low to high) and data input from either single or many data sensors and data sources

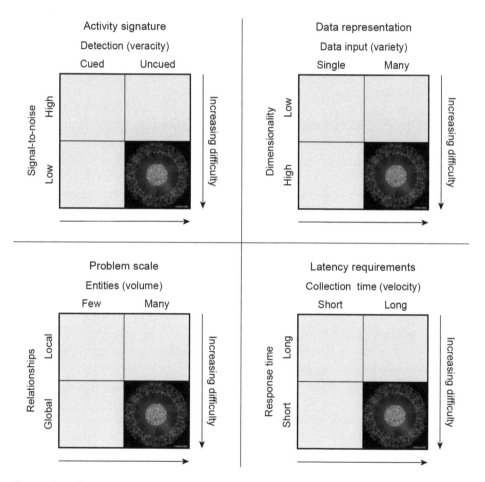

Figure 3.2   The 4 "Vs" characterizing big data complexity.

- **Velocity**: A characteristic of data speed, or equivalently driving the latency requirements as a function of collection time versus response time
- **Volume**: A scalability across relationships (global to local) and from few to many entities
- **Veracity**: Characterizes levels of data integrity and usefulness with respect to signal-to-noise and the ability for the ML algorithms to extract (or detect) the content of interest existing in the data (e.g., to perform prediction, classification, clustering, and other actions)

In the past two decades, a lot has been written about the benefits, impact, and challenges with big data [5]. There has also been an emphasis—and rightly so—on organizations needing to treat data as a strategic asset [6]. Unfortunately, many organizations continue to struggle with how to benefit from the availability of both internal and external data. There are many reasons for these obstacles, but at the core are people, infrastructure, and the associated culture [7, 8]. As Redman and Davenport point out, "if it costs $1 to develop a [machine learning] model, it costs approximately $100 to deploy it" [7].

At the macro level, Erik Brynjolfsson (at Stanford) and colleagues have discussed how general purpose technologies such as AI require other complementary investments to occur before AI can be used to its full potential. Some examples of such investments are in human capital (people), processes, infrastructure, and business models [9]. The lack of these complementary investments will continue to be a limiting factor for organizations in successfully transitioning AI products and services into operations. Rackspace Technology, a cloud service provider, found that in a survey of 1,870 organizations in a variety of industries, only 20 percent of these companies have mature AI initiatives. The rest are still trying to figure out how to make it work [10].

In addition to the complementary investments discussed earlier, in the context of organizations needing to treat data as a strategic asset, senior leaders must make the necessary investment to undergo the digital transformation required for enabling access to data inside and outside the organization. Data is a force multiplier to achieve competitiveness, improve efficiency, and deliver greater value to the company's stakeholders. As we discuss in the next section, a digitally transformed company must evolve through a number of stages, starting with data being fully digitized (digitization stage) and moving ultimately to a digitally transformed organization. Business-unit silos can be eliminated (or at least minimized, demonstrating thin organizational membranes), so the organization can rapidly respond to changes in the marketplace.

3.2   Digital Transformation

Many organizations struggle in the transitioning and scaling of AI products and services. As we discussed in the previous section, one challenge for organizations facing this barrier is the lack of attention to their digital infrastructure and associated capability maturity levels. We can think of a three-tier approach to address this challenge:

1. Undergo a digital transformation.
2. Create a data-driven culture.
3. Implement AI capabilities based on a data-centric discipline.

In this section, we formulate a systematic approach to undergoing digital transformation [11]. Regardless of organization size, the business must invest in transforming its infrastructure environment by eliminating silos and sustaining a digital ecosystem. Professor Jeanne Ross (an American organizational theorist and former principal researcher at the MIT Center for Information Systems Research) emphasizes that a digitized business is not the same as a digitally transformed business [12]. The distinction might be subtle, but it is profound to successfully create a data-driven organizational culture.

Figure 3.3 illustrates five digital transformation milestones and the associated capability maturity levels. An organization at maturity level 1 complies with a digitized infrastructure. A business is considered to be digitized when most, if not all, of its internal infrastructure conforms to digitized operations. In the context of a data-driven organization, the implementation of the AI system architecture shown in figure 1.3 requires that business silos be eliminated by enabling the access and sharing of all data within each of the architecture building blocks (e.g., data at the output of the data-conditioning stage).

We refer to a business unit—for example, a line of business, a subsidiary, a division, or a group—as any entity within the organization responsible for using AI in the development and demonstration of capabilities (i.e., products and services). An AI organization can be structured as either a multidivisional business or a functional organization. For example, Apple went through a reorganization to consist of cross-functional teams that can adapt more rapidly to technological change [13]. The digital transformation discussed herein applies to both types of organizational structure.

Organizations confronting a barrier in developing and transitioning AI capabilities exhibit an "as is" infrastructure dominated by business silos. The goal is to evolve into the "to be" organization infrastructure shown in figure 3.4 [12]. This level 1 milestone is necessary to successfully develop AI capabilities, employing both internal and external organizational resources across the triad of people-process-technology discussed in chapter 2.

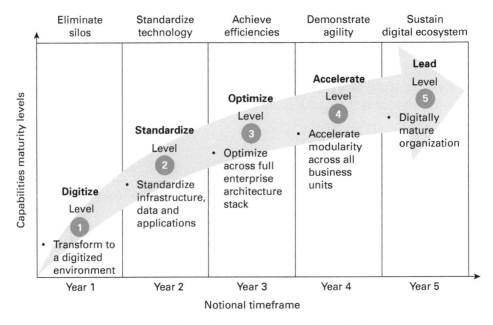

Figure 3.3   Digital transformation milestones and capability maturity levels.

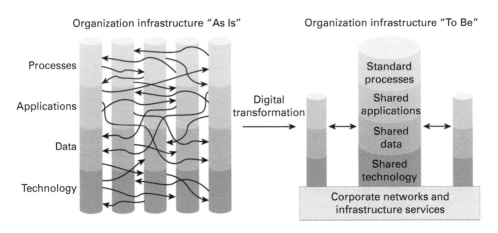

Figure 3.4   Organization infrastructure evolution from "as is" (many silos) to the "to be" state.

Once an organization has successfully reached level 1, the next logical step is to standardize all processes, and with common infrastructure standards, the business units can then share applications and data. This is level 2, shown in figure 3.3. Although not explicitly shown, level 2 also assumes that there is a well-established organizational governance monitoring compliance with enterprise standards.

Since it is imperative for all AI organizations to rapidly iterate between development (prototyping) and operations, as discussed in chapter 2, organizations at level 3 can achieve efficiencies across the full architecture pipeline shown in figure 1.3. "Efficiency" in this context means that organizations are able to use data, ML models, and computing infrastructure, as well as maintaining intermediate data products from each of the processing stages of the AI system architecture, from one AI product or service to updated or new offerings.

Maturity level 4 (accelerating development and deployment) implies that the AI organization is agile and modular in developing AI capabilities and deploying them in production. More specifically, referring back to figure 2.2, as the stakeholders' business goals change, the AI data-driven organization is able to adapt rapidly to new market opportunities. An organization at level 4 is also able to more aptly evolve its AI strategy through the step-by-step AI strategic development model discussed in chapter 2 and shown in figure 2.4.

Finally, at level 5, AI organizations are able to sustain their position as a mature, digitally transformed business. At this level of maturity, the organization infrastructure becomes a development and operational platform. If the AI capabilities are to be used internally within the organization, the operational platform must be a high-fidelity representation of the ultimate deployment environment. Similarly, if the AI product or service is targeted at external stakeholders, the operational platform must also be a high-fidelity representation of how the user or consumer will be employing the AI capability.

The journey from level 1 to level 5 does not imply to senior leaders that their organizations must go through a clean-slate transformation. In fact, we advocate the opposite. An organization must adapt its AI strategy and infrastructure by bridging between the existing organization infrastructure, with the ultimate goal of being a data-driven organization enabled by digital transformation. As discussed by Furr et al., digital does not have to be disruptive, meaning implementing adaptation rather than reinvention [14].

That said, digital transformation is imperative to avoid stagnation and limiting the ability of the organization to innovate. Melvin Conway said it well, and today his quote is commonly referred to as Conway's Law: "Any organization that designs a system (defined broadly) will produce a design whose structure is a copy of the organization's communication structure" [15]. An AI organization at level 5—that is, a digitally transformed enterprise—will eliminate the communication barriers, resulting in a more flexible and lean organization.

In this section, we presented a systematic approach for AI organizations to follow, as they embark on a digital transformation journey. We believe that a digitally transformed organization will be able to reduce barriers to the development and deployment of AI capabilities. At level 5 of this journey, organizations will be able to more effectively treat data as a strategic asset. In the next section, we elaborate further on databases (management and evolution) that use this organization's platform infrastructure.

3.3 Databases: Management and Evolution

An AI organization undergoing a digital transformation from level 1 through level 5, as illustrated in figure 3.3, will have the foundational platform to perform data conditioning on structured and unstructured data. As shown in figure 3.4, data and associated technologies must comply with standard processes.

In this section, we discuss different database management standards and their associated evolution. However, it is imperative that AI data scientists and data engineers begin by addressing the stakeholders' (users/consumers' or customers') requirements and expected results—this means data needed to achieve the desired insights, the computing platform environment (batched and/or streaming), and success criteria (measures of effectiveness).

Table 3.1 enumerates a set of questions that must be asked at the outset of the AI system architecture implementation framework shown in figure 2.2 in chapter 2. The AI team must be customer focused. Answers to these questions, at the outset of the AI development stage, are very important to achieve success prior to the start of the data-conditioning function. Business understanding for creating value out of AI capabilities is intrinsic to the key phases of the cross-industry standard process for data mining (CRISP-DM) process [16].

Since the advent of big data, there has been a surge in database types and formats. For many years, the traditional database was based on a need for transactional processing. Most transactional processing adhered to a row-column format encapsulated in a relational Structured Query Language (SQL) database. One of the early pioneers in relational databases was Edgar Codd at the IBM Research Laboratory; his seminal paper was originally published in 1970 [17].

Later, Chang et al. designed a new distributed storage system for structured data known as Bigtable [18]. Bigtable was foundational, and it was made into an open-source

Table 3.1   Key questions prior to starting the data-conditioning stage

Key Questions	Comments
What insight does the customer need?	Work from the expected result back to the needed data.
What data input is required to achieve the desired insight?	Make sure that we have curated data.
Are the incoming data and required processing batched or streaming?	Forwardly deployed applications require batched processing in the cloud, streaming, or both.
Can we solve the AI problem using ML techniques other than deep learning?	Deep learning is very powerful when we have data that is representative (statistically) of the population, scenarios, or both.
Do we need edge computing (in real time)?	We might need to operate under low-capacity and intermittent data links.
How do we protect our AI system from adversarial AI (e.g., data poisoning)?	AI systems are very fragile and easy to fool.
What measures of effectiveness (MoEs) are critical?	Many measures of performance (MoPs) are available. The challenge is to aggregate MoPs into MoE metrics

distributed structured data table referred to as HBase. HBase (an Apache open-source standard), together with MapReduce (a distributed processing model) plus the Hadoop Distributed File System (HDFS), became accepted standards for their scalability features and ability to enable rapid ingestion of streaming data [19]. The new approach and standard, for managing distributed files across a large number of computing nodes, became known as Not-only Structured Query Language (NoSQL).

Today, the Apache Hadoop open-source project develops and maintains the library necessary for managing large data volumes and ingesting data at high speeds. Hadoop consists of common utilities, such as HDFS, MapReduce, and Hadoop YARN (a framework for job scheduling and cluster resource management; which refers to yet another resource negotiator) [20]. This library set is popular and well accepted for implementing data processing in a large cloud computing environment.

Database designers strive to have the standards meet a set of properties referred to as atomicity, consistency, isolation, and durability (ACID). Table 3.2 defines these properties, including the benefits and challenges of each. The goal with relational databases using a SQL format was to meet these properties, but unfortunately, as data grew in

Table 3.2   ACID properties

Property	Benefits	Challenges
Atomicity	Maintains data integrity	Expensive to set up, maintain, and preserve as data grows
Consistency	Reliable and easy to read and interpret	Hard to integrate when ingesting unstructured data
Isolation	Manageable and can be changed in modules	Schema changes take time to implement and depend on specific use cases
Durability	Robust in tracking and storing	Slow when processing data

Table 3.3   SQL and NoSQL comparison

SQL	NoSQL
Relational database with fixed columns and rows typically found with structured data	Nonrelational database typically found with unstructured data
Follows a specific schema; for example, last name (row) followed by first name, home address, phone number, email address, and other information	Relational data (structured data) can be stored, but not in a row-column pair. Instead, data is nested in a single data structure
Much easier to comply with the ACID properties	Harder to meet the ACID properties
Difficult to scale to large volume and velocity	Easier to scale; optimized for developer productivity; runs well on distributed clusters (cloud computing)
Originated in the early 1970s with the need for transactional processing	Originated in the mid-2000s with the need for rapid ingestion (e.g., internet searches)

volume, the SQL format did not scale well. In contrast, NoSQL does not meet all these properties, but it is better matched to a large volume of unstructured data. Gessert et al. discussed a number of NoSQL databases and offered a toolbox for helping AI practitioners decide on the most applicable choice depending on the application use cases, including an assessment of database properties [21].

Table 3.3 compares SQL and NoSQL database formats. NoSQL can ingest both structured and unstructured data. These database formats, together with the required

- **Key-value databases**
  (shown at right)
  Key-value stores are the
  simplest NoSQL. For
  example, Key can be
  Customer ID#, and Value contains
  attributes about the customer
  (e.g., Amazon DynamoDB)

- **Document databases**
  Key is an document identifier. Value
  contains the hierarchical tree data
  structures containing the document
  (e.g., MongoDB)

- **Wide-column databases**
  Each column family can be compared
  to a container of rows in an RDBMS
  (Relational Database Management
  System) table. Rows do not have
  to have the same columns (e.g.,
  Facebook Cassandra)

- **Graph databases**
  Data is stored in nodes and edges.
  Nodes can be for example people
  or places, with edges identifying
  relationships like in social networks
  (Neo4J, OrientDB)

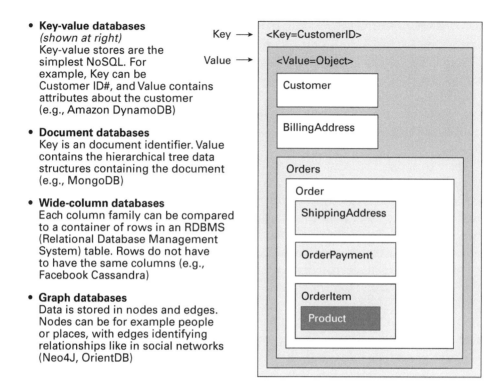

Key ⟶ <Key=CustomerID>
Value ⟶ <Value=Object>

Figure 3.5   NoSQL database types [22].

compute environment such as batch or stream processing, are standards that must be part of the digitally transformed platform infrastructure discussed earlier. The platform infrastructure facilitates the appropriate standards and enables the sharing of applications, data, and associated technologies, plus staff expertise—for the data-conditioning stage and required management done on the premises, through a cloud computing service provider, or a hybrid of these two environments.

There are different types of NoSQL databases, such as key-value stores, document databases, wide-column databases, and graph databases. Figure 3.5 provides a short description of each, as well as examples of database management tool providers, plus an example of the key-value store format.

In addition to database management and formats, there are different types of data storage environments. Currently, two of the most common repository environments are data warehouses and data lakes. There are commercial companies that offer a convergence of data storage environments employing the best of both data warehouses and data lakes (e.g., Snowflake and Databricks). At present, both Snowflake and Databricks were domi-nant in their specialties, with a market capitalization of tens of billions of dollars, which gives an indication of how significant data management and storage are to modern data processing and AI.

Databricks uses something that it calls the "data lakehouse," a platform meant to com-bine the best of data warehouses and data lakes [23]. Data warehouses primarily store data in structured format, whereas data lakes predominantly store data in unstructured format [24–26].

There is an ongoing trend to use AI in business intelligence, as well as integrating structure, semistructured, and unstructured data into the decision support workflow for a modern AI organization. This trend is very healthy for the AI community to avoid looking at databases and types as uniquely distinct environments. The digital transfor-mation platform, discussed earlier, should incorporate these databases and data types into an integrated data environment [27].

An AI organization that has implemented a digital transformation, owning a modern platform infrastructure, can create value by either performing the process of Extract-Transform-Load (ETL) or Extract-Load-Transform (ELT), across data storage environ-ments (e.g., data warehouses, data lakes, or data lakehouses). In the data-conditioning stage, ETL is employed in the classical data manipulation flow using the data and data-base management tools described earlier. ELT, in addition to reversing the operation of load and transform, is becoming the accepted data manipulation flow for a large volume of unstructured data, with data processing occurring closer to where the data is stored (minimizing latency). An excellent resource to stay abreast of changes in the data and AI ecosystem is by Matt Turck (a venture capitalist at FirstMark), who regularly writes a comprehensive post with relevant updates [28].

So far, we have articulated the need for an AI organization to undergo a digital trans-formation in order to achieve a modern platform infrastructure. This platform infrastruc-ture facilitates the use of the database management tools and techniques necessary to implement the data-conditioning stage. The next steps, after these foundational pieces are in place, are to perform data quality assessment, cleaning, and preparation prior to the ML stage. The next section elaborates on these data-processing functions in more detail.

3.4   Data Quality, Cleaning, and Preparation

The third-tier component of our earlier discussion of the transitioning and scaling of AI products and services is to focus on data-centric AI instead of putting the majority of the emphasis on model-centric AI. "Data-centric AI" simply refers to iterating between the data-conditioning stage and the ML stage shown in figure 2.2, where the emphasis is on enhancing data (through data engineering) to achieve better performance from the ML algorithms [29]. Model-centric AI, in contrast to data-centric AI, focuses on improving the ML algorithms while primarily doing minor tweaks to the initial data used in training/cross-validation and testing.

Effective implementation of data-centric AI relies on careful attention to the data-conditioning stage. There are many references in the literature that present an in-depth discussion on data conditioning [16, 30–33]. For our purposes, in this chapter we highlight some of the important data-conditioning exploration details and issues that data scientists and data engineers must address.

Table 3.4 breaks the data conditioning exploration into the following categories:

- Data quality: Initial analysis of the available data.
- Data cleaning: Addressing corrections to the existing data, and determining if additional data collection is needed.
- Data preparation: Organizing the data prior to the ML stage.

These categories must be undertaken to convert raw data into useful information, leading to effective extraction of knowledge from the ML stage. It is also important to emphasize that this exploration is done iteratively with the ML step. The iterative process requires that the performance of the ML algorithm be evaluated through a set of performance metrics such as accuracy, confusion matrix, area under the curve (AUC), as discussed in more detail in chapter 4.

AI leaders and practitioners must have capable data scientists and data engineers on their AI teams. It is one of the most critical roles for the successful development and deployment of AI capabilities [34]. Data scientists are typically responsible for analyzing the data, performing the data exploration functions enumerated in table 3.4. Data engineers complement the data scientists, and they are pivotal members of the team, with the responsibility for acquiring and maintaining robust and scalable database management tools. They also must play the stewardship roles for evolving the platform infrastructure discussed in sections 3.2 and 3.3.

The role of data scientists and data engineers is to do an initial data quality exploration. After understanding the stakeholders' business needs, the set of questions posed in table 3.1 must be answered. From these answers, one can perform a data quality assessment to determine if there is missing or incomplete data. For example, the output of ML could be either classification, predictions, or clustering estimates. Input data must be sufficiently heterogenous to span the ML output. For example, to overcome this limitation, more data might need to be collected.

Table 3.4   Data exploration and issues

Data Exploration	Issues
Data quality	Missing or incomplete data
	Outlier detection
	Duplication
	Bias versus variance trade-off
Data cleaning	Imputation
	Rescaling
	Pruning
	Feature selection
Data preparation	Curse of dimensionality
	Dimensionality reduction
	Data augmentation
	Data labeling
	Data drift

Data might also contain outliers, which can result in the poor performance of an ML model (e.g., underfitting). Data duplication must also be eliminated from the available data, as it can lead to data bias, discussed next.

There is always a trade-off between data bias and data variance. In this context, data bias is the difference between the ML output and the expected output. High bias results in a model that underfits the training data. A solution is to try a different model or a more complex model (i.e., additional hyperparameters). Data variance is an error resulting from a high degree of sensitivity to small changes or fluctuations in the training data. High variance creates overfitting to the training data (e.g., when there are too many features but not enough training samples or the model is too complex). Employing a regularization term in the model's objective function can lead to higher bias but potentially lower variance, as well as a simpler model. We should also point out that model bias in this context is different than data bias, discussed in chapter 8 in the context of responsible AI (RAI), such as gender, age, and ethnic/race biases.

Once an initial data exploration is performed, under a data quality assessment, the data cleaning involves multiple efforts such as the following:

- Imputation: Include representative values to enrich the data set relative to the desired features of interest. This could be done through simulation or emulation using high-fidelity representations of the real-world environment.

- Rescaling: Avoid large dynamic range in the absolute data values. This can be addressed by normalizing the value range.
- Pruning: Excise training samples that are outliers or noisy from the data set. Data pruning helps in improving generalization and the resulting accuracy [35, 36]. There are also data-pruning techniques that focus on achieving acceptable performance with smaller data sets, while also reducing computation cost [37].
- Feature selection: Remove irrelevant or redundant features that would cause overfitting, increase computational complexity, or both [38]. This step also helps with reducing the curse of dimensionality (as addressed next).

Data quality assessment and data cleaning permit data scientists and data engineers to move to the next step of data preparation. During data preparation, it is conditioned to be well matched to the ML stage. For example, one must avoid the effect known as the "curse of dimensionality," which occurs when additional features are incorporated into the model but there is not enough data to support the desired output from the model. Let us look at a use case as an example. In the application of real estate housing value prediction, if we add more representative features (e.g., more bedrooms or square footage), we must also increase the number of available training samples to span the additional features. If we do not do so, the high-dimensionality increase causes no training samples to occupy that increased high-dimensional space. One approach to avoid this effect is employing dimensionality reduction, such as principal component analysis (PCA).

Data augmentation is most often used when there is underfitting in the model to perform a function such as computer vision. The training set can be augmented by rotating the input image. These image rotations can be random geometric transformations without actually changing the class object of interest (i.e., the target image).

Another important step in data conditioning is to label the data prior to employing a supervised learning algorithm. There are data sets that are already labeled, which are often used by algorithm developers and researchers, as discussed in the next section.

Finally, data conditioning is not an end state. Not only does it need to be done in an iterative fashion with the ML stage, but it is imperative that during deployment, the AI system must be checked for data drift. Data drift means that the original ML model no longer matches well to the new real-world scenario and incoming data. This effect is not uncommon as AI is applied to rapidly changing applications. The monitoring stage, shown in figure 2.2 in chapter 2, might cause us to return to the data-conditioning stage to collect or perform additional data quality assessment, data cleaning, or data preparation.

Sambasivan et al. performed a study in which they interviewed fifty-three AI practitioners, from several countries (e.g., India, East and West Africa, and the US) [39]. They found empirical evidence of what they referred to as "Data Cascades" effects. Data cascades effects mean compounding events that cause negative downstream data errors. These compounding effects are detrimental to the final ML results, particularly in high-stakes AI applications.

Before we finish this section, it is important to mention other critical, perhaps even existential issues in developing AI system capabilities, such as attention to data governance, access controls, privacy, and confidentiality. Data governance is still a challenge in terms of who has the responsibilities for effective and compliant use of data for AI (such as data ownership, maintaining high data quality standards, and data stewardship), and adapting to data standards as the system architecture evolves [27, 32].

Many advances in narrow AI have been achieved, as we have discussed in earlier chapters. This chapter has presented a set of tools and approaches for reducing risk and likely achieving success in developing and deploying AI capabilities by performing a more thorough job at the data-conditioning stage prior to the ML step. As a result of the wide availability of significant open-source data, AI practitioners do not need to start from scratch. The next section presents a sample of curated data sets, as well as some of their attributes that are relevant to designing ML algorithms.

3.5   Curated Data Set Examples and Attributes

In previous sections, we have elaborated on different topics under the rubric of data conditioning, beginning with the need of modern AI organizations to develop and sustain a platform infrastructure by implementing a digital transformation. We then discussed database management and techniques as part of the role of a data-driven organization to conform with established standards and formats. In section 3.4, we delved into data quality, cleaning, and preparation in the context of data-centric AI.

In this section, we take more of a micro-level view and present a small but representative set of open-source data sets, their key attributes, and their implications to ML algorithm development. Within the data-centric AI discipline, it is important to evolve to a mindset of creating data products instead of just collecting data. Data products become a strategic asset for any AI organization. Thus, the open-source data sets can be a starting place as being available and curated. However, AI organizations must adapt the algorithms to their own specific applications and create data and preserve products across the full pipeline shown in figure 1.3 in chapter 1.

The logical progression in this chapter from a macro-level view (i.e., digital transformation) to more of a micro-level discussion of data sets is driven by the exponential data growth discussed in section 3.1. In figure 1.3, we illustrated the types of data inputs, such as structured or unstructured data. There is also semistructured data, which is similar to unstructured data but contains additional characteristics. Examples of semistructured data are images annotated with geotags, data-time, coloration, brightness, and contrast.

In a simplistic way, we can think of the data-conditioning functional block as the step to convert raw data into a well-defined schema, accompanied by metadata incorporated into a data catalog. The metadata catalog also facilitates maintaining track of data provenance. The data attributes would also include data labels necessary for the supervised learning algorithm, which will be discussed in chapter 4.

A systems engineering approach to data-centric AI must incorporate, within its discipline, a golden data set for performing benchmark measurements as part of the integration, verification, and validation shown in figure 2.1 in chapter 2. We elaborate further on the importance of performing AI benchmarks in chapter 10, in the context of an AI ecosystem.

The fast acceleration that the AI community has experienced for over a decade is credited, in a significant way, to advances in ML algorithms because of the availability of open-source data sets. For example, LeCun et al. demonstrated the ability to apply ML classifiers to recognize handwritten digits using the Modified National Institute of Standards and Technology (MNIST) data set. As shown in figure 3.6, the data set is well accepted as a standard for initial algorithm performance evaluation [40].

The MNIST data set consists of 60,000 labeled samples, and another 10,000 unlabeled test samples. In the appendix of this book, we show a multilayer perceptron (MLP) and a convolutional neural network (CNN) employing the MNIST Fashion data set. The MNIST Fashion data set also consists of 60,000 labeled images and 10,000 unlabeled test images. Each image is centered on a 28×28 set of pixels [41]. For either the MNIST handwritten or fashion data set, it is common to use 55,000 labeled images for training, 5,000 for cross-validation, and the full 10,000 images for testing.

Another iconic data set developed by Li et al. is ImageNet, used as the gold standard for the evaluation of visual object recognition algorithms [42]. It served as the data set in the now-famous ImageNet Large-Scale Visual Recognition Challenge (ILSVRC), a competition to find the best algorithms for object detection and image classification at large scale. During the earlier phases of the ILSVRC competition, the data set consisted of approximately 1.5 million labeled images, under 1,000 classes (i.e., images ontology with 1000 images per class), presently the ImageNet data set contains approximately 15 million labeled images. In 2012, G. Hinton and his students at the University of Toronto demonstrated a watershed moment with the AlexNet CNN, with an almost 11 percent lower error rate performance than the best image classification algorithm demonstrated in prior ImageNet competitions—a top-five 15.3 percent error rate [43]. Only three

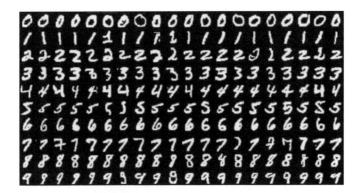

Classifier	Description	Error rate (%)
Linear classifier	Pairwise linear classifier	7.6
K-Nearest Neighbors	K-NN with non-linear deformation (P2DHMDM)	0.52
Boosted Stumps	Product of stumps on Haar features	0.87
Non-linear classifier	40 PCA + quadratic classifier	3.3
Support-vector machine	Virtual SVM, deg-9 poly, 2-pixel jittered	0.56
Deep neural network	2-layer 784-800-10	1.6
	2-layer 784-800-10 (elastic distortions)	0.7
	6-layer 784-2500-2000-1500-1000-500-10	0.35
Convolutional neural network	6-layer 784-40-80-500-1000-2000-10	0.31
	6-layer 784-50-100-500-1000-10-10	0.27
	Committee of 35 CNNs, 1-20-P-40-P-150-10	0.23
	Committee of 5 CNNs, 6-layer 784-50-100-500-1000-10-10	0.21

Figure 3.6   MNIST handwritten data set and algorithm test error rate performance [40].

years later (ILSVRC 2015), Microsoft researchers demonstrated the ResNet CNN algo-rithm with a top-five 3.57 percent error rate. The ResNet CNN algorithm used 1.28 million labeled images for training, 50,000 images for cross-validation, and 100,000 test images [44].

With great gratitude to several AI researchers, these major leaps in algorithm performance have been possible only because of the advent of curated open-source data sets, supported by computing engines such as graphics processing units (GPUs). Although the MNIST and ImageNet data sets continue to serve as benchmarks, there are many other curated data sets that are commonly used by ML algorithm developers to demonstrate algorithm advances.

In table 3.5, we outline a few representative data sets and their associated attributes. As examples of knowledge derived from ML outputs, ML developers have continued to address not just image classification, but also scene understanding, image segmentation, and video activities. For example, the COCO data set, sponsored by several major AI companies and developed with a number of collaborators, can be used for object detection, segmentation, and captioning [45]. MIT CSAIL developed the Places data set with the goal of advancing scene classification and understanding. The data set contains 10 million images representing real-world scenarios in context, labeled via Amazon Mechanical Turk (AMT) [46]. AMT continues to be used as a mechanism to achieve data labeling. However, careful attention must be exercised with this data-labeling approach via crowdsourcing to avoid mislabeling errors. The use of humans in the loop as part of a digital assembly line, performing a set of tasks at low cost, is becoming a significant component of the worldwide economy, as discussed by Gray and Suri [47].

Another very useful data set was also developed by MIT CSAIL, containing videos instead of images. The Moments in Time data set integrates 1 million 3-second videos, also labeled by humans via AMT and exemplifying the diversity in activities produced by humans, animals, objects, or nature [48].

In table 3.5, we also show a data set created by Mozilla's common voice initiative [49]. The intended use of the common voice multilingual data set is to advance automatic speech recognition technologies. The massive speech collection consists of thirty-eight languages amounting to 2,500 hours of speech. This data set consists of both common languages and low-resource languages.

The last data set that we decided to include in table 3.5, is the Enron Email data set. It is often used in the analysis and development of graph neural networks representing a social network [50, 51]. The data consists of about 150 people (mostly senior managers) who were communicating via 0.5 million email messages leading up to Enron's collapse in December 2001. The data set was redacted in order to shield real names and make the information otherwise accessible to the public by the US Federal Energy Regulatory Commission.

For readers interested in exploring additional data sets, there are a large number of excellent resources containing curated data sets applicable to a variety of industries and applications. Here are a few of these resources and associated links:

- Kaggle: https://www.kaggle.com/datasets
- University of CA Irvine (USI) Machine Learning Repository

Table 3.5 Small sample of curated data sets and their respective attributes

Data Set Name	Attributes	Comment/Availability
MNIST Digit Recognizer	60,000 training images; 10,000 test images	http://yann.lecun.com/exdb/mnist/
MNIST Fashion	60,000 training images; 10,000 test images	https://www.kaggle.com/zalando-research /fashionmnist
ImageNet	1.2 million images for training; 50,000 cross-validation images; 100,000 test images	https://image-net.org/index.php
COCO (Common Objects in Context)	91 object types; 2 million labeled instances in 328,000 images	Object detection, segmentation, and captioning https://cocodataset.org/#home https://arxiv.org/pdf/1405.0312.pdf
Places	10 million images; 434 scene categories; 5,000 to 30,000 training images per class	Visual object and scene recognition http://places2.csail.mit.edu/
Moments in Time Videos	1 million 3-s videos; 800,000 labels in training; 339 classes; more than 2,250 videos per class	Special-audio-temporal dynamics http://moments.csail.mit.edu/
Common Voice Corpus	38 languages; 2,500 hours of speech	Automatic speech recognition https://commonvoice.mozilla.org/en
Enron Emails	Redacted data set to anonymized personnel; 150 people and 0.5 million email messages	Social network analysis https://www.kaggle.com/wcukierski/enron -email-dataset https://www.cs.cmu.edu/~./enron/

- Stanford Future Data Systems: https://github.com/stanford-futuredata
- Medical Imaging and Data Resource Center (MIDRC): https://data.midrc.org/
- MIMIC III (Medical Information MART Intensive Care): https://physionet.org/content/mimiciii/1.4/
- NIST Artificial Intelligence: https://www.nist.gov/artificial-intelligence
- AI index annual report by Stanford University: https://aiindex.stanford.edu/report/

In the next section, we discuss some of the challenges that AI practitioners must take into account when addressing data conditioning, writ large, across the spectrum from digital transformation, a data-driven culture, and data-centric AI. We also include approaches to ascertain compliance with existing data protection regulations.

3.6   Data Conditioning Challenges

At the start of this chapter, we emphasized the importance of having a robust and sustainable AI platform infrastructure, achieved through digital transformation. For data-driven organizations, it is imperative that this AI platform infrastructure supports a broad range of data types (e.g., structured and unstructured/semistructured data), and different database formats, such as those standards established by data warehouses and data lakes. This foundational platform infrastructure is necessary to successfully implement the data conditioning functional building block shown in figure 1.3 of chapter 1.

Important challenges in undertaking digital transformation in an organization, can be summarized as follows:

- Proceeding diligently through each of the capability's maturity levels shown in figure 3.3
- Lack of support and patience from upper management
- Difficulty in breaking up organizational silos
- Recognizing that a digital transformation includes the business operations, AI development, and transitioning of AI products and services into operations
- Integrating transformational changes within an existing infrastructure

A data-driven organization must use their platform infrastructure, resulting from their digital transformation journey, and incorporate and sustain advances in database management and its continuing evolution, as discussed in section 3.3. The most significant challenges for a data-enabled AI organization are

- Organizing data according to the best matched database environment, while still preserving some of the desired ACID properties

- Able to ingest SQL, NoSQL, and streaming data
- Enabling efficient and effective data discovery and exploitation across multiple database environments
- Preserving the same development infrastructure regardless if the AI implementation is done in a commercial cloud-service center, on the premises, or a combination of both
- Integrating third-party commercial tools and making sure that they interoperate among them
- Clearly identifying the AI organizational entity responsible for data ownership and stewardship

These digital transformation and data-driven organization challenges are very tall orders. These challenges are at the macro-level, and they present potential barriers to successfully developing and deploying AI capabilities. However, AI practitioners must also attend to more specific challenges at the data conditioning functional building block in terms of data quality, cleaning, and preparation [52].

Both at the input of the data-conditioning stage and at its output, we must ascertain compliance with fairness, accountability, safety, transparency, ethics, privacy, and security (FASTEPS), discussed in more detail in chapter 8. Professor James Zou and colleagues have posed the following question (in the context of data-centric AI): Who is responsible when an ML algorithm underperforms on a subpopulation at test time? They argue that it is a joint process iterating both on the data and the ML algorithm, and they present a framework for undertaking the analysis [53].

There are also important challenges in ascertaining that data has been labeled correctly. Northcutt et al. have demonstrated the impact of label errors on ML performance and benchmark results using ten of the most commonly used computer vision, natural language, and audio data sets [54]. There has been research on more optimally performing approximate selections for labeling objects and events of interest while using oracles (i.e., human labelers), constrained by a set of performance metrics, including precision and recall [55]. This approach has the promise to reduce the cost and effort in labeling data by using cheap proxy models, such as image classifiers, and identifying an approximate set of data points satisfying a data selection filter. Koch et al. have also addressed the use of benchmark data sets and their use in ML research assessment across time (specifically a timeframe from 2015 to 2020) [56].

To ascertain not only acceptable ML algorithm performance, but also robustness across the full spectrum of operations throughout all the building blocks shown in figure 1.3, the AI community must employ reliable and well-formulated benchmarks. There have been several efforts to formulate a set of AI benchmarks [57, 58]. Kiela et al. proposed a more dynamic approach to benchmarking instead of today's static evaluations [59]. Dynabench, as developed by Kiela et al., avoids the ML algorithm becoming static in reaching optimum performance (i.e., benchmark saturation) by continuing assessing improvements through iterating among data creation, model development, and model assessment. Another very innovative advancement in benchmarking ML algorithms is the work of Gordon et al. [60]. Their approach integrates the technical performance assessment (e.g., precision and recall) with human assessments. Performing an assessment integrating the output of the algorithm and the agreement or disagreement between human labelers leads to a more realistic ML algorithm performance, making the coupling between the ML model development stage and the human-machine teaming (HMT), shown in figure 1.3, more robust.

In table 3.6, we offer a set of recommendations for additional research and development (R&D) study areas across the three horizons discussed in chapter 1. These recommended study areas would help the AI community in addressing some of the limitations and challenges highlighted earlier.

Table 3.6 Additional data conditioning challenges and recommendations

Challenge Area	Data Conditioning
Horizon 1: Content-Based Insight (1–2 years)	• Automate data labeling • Create benchmarks serving as the gold standard, including test harness (referring to the test infrastructure), data sets, and reproducible code • Perform end-to-end AI capability performance assessments • Assess compliance with FASTEPS
Horizon 2: Collaboration-Based Insight (3–4 years)	• Aggregate real data, simulated plus emulated data, and prior knowledge • Access intermediate data products across the AI architecture pipeline • Iterate among data collection, model development, and HMT
Horizon 3: Context-Based Insight (5+ years)	• Develop ML algorithms incorporating perceived world models • Build context into the model development by exploiting a richer set of data constraints (e.g., what needs to be estimated together with spatial-time data information) • Operate in degraded environments while accessing real-time data

In terms of data privacy and ethics, several countries are establishing policies. Examples of these are the General Data Protection Regulation (GDPR) in Europe and the California Privacy Rights Act (CPRA). In the US, the White House Office of Science and Technology Policy (OSTP) has developed principles under a proposed AI Bill of Rights to guard against technologies that disproportionately affect marginalized individuals and communities [61]. There is also a need for the broader AI community to share data sets, models, and results to more rapidly advance the state of the art and ensure compliance with FASTEPS principles, as introduced in chapter 8. This sharing must be done without stifling innovation and collaboration [62, 63].

Several AI practitioners have recommended the need to formally establish a standardized process for documenting ML data sets. More specifically, Timnit Gebru and colleagues have proposed datasheets for data sets [64] that would be important for both the creators and consumers of data sets. The creation of datasheets for data sets is an excellent approach to avoid mismatches that potentially could contribute to critical consequences in the use of ML models in high-stakes applications.

3.7   Main Takeaways

The exponential growth in data experienced in the past decade has contributed to remarkable advances in AI capabilities. For AI organizations to be well positioned to take advantage of these advances, as well as to continue to develop and transition AI products and services into the market place, we recommended a three-tier approach:

1. Undertake a digital transformation journey to build and sustain a modern platform infrastructure.

2. Become a data-driven organization integrating, conforming, and deploying AI capabilities consistent with database management standards.

3. Follow a set of data-centric AI principles, including diligent attention to data quality assessment, data cleaning, and data preparation.

One important benefit of evolving into a digitally transformed enterprise is to break the organizational silos that hinder the proper implementation of the AI pipeline shown in figure 1.3 in chapter 1, starting with the data-conditioning stage. We highlighted five major capability maturity levels that an AI organization must achieve as they progress through a digital transformation journey. These levels are digitize, standardize, optimize, accelerate, and lead by sustaining a competitive digital ecosystem.

A digital transformation enables organizations to build and sustain a modern platform infrastructure that permits proper database management and compliance with data format and schema standards. In the past few years, there has been a need to advance from transactional data processing, using structured data, to more unstructured data using a NoSQL data schema.

Commercial organizations are also making significant strides in providing repository environments supporting both structured data and unstructured/semistructured data. Examples of data repository environments are data warehouses and data lakes. There is a trend for these commercial providers to converge in their ability to conform to SQL, NoSQL, and streaming plus processing at the data source.

Digital transformation and the establishment of a data-driven organizational culture are necessary but not sufficient for successfully developing and deploying AI capabilities. We must also attend to a third tier—adhering to a set of data-centric AI principles.

The data-centric AI approach (iterative with the ML stage) requires diligent attention to data quality assessment, data cleaning, and data preparation within the data conditioning functional block. We discussed typical functions performed during each of these steps. We also emphasized the importance of understanding the stakeholders' requirements prior to undertaking these specific data conditioning processing functions in order to make sure the data is suitable for the ML algorithm.

The rapid and remarkable acceleration that the AI community has experienced for over a decade is credited to the availability of well-curated open-source data sets, such as MNIST and ImageNet, complemented by advances in computing technologies. There were several watershed moments where dramatic improvements were demonstrated in object classification. Since then, there have been additional efforts to create curated data sets applicable to different modalities including additional image ontologies, videos, text, and speech.

In this chapter, we outlined a number of challenges for each of the three tiers—digital transformation, data-driven culture, and data-centric AI—that must be addressed to develop and deploy AI products and services successfully.

There are many ML metrics. However, the AI community needs to attend to performance metrics at all levels of the AI pipeline shown in figure 1.3, including reliable and trusted data-set labels. The recommended approach is to institute a set of data-centric AI benchmarks, with a requisite test harness, gold-standard data sets, and reproducible code. Each data set must also include a datasheet describing a metadata-level description of what is in it. Data sets must also conform to a set of regulations without stifling technology innovation.

3.8   Exercises

1. Identify and elaborate on one important benefit of undertaking a digital transformation journey.
2. Highlight the main differences between structured and unstructured data.
3. What are the "four Vs" characterizing big data? Explain what each "V" means.
4. Describe the five capability maturity levels that an organization must achieve during a digital transformation journey.
5. Is an SQL schema well matched to unstructured data?
    a. Yes
    b. No
6. Describe the main components of the Apache Hadoop open-source tool set.
7. Elaborate on the main differences between a data warehouse and a data lake.
8. Choose an AI application and explain how the curse of dimensionality can be a significant issue in achieving the desired ML algorithm performance.
9. What are typical functions performed during data quality assessment, data cleaning, and data preparation?
10. Explain the data bias versus data variance trade-off.
11. How can data scientists overcome the curse of dimensionality?
12. Describe examples of data-centric benchmarks and some of their challenges.
13. What are datasheets for data sets?
14. The trend is to incorporate multimodal data (e.g., text, video, and images). How would that help in gaining better context with the results of AI?
15. Address challenges, at the data-conditioning stage, in incorporating multimodal data. Hint: Refer back to table 3.4.

3.9   References

1. Ng, A., *NeuroIPS Data-Centric AI Workshop*. 2021.
2. Hulten, G., *Building Intelligent Systems: A Guide to Machine Learning Engineering*. 2018, Apress.
3. DOMO, *Ninth Annual "Data Never Sleeps"*. 2021. https://www.domo.com/learn/infographic/data-never-sleeps-9.
4. Datareportal, *Digital around the World*. 2021. https://datareportal.com/reports/a-decade-in-digital.

5. Lee, I., Big data: Dimensions, evolution, impacts, and challenges. *Business Horizons*, 2017. 60(3): 293–303.

6. Short, J. and S. Todd, What's your data worth? *MIT Sloan Management Review*, 2017. 58(3): 17.

7. Redman, T. and T. Davenport, Getting serious about data and data science. 2020, *Recuperado de MIT Sloan Management Review*: https://sloanreview.mit.edu.

8. Redman, T. C., What's holding your data program back? *MIT Sloan Management Review*, 2021. 63(1): 1–10.

9. Brynjolfsson, E., D. Rock, and C. Syverson, The productivity J-curve: How intangibles complement general purpose technologies. *American Economic Journal: Macroeconomics*, 2021. 13(1): 333–372.

10. Dickson, B., Why most machine learning strategies fail—TechTalks (bdtechtalks.com), 2021.

11. Burianek, D. and R. Solis, Personal communication on digital transformation maturity levels. MIT Lincoln Laboratory.

12. Ross, J. W., C. M. Beath, and M. Mocker, *Designed for digital: How to architect your business for sustained success*. 2019, MIT Press.

13. Podolny, J. M. and M. T. Hansen, How Apple is organized for innovation. *Harvard Business Review*, 2020. 98(6): 86–95.

14. Furr, N. and A. Shipilov, Digital doesn't have to be disruptive: The best results can come from adaptation rather than reinvention. *Harvard Business Review*, 2019. 97(4): 94–104.

15. Conway, M. E., How do committees invent? *Datamation*, 1968. 14(4): 28–31.

16. Kelleher, J. D., B. Mac Namee, and A. D'arcy, *Fundamentals of machine learning for predictive data analytics: Algorithms, worked examples, and case studies*. 2020, MIT Press.

17. Codd, E. F., A relational model of data for large shared data banks, *Communications of the ACM*. https://dl.acm.org/doi/pdf/10.1145/362384.362685.

18. Chang, F., J. Dean, S. Ghemawat, et al., Bigtable: A distributed storage system for structured data. *ACM Transactions on Computer Systems (TOCS)*, 2008. 26(2): 1–26.

19. Lu, H., C. Hai-Shan, and H. Ting-Ting. Research on Hadoop cloud computing model and its applications, In *2012 Third International Conference on Networking and Distributed Computing*, 59–63. 2012. IEEE.

20. Apache-Software-Foundation. *Apache Hadoop*. 2022; https://hadoop.apache.org/.

21. Gessert, F., W. Wingerath, S. Friedrich, and N. Ritter, NoSQL database systems: A survey and decision guidance. *Computer Science-Research and Development*, 2017. 32(3): 353–365.

22. Sadalage, P., *NoSQL databases: An overview*. 2014. https://www.informit.com/articles/article.aspx?p=2266741

23. Inmon, B., M. Levins, and R. Srivastava, *Building the data lakehouse*. 2021, Technics Publications.

24. Ballou, D. P. and G. K. Tayi, *Enhancing data quality in data warehouse environments. Communications of the ACM*, 1999. 42(1): 73–78.

25. Eichler, R., C. Giebler, C. Gröger, et al., HANDLE—a generic metadata model for data lakes, in *International Conference on Big Data Analytics and Knowledge Discovery*. 2020, Springer. https://www.ipvs.uni-stuttgart.de/departments/as/publications/eichlera/DaWaK2020-HANDLE_Pre-Print.pdf.

26. Giebler, C., C. Gröger, E. Hoos, et al., Leveraging the data lake: Current state and challenges, in *International Conference on Big Data Analytics and Knowledge Discovery*. 2019, Linz, Austria, Springer.

27. Gröger, C., There is no AI without data. *Communications of the ACM*, 2021. 64(11): 98–108.

28. Turck, M. *The 2021 Machine Learning, AI and Data (MAD) Landscape*. 2021, https://mattturck.com/data2021/.

29. Ng, A. *Data-Centric AI*. 2022, https://datacentricai.org/.

30. Chu, X., I. F. Ilyas, S. Krishnan, and J. Wang, Data cleaning: Overview and emerging challenges, in *Proceedings of the 2016 International Conference on Management of Data*, San Francisco. 2016. https://bpb-us-w2.wpmucdn.com/sites.gatech.edu/dist/b/1653/files/2020/10/data-cleaning-sigmod-tutorial.pdf.

31. Grus, J., *Data science from scratch: First principles with Python*. 2019, O'Reilly Media.

32. Gudivada, V., A. Apon, and J. Ding, Data quality considerations for big data and machine learning: Going beyond data cleaning and transformations. *International Journal on Advances in Software*, 2017. 10(1): 1–20.

33. Roh, Y., G. Heo, and S. E. Whang, A survey on data collection for machine learning: A big data-AI integration perspective. *IEEE Transactions on Knowledge and Data Engineering*, San Diego, 2019. 33(4): 1328–1347.

34. Cao, L., Data science: A comprehensive overview. *ACM Computing Surveys (CSUR)*, 2017. 50(3): 1–42.

35. Angelova, A., Y. Abu-Mostafam, and P. Perona. Pruning training sets for learning of object categories, in *2005 IEEE Computer Society Conference on Computer Vision and Pattern Recognition (CVPR'05)*, San Diego. 2005. IEEE.

36. Thundyill Saseendran, A., V. Chhabria, A. B. Roy, et al., *Impact of Data Pruning on Machine Learning Algorithm Performance*. arXiv e-prints, 2019: p. arXiv: 1901.10539.

37. Sorscher, B., R. Geirhos, S. Shrekhar, et al., *Beyond neural scaling laws: Beating power law scaling via data pruning.* arXiv preprint arXiv:2206.14486, 2022.

38. Chandrashekar, G. and F. Sahin, A survey on feature selection methods. *Computers & Electrical Engineering*, 2014. 40(1): 16–28.

39. Sambasivan, N., S. Kampania, H. Highfill, et al., "Everyone wants to do the model work, not the data work": Data cascades in high-stakes AI, in *Proceedings of the 2021 CHI Conference on Human Factors in Computing Systems,* Yokohama, Japan. 2021.

40. LeCun, Y., C. Cortes, and C. J. C. Burges, Gradient-based learning applied to document recognition. *Proceedings of the IEEE*, 1998. 86(11): 2278–2324. http://yann.lecun.com/exdb/mnist./

41. Xiao, H., K. Rasul, and R. Vollgraf, *Fashion-MNIST: A novel image dataset for benchmarking machine learning algorithms.* arXiv preprint arXiv:1708.07747, 2017. https://github.com/zalandoresearch/fashion-mnist.

42. Fei-Fei, L., J. Deng, and K. Li, ImageNet: Constructing a large-scale image database. *Journal of Vision*, 2009. 9(8): 1037–1037. https://image-net.org/index.php.

43. Krizhevsky, A., I. Sutskever, and G. E. Hinton, ImageNet classification with deep convolutional neural networks. *Advances in Neural Information Processing Systems*, 2012. 25.

44. He, K., X. Zhang, S. Ren, and J. Sun, Deep residual learning for image recognition, in *Proceedings of the IEEE Conference on Computer Vision and Pattern Recognition,* Las Vegas. 2016.

45. Lin, T.-Y., M. Maire, S. Belongie, et al., Microsoft COCO: Common objects in context, in *European Conference on Computer Vision,* Zurich. 2014. Springer.

46. Zhou, B., A. Lapedriza, A. Khosla, et al., Places: A 10 million image database for scene recognition. *IEEE Transactions on Pattern Analysis and Machine Intelligence*, 2017. 40(6): 1452–1464.

47. Gray, M. L. and S. Suri, *Ghost work: How to stop Silicon Valley from building a new global underclass.* 2019, Eamon Dolan Books.

48. Monfort, M., A. Andonian, B. Zhou, et al., Moments in Time dataset: One million videos for event understanding. *IEEE Transactions on Pattern Analysis and Machine Intelligence*, 2019. 42(2): 502–508.

49. Ardila, R., M. Branson, K. Davis, et al., *Common voice: A massively-multilingual speech corpus.* arXiv preprint arXiv:1912.06670, 2019.

50. Shetty, J. and J. Adibi, *The Enron email dataset database schema and brief statistical report.* Information Sciences Institute Technical Report, University of Southern California, 2004. 4(1): 120–128.

51. Shetty, J. and J. Adibi. Discovering important nodes through graph entropy: The case of Enron email database, in *Proceedings of the 3rd International Workshop on Link Discovery*, Chicago. 2005.

52. Gadepally, V., S. Madden, and M. Stonebraker, *Managing heterogeneous data and polystore databases*. MIT Lincoln Laboratory Series. MIT Press, forthcoming.

53. Yona, G., A. Ghorbani, and J. Zou. Who's responsible? Jointly quantifying the contribution of the learning algorithm and data, in *Proceedings of the 2021 AAAI/ACM Conference on AI, Ethics, and Society*. 2021.

54. Northcutt, C. G., A. Athalye, and J. Mueller, *Pervasive label errors in test sets destabilize machine learning benchmarks*. arXiv preprint arXiv:2103.14749, 2021.

55. Kang, D., E. Gan, P. Bailis, et al., *Approximate selection with guarantees using proxies*. arXiv preprint arXiv:2004.00827, 2020.

56. Koch, B., E. Denton, A. Hanna, and J. G. Foster, *Reduced, reused and recycled: The life of a dataset in machine learning research*. arXiv preprint arXiv:2112.01716, 2021.

57. Coleman, C., D. Narayanan, D. Kang, et al., DAWNbench: An end-to-end deep learning benchmark and competition. *Training*, 2017. 100(101): 102.

58. Mattson, P., V. J. Reddi, C. Cheng, et al., MLPerf: An industry standard benchmark suite for machine learning performance. *IEEE Micro*, 2020. 40(2): 8–16.

59. Kiela, D., M. Bartolo, Y. Nie, et al., *Dynabench: Rethinking benchmarking in NLP*. arXiv preprint arXiv:2104.14337, 2021.

60. Gordon, M. L., K. Zhou, K. Patel, et al., The disagreement deconvolution: Bringing machine learning performance metrics in line with reality, in *Proceedings of the 2021 CHI Conference on Human Factors in Computing Systems*, Yokohama, Japan. 2021.

61. Yoo, C. S. and A. Lai, Regulation of algorithmic tools in the United States. *Journal of Law and Economic Regulation*, 2020. 13: 7.

62. Mayer-Schonberger, V. and T. Ramge, A big choice for big tech: Share data or suffer the consequences. *Foreign Affairs*, 2018. 97: 48.

63. Nonnecke, B. and C. Carlton, EU and US legislation seek to open up digital platform data. *Science*, 2022. 375(6581): 610–612.

64. Gebru, T., J. Morgenstern, B. Vecchione, et al., Datasheets for datasets. *Communications of the ACM*, 2021. 64(12): 86–92.

4

Machine Learning

I believe there is no deep difference between what can be achieved by a biological brain and what can be achieved by a computer. It, therefore, follows that computers can, in theory, emulate human intelligence—and exceed it.
—Stephen Hawking, English theoretical physicist, cosmologist, author, and former director of research at the Centre for Theoretical Cosmology at the University of Cambridge (1942–2018)

Recent years have witnessed remarkable progress in machine learning (ML) due to the convergence of big data, advanced algorithms, and modern computing technologies. The proliferation of large volumes of data, known as "big data," is largely attributed to the prevalence of smart phones, Internet of Things (IoT), and social media, which have collectively contributed to over 2.5 million terabytes of data being generated daily and to over 90 percent of the world's data that has been generated over the last two years [1]. Along with big data, advancements in modern computing technologies such as graphics processing units (GPUs) and tensor processing units (TPUs) have made it possible to efficiently process data and large-scale computations, which is crucial for the computationally intensive task of model training in ML, as will be discussed in chapter 5. Finally, breakthroughs in new algorithms and ML techniques have been key to efficiently and effectively extracting actionable insights from massive volumes and varieties of data in ways that were previously unimaginable.

ML—one of today's most rapidly growing technical fields—is a branch of artificial intelligence (AI) focused on employing data and algorithms to extract knowledge and insights without explicit instructions. At its core, ML is all about designing algorithms and models that can learn from data without being explicitly programmed. It is used to recognize patterns, make classifications or predictions, and even learn to perform tasks once thought to be exclusive to human intelligence, such as language and image under-standing and generation. These outputs ultimately enable human-machine teams to derive insights used for decision-making in domains such as healthcare, law enforcement, manufacturing, education, and financial services. Over the past decade, advances in the field of deep learning—a technique that uses artificial neural networks (ANNs) to learn from large volumes of data by mimicking the way that the human brain functions—have most notably led to breakthroughs in computer vision, speech recognition, and natural-language processing (NLP) [2]. As noted by Stephen Hawking's quote at the start of this chapter [3], these new techniques attempt to achieve, and in many cases exceed, the abilities of a biological brain in various object detection or language-understanding tasks. As a result, these technologies have been applied to many products that we use and interact with every day such as virtual personal assistants (e.g., Siri, Alexa, Google Home), entertainment and shopping recommendation algorithms (e.g., Spotify, Netflix, Amazon), self-driving cars (e.g., Tesla), and many more. However, it is important to note that many capabilities achieved by modern ML systems fall under the umbrella of narrow AI. There are many limitations to modern day ML from achieving artificial general intelligence (AGI), which would be key to emulating general human intelligence, and even exceeding it.

In this chapter, we provide an overview of the field of ML to equip readers with a foundational understanding of key ML concepts, techniques, applications, and performance metrics. Those looking to gain a deeper understanding of ML topics are recommended to consider textbooks with a more comprehensive coverage of the fundamentals, such as Kevin Murphy's *Machine Learning: A Probabilistic Perspective* and *Probabilistic Machine Learning: An Introduction*, or Christopher Bishop's *Pattern Recognition and Machine Learning* and *Neural Networks for Pattern Recognition* [4–7]. We begin by exploring the three broad classes of ML—unsupervised learning, supervised learning, and reinforcement learning. We review each of these in more detail, highlighting the motivations for each approach and potential use cases. After introducing these three classes of ML, we discuss a set of common measures of performance for evaluating ML models, such as precision, recall, and area under the curve (AUC). With a foundation of classical ML techniques and ML evaluation metrics, we discuss deep learning, which has been the primary technique used in the past decade for achieving unprecedented results. More specifically, we discuss neural networks, providing a brief review of the historical and biological underpinnings of neural networks and discussing backpropagation—a widely used algorithm and a key innovation to the training of neural networks [8, 9]. We finally end with a discussion of one type of supervised deep learning model—convolutional neural networks (CNNs)—which has been widely adopted for analyzing

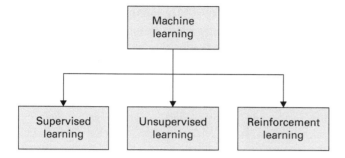

Figure 4.1   The three major classes of machine learning.

visual imagery in computer vision tasks. We end the chapter with a discussion of ML challenges and future opportunities, and finally give the main takeaways and a set of exercises for the reader.

4.1   Machine Learning Classes

ML is a field of technology that focuses on developing computer programs capable of automatically adapting and learning from data without being explicitly programmed to do so. ML, as illustrated in figure 4.1, is broadly categorized into supervised learning, unsupervised learning, and reinforcement learning [2]. These classes vary according to the type of input data provided and the desired output. However, a consistent and core requirement across all ML techniques is the need for model generalizability—a trained model's ability to properly "generalize," or adapt to new, unseen data drawn from the same distribution as the one used to train the model [10]. This generalizability is key to ensuring the model's practical applicability.

Supervised learning, the most widely used ML method, uses labeled data to explicitly train a model to classify or make predictions about new, unseen data. The notion of labeled data simply refers to a collection of prelabeled (x, y) pairs where each input x is annotated with some ground truth output class or feature y by human annotators, special devices, or a physical experiment. An example of a labeled data set is a set of images of cats and dogs that have been annotated by humans with a label of the animal in the image (i.e., cat or dog). Such a data set can be used to train an ML algorithm to classify the animal in a new, unseen image as either a cat or dog. Similarly, another example of labeled data is a historical set of daily temperatures for a given city over the past five years. Such a data set can be used to train an ML algorithm to predict the temperature for that city on a future date. Labeled data is key to enabling supervised ML, and the importance of it has given rise to new job markets and crowdsourcing marketplaces for data labeling such as Amazon Mechanical Turk and start-ups such as LabelBox, ScaleAI, and Sama. In addition to using these marketplaces to facilitate the creation of new labeled data sets, sites like Google Dataset Search, Kaggle, and Data.gov have served as repositories for data sets and hubs for students, researchers, and practitioners to explore, analyze, and share quality data [11–13].

Unsupervised learning, a second major class of ML, employs unlabeled data to identify patterns and clusters in the data. Rather than learning a function from input to output based on a prelabeled set of (x,y) input-output pairs, unsupervised learning seeks to find some inherent structure and pattern in the given data set.

Reinforcement learning, a third major class of ML, is an approach in which an algorithm learns to make decisions by interacting with an environment and receiving feedback in the form of rewards and penalties. The algorithm uses different policies to make a sequence of decisions in a way that maximizes the total reward over time by learning which actions lead to desirable outcomes. In this section, we will discuss each of these major classes of ML in greater detail, highlighting their motivations, example algorithms, and applications. In the appendix of this book, we present use cases demonstrating how the ML techniques are key technology enablers in the AI system architecture as part of the AI value proposition.

Supervised Machine Learning

In supervised learning, the most common class of ML, a model is trained on labeled data to make predictions using a learned mapping $y = F(x)$, enabling the model to eventually predict output labels for new, unseen inputs [2]. More formally, training data D_n takes the form of input-output pairs $[(x^{(1)}y^{(1)}), (x^{(2)}y^{(2)}), \ldots, (x^{(n)}, y^{(n)})]$, where each of the input variables $x^{(i)}$ are d-dimensional vectors of real and/or discrete values and represent an input variable to be classified or predicted. The output variables $y^{(i)}$ represent an output label or continuous value associated with a given input $x^{(i)}$. In supervised learning, the target output variables $y^{(i)}$ are specified for each of the training examples $x^{(i)}$. Thus, the goal is to identify common patterns and trends in the training data that would allow the model to learn a function from input to output, and ultimately predict the output variable $y^{(n+1)}$ for a new, unseen input value, $x^{(n+1)}$.

Supervised learning primarily takes one of two forms—classification and regression. These two supervised learning approaches are visually depicted in figure 4.2. "Classification" is the task of predicting a discrete class label. It uses training data comprised of

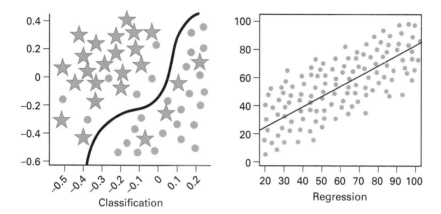

Figure 4.2 Supervised learning approaches: Classification versus regression.

input data annotated with discrete, categorical labels to predict a class label $y^{(i)}$ for new, unseen data $x^{(i)}$. A classification task is binary if $y^{(i)}$ is drawn from a set of two possible values (i.e., hot or cold); otherwise, the task is a multiclass classification (i.e., hot, warm, or cold). For example, classification techniques are used in the medical domain to detect and classify tumors via X-ray and cathode-ray-tube images [14]. Regression is the task of predicting a continuous value. It uses training data to predict the values of an output (dependent) variable $y^{(i)}$, such that $y^{(i)} \in R$ based on the value of a single- or multiple-input predictor (independent) variables. For example, a business organization might use a multiple regression model using predictor variables such as price, geographic location, weather, and consumer demographics to forecast sales. Such a model can help the business better understand how certain factors affect sales performance and enable better predictions for how sales would react to different changes in those input predictor variables (e.g., price, weather).

The anatomy of a supervised ML technique—including both classification and regression—involves a set of six key steps, as shown in figure 4.3. Next, we discuss each in greater detail. Part of this material is adapted from a lecture by Dr. Danelle Shah in the "Artificial Intelligence: A Systems Engineering Approach Course," April 13–27, 2020.

1. **Gather and preprocess data:** Supervised learning algorithms require labeled data. The first step is to gather labeled training data D_n in the form of input-output pairs $[(x^{(1)}y^{(1)}), \ldots ,(x^{(n)},y^{(n)})]$. Once the labeled data is gathered, the data set must be conditioned and preprocessed in accordance with the key data-conditioning principles and practices described in chapter 3.

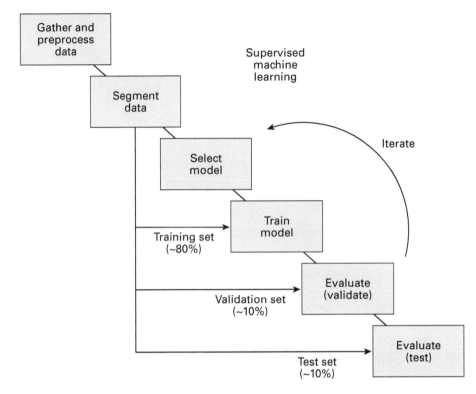

Figure 4.3   Anatomy of a supervised ML technique.

2. **Segment the data:** Once labeled data is gathered and preprocessed, the labeled data set must be appropriately split into training, validation, and testing sets. Validation sets are used to iteratively validate model performance and ensure that the model is not overfit to the training set, while test sets are used to get an evaluation of final model performance that is indicative of how the model would perform in a production setting. An example of a common split is 80 percent training, 10 percent validation, and 10 percent testing, but model owners can opt for different data split ratios (e.g., 70/15/15 or 60/20/20). There are many ways to split the data, but the most common approach is to use some version of random sampling (i.e., each sample has an equal probability of selection), as it is efficient and easy to implement [10]. Whatever the data split ratio or sampling strategy may be, it is important that a segment of data be set aside as a test set so the model can be evaluated on a blind set.

3. **Select the model and train:** Once labeled data has been gathered, processed, and segmented, a decision must be made on the best supervised ML technique for the given task. There are a range of options, including decision trees, linear regression, logistic regression, random forest, gradient boosting, and k–nearest neighbors. Cunningham et al. provided an overview of key supervised learning techniques [15]. The type (regression versus classification) and complexity of the task, the volume and variety of training data, computational cost, and performance trade-offs should be considered when determining the best ML technique for the given task. Once a model of interest is defined and selected, it is trained with the training data set to learn a mapping $y = F(x)$ based on the input-output pairs in the training data set that would allow the model to eventually predict output labels for new, unseen input variables. This mapping is learned by minimizing a defined loss function. A loss function is a measure of the error between the prediction $F(x)$ and the ground truth y. A good prediction will result in a lower error or loss, while a bad prediction will result in a higher one. By defining a loss function, we approach the task of learning as an optimization problem—optimizing our model by minimizing the loss function. Wang et al. provide a comprehensive survey of common loss functions for both classification and regression ML techniques [16].

4. **Evaluate (validate):** Once the target model is selected and trained, the model's performance must be evaluated using some of the common measures of performance described in section 4.2 on ML performance metrics. Most important, model owners must ensure that the final model avoids overfitting or underfitting to the training set. "Underfitting" refers to a modeling error in which a trained model can neither model the training data nor generalize well to new data. On the other hand, "overfitting" refers to a modeling error in which a trained model is too closely aligned to the training set and thus does not generalize well to new data. You can see these two phenomena presented in figure 4.4.

   When training an ML model, recall that a core requirement is generalizability, such that our model adapts well to new, unseen data. The desired state fits well but is generalizable with a good-fit model, also illustrated in figure 4.4 (middle). A model's generalizability can be estimated on a held-out validation set. Evaluation metrics should be used to iterate and fine-tune a model's hyperparameters until the desired level of model performance is achieved. However, as setting aside a portion of the data for validation drastically reduces the data that can be used for training the model, a cross-validation process can instead be used to evaluate model performance. A basic approach shown in figure 4.5, called k-fold cross-validation, splits the training data

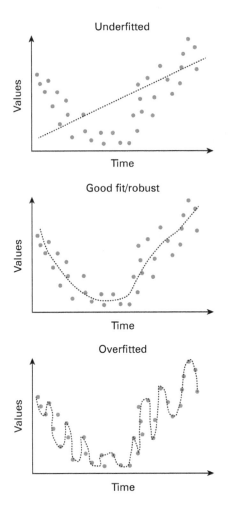

Figure 4.4   Overfitting, good fit, and underfitting.

into k smaller sets. For each of the k folds, a model is trained using $k-1$ of the folds as training data, and the final estimator is validated on the remaining fold [17]. The performance measure using a k-fold is averaged across all values, where each fold is used for validation. While this process can be computationally expensive, it provides the added advantage that more data is freed up to be used for model training. Cross-validation techniques are used to ensure good generalization and avoid overfit or underfit models [10].

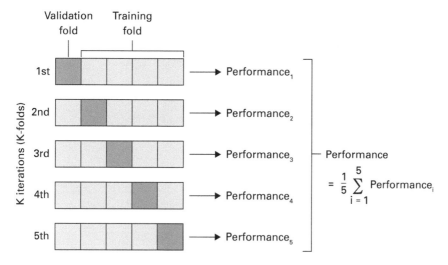

Figure 4.5   Cross-validation process.

5. **Evaluate (test):** After a sufficient convergence is reached in terms of model performance, the model is evaluated on a blind test set. We discuss important measures of performance that should be considered in the evaluation of a model in the following section.

Supervised learning can be a powerful tool for turning data into real, actionable insights. Many classical supervised ML techniques, such as linear regression, logistic regression, support vector machines (SVMs), and random decision forests, are used across a wide range of industries for tasks such as sentiment analysis, recommendation systems, medical diagnosis, and forecasting.

Unsupervised Machine Learning

Unsupervised ML describes a class of ML techniques that seek to analyze and identify patterns in unlabeled data sets. Unlike supervised learning algorithms, which require labeled data, unsupervised learning algorithms are used to identify hidden patterns or clusters without the need for an explicit data labeling process. Two of the main techniques in unsupervised learning are clustering and dimensionality reduction [2], as illustrated in figure 4.6.

Clustering is used to identify groups or clusters in data based on similarities or differences. It identifies structures or patterns in the information that would be difficult

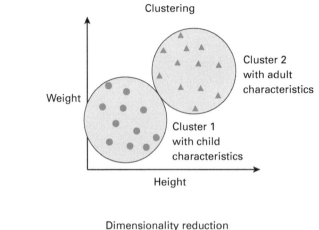

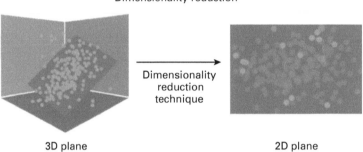

Figure 4.6   Unsupervised ML approaches: Clustering and dimensionality reduction.

for humans to explicitly identify. Given samples $x^{(1)}$, $x^{(2)}$,..., $x^{(n)}$, the goal is to identify clusters that group samples based on similarities. There are many types of clustering algorithms that depend on the definition of similarity among the samples and
criteria used (e.g., minimizing the pairwise distance of samples within the same cluster). Xu et al. provide a comprehensive survey of clustering algorithms [18]. Clustering is used to find useful structures in large data by identifying similar segments in the
data or detecting anomalies by identifying outliers in the data. For example, in marketing, user data on social media activity and past spending habits can be used to
cluster customers into groups with similar purchasing behaviors and ultimately target
advertisements to those groups [19].

Dimensionality reduction is another unsupervised learning technique for reducing the
dimensionality and subsequent complexity of high-dimensional data with the goal of

increasing interpretability while minimizing information loss. This is achieved by re-representing the data in a lower-dimensional space by reducing noise, redundancy, and computational complexity. Namely, given samples $x^{(1)}, x^{(2)}, \ldots, x^{(n)} \in R^D$, the goal of dimensionality reduction is to re-represent the samples as points in a d-dimensional space such that $d < D$. In doing so, we seek to retain as much information as possible such that we can discriminate elements of one class from another while reducing the complexity of high-dimensional data [20]. Dimensionality reduction is used, for example, for data visualization and understanding of high-dimensional data. Dimensionality reduction is also often used as a first step to reduce uninformative features and elucidate the features that are most important or informative before performing regression or classification. This is because dimensionality reduction can mitigate the curse of dimensionality, discussed in chapter 3, by excluding irrelevant features, which can often be a cause for confusion, complexity, and overfitting for ML algorithms.

Some examples of unsupervised learning techniques include principal component analysis (PCA) and k-means clustering. These techniques are used for tasks like data visualization and data clustering and can be applied to effectively visualize the salient features of high-dimensional data, and they also have practical applications such as customer segmentation and anomaly detection.

The decision to use supervised or unsupervised learning depends on the availability of labeled data and the desired task and inferences to be drawn. For example, unsupervised learning is better suited for identifying natural patterns and clusters in a data set. More recently, there has been research combining these two learning techniques, starting with unsupervised learning to reduce the dimensionality of a high-dimensional data set (e.g., using PCA) before using a supervised learning algorithm. In addition, semisupervised learning can be used to combine the benefits of both techniques using a small amount of labeled data and a large amount of unlabeled data, where the labeled data provides some constraints to the learning algorithm during the unsupervised training process [21]. This approach helps avoid the challenges of finding or generating large amounts of labeled data, making this technique particularly useful in tasks where data labeling is difficult or expensive.

Reinforcement Machine Learning

Reinforcement ML describes a class of ML techniques that use a reward-based learning process. Unlike supervised learning techniques, in which training examples are used to indicate a ground truth output for a given input, the training data in reinforcement learning is assumed to provide an indication of whether a taken action is correct. If the action

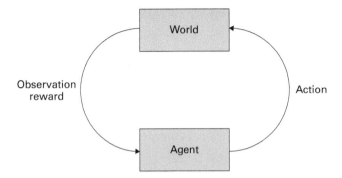

Figure 4.7   Reinforcement ML.

is incorrect, there still remains the problem of finding the right or correct action [2]. As such, these techniques are trained to learn by using trial and error, such that an agent interacts with the environment and makes decisions and choices to ensure that some overall notion of reward is maximized. The agent is provided with rewards based on how well the decisions and choices made by the agent resulted in achieving the desired goal.

More formally, the agent observes the current state $x^{(0)}$. Based on that observed state, it makes a decision, $y^{(0)}$. The agent is provided a reward, $r^{(0)}$, which typically depends on $x^{(0)}$ and possibly $y^{(0)}$. The environment transitions probabilistically to a new state, $x^{(1)}$, with a distribution that depends only on $x^{(0)}$ and $y^{(0)}$. The agent now observes the new current state, $x^{(1)}$, and the process continues. Ultimately, a sequence of successful outcomes is reinforced, while unsuccessful outcomes are discouraged to develop the best recommendation or policy for a given problem and environment. The goal is to find the policy that maps states to a decision ($x^{(0)}$ to $y^{(0)}$) such that some long-term sum or average of rewards $r^{(n)}$ is maximized. It is worth noting that deep learning has similarly accelerated progress in reinforcement learning, with applications of deep learning algorithms in reinforcement giving way to the field of deep reinforcement learning [22].

Reinforcement learning is notably applied to train computer systems to play games like Go or chess, robotics, and motion planning for self-driving cars [23–25].

4.2   Common Measures of Performance

A core requirement of any ML technique is generalizability—a trained model's ability to properly adapt to new, unseen data [10]. Once we identify the most appropriate ML technique for a given task, based on the constraints of the available data and desired outputs, it is essential to define a criterion for evaluating our predictions and the overall performance of our model.

To asses a model's generalizability, it is crucial to evaluate its performance on a blind test set that was not used in the training process. Using a blind test set ensures that evaluations of the model are a good measure of how well the model will perform when it is deployed to production, assuming that the test set follows the same distribution as the set used for training and the target population where the ML system will be deployed. But what defines a good model, and what metrics should be used to evaluate its performance? Fortunately, there are several common measures of performance; in this section, we will define key metrics that can be used to evaluate the performance of a ML model.

Before discussing ML evaluation metrics, it's important to make a distinction between loss functions and evaluation metrics. Loss functions are error functions used to measure how much the predicted value of a learning function deviates from the ground truth. Formally, loss function $L(p, t)$ defines the error penalty during training for making prediction p, given some ground truth t. During training, the loss function is minimized to optimize the ML model. Wang et al. provide a comprehensive summary and analysis of thirty-one common loss functions in ML, providing their formulations and describing the loss functions from the perspectives of traditional ML (i.e., classification, regression, and unsupervised learning) and deep learning [16].

While loss functions are implemented and used to optimize a model during training, evaluation metrics are used after training to assess the predictive performance of the model, independent of the training process. Table 4.1 presents several common measures of performance for evaluating the performance of ML models [26]. Many of the defined metrics are used to evaluate classification problems. There is also a set of metrics used to measure the performance of regression models, such as mean squared error (MSE), mean absolute error (MAE), and R^2.

Many of the key metrics defined here for evaluating classification performance can be clarified further through an understanding of a "confusion matrix," a tabular visualization of the performance of a classification model's predictions on a set of data for which the true values are known. The rows of the confusion matrix correspond to the ground truth, or "target" labels, while the columns correspond to the predicted labels. The diagonal elements of the matrix represent samples correctly classified for each class, while the off-diagonal elements represent the misclassifications between classes. The table is an $N \times N$ matrix—where N is the number of classes/outcomes being predicted, capturing true positive, true negative, false positive, and false negative outcomes. By simply comparing predicted positive and negative data to true positive and true negative data in a test sample, a variety of summary statistics can be computed. Namely, three important metrics can be

Table 4.1 Common evaluation metrics for ML models

Metric	Description	Formula
False positive (FP)	When the actual class is negative (0), but the predicted class is positive (1)	FP = Predicted Positive, Actual Negative
False negative (FN)	When the actual class is positive (1), but the predicted class is negative (0)	FN = Predicted Negative, Actual Positive
True positive (TP)	When the actual class is positive (1) and the predicted class is positive (1) as well	TP = Predicted Positive, Actual Positive
True negative (TN)	When the actual class is negative (0) and the predicted class is negative (0) as well	TN = Predicted Negative, Actual Negative
Accuracy	A measure of the rate of predictions that our model got right from the total number of predictions	$Accuracy = \dfrac{TP+TN}{TP+FP+FN+TN}$
Misclassification rate	A measure of the rate of predictions that were wrong, regardless of whether they were positive or negative (1 − Accuracy)	$Misclassification\ rate = \dfrac{FP+FN}{TP+TN+FP+FN}$
Precision	A rate of how often the model makes a correct prediction among all predicted positives	$Precision = \dfrac{TP}{TP+FP}$
Recall (sensitivity)	A TPR telling how confident we can be that the model found instances of a positive target level	$Recall = \dfrac{TP}{TP+FN}$
F1 score	Harmonic mean of precision and recall	$F1\ score = 2 \times \dfrac{Precision \times Recall}{Precision + Recall}$
MSE	The average squared error between the predicted values and the actual values	$MSE = \dfrac{1}{N}\sum_{i=1}^{N}(y_i - \hat{y}_i)^2$

MAE	The average absolute distance between the predicted values and the actual values	$\mathrm{MAE} = \dfrac{1}{N}\displaystyle\sum_{i=1}^{N}\lvert y_i - \hat{y}_i\rvert$
Coefficient of determination (R^2)	A statistical measure of the proportion of the variation in the dependent (output) variable that is predictable from the independent (input) variables	$R^2 = 1 - \dfrac{RSS}{TSS}$ * RSS = Sum of the squares of residuals * TSS = Total sum of the squares

Table 4.2   2×2 confusion matrix

	Prediction Positive	Prediction Negative	
Target Positive	TP	FN	$Sensitivity = \dfrac{TP}{TP+FN}$
Target Negative	FP	TN	$Specificity = \dfrac{TN}{FP+TN}$
	Positive Predicted Value $=\dfrac{TP}{TP+FP}$	Negative Predicted Value $=\dfrac{TN}{TN+FN}$	$Accuracy = \dfrac{TP+TN}{TP+TN+FP+FN}$

computed from this confusion matrix: precision, recall, and accuracy. Precision seeks to determine the proportion of identified positives (TP+FP) that were actually positive (TP). Precision is also known as positive predicted value, as shown in table 4.2. Recall seeks to determine the proportion of actual positive (TP+FN) that were identified correctly (TP). Recall is also known as "sensitivity." Finally, accuracy is a measure of the proportion of correctly classified predictions among all predictions. Table 4.2 presents a 2×2 confusion matrix for a simple binary classifier. It's important to note that for an N-class classification, our confusion matrix would be of N×N dimensions.

It is important to consider that the consequence of actions resulting from a false positive versus a false negative prediction will depend on the use case. Therefore, it is imperative that model owners make mindful trade-offs between the false positive and false negative rates of their model (see the discussion in chapter 6 on confidence in the machine making the decision versus the consequence of actions). For example, in the medical field, diagnostic medical tests may produce false positive and false negative errors. A false negative may cause a sick patient to remain untreated, while a false positive may lead to unnecessary treatment of a healthy patient. While efforts can be made to minimize errors by collecting more data, adjusting sensitivity and specificity thresholds, and further testing the predictive model, reducing one type of error may often come at the expense of increasing the other type [27]. In such cases, model owners must decide which error is preferable and tolerable based on an evaluation of the consequence of false action.

The AUC and receiver operating characteristic (ROC) curve are important measures to evaluate the performance of a binary classification model in ML and are extensions of some of the metrics presented in table 4.1. ROC is a visual representation of the trade-off

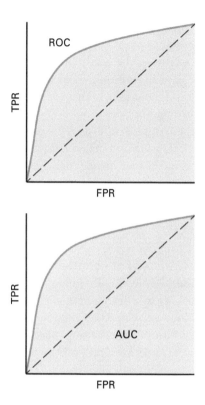

Figure 4.8 The ROC and AUC.

between a true positive rate (TPR) and a false positive rate (FPR) at various classification thresholds. The curve is plotted by computing the FPR and TPR at all classification thresholds, as shown in figure 4.8. AUC is an aggregate measure of the performance of the model and is calculated as the area underneath the ROC curve. The AUC is threshold invariant and can range from 0 to 1. For a random guess classifier, the AUC is 0.5 (dotted line), while a perfect classifier has an AUC of 1. A large AUC indicates a better-performing model overall, as the model can better distinguish between the two classes regardless of the chosen threshold. ROC and AUC are important metrics that provide a single value to assess the performance of a binary classification model across classification thresholds.

Developing an ML solution involves significant investments of time and resources. Model training can take a long time depending on the size and complexity of the data and the model, and often specialized, and costly hardware and software are required to train and run these models in production. As a result, it is important not only to evaluate a model's performance using standard performance metrics, but also to articulate the value proposition of the ML solution in justifying the costly investments. This involves considering the potential performance gains in relation to the associated risks, time, resources, and financial costs to ensure that the investment in an ML solution is justified and delivers significant benefits to the organization.

With a foundation of classical ML techniques and common measures of performance, we turn our attention in the following section to deep learning and neural nets, a subfield of ML that has been the primary technique used in the past decade for achieving unprecedented results.

4.3   Introduction to Deep Learning and Neural Nets

Deep learning was largely motivated by the performance limitations of classical ML algorithms, which relied heavily on human domain knowledge and feature engineering [8]. In classical ML techniques such as linear regression, decision trees and random forests, domain knowledge, and human intervention are necessary to extract and engineer features from raw data prior to training a model [8, 28]. However, with representation learning, deep learning has eliminated the need for manual feature preprocessing, making it possible for deep learning models to extract important features from raw, unstructured, and unprocessed data such as text and images with minimal human intervention [28].

Representation learning is a set of methods that enable a machine to start with raw input data and learn representations of data that are necessary for detection or classification tasks. The ability to learn representations at higher and more abstract levels has enabled deep learning techniques to be very good at discovering intricate structures in high-dimensional data, enabling great success in solving complex tasks such as language and image understanding [8]. Deep learning has had and will continue to demonstrate great success due to its ability to learn multilevel complex representations with very little to no human engineering. Furthermore, as illustrated by Andrew Ng in figure 4.9, deep learning has shown great promise in demonstrating better performance gains with increases in the amount of available computation power and data compared to traditional ML techniques [8, 29].

Deep learning—inspired by the anatomy and function of the human biological brain—is a subfield of ML that uses representation learning to train a computer to learn from large amounts of examples. ANNs, also referred to as "simulated neural networks," are the foundation for deep learning and date back to the 1940s [30]. The design of ANNs is motivated and inspired by the biological structure of the human

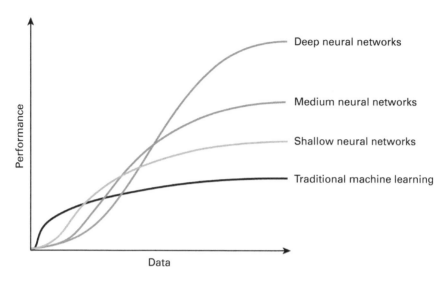

Figure 4.9   Scale driving deep learning progress [29].

brain and seeks to mimic the way that biological neurons fire and signal to one another in the human brain. By design, ANNs are a collection of units or nodes that model the neurons in a biological human brain, as illustrated in figure 4.10. Each of these nodes are connected by edges to transmit signals to one another, akin to synapses in a biological brain [31]. Ultimately, ANNs seek to capture and achieve the computational power of biological brains by linking together many artificial neurons.

A neural network is composed of nodes, or artificial neurons, which are arranged in a series of layers. The first layer, or the input layer, is responsible for inputting the various features or attributes of the input data. These features define the input data that the network uses to learn. From the input layer, the data goes through one or more hidden layers. At each layer, mathematical operations are carried out to transform the input into a signal that is propagated throughout each layer and finally delivered to the final output layer. As shown in figure 4.11, a neural network with more than one hidden layer (a total of four or more layers, including input and output layers) is referred to as a "deep neural network (DNN)" or "deep learning."

Every node in every layer of the network is connected by edges, as shown in figures 4.10 and 4.11. Each neuron receives input signal or signals x_i by way of this connection and outputs signal y_i to other nodes it is connected to. Each of these connections are given weight w_i, where the greater the weight of a connection, the greater the influence

Biological neuron versus artificial neural network

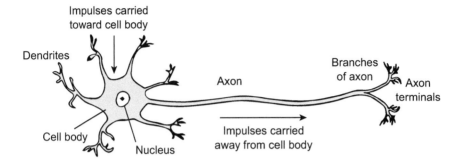

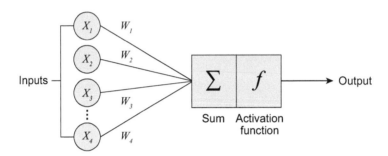

Figure 4.10 Biological neuron versus ANN. Based on Jain, Mao, and Mohiuddin (1996).

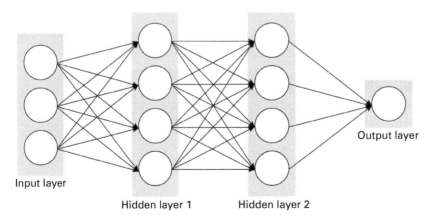

Figure 4.11 DNN with two hidden layers.

the first node has on the second node, ultimately reflecting how important the first node is. The output signal y_i of a node is determined using a weighted sum of its inputs, called the "activation," and a bias (constant) term b added to the activation, $b + \sum_{i=1}^{n} (w_i * x_i)$. The role of the bias term is simply to shift the value produced by the activation, akin to the role of a constant in a linear function. The computed value is fed into an activation function F_a, which produces a single output signal $y = F_a(b + \sum_{i=1}^{n} (w_i * x_i))$ that is sent to all other connected nodes.

Currently, the most widely used activation function is the rectified linear unit (ReLU), but figure 4.12 outlines some other notable activation functions commonly used in neural networks. Nwankpa et al. provide a comprehensive survey of the existing activation functions used in deep learning applications and the current trends in the applications and usage of these functions [32]. Under this process, data is propagated through every layer until the final layer. The outputs of the neurons in the final output layer accomplish the initial desired task, which may have involved either classifying an object in an image or predicting some value.

There are numerous types of neural network architectures developed using different types of layers. Each type of layer performs different transformations on the inputs, and as a result, has a different use case. Figure 4.13 outlines common neural network layer structures. For example, a fully connected layer, also known as a "dense layer," is a general-purpose layer where every neuron in the current layer is connected to every neuron in the previous layer. As such, each neuron in the current layer takes as input a vector of values representing the outputs of all neurons in the previous layer. Convolution layers are key to CNNs, which are commonly used to analyze and detect features in images. Some other common neural network layers include Deconvolutional, MaxPool, and Dropout.

Now that we have established the motivation behind deep learning and the biological underpinnings and the anatomy of ANNs, we can proceed to discuss the process of training a neural network. Namely, we provide a deep dive into backpropagation, the key enabling algorithm by which DNNs are able to learn and converge toward optimal outputs.

4.4   Training Neural Networks with Backpropagation

The goal of any supervised deep learning algorithm is to find a generalizable function that best maps a set of inputs to their correct outputs. During training, a neural network

Activation functions

Rectified linear unit (ReLU):

$$f(x) = max\,(0,x)$$

Sigmoid function:

$$f(x) = \frac{1}{1 + e^{-x}}$$

Tanh function:

$$f(x) = tanh\,(x)$$

Step function:

$$f(x) = \begin{cases} 0, & x < 0 \\ 1, & x \geq 0 \end{cases}$$

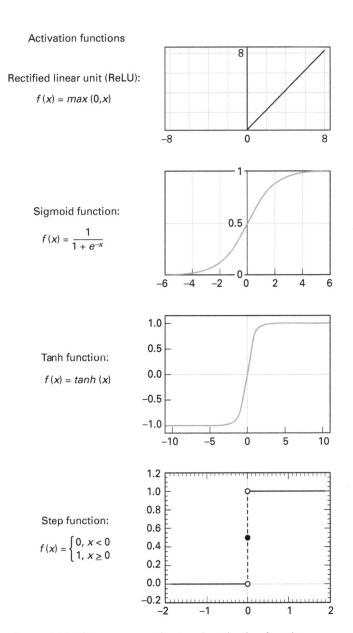

Figure 4.12   Common neural network activation functions.

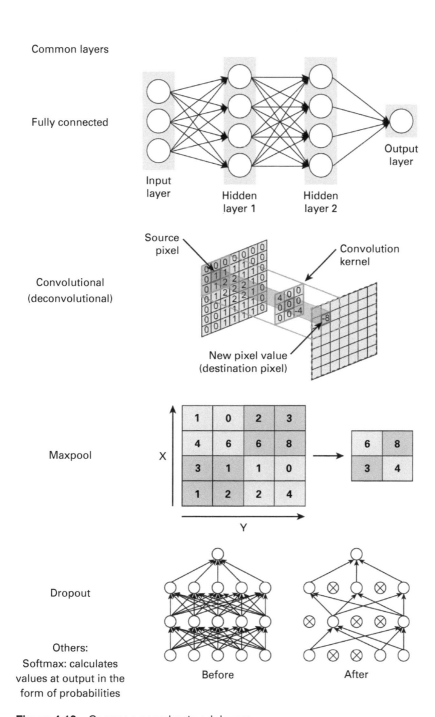

Common layers

Fully connected

Input layer
Hidden layer 1
Hidden layer 2
Output layer

Convolutional (deconvolutional)

Source pixel
Convolution kernel
New pixel value (destination pixel)

Maxpool

Dropout

Before

After

Others:
Softmax: calculates values at output in the form of probabilities

Figure 4.13 Common neural network layers.

is initialized with a set of weights and biases (networks are often initialized with a set of random weights, and biases are set to 0) and is fed large amounts of training data. Training data is passed through each layer of the network, and the network's final prediction output is compared to the ground truth label. Recall that when training, we are seeking to minimize some predefined loss function, which represents the cost of a wrong prediction [16]. In this section, we discuss the backpropagation algorithm, a key processing step that is used to train neural networks by minimizing the loss function by efficiently updating the network's weights and biases.

Backpropagation was notably popularized as the preferred learning procedure to train neural networks in a 1986 paper in *Nature* titled "Learning Representations by Back-propagating Errors," by Rumelhart et al. [9]. At a high level, the backpropagation algorithm indicates how a model should update its internal parameters (weights and biases), which are used to compute the representation in each layer based on the representation in the previous layer [8]. This algorithm is the workhorse behind how DNNs learn and discover intricate structures in large data sets.

Our goal with backpropagation is simple—to adjust the weights and biases of our network by minimizing our loss function. It works by computing the gradient of the loss function with respect to the weights of the network and then using the gradient to update the weights in a way that reduces the loss. We describe some of the key elements of the backpropagation algorithm using a simple network with only one neuron per layer.

We start by describing some relevant equations. First, we define the activation, $a^{(L)}$, of the neuron at the final output layer L. The activation is computed by feeding $z^{(L)}$, the weighted sum of all input activations plus a bias term, into an activation function F_a:

$$z^{(L)} = w^{(L)} * a^{L-1} + b$$
$$a^{(L)} = F_a(z^L)$$

For a given predicted output and its associated true value y, we compute the cost, or loss, of our prediction with respect to the predefined cost function F_c:

$$C = F_c(a^{(L)}, y)$$

Given the cost function, the task now is to understand how sensitive the cost function is to the different weights and biases in the network. Our goal during training is to minimize the cost of the network's predictions. A better understanding of the relationship among different components of the network and the overall cost will allow us to effectively identify and update the weights and biases that will minimize the cost.

Formally, we will need to compute the derivative of cost C with respect to $w^{(L)}$. Note that since $w^{(L)}$ is not directly found in the cost function, we start by considering the change of $w^{(L)}$ in $z^{(L)}$. Next, we consider the change of $z^{(L)}$ in $a^{(L)}$, and then finally the change of $a^{(L)}$ in C. By doing so, we can measure the change of C in relation to a particular weight $w^{(L)}$. More formally, we use the chain rule to compute the derivative of C with respect to $w^{(L)}$:

$$\frac{dC}{dw^{(L)}} = \frac{dC}{da^{(L)}} \frac{da^{(L)}}{dz^{(L)}} \frac{dz^{(L)}}{dw^{(L)}}$$

Multiplying these three ratios gives us the sensitivity of C to changes in $w^{(L)}$. Note that we must modify only slightly the partial derivative given here to get the sensitivity of C to changes in $b^{(L)}$:

$$\frac{dC}{db^{(L)}} = \frac{dC}{da^{(L)}} \frac{da^{(L)}}{dz^{(L)}} \frac{dz^{(L)}}{db^{(L)}}$$

While we can't directly influence activations, we can slightly modify the partial derivative here to also get the sensitivity of C to changes in the previous activations in our network $a^{(L-1)}$:

$$\frac{dC}{da^{(L-1)}} = \frac{dC}{da^{(L)}} \frac{da^{(L)}}{dz^{(L)}} \frac{dz^{(L)}}{da^{(L-1)}}$$

However, what we have demonstrated in this example takes only output layer L into consideration, indicating the list of updates that we want to happen to weights, biases, and activations of the previous layer. Once we have those, we simply recursively apply the same backpropagation process to the relevant weights and biases that determine those activations for the same previous layer until the first layer is reached. We do this by chaining more partial derivatives to find the weights and biases for those layers, moving backward through each layer of the network. So, to calculate the derivate of the cost function with respect to the weights of hidden layer $L-1$, we must build on the previous calculations for layer L:

$$\frac{dC}{dw^{(L-1)}} = \frac{dC}{da^{(L)}} \frac{da^{(L)}}{dz^{(L)}} \frac{dz^{(L)}}{dw^{(L)}} \frac{dw^{(L)}}{da^{(L-1)}} \frac{da^{(L-1)}}{dz^{(L-1)}} \frac{dz^{(L-1)}}{dw^{(L-1)}}$$

Now we continue this process, working our way backward through the network and adding more partial derivatives for each extra layer.

Note that thus far we have only computed the derivative of C with respect to w for just one training example. The full cost of our model is the average of all the individual costs for each training example, so a true gradient descent step would require computing the derivative for all n training examples and averaging the desired changes that we would get for each weight and bias, as follows:

$$\frac{dC}{dw} = \frac{1}{n}\sum_{k=1}^{n}\frac{dC_k}{dw_k}$$

$$\frac{dC}{db} = \frac{1}{n}\sum_{k=1}^{n}\frac{dC_k}{db_k}$$

However, this approach of computing the influence of every training example for every single gradient descent step would be a slow and computationally expensive task, given training sets of large sizes. So instead, we randomly subdivide the data into "mini-batches" (often sixteen or thirty-two samples) and compute the average output for each weight and bias, which becomes the step, or output, of the gradient. Repeatedly going through all the mini-batches in the training set and making adjustments to the network's weights and biases allow us to converge to a pretty good estimation of the global minimum of the cost function. This is a pretty good approximation of the global minimum that we would achieve if we used the full training set to compute every single gradient descent step. This technique is referred to as "stochastic gradient descent (SGD)," and the process of SGD steps that minimize the cost function toward a local minimum are depicted in figure 4.14.

Each of the partial derivatives from the weights and biases are saved in a gradient vector. That vector points in the direction of the greatest increase in cost and loosely

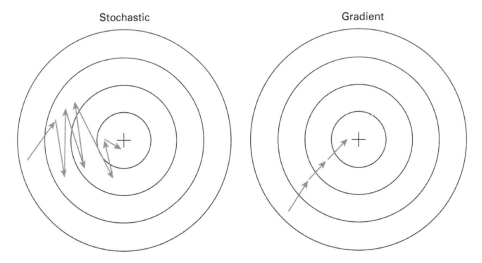

Figure 4.14 SGD versus gradient descent.

represents how we need to change each of the weights and biases in the network to most efficiently decrease the cost. By taking the negative of the gradient vector, we point in the direction of the steepest descent, as shown here:

$$-\nabla C(w_1, b_1, \ldots, w_n, b_n) = \begin{bmatrix} \dfrac{dC}{dw_1} \\[2mm] \dfrac{dC}{db_1} \\[2mm] \vdots \\[2mm] \dfrac{dC}{dw_n} \\[2mm] \dfrac{dC}{db_n} \end{bmatrix}$$

The backpropagation algorithm is key to enabling neural networks to learn. With enough input-output training data pairs and many iterations and epochs, our network eventually converges to an optimum that minimizes our cost function. For an audiovisual review of both the intuition and the calculus of backpropagation, we highly recommend the set of audiovisual materials developed by 3Blue1Brown (available on their YouTube channel) on their series on deep learning and backpropagation [33, 34]. With a foundational understanding of the key enabling algorithm for how neural networks can learn, we now discuss important responsibilities and decisions in the design of neural networks.

4.5 Designing a Neural Network

Neural networks have undoubtedly demonstrated great promise in learning intricate details from high-dimensional data. While the greatest appeal of neural networks is their ability to learn from raw, unstructured data with little human intervention and feature engineering, there are still many roles and responsibilities that a human must play in enabling the application and use of neural networks [8]. To successfully deploy a ML solution, humans must still do all the following:

1. Frame the problem area and ML task.
2. Identify, acquire, and prepare the data.

3. Select the appropriate learning algorithm.

4. Define and tune hyperparameters during training.

5. Validate the results, making trade-offs and determining the readiness of the solution for deployment and production use.

Other than the core responsibility of training a model, these human tasks are integral and of great importance to the successful design, development, and deployment of an ML solution. In this section, we briefly discuss some of the key responsibilities in selecting the appropriate ML algorithm and defining the right hyperparameters during training. In the appendix of this book, we discuss the design of a multilayer perceptron neural network and a CNN algorithm, going through the steps discussed in this chapter.

Advances in the deep learning community have led to the innovation of new neural network architectures. Different architectures have been developed to address specific use cases by better emulating human biological processes for tasks like image and language understanding. As such, these new techniques, coupled with the advances in computing technologies and big data, have proved to be highly successful when applied to specific tasks. For example, CNNs are primarily used for image processing, but they also can be used for other types of input such as audio. Composing a complete and comprehensive list of all existing network architectures is practically impossible, as new architectures are being developed all the time, and there is no comprehensive summary of all existing neural network architectures. Liu et al. discuss some widely used deep learning architectures and their practical applications, providing a deeper discussion of autoencoders, CNNs, deep belief networks, and restricted Boltzmann machines [35]. While it is by no means a comprehensive list, Leijnen et al., from the Asimov Institute, also identify and describe some salient neural network architectures. These architectures are captured in figure 4.15 [36]. Our objective is to expose the reader to the variety of neural network architectures that exist today and the different functions they serve, not to give an exhaustive list.

One very important advancement in neural networks (captured in figure 4.15) is the development of general adversarial networks (GANs). GANs, introduced by Goodfellow et al. in 2014, are a class of ML frameworks in which two neural networks compete in a zero-sum game [37]. One neural network, called the "generator," creates new instances of data with the same statistics as the training set, while the other neural network, called the "discriminator," evaluates the authenticity of those new instances of data. The training objective of the discriminative network is to evaluate the data as authentic (i.e., belonging to the training set) or not, while the objective of the generative network is to increase the error rate of the discriminative network. Both discriminator and generator seek to

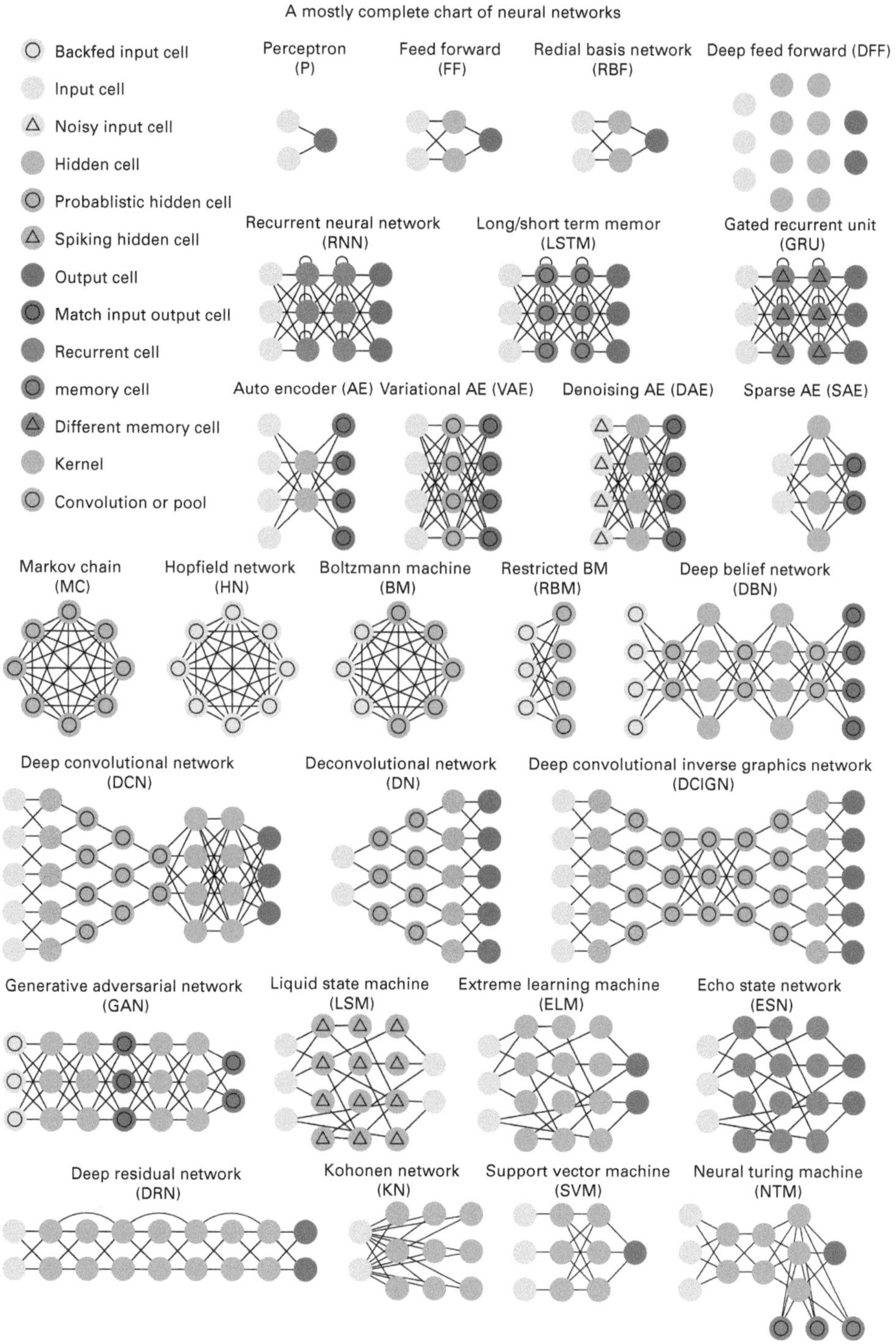

A mostly complete chart of neural networks

Legend:
- ○ Backfed input cell
- Input cell
- △ Noisy input cell
- Hidden cell
- ◎ Probablistic hidden cell
- △ Spiking hidden cell
- Output cell
- ◉ Match input output cell
- Recurrent cell
- ◎ memory cell
- △ Different memory cell
- Kernel
- ○ Convolution or pool

Architectures: Perceptron (P), Feed forward (FF), Redial basis network (RBF), Deep feed forward (DFF), Recurrent neural network (RNN), Long/short term memor (LSTM), Gated recurrent unit (GRU), Auto encoder (AE), Variational AE (VAE), Denoising AE (DAE), Sparse AE (SAE), Markov chain (MC), Hopfield network (HN), Boltzmann machine (BM), Restricted BM (RBM), Deep belief network (DBN), Deep convolutional network (DCN), Deconvolutional network (DN), Deep convolutional inverse graphics network (DCIGN), Generative adversarial network (GAN), Liquid state machine (LSM), Extreme learning machine (ELM), Echo state network (ESN), Deep residual network (DRN), Kohonen network (KN), Support vector machine (SVM), Neural turing machine (NTM)

Figure 4.15 Neural network architectures. Source: Lejnen and van Veen (2020).

optimize different and opposing objective functions, which enables the model to ultimately learn in an unsupervised manner. GANs have many practical applications, such as image, video, and voice generation.

Similarly, the introduction of the transformer architecture has been revolutionary to the recent advancements in the fields of natural-language generation. The architecture was first introduced in 2017 by a team at Google in a paper titled "Attention Is All you Need," and it has since become a state-of-the-art technique in the field of natural-language processing (NLP) [38]. The transformer network has an encoder and decoder structure that uses a self-attention mechanism to capture long-term dependencies in the input data, making it possible to process longer sequences efficiently and effectively. This approach achieves superior quality and performance over previous models such as recurrent neural networks (RNNs) and long short-term memory (LSTM) networks. Another benefit of this new transformer approach is that it lends itself well to a parallel architecture, suitable for GPU architectures. The novel architecture for processing sequential data, such as natural-language text, has motivated many variants and extensions, such as BERT, GPT-3, and GPT-4 [39, 40].

Upon selecting a specific neural network architecture for a given task, model owners have several "knobs" available to them that must be defined before training the neural network, and ultimately must be tuned during the iterative process of training and validating the performance of the model. These knobs, formally known as "hyperparameters," are the configuration variables that govern the training process itself; they include things like the number of layers, number of nodes, activation function, epochs, iterations, and batch size of your model. Decisions on these configuration variables will have a subsequent impact on how efficiently and effectively your model is able to learn. The process of training a model involves initializing and tuning these hyperparameters to ensure that the model learns the optimal parameters (i.e., weights and biases) that will allow it to achieve the desired predictive performance. "Tuning" simply refers to the process of adjusting these variables until the desired model performance is achieved. This will require an iterative process of running your training job, looking at the aggregate metrics, and adjusting the configuration variables to improve model performance and arrive at the optimal values.

It's important to note the differences between parameters and hyperparameters. Hyperparameters are not learned by the model itself from the data, and thus they must be explicitly defined and iteratively tuned by the model owner before training begins. On the contrary, parameters are internal to the model and represent the attributes that are learned or estimated by the model from the training process, such as the weights and biases of the neural network. In table 4.3, we present and define some of the most

Table 4.3 Hyperparameters when designing a neural network

Hyperparameter	Definition
Learning rate	The rate at which the gradient descent searches for a global minimum.
Epoch	A pass through the full training or validation data forward and backward.
Batch size	The subset of data used in the training or validation stages. The larger the batch size, the more the processing can be done in parallel using GPUs.
Iteration	The number of batches to complete one epoch.
Activation function	A function used to determine the output of a node given an input or set of inputs
Cost or loss function	A function used to determine the penalty for an error
Train/test split ratio	The split of data used for training and testing a model. A common split strategy might be 90% train, 10% test, but could vary (e.g., 70/30, 80/20, or 85/15).
Network depth	A measure of the number of hidden and visible layers in a neural network.
Dropout	Refers to the dropping-out of units (hidden and visible) in a neural network, usually done to prevent overfitting.
Initialized weights and biases	While weights and biases are parameters learned by the model during the training process, model owners must initialize weights and biases. It is common for weights to be initialized randomly and biases to be set to 0.

important general purposes hyperparameters that model owners should be mindful of when designing a neural network.

It's important to note that specific network architectures may introduce their own sets of additional hyperparameters that must be initialized and tuned. For example, we will see in the discussion of CNNs in section 4.6 that we will have to initialize and tune hyperparameters specific to the convolutional layer of the network, such as the kernel size and stride. As the complexity of the network increases, the task of defining and tuning hyperparameters becomes more challenging. While there are best practices and guidelines that can help improve the tuning process, designing a neural network and tuning hyperparameters is a complex task that requires both scientific understanding and experience.

In the following section, we provide a short introduction to CNNs, which notably are used for image and object recognition tasks. In the appendix of this book, we walk the reader through a hands-on design of a CNN using the MNIST Fashion data set.

4.6   Introduction to Convolutional Neural Networks

One of the biggest and most compelling applications of deep learning is in computer vision, where neural networks are trained to detect, segment, and recognize instances of semantic objects in digital images and videos [8]. These capabilities have been essential to computer vision tasks such as face recognition, object detection, and image annotation, which have many practical applications to areas such as spatial mapping for robotics and autonomous vehicles, video surveillance for law enforcement, and image segmentation and analysis of X-rays for medical diagnoses. Much of the massive progress and success in computer vision over the past decade have been attributed to CNNs, which serve as the backbone and enabling neural network architecture for many modern high-performing computer vision systems. In this section, we introduce CNNs, discussing their important building blocks and helping the reader develop an intuitive foundation for how a CNN is able to "see."

The introduction of CNNs was formalized by LeCun et al. in their 1998 paper, "Gradient-Based Learning Applied to Document Recognition." They demonstrated that a CNN model, capable of aggregating simpler features progressively into more complicated features, can be successfully used to recognize handwritten characters. Using the MNIST database of handwritten digits, LeCun et al. showed that CNNs outperform all other techniques at doing the task of handwritten character recognition [41].

CNNs, also known as "ConvNets" or "deep CNNs," were largely forsaken by the ML community since the late 1990s. That changed in 2012 with the ImageNet Large Scale Visual Recognition Challenge (ILSVRC), a visual recognition challenge using ImageNet, a data set with nearly 1.5 million high-resolution images that contained 1,000 classes [8]. At the ILSVRC-2012 challenge, CNNs saw a huge resurgence in popularity after a CNN architecture network called AlexNet achieved a better than previous state-of-the-art performance labeling ImageNet images. The AlexNet CNN, designed by Alex Krizhevsky, a PhD student at the time, and his colleagues, was able to achieve a top-five error of 15.3 percent, compared to the 26.2 percent achieved by the second-best entry [42]. The creators of AlexNet attributed the depth of the model as being essential for its high performance. The model contained eight layers: five convolutional layers and three fully connected layers, totaling 60 million parameters and 650,000 neurons. It took five to six days to train [42].

So, what is a CNN? How is it able to detect, segment, and recognize items in an image, and what are some of the key building blocks of the architecture? CNNs are feed-forward neural networks that use a mathematical operation called a "convolution." A typical CNN is structured as a series of stages. Each layer of a CNN is able to "learn" different levels of abstractions within a given image, as shown in figure 4.16. The first layers learn basic features within an image, including edges and corners. That learning is further abstracted in the middle layers, where the model can detect shapes, parts of objects, and textures. Finally, the last layers can learn higher representations until the network finally identifies the intended object [8]. CNNs consist of four main types of layers that enable this representation learning:

- *Convolution layer:* The convolution layer applies a filter (formally known as a "kernel") to the input image to detect what features, such as edges, are present throughout the image. The kernel slides across the height and width of the image, using the kernel to apply a convolution operation to the input image. The sliding size of the kernel is called a "stride," and the convolution operation converts the pixels in its receptive field into a single value by taking a dot product calculation between the input pixels and the kernel and feeding the result into an output array. The kernel shifts by a stride and repeats the process until the kernel has swept across the entire image. The final output is known as a "feature/activation map."

- *Pooling layer:* The pooling layer is used to down sample data by reducing the spatial size of the input vector (dimensionality reduction). The pooling operation sweeps a filter across the entire input, applying an aggregation function to the values within the receptive field and populating the output array. There are two types of pooling: (1) max pooling, where the filter moves across the input and selects the pixel with the maximum value to send to the output array; and (2) average pooling, where the filter moves across the input and calculates the average value within the receptive field to send to the output array.

- *Dense/fully connected layer:* The dense/fully connected layer has full connectivity with all neurons in the preceding and succeeding layers.

- *Dropout layer:* The dropout layer is used as a regularization technique to prevent overfitting. As the name suggests, "dropout" refers to the dropping-out of units (both hidden and visible) within a network.

There are numerous salient CNN architectures beyond AlexNet, such as VGGNet, GoogLeNet, and ResNet, to name a few. CNNs have served as an invaluable breakthrough in the field of computer vision and in enabling computers and systems to derive meaningful information and details from digital images, videos, and audio input. The

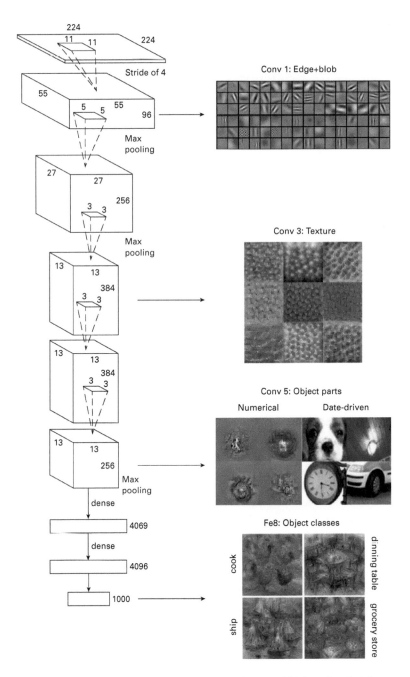

Figure 4.16 How a CNN is able to "see." Source: Krizhevsky, Sutskever, and Hinton (2012).

advances in computer vision, largely due to the progress that has been achieved with CNNs, have led to practical applications of computer vision to a range of industries from healthcare and retail to automotive and legal services.

4.7 Machine Learning Challenges

The proliferation of data, advancements in computing technologies, and breakthrough in ML algorithms and architectures have led to great success within the ML community. With the advent of deep learning and DNNs, we have especially seen the added performance gains that have come with larger networks incorporating more data, as shown in figure 4.9. This has prompted a paradigm shift with the rise of large-scale foundation models such as BERT, DALL-E, and GPT-3, all of which are trained on broad data at scale and are adaptable through fine-tuning for many downstream applications [38, 43]. Vaswani et al. (from Google and the University of Toronto) published the seminal paper on a new simple network architecture referred to as the "Transformer"—this work spun off a whole field initially applied to natural-language processing (NLP) [38]. Bommasani et al. provided a commentary on the opportunities and risks of foundation models, ranging from their capability to their applications and societal impact [43].

Nonetheless, these foundation models can be a key step toward democratizing AI capabilities and allowing individuals and organizations, small and large, to benefit from the scale, power, and size of these foundational models, which can be fine-tuned for different downstream tasks. The future holds great promise, with recent work and breakthroughs around generative models and multimodal models. In table 4.4, based on these opportunities, we outline a set of challenges and recommendations using the same horizon format as in previous chapters. Alongside many of these advances and successes, though, are real threats around robustness of AI systems and issues of responsible AI (RAI), which have become pervasive topics of recent discussion. These topics are covered in greater detail in chapters 7 and 8, with a review of harms and challenges, case studies, and mitigations.

4.8 Main Takeaways

In this chapter, we discussed important fundamental building blocks for ML. We began with a discussion on the three classes of ML—supervised learning, unsupervised learning, and reinforcement learning. Supervised learning uses labeled (annotated) training data to learn a mapping between inputs and outputs. On the other hand, unsupervised learning tries to explore and identify patterns, similarities, and outliers to find inherent

Table 4.4   ML challenges and recommendations

Horizon	Opportunities
Horizon 1 (1–2 years): Content-based insight	*Apply AI to gain insight into the content of interest in disparate types of data:* • Blend supervised and unsupervised learning • Demonstrate adversarial learning
Horizon 2 (3–4 years): Collaboration-based insight	*Extend AI roles to include multiple human-machine teams working together:* • Advance algorithm accuracy through human machine collaboration • Exploit physics and causal relationships to improve model
Horizon 3 (5+ years): Context-based insight	*Incorporate semantic context into the application of AI:* • Train with limited data • Converge on universal (all languages) multimodality models • Low-shot or one-shot learning • Context-aware learning • Advanced research on goal reasoning

structures and clusters in unlabeled data. Finally, reinforcement learning follows a trial-and-error method to maximize some long-term notion of reward for a good action. As supervised learning algorithms represent the most widely used ML method, we provided a deep dive into the anatomy of a supervised learning procedure. Consistent across all these three classes of ML, however, is the need for generalizability—the ability for a model to adapt well to new, unseen data.

With a foundational understanding of the three classes of ML, we reviewed common measures of performance, such as precision, recall, ROC and AUC, and MSE/MAE. We made a distinction between performance metrics and loss functions, defining loss functions as error penalties that are used to evaluate and optimize an ML model during training. After discussing some key ML performance metrics, we defined the general procedure for how validation and test sets can be used to assess and measure the quality of a trained model. With a background on classical ML techniques and common measures of performance, we turned our attention to deep learning and neural networks in the second half of the chapter—an area of ML where there have been great advances in the last decade.

We provided a review of the historical and biological underpinnings and motivations for deep learning and neural networks. We described representational learning as the key driver for how neural networks are able to learn intricate details and abstract topics/concepts from high-dimensional data. To better understand how neural networks work, we discussed key building blocks of neural networks, including nodes, different types of layers, and activation functions, as well as how signals are transmitted throughout a network. We then explained the task of training neural networks. To do this, we provided a deep dive into the backpropagation algorithm, the workhorse behind how neural networks are able to learn.

While the greatest motivation behind and appeal of neural networks, as compared to classical ML techniques, is the limited need for feature engineering and human intervention, we described some of the core designer roles and responsibilities in identifying the right model for a given task, and the task of initializing and tuning hyperparameters to achieve the best performance. We finally ended the chapter with an introduction to CNNs, one of the key enablers for computer vision systems.

In chapter 5, we discuss modern computing technologies, which constitute the third area contributing to the remarkable success of AI. Modern computing, a key subcomponent of the AI system architecture, has enabled the computationally intensive task of ML. It provides a review of the computational needs for neural network training and inference, some of the computational drivers for ML algorithms, and some future challenges and opportunities in the modern computing domain.

4.9 Exercises

1. What are the three classes of ML, and what are some of their key differentiators?
2. What are the primary drivers that dictate whether you should pursue a supervised or unsupervised learning approach?
3. Describe how loss functions are different from evaluation metrics.
4. A regression supervised learning technique can be evaluated using a confusion matrix.
   a. True
   b. False
5. Identify a scenario in which higher sensitivity is more important than higher specificity. Now, identify a scenario where higher specificity is more important than higher sensitivity.

6. What is meant by a stochastic gradient descent (SGD) and its importance in calculating a loss function?

7. _____ learning is a set of methods that enable a machine to start with raw, unstructured input data and automatically learn representations of data necessary for detection or classification tasks.

   a. Transfer

   b. Representation

   c. Deep

   d. None of the above

8. Backpropagation is an algorithm used during training with the goal of adjusting the neural network weights and biases by minimizing a cost function to get the desired result at the output of the network.

   a. True

   b. False

9. As a model owner, you are responsible for tuning model parameters to achieve optimal results.

   a. True

   b. False

10. Elaborate on what one epoch means and how batches and iterations relate to it.

11. The primary mathematical operation applied to an input image within the convolutional layer of a CNN is called _____.

12. What are foundation models?

13. Explain the concept of overfitting in ML, and provide two techniques to prevent it.

14. Use a public generative AI product (such as ChatGPT, Bing AI, or Google Bard) to prompt and perform a custom task or play a custom game. Before starting, do some research on prompt engineering best practices and few-shot learning to understand how to best frame your task/game prompt in a way that the model can understand and learn. Then, train the model using only a few examples of the task or game, and see how well it can generalize and perform on new examples. Experiment with various prompts and few shot examples and training strategies to see what works best.

15. What are some of the limitations of large language models?

16. Research a specific application of ML (i.e., large language models) and the recent advances in the field. Write a short report summarizing some of the major research advances and reflecting on the potential implications of the research for future developments in the field.

4.10 References

1. DOMO, Data never sleeps. https://www.domo.com/solution/data-never-sleeps-6.
2. Jordan, M. I. and T. M. Mitchell, Machine learning: Trends, perspectives, and prospects. *Science*, 2015. 349(6245): 255–260.
3. Hawking, S., *Launching of the leverhulme centre for the future of intelligence.* 2016.
4. Bishop, C. M., *Neural networks for pattern recognition.* 1995, Oxford University Press.
5. Bishop, C. M. and N. M. Nasrabadi, *Pattern recognition and machine learning.* Vol. 4. 2006, Springer.
6. Murphy, K. P., *Machine learning: A probabilistic perspective.* 2012, MIT Press.
7. Murphy, K. P., *Probabilistic machine learning: An introduction.* 2022, MIT Press.
8. LeCun, Y., Y. Bengio, and G. Hinton, Deep learning. *Nature*, 2015. 521(7553): 436–444.
9. Rumelhart, D. E., G. E. Hinton, and R. J. Williams, Learning representations by back-propagating errors. *Nature*, 1986. 323(6088): 533–536.
10. Reitermanova, Z. Data splitting, in *WDS*. 2010.
11. Google. *Google data set search.* https://datasetsearch.research.google.com/.
12. Kaggle. *Kaggle datasets.* https://www.kaggle.com/datasets.
13. Data.gov. *Data.gov.* https://data.gov/.
14. Cruz, J. A. and D. S. Wishart, Applications of machine learning in cancer prediction and prognosis. *Cancer Informatics*, 2006. 2: 117693510600200030.
15. Cunningham, P., M. Cord, and S. J. Delany, Supervised learning, in *Machine learning techniques for multimedia,* 21–49. 2008, Springer.
16. Wang, Q., Y. Ma, K. Zhao, and Y. Tian, A comprehensive survey of loss functions in machine learning. *Annals of Data Science*, 2022. 9(2): 187–212. https://link.springer.com/article/10.1007/s40745-020-00253-5.
17. Refaeilzadeh, P., L. Tang, and H. Liu, Cross-validation. *Encyclopedia of Database Systems*, 2009. 5: 532–538.

18. Xu, D. and Y. Tian, A comprehensive survey of clustering algorithms. *Annals of Data Science*, 2015. 2(2): 165–193.

19. Huang, J.-J., G.-H. Tzeng, and C.-S. Ong, Marketing segmentation using support vector clustering. *Expert Systems with Applications*, 2007. 32(2): 13–317.

20. Hinton, G. E. and R. R. Salakhutdinov, Reducing the dimensionality of data with neural networks. *Science*, 2006. 313(5786): 504–507.

21. Zhu, X. J., *Semi-supervised learning literature survey*. 2005. University of Wisconsin–Madison.

22. Arulkumaran, K., M. P. Deisenroth, M. Brundage, and A. A. Bharath, A brief survey of deep reinforcement learning, 2017. arXiv preprint arXiv:1708.05866.

23. Ibarz, J., J. Tan, C. Finn, et al., How to train your robot with deep reinforcement learning: Lessons we have learned. *International Journal of Robotics Research*, 2021. 40(4–5): 698–721.

24. Kober, J., J. A. Bagnell, and J. Peters, Reinforcement learning in robotics: A survey. *International Journal of Robotics Research*, 2013. 32(11): 1238–1274.

25. Silver, D., T. Hubert, J. Schrittweiser, et al., A general reinforcement learning algorithm that masters chess, shogi, and Go through self-play. *Science*, 2018. 362(6419): 1140–1144.

26. Sokolova, M. and G. Lapalme, A systematic analysis of performance measures for classification tasks. *Information Processing & Management*, 2009. 45(4): 427–437.

27. Hanley, J. A. and B. J. McNeil, The meaning and use of the area under a receiver operating characteristic (ROC) curve. *Radiology*, 1982. 143(1): 29–36.

28. Wang, H. and B. Raj, On the origin of deep learning, 2017. arXiv preprint arXiv:1702.07800.

29. Ng, A., *Nuts and bolts of building AI applications using deep learning*. NIPS Keynote Talk, 2016.

30. McCulloch, W. S. and W. Pitts, A logical calculus of the ideas immanent in nervous activity. *Bulletin of Mathematical Biophysics*, 1943. 5(4): 115–133.

31. Jain, A. K., J. Mao, and K. M. Mohiuddin, Artificial neural networks: A tutorial. *Computer*, 1996. 29(3): 31–44.

32. Nwankpa, C., W. Ijomah, A. Gachagan, and S. Marshall, Activation functions: Comparison of trends in practice and research for deep learning, 2018. arXiv preprint arXiv:1811.03378.

33. 3blue1brown, What is backpropagation really doing? | *Chapter 3, Analyzing our neural network*. https://www.3blue1brown.com/lessons/neural-networks.

34. 3blue1brown, Backpropagation calculus | *Chapter 4, Deep learning.* https://www
     .3blue1brown.com/lessons/neural-networks.

35. Liu, W., Z. Wang, X. Liu, et al., A survey of deep neural network architectures
     and their applications. *Neurocomputing*, 2017. 234: 11–26.

36. Leijnen, S. and F. van Veen. The neural network zoo, in *Multidisciplinary digital
     publishing institute proceedings.* 2020.

37. Goodfellow, I., J. Pouget-Abadie, M. Mirza, et al., Generative adversarial nets.
     Advances in Neural Information Processing Systems, 2014.

38. Vaswani, A., N. Shazeer, N. Parmar, et al., Attention is all you need. *Advances in
     Neural Information Processing Systems*, 2017.

39. Devlin, J., M.-W. Chang, K. Lee, and K. Toutanova, BERT: Pre-training of deep
     bidirectional transformers for language understanding, 2018. arXiv preprint
     arXiv:1810.04805.

40. OpenAI, *GPT-4 technical report.* 2023. https://arxiv.org/abs/2303.08774.

41. LeCun, Y., L. Bouttou, Y. Bengio, et al., Gradient-based learning applied to
     document recognition. *Proceedings of the IEEE*, 1998. 86(11): 2278–2324.

42. Krizhevsky, A., I. Sutskever, and G. E. Hinton, ImageNet classification with
     deep convolutional neural networks. *Advances in Neural Information Processing
     Systems*, 2012.

43. Bommasani, R., D. A. Hudson, E. Adeli, et al., On the opportunities and risks
     of foundation models, 2021. arXiv preprint arXiv:2108.07258.

5

Modern Computing

Things don't have to change the world to be important.
—Steve Jobs, former Apple CEO

In prior chapters, we have discussed two of the impactful enablers of the rapid advancements in artificial intelligence (AI)—data and machine learning (ML). The third is modern computing, which is the focus of this chapter.

Computing technologies have been one of the foundational pillars revolutionizing the information age (also referred to as the "digital age"). It is the confluence of an immense amount of data, coupled with data-driven algorithms (i.e., ML), and advances in computing hardware and software that have led to useful AI products and services.

The exponential data growth that we are experiencing, due to the advent of interconnected devices, will continue to demand augmented human intelligence—enabled by AI facilitating timely data-driven decision making. Globally, it is predicted that by 2025, there will be more than 400 exabytes of data (1 exabyte = 10^{18} bytes) created every day. Erik Brynjolfsson, director of the Stanford Digital Economy Lab, makes the important point that AI is a general purpose technology, and as such, these technologies must evolve with many other complementary innovations [1]. Economic benefit (i.e., value) from AI will take time to develop, as the technology initially goes through a stage of small contributions to productivity, but then increases in economic value as other parts of the economy that complement AI as a GPTe mature [2].

There are many capabilities needed that complement AI. They fit nicely into the people-process-technology triad that was discussed earlier in chapter 1. Within this triad, an example of people is the need for human capital, meaning talent at all levels across academia, government, and industry. An example of process is the need for a digital transformation, as explored in chapter 3. In this chapter, we elaborate on modern

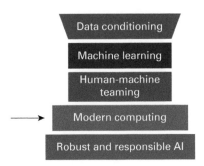

Figure 5.1   A simplified AI system architecture emphasizing modern computing.

computing as an enabling technology. It is important for us to emphasize that under the rubric of modern computing, we include both hardware and software. Hardware and software must work in unison supporting the full end-to-end AI system architecture.

Figure 5.1 depicts a simplified AI system architecture with an emphasis on modern computing technologies. Modern computing is the bedrock that enables the processing performed in each of the other subcomponents of the architecture. In subsequent sections, we discuss in more detail the various classes of computing engines, including central processing units (CPUs), graphics processing units (GPUs), tensor processing units (TPUs), and field-programmable gate arrays (FPGAs).

Steve Jobs's quote at the start of this chapter applies well to what we discuss later in the chapter. One example of a thing that had not "changed the world" but has been very important to AI is the use of domain-specific architecture (DSA) such as GPUs and TPUs. GPU and TPU hardware designers exploit variable precision in these microprocessor architectures by mapping integer and floating-point arithmetic to the specific kernels internal to an ML algorithm, resulting in very high throughput at low power.

For example, in the development of a neural network, we often need high precision at the training stage. However, during the application of the weights (i.e., at the test stage), operations can be done using integer precision with a lower number of bits. In production, the ability to perform ML at the edge—which has become a required and a very important AI capability—is now feasible by exploiting this type of hardware property, resulting in phenomenal capabilities, such as automatic speech recognition, computer vision, and scene understanding, to name just a few examples.

AI hardware is one important technological advance. But AI tools and implementation environments are very important innovations as well, reducing the barrier to entry. Figure 5.2 illustrates examples of popular ML development environments. As expected,

Easier
to use

K Keras

- Application programming interface (API) for tensorflow and other AI packages
- Suitable for rapid prototyping on smaller data sets
- Relatively slow and inflexible
- Python

Ů PyTorch

- Lies between keras and tensorflow in performance and ease of use
- Built by Facebook
- Python

More
powerful

TensorFlow

- Most powerful and flexible option
- Built by Google
- C++ with APIs compatible with python

Figure 5.2   Popular ML software development environments.

some are easier to use than others. For example, TensorFlow, built by Google [3], is powerful but requires domain knowledge of the AI application and ML algorithm workflow. PyTorch, built by Facebook researchers in collaboration with others [4], enables ML algorithm developers to make them easier to use, and it is well matched to prototyping early in the development cycle. Keras serves as an application programming interface (API) to TensorFlow and other packages, developed for fast experimentation and ease of use [5]. In the appendix of this book, we describe the implementation of several ML algorithms using Keras, TensorFlow, and other ML packages (all open-source).

In the next section, we present a short history of computing technologies as they relate to artificial intelligence. Over seventy years of computing evolution has led to AI systems with unprecedented performance at the enterprise levels—where size, weight, and power (SWaP) are acceptable. AI at the edge, where SWaP is scarce, has limitations on what is feasible. We discuss this topic in section 5.2, where we compare enterprise against edge computing.

In sections 5.3 and 5.4, we elaborate on the computational kernels that drive many of the design requirements. Several of the computational functions are not new. In fact, there is a vast amount of foundational research and development (R&D) dating to concepts in linear algebra and statistical signal processing. However, the most recent innovations have been in designing microprocessors suitable for exploiting the data and algorithm parallelism and arithmetic precision, while at the same time minimizing communication and latency overhead. As Moore's Law, reflecting computing advances, continues to slow down, chip architects and designers have found ways to deliver exponential growth by using properties inherent in the ML computational kernels.

With an understanding of computational kernels and arithmetic precision, in section 5.5, a number of well-known deep neural network (DNN) algorithms, their respective computation times, and the hardware are discussed. In section 5.6, we discuss advancements in DSA and software—specifically, the characteristics that make GPUs and TPUs so well matched to operating on the ML computational kernels. Section 5.7 highlights contemporary processing engines and computing platforms in terms of computation throughput and power usage.

A very interesting way to determine improvements in microprocessors, in terms of memory bandwidth and computation throughput, is by assessing attainable floating-point operations per second (FLOPS) versus operational intensity (meaning FLOPS per byte). This metric is referred to as the "Roofline," and it is discussed very eloquently by Jouppi et al. from Google [6]. In section 5.8, we overview the Roofline metric and its implication on assessing computing engine performance.

The vulnerability of modern computing to cyberattacks is very concerning. Therefore, in section 5.9, we highlight proposed approaches for making computing systems more secure. We also address the use of AI for improving the cybersecurity domain.

We conclude the chapter with a discussion on modern computing challenges, the main takeaways, and a set of exercises. Among several important challenges, one that is of paramount importance for our AI community to address is the ever-increasing significance of required computing power generating exorbitant amounts of carbon dioxide (CO_2). Today's foundation models, such as BERT, GPT-3, and DALL-E 2, deliver remarkable capabilities, demanding computational throughputs at the expense of very large carbon footprints [7].

5.1   A Short History of Computing Technologies

In this section, we provide a short history of computing technologies to help put in context today's modern computing technologies as they relate to AI. For a very elegant and in-depth treatment of the history and achievements, we direct the reader to Nils Nilsson's book *The Quest for Artificial Intelligence,* especially chapter 1 [8].

The field of computing is fascinating, and one iconic figure is Alan Turing, on the basis of his work producing a computer that could interpret a program encoded in a tape representing a general logic unit [9, 10]. The Electronic Numerical Integrator and Computer (ENIAC), developed at the University of Pennsylvania's Moore School of Electrical Engineering between 1945 and 1955, was the first general-purpose programmable electronic computer [11, 12]. John von Neumann is recognized as a forefather of computer architecture whose work has stood the test of time [13]. Working at the University of Pennsylvania, von Neumann designed the successor to the ENIAC, called the Electronic Discrete Variable Automatic Computer (EDVAC). In so doing, he presented a new approach to address computer memory containing both the data and the program instructions, now known as the "von Neumann architecture." During that early work, it is not clear if von Neumann's computer architecture concepts evolved from Turing's initial computer concepts, even though von Neumann recognized Turing's logic machine (i.e., the universal machine) as being well matched to arithmetic functions [14].

We recount this initial work since in von Neumann's seminal report, he points out the ability to represent a logic unit described by McCulloch and Pitts [15]. As discussed in chapter 1, McCulloch and Pitts were the first to formulate a set of logic structures representing mathematical functions performed by analog neural networks consisting of neurons and interconnections (i.e., synapses) [16].

Claude Shannon, in his MIT master's thesis, *A Symbolic Analysis of Relay and Switching Circuits*, set the foundation for Boolean algebra operators used in all of today's classical digital computers [17]. Claude Shannon, John McCarthy, Oliver Selfridge, Marvin Minsky, and others were early pioneers present at what is considered the dawn of AI—the 1956 Dartmouth College Summer Research Project—that was highlighted in chapter 1.

Another important work of the 1950s era that greatly influenced computers, was by Oliver Selfridge (MIT Lincoln Laboratory), associate director of Project MAC [18]), who was considered the father of perception by virtue of his famous paper on a paradigm for learning and simulating the process in digital computers [19–21]. In 1955, Wes Clark (MIT Lincoln Laboratory) and Ken Olsen (founder of the Digital Equipment Corporation and formerly at MIT Lincoln Laboratory) designed and developed the TX-0 (whose name stands for "Transistorized experimental computer—zero"). That was a demonstration of an 18-bit transistorized computer with a 16-bit address range in a magnetic core memory. TX-0 was instrumental during the early research in AI and computer technologies at the MIT Artificial Intelligence Laboratory.

There were influential innovations in AI and computer science that spun off from the original MIT AI Group (founded by Marvin Minsky and John McCarthy in 1958), Project MAC, and the MIT Artificial Intelligence Lab. A couple of notable examples were the TX-0 being used to implement interactive time-sharing and to demonstrate interactive code debugging. For readers interested in the early history of MIT computing and its influence on the formation of several commercial computer companies that followed, Chiou et al. chronicles these developments [22].

From the 1950s to the 1980s, there were many advances coupling innovations in AI algorithms and computer systems. For example, Ed Feigenbaum and colleagues, demonstrated a functional expert system, the DENDRAL project, which was applied to help with the identification of organic molecules from their spectra in organic chemistry. The project began in the mid-1960s, and by the 1980s, it showed impressive results of using an expert system based on a rule-based decision process [23, 24].

In 1965, a famous and seminal paper by Gordon Moore laid the microprocessor path for the whole semiconductor industry—initially, he predicted that transistors in an integrated circuit would double every twelve months [25]. Carver Mead, a professor at Caltech who revolutionized the semiconductor industry based on his seminal work on very-large-scale integration (VLSI), is credited with coining the term "Moore's Law," in reference to Moore's 1965 findings. At the Computer History Museum in 2005, Moore and Mead assessed the state of the industry in a talk celebrating the fortieth anniversary of Moore's Law [26]. In 1975, Moore updated his prediction to transistor density doubling every two years—since then, his prediction has stayed accurate for decades (indeed, it is closer to doubling every eighteen months).

The Association of Computing Machinery (ACM) gave the A. M. Turing Award (considered the Nobel Prize in computer science) in 2017 to two iconic scholars in the computer industry: John Hennessy, a former Stanford University president, and David Patterson, a retired professor at the University of California, Berkeley, and a distinguished engineer at Google [27]. Both Hennessy and Patterson pioneered a systematic, quantitative approach to the design and evaluation of computer architecture, including their research and development (R&D) of Reduced Instruction Set Computer (RISC) microprocessors [28]. They built computer systems that showed faster speeds by exploiting the simpler RISC instruction set. The RISC architecture enabled computer systems to rely on Moore's Law ($2x/1.5$ years) for a couple of decades (about 1984 to 2004).

Moore's Law has affected end-to-end computer systems performance at approximately the same rate as microprocessor performance from the mid-1980s to the mid-2000s. This has been possible because of the joint architectural optimizations between hardware and software. However, Moore's Law has slowed significantly since approximately 2004; which has led to additional computer architecture innovations using a larger number of cores embedded in DSA. We will elaborate in section 5.6 on types of DSA engines and their relevance to modern computing AI systems.

There are many other factors influencing overall computing performance such as clock frequency, interconnect technology, power dissipation, memory access speeds, data parallelism, and algorithm parallelism [29]. Two of these factors, which limit continuing computer performance growth, are known as Amdahl's Law [30] and Dennard Scaling [31]:

- *Amdahl's Law (1967):* With a computer system executing functions in parallel, the maximum theoretical speed-up is limited by the sequential portion of the code.

- *Dennard Scaling (1974):* The power density (i.e., power dissipation relative to circuit area reduction) of a metal-oxide-silicon transistor stays constant regardless of the transistor's feature size if the current and voltage can scale correspondingly to the transistor area.

These limitations contribute to a further slowing of Moore's Law. As circuit designers of microprocessors were able to put more cores into a single integrated chip, their throughput was limited by such elements as parts of the code that were not parallelizable (i.e., limited by Amdahl's Law). Similarly, clock frequencies and associated voltages are no longer able to scale according to Dennard Scaling. In fact, clock frequencies and voltages have reached a wall causing power dissipation limits, and thus reducing microprocessor performance. Hennessy and Patterson, in their Turing lecture at the 2018 ISCA conference, emphasized the need for domain-specific chip architectures to achieve increased microprocessor performance, with the ability to deliver throughputs commensurate with the requirements of ML algorithms [27].

William Dally, the chief scientist and senior vice president of research at NVIDIA Corporation, and former chair of computer science at Stanford University, developed and demonstrated several computer architecture approaches that are found in modern computing AI systems. While at Stanford, Dally and his students designed the Merrimac scientific computing system based on the concept of streams tailored to exploit parallelism and data locality close to where arithmetic operations were executed. As Dally points out: "Modern semiconductor technology makes arithmetic inexpensive and [communication] bandwidth expensive." [32] Therefore, it is best to optimize arithmetic intensity (i.e., increase the ratio between arithmetic operations to bandwidth; or, equivalently, the arithmetic operations per byte accessed from memory). These earlier concepts led to architectural principles found in the NVIDIA GPUs.

Yann LeCun, in his 2019 paper, provided a perspective on the evolution of deep learning hardware in much more detail than we have done in this section [33]. We have limited the scope of this section to a much smaller group of past and present pioneers to introduce the reader to the progression from early computing hardware to today's trend toward contemporary computing. Having a general-purpose compute engine (like a CPU) and still achieving the high-performance processing required of ML is no longer feasible because of the limitations discussed earlier.

In the subsequent sections of this chapter, we look in more detail not just at computing hardware—which has been the primary focus of this section—but also advances in software. It is the joint optimization of modern computing hardware and software that will continue to deliver the requisite processing throughputs and performance in AI systems. Next, we discuss computing at the enterprise and computing at the edge, including similarities and differences in their respective environments.

5.2   Computing at the Enterprise versus Computing at the Edge

The evolution of AI has led to capabilities requiring deployment at the enterprise level, at the edge, or both. For example, AI applied to deciding to whom to grant a bank loan can be processed in a data center (e.g., a cloud compute environment) residing at the enterprise level. In contrast, the large increase of Internet of Things (IoT) devices has led to the AI computation taking place closer to the physical devices. For example, smart meters, smart medical devices, drone navigation, robotics, and mobile phones, to name just a few instances, depend on low latencies to be useful; and reaching back to a distant enterprise will increase the latency.

In this section, we address similarities and differences between computing at the enterprise versus computing at the edge. For enterprise computing, we include cloud computing offered by commercial cloud providers such as Amazon Web Services (AWS), Microsoft Azure, and Google Cloud, and/or on-premise computing.

There is a relatively new paradigm referred to as "fog computing" [34]. Fog computing fits between an enterprise (residing in a cloud computing environment) and edge computing. There are benefits to fog computing, particularly in delivering services (like AI capabilities) closer to the physical devices with lower latency. However, there are also challenges. Fog computing challenges include physical security and cybersecurity, maintenance, and amortization of the cost among the number of users. Commercial cloud providers have been wrestling with these issues for many years. In this discussion, we subsume fog computing within the context of cloud computing, but with its location closer to the physical devices at the edge; minimizing latencies.

We highlight the definitions of these concepts as shown here. More specifically, we make a distinction in terms of number of locations, as follows:

- *Enterprise/cloud computing (limited number of locations):* AI processing is performed at a commercial cloud provider, on-premise data center, or closer to the point of service.

- *Edge computing (thousands to millions of devices):* AI processing is performed on the physical device (i.e., sensors in a smart city, in a smart phone, or on a personal assistant device). or the processing is carried out over a small hop with proximity to the physical devices, while still meeting low latency standards.

In any of these computing paradigms, there is a balance among computing resources, storage, and communication (i.e., network) bandwidths. The selection depends on many factors, such as latency, data privacy, real-time performance, security, and cost. As the AI capabilities are formulated, architects, designers, developers, and implementers must define the AI business needs and value early on, with inputs from the stakeholders, as discussed in chapter 2.

Figure 5.3 illustrates the ML stages in the process of developing a neutral network model. The ML stages help us understand the computing demands in an enterprise versus an edge environment. Table 5.1 outlines key similarities and differences between AI at the enterprise compared to AI at the edge. In this comparison, we make two key assumptions, relative to computing capabilities today. The first assumption is that, in most cases, training (including cross-validation) and tests are done at the enterprise level during model development (either in a commercial cloud or on-premise). The second assumption is that once the model is developed (with acceptable performance), then it is deployed at the edge for the inference processing.

Referring to table 5.1 (observing the similarities between computing at the enterprise versus at the edge), in both cases we face the need for high dimensionality (i.e., many sensors and sources of data) and large data volume for training to be done in the cloud (i.e., at the enterprise). Inference can be performed either at the enterprise level or at the edge (for low-latency applications). Both environments need computational efficiency, which means to have compute resources and memory fully used in completing the ML task (i.e., low overhead).

A low carbon footprint in data centers at the enterprise level is a major challenge. We will address this topic in section 5.10. SWaP plus cost is challenging for deployed

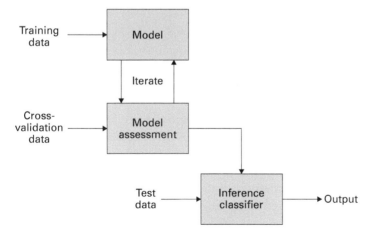

Figure 5.3 ML algorithm stages for training, cross-validation, and test for a neural network classifier.

Table 5.1   Characteristics of AI deployed at the enterprise and/or at the edge

Enterprise AI	Edge AI
High dimensionality (i.e., variety).	High dimensionality (i.e., variety).
Large data volume (labeled data for training and unlabeled for inference) and high velocity.	Large data volume (often unlabeled) and high velocity for inference.
Processing performed in the cloud for training and/or inference.	Need for cloud (e.g., training) followed by edge computing for inference.
Need for computational efficiency and low carbon footprint.	Need for computational efficiency. Maintain low SWaP and cost.
Adversarial AI is a concern.	Adversarial AI is a concern
Availability of high-capacity datalinks.	Must continue to operate under low capacity and/or intermittent datalinks.
Batch processing and less stringent latency requirements.	Need for real-time with low latencies.
Human-machine teaming is tolerant to errors (e.g., speech-based virtual assistants).	Human-machine teaming is less tolerant to errors.
Pay-for-service and some free apps.	Continuous operations.
Need for monitoring after deployment.	Need for monitoring after deployment.

sensors at the edge. However, there has been a continuing increase in cloud computing costs. Thus, AI architects, designers, and implementers must treat cost as a variable when addressing the development of AI products and services relative to the business value.

Adversarial AI is of concern for both at the enterprise level or at the edge, as discussed in chapter 7. Communication links are more limited at the edge and can be intermittent. At the edge, low latencies are critically important.

Since many AI edge applications must be done in or near real time, there is little margin for error. In contrast, at the enterprise, if the human suspects that the answer is incorrect, the AI system can be queried again. Edge computing also depends on continuous operations to achieve real-time performance. Finally, both at the enterprise and at the edge, we must employ AI system monitoring to address data drifts, algorithm drifts, or adapting to new user needs, as discussed in chapter 11.

Let us now walk through a couple of examples of employing enterprise and edge computing as discussed here. The first involves using a Google android phone, a Google

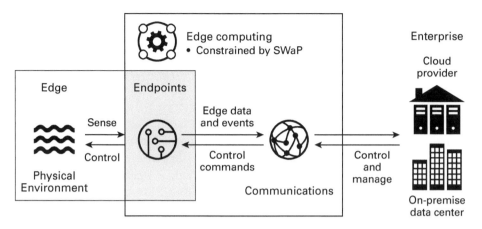

Figure 5.4   Enterprise and edge computing environments. Source: Gartner, Inc.

hub, or a Nest thermostat. Google does the model training at the enterprise in its large data centers, deploys its model to the edge devices, and collects data, which gets aggregated back at the data center. In the second example, Apple does the training for its iPhones in its enterprise data centers via the Core ML modeling tool. It then deploys the models to run in its Apple-embedded neural engine. Apple also collects and aggregates data for improving the models at the enterprise data centers.

As shown in figure 5.4, an enterprise compute environment and computing at the edge are often coupled (as demonstrated with the two examples discussed earlier) where some computation is done at the enterprise (e.g., in a commercial cloud provider and/or on-premise), and other processing is done at the edge. Therefore, in addition to focusing on computational resources, we need to pay attention to communication links and command/control functions between the two environments.

In the next section, we elaborate on the key computational kernels driving the complexity of ML algorithms. Computational kernels are relevant to computing either at the enterprise or at the edge. Of course, compute engines (i.e., GPUs, TPUs), memory, and interconnects are most constraining in an edge computing environment. For enterprise computing (i.e., carbon footprint) and edge computing (i.e., SWaP), we have to address the power consumed and also the respective energy efficiencies.

5.3   Neural Network: Key Computational Kernels

In this section, we introduce the key computational kernels within a neural network. As we have discussed in chapter 4, neural networks are a class of algorithms within the larger class of ML techniques. However, we limit our discussion to neural networks since other ML techniques share similar linear algebra operands. Our primary objective is to give the reader an appreciation of the type of mathematical functions used and an understanding of why these functions are so well matched to high-performance, modern compute engines such as GPUs and TPUs.

In chapter 4, we discussed the neural network parameters that are either computed internally to the algorithm or are set by the model designer. In the latter case, these are referred to as "hyperparameters" (e.g., number of nodes, number of layers, activation function, size of a batch, number of epochs). These hyperparameters drive the total number of algorithm parameters, which in turn drive the number of arithmetic operations. Thus, the neural network algorithm designer must balance between number of algorithm parameters—directly affecting algebraic operations—and the level of algorithm performance. As expected, this balancing is an iterative process.

The reason why GPUs, for example, are so well matched to the algorithm kernel functions is because many of these operations can be done in parallel, across a large number of microprocessor cores, with enough local memory to optimize operations per byte (i.e., reads and writes) [35]. However, the number of parameters is growing at an exponential rate. As we discussed in chapter 2 and illustrated in figure 2.3, starting around 2012, the algorithm computation increased $2x$ every 3.4 months (or, approximately $8x$ every year). This is a result of an increase in number of model parameters (approximately doubling every 2.3 months). Model performance accuracy often improves as the number of parameters in the model increases, and as the training data size increases; requiring a very large number of arithmetic operations.

Generations of the Generative Pretrained Transformer (GPT) algorithm, developed by OpenAI [36], have grown in number of parameters as follows:

- GPT-1 (2018): 117 million parameters
- GPT-2 (2019): 1.5 billion parameters
- GPT-3 (2020): 175 billion parameters

A seminal paper on transformers was published by Google researchers in 2017 [37]. The GPT versions are a class of transformers with impressive performance in classes of natural-language processing (NLP) applications. OpenAI announced GPT-4 in 2023. Although the specific number of parameters were not disclosed in the mid-March 2023 unveiling, GPT-4 demonstrated a significant improvement over GPT-3 and GPT-3.5, likely due to the orders of magnitude increase in number of parameters and its continuing to employ reinforcement learning.

As an example, it is useful to look briefly at the number of computations required in the training stage of the GPT-3 transformer. GPT-3 had available for its training data

499 billion tokens (from many corpora of data, including crawling the web and Wikipedia). It used 300 billion tokens in training in order to achieve the lowest cross-entropy loss (see figure 4.1 in [38]). Tokens are often words, but they can also be characters or other meaningful units depending on the task.

Since generative AI algorithms, such as the GPT family and others, are starting to be used by commercial companies to improve their products and services, we encourage readers to consult the writings of Stephen Wolfram [39]. Wolfram explains in simple and understandable terms (supported by representative code) how ChatGPT creates human-readable text. However, we also caution that these technologies are in their infancy, sometimes (and at times often) resulting in significant errors in the created text (referred to as "hallucinations"). Therefore, we encourage AI architects, designers, developers, implementers, and users to follow a rigorous approach in their development, deployment, and use of what are likely to become revolutionary capabilities.

Brown et al. describe in detail the GPT-3 transformer algorithm, including its architecture, model performance, data sets, and numerical computations [38]. The total number of computations reported, with 175 billion parameters, was 3.14×10^{23} floating-point operations (FLOPS), or equivalently 3.64×10^3 peta-FLOPS-days. OpenAI implemented the model training on a high-bandwidth cluster from Microsoft with NVIDIA V100 GPUs. Because of the number of parameters, batch size, learning rate, and other factors, the model designers had to implement model-level parallelism. We discuss computational modes of parallelism later in this chapter.

The neural network operations required to perform the processing stages shown in figure 5.3 (e.g., in a computer vision implementation) are dominated by a set of linear algebra operations, which constitute the key computational kernels. The dominant computation in many of the DNN algorithms, including NLP, is GEneral Matrix Multiplication (GEMM) operator [40]. For a forward-propagation, fully connected network, the kernels are as follows:

- Vector-Vector adds.
- Matrix-Vector multiplies and adds.
- Activation function operand.
- Matrix-Matrix multiplies and adds.

Let us work through a simple example for illustration purposes, consisting of an input layer, three hidden layers, and an output layer, as shown in figure 5.5. The input layer consists of eight nodes; followed by nine nodes at each of the three hidden layers and four nodes at the output; these are a subset of the hyperparameters that a model designer chooses. The activation function, in the example, is a rectified linear unit (ReLU).

Table 5.2 Algebraic operations for a simple neural network

	Hidden Layer 1	Hidden Layer 2	Hidden Layer 3	Output Layer 4	Total Operations
Matrix-Vector operations	72 multiplies	81 multiplies	81 multiplies	36 multiplies	270 multiplies
	63 adds	72 adds	72 adds	32 adds	239 adds
Vector-Vector operations	9 adds	9 adds	9 adds	4 adds	31 adds
ReLU activation	9 max	9 max	9 max	4 max	31 max
	comparisons	comparisons	comparisons	comparisons	comparisons
Total operations	153	171	171	76	571

The total numerical operations are dominated by the Matrix-Vector multiplies and adds needed to calculate the input to each of the network nodes. These inputs are then passed through the ReLU activation function. The total number of operations, for this multilayer perceptron (MLP) neural network, is 571.

This simple network, shown in figure 5.5, consists of four sets of weight matrices plus four bias vectors; representing a subset of the parameters. As shown earlier, today's modern neural network models can exceed hundreds of billions of parameters. Therefore, in addition to exploiting modes of computational parallelism, algorithm designers and implementers must employ mixed-level precision to achieve computation efficiency. In the next section, we discuss arithmetic precision in modern computing.

5.4   Arithmetic Precision

There are several important factors that designers of modern computing engines must take into account in order to achieve high throughput, minimum latency, lower SWaP, improved computation and energy efficiencies, fast reads and writes to memory (i.e., operations per byte), and other benefits. One example of a high-performance accelerator from IBM—meeting some of these characteristics for training and inference—is trillion operations per second per watt (TOPS/W) power efficiency, designed in 14-nanometer feature size technology (in the 2018 timeframe) [41]. These performance metrics are important for today's large ML models. Because of these stringent metrics, designers have achieved major gains by using mixed-precision, in addition to other hardware architectural features such as parallelism and high-speed memory access. Since arithmetic precision is one of the knobs that microprocessor designers exploit, we discuss important issues affecting ML training and inference.

We begin with an arithmetic precision primer. Table 5.3 shows several examples of arithmetic precision (floating-point and integer formats) in terms of the total number of bits. Other levels of precision are also used by simply scaling the number of bits from the set in the table. A general rule, referring to figure 5.3, is that ML algorithms require greater precision at training (i.e., during the development of the model) as opposed to the application of the model weights (e.g., during the inference stage). The trade-off, in terms of arithmetic precision formats, comes down to the number of precision bits (i.e., mantissa bits) and the number of bits of dynamic range (i.e., exponent bits).

There are several important characteristics worth highlighting. Because of differences in floating-point implementations among computer system architectures, in 1985 a standard was established for the IEEE-754 floating-point standard. The fp32 and fp16 formats, shown in table 5.3, are implemented in hardware using the IEEE-754 standard.

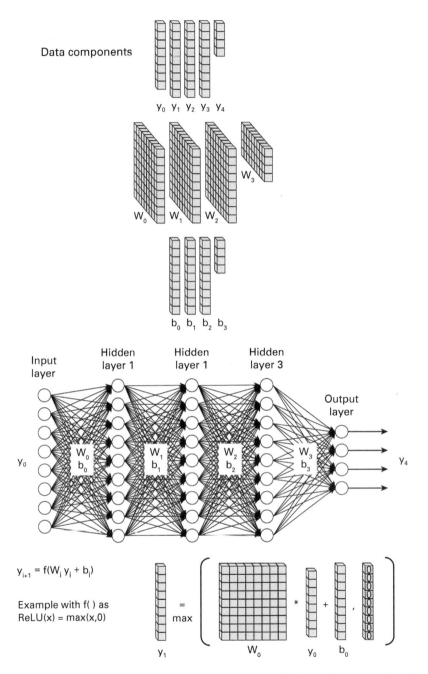

Data components

y_0 y_1 y_2 y_3 y_4

W_3

W_0 W_1 W_2

b_0 b_1 b_2 b_3

Input
layer

Hidden
layer 1

Hidden
layer 1

Hidden
layer 3

Output
layer

y_0

W_0
b_0

W_1
b_1

W_2
b_2

W_3
b_3

y_4

$y_{i+1} = f(W_i\, y_i + b_i)$

Example with f( ) as
ReLU(x) = max(x,0)

y_1

=
max

W_0

*

y_0

+

b_0

Figure 5.5 Simple example of a neural network with forward propagation. Source: Adapted from: http://www.rsipvision.com/exploring-deep-learning/.

Table 5.3 Examples of arithmetic precision formats

Formats	Exponent Bits Plus Sign Bit	Mantissas Bits	Numerical Ranges
Floating-point representations			
Floating-point 32-bits (single-precision fp32)	9	23	+/− (about 1×10^{-38} to about 3×10^{38})
Floating-point 16-bits (half-precision fp16)	6	10	+/− (about 6.1×10^{-5} to 65,504)
Brain floating-point (bfloat16)	9	7	+/− (about 1×10^{-38} to about 3×10^{38})
Integer Representation	**Sign Plus Binary Bits**		
Integer 32 (sign int32)	Sign plus 31 bits		-2^{31} to $2^{31} -1$
Integer 16 (sign int16)	Sign plus 15 bits		-2^{15} to $2^{15} - 1$
Integer 8 (sign int8)	Sign plus 7 bits		-2^{7} to $2^{7} - 1$

Google invented a third format called the "brain floating-point." As shown in the table, relative to exponent versus mantissa bits, Google designers accepted sacrificing some amount of precision to permit a much larger numerical range than is available with fp16—but still maintain a word length of 16 bits.

The IEEE-754 floating-point standard incorporates a format structure that enables more efficient numerical representations. For example, the exponent sign is implicit by incorporating a bias of 127 for fp32. Similarly, the fp16 format incorporates a bias of 15. The binary representation of the mantissa starts with an implicit first bit. This bit is not stored in memory. As shown by Oser [42], the numerical ranges can be calculated by the following formula:

$$\text{Numerical decimal value} = (-1)^s \times 2^{(e\text{-bias})} \times (1 + \text{fraction})$$

where **s** is either a 0 or a 1 (depending on the sign bit); **e** is the value of the exponent (converted from binary to decimal); the **bias** is either 127 for fp32 or 15 for fp16; and **fraction** is the numerical representation of the mantissa value (converted from binary to a fractional decimal). The representation (1 + fraction) is determined by taking the original decimal number (N) and calculating a factor fitting within $(1 \leq N < 2)$ times a power of 2 (i.e., 2^{power}). "Power" is the decimal representation of the exponent term (i.e., $\mathbf{e}$)—the bias term. Because the first term of (1+fraction) is always 1, the 1 is implicit (hidden) and not stored—saving an extra bit.

There are additional formats in the IEEE-754 floating-point standard, including higher precision, such as double-precision with 64 bits (fp64). Here, we discuss only two examples, fp32 and fp16, to illustrate the main differences and to simplify the discussion.

Let us now work through an example for fp16. For the largest positive value $\mathbf{s}=0$; $\mathbf{e}=11110_2$ (converted to decimal$=30$); bias$=15$; or fraction f$=1111111111_2$ (converted to decimal$=2^{-1}+2^{-2}+2^{-3}+\ldots+2^{-10}$), resulting in the largest positive normal value$=2^{15}\times(1+1023/1024)=65{,}504$. For the smallest positive value $\mathbf{s}=0$; $\mathbf{e}=00001_2$ (converted to decimal$=1$); bias$=15$; or fraction f$=0000000000_2$ (converted to decimal$=1$), resulting in the smallest positive normal value$=2^{-14}=6.1\times10^{-5}$.

We elaborated on these representations, and this example, to emphasize the importance of precision versus the maximum and minimum range of values as a function of arithmetic formats. The larger the number of bits, the higher the range of numerical values and precision achievable, but at the expense of more complex hardware requiring more power to perform the add and multiply operations, as well as the need for increased memory storage. Modern computing hardware designers recognized the importance of mixed precision such that, during model training, the computation is done at higher precision. In contrast, for inference, the computation can be done with a smaller number of bits, resulting in less power consumption and reduced storage requirements.

There is ongoing research to find ways to improve power efficiencies without a significant sacrifice in model accuracy relative to the IEEE-754 standard format [43, 44]. Also, Zhao and colleagues have demonstrated the use of a logarithm numerical system in training and weight updates at lower precision [45]. The benefit of the logarithm approach is that multiplications are performed through logarithmic additions, and add operations can be efficiently implemented via lookup tables. Zhao et al. demonstrated a reduction in energy consumption of over 90 percent on several well-known computer vision and NLP algorithms, compared to high-precision floating-point techniques.

In the next sections, we discuss the confluence of algorithm improvements enabled by advances in computing technology. We also explore various types of DSA, such as GPUs and TPUs. DSAs exploit arithmetic mixed precision, among other hardware design options, such as improved memory, fast interconnects, and parallelism, making them well suited to ML algorithm computations.

5.5 Confluence of ML Algorithm Improvements and Computing Technology

The rapid AI advancement, across many industries, will continue to have a profound impact on businesses, societies, and helping to solve worldwide problems (i.e., climate change, food shortages, health, education, and other issues). Economists predict that AI will undergo the typical S-curve, with initially slow acceptance, followed by rapid assimilation into business in the form of AI products and services, and plateauing once AI reaches maturity. We are at the beginning of this S-curve.

The principal factors in AI reaching adoption and value hinges on many factors discussed throughout this book. Four of the most important technical factors are (and continue to be):

1. Data availability (i.e., good and reliable data), especially across multimodalities, as discussed in chapter 3.
2. ML algorithms, as discussed in chapter 4.
3. Modern computing, which enables ML algorithm developers to exploit hardware features with software developments. The most benefit will come from using innovative microprocessor architectures, making microprocessor hardware well matched to the ML parallel structure and complementing advances in software, as discussed in this chapter.
4. AI implementation and deployment based on a rigorous systems engineering discipline, improving business value. We elaborate further on this topic in chapters 10 and 11.

In terms of the third factor, let us briefly look at ML algorithms and the hardware used to achieve world-record performance. In table 5.4, we itemized several contemporary DNN algorithms, their architecture depth in terms of the number of layers, the Top-5 error in percent, the training time, and the GPU hardware.

The ILSVRC competition was a phenomenal demonstration of the rapid advancements experienced by DNN algorithms from 2012 through 2015, specifically in the computer vision domain. Many of the same reasons that led to these advances in computer vision are just as applicable to other domains (e.g., transformers, automatic speech recognition, and image and video scene segmentation) From table 5.4, we can identify several important takeaways:

- The AlexNet DNN was one of the first demonstrations employing the power of advances in GPU technology. It used, among several innovative algorithm

Table 5.4   Representative DNN algorithms, training time, and hardware used

DNN Algorithms [46]	Architecture Depth	Top-5 Error	Training Time	GPU Hardware
AlexNet [47]; first place in ILSVRC 2012	8 layers	15.3%	6 days	2x NVIDIA GTX 580
VGG16 [48]; second place in ILSVRC 2014	16 layers	7.3%	2–3 weeks	4x NVIDIA Titan Black
GoogLeNet [49]; first place in ILSVRC 2014	22 layers	6.67%	Estimate: 1 week	Used CPUs in training; estimate was <1 week if using contemporary GPUs
Google Inception V3 [50]; published in 2015	42 layers	5.6%	2 weeks	8x NVIDIA Tesla K40
ResNet50 [51]; first place in ILSVRC 2015 (with 152 layers; 3.57% Top-5 error)	50 layers	7.02%	5 days	4x NVIDIA Tesla M40 [52]

ILSVRC = ImageNet Large Scale Visual Recognition Challenge, which contained about 1.2 million training images, 50,000 validation images, and 100,000 test images

techniques, a dual parallel path employing model parallelism—each of two NVIDIA GPUs processed a subset of the model's neurons, with the GPUs communicating at each stage of the processing flow—the dual-path results were combined at the end.

- All other subsequent DNN algorithms that either won the competition or scored well also used GPUs.

- There was an unprecedented improvement in the reduction of the Top-5 error rate. The ResNet-152 DNN algorithm was able to demonstrate a 3.57 percent image classification error, with a neural network depth of 152 layers—about $5x$ improvement in classification error performance from 2012 to 2015.

- The advancement in GPU technologies was also very remarkable. An NVIDIA GTX 580 had about 3 billion transistors in 40-nanometer process technology. In contrast, the NVIDIA Tesla M40 had about 8 billion transistors in 28-nanometer technology. The NVIDIA Tesla M40 showed more than $4x$ improvement of computation throughput relative to the GTX 580, at about the same peak power (based on the manufacturer's specifications).

• Execution time to train the algorithms, using several tens of millions of parameters, took only a few days for both AlexNet and ResNet-50—albeit, with different GPUs, as modern computing technology advanced.

These achievements were possible by many improvements in the algorithm techniques, implementation approaches using microprocessor enhancements, fabrication chip technology, optimized deep learning libraries, and other elements. For example, in the category of microprocessor enhancements, the high-speed intercommunication across GPUs enables exploiting model parallelism [53].

Another significant improvement came from the large number of compute cores available within a GPU. Because of several limitations—such as the slowing of Moore's Law, the limitation due to the serial part of the code as per Amdahl's Law, and the inability to dissipate heat as per the end of Dennard Scaling—the increase in the number of transistors was used to integrate a larger number of low-power cores per chip (as the fabrication process technology decreases) [54]. For example, the NVIDIA Tesla M40 had a total of 3,072 Compute Unified Device Architecture (CUDA) cores. In contrast, one of the latest NVIDIA GPUs, the A100 Tensor Core, has a total of 6,912 CUDA cores [55].

The year after the ILSVRC competition ended, after the ML techniques performed better—on average—than a human on the top-five error rate, another watershed moment happened in AI. In 2016, the AlphaGo algorithm from DeepMind beat the world champion Lee Sedol at the game of Go. A DNN was trained from human expert moves, together with reinforcement learning through self-plays [56]. AlphaGo used 1,202 CPUs and 48 TPUs, distributed across many machines. A year later, in 2017, AlphaGo Zero defeated AlphaGo 100 games to 0 employing only 4 TPUs, on a single machine running on the Google Cloud. Figure 2.3, in chapter 2, illustrates the impressive exponential growth in peta-FLOPS-days in computation requirements, including AlexNet, VGG, ResNets, and AlphaGo Zero, among other ML algorithms.

These events were enabled by rapid advances in ML models, together with advances in domain-specific designs, represented, for example, by GPUs and TPUs. Jeff Dean, chief scientist at Google, explained why specialized hardware makes sense for deep learning models [57], pointing out three main reasons for this:

1. There is a tolerance of reduced-precision computation (see section 5.4, on arithmetic precision).

2. The computations performed by most models are a small handful of linear algebra operations, but they are required to achieve very high throughputs (see section 5.3, on computational kernels).

3. The chip complexity of general-purpose CPUs is not required for ML operations (section 5.6, later in this chapter, elaborates on DSAs).

This shift to specialized hardware is not unprecedented. Back in the 1980s and 1990s, the advent of digital signal processors led to specialized DSP chips dedicated to signal processing operations, enabling real-time, low latencies, and at low power [58]. In the next section, we look at domain-specific hardware and software in more detail.

5.6   Domain-Specific Hardware and Software

The demands of ML (e.g., DNNs) are driving computer architects to look for more efficient designs because the prior advances credited to Moore's Law are no longer sustainable. Fortunately, as discussed earlier, ML models have inherently a lot of structure that can be exploited with domain-specific hardware supported by domain-specific software.

For decades, it has been well recognized that application-specific hardware such as application-specific integrated circuits (ASICs) outperforms general-purpose CPUs. CPUs implement architectural features (e.g., branch prediction, out-of-order execution, a translation lookaside buffer, hierarchy levels of caches, and instruction-level parallelism), adding chip complexity, more energy usage, and increasing area per a given number of transistors. These general-purpose CPU features are details hidden from the software developers, but they are no longer acceptable when performance (such as low latencies to deliver an answer) are of utmost importance to ML implementations. Thus, today's computer architects have opted instead to design domain-specific hardware and software.

A type of domain-specific hardware and software design is the TPU developed by Google to accelerate ML workloads, while reducing the energy use in their data centers [6, 59]. The first-generation TPU was designed to improve the performance-cost-energy of the neural network's inference stage. TPUs are an example of a DSA [60], and the latest generations of them can also be used for both training and inference processing (see the neural network processing stages illustrated in figure 5.3). The TPU enables hardware parallelism to be exploited by the software via domain-specific software, such as Google's TensorFlow [3].

In this section, we use DSA terminology to describe hardware, supported by software, designed for a special-purpose domain, which performs a narrow range of tasks extremely well [27]. The DSA designs permit several efficient uses of the hardware, while at the same time performing more energy-efficient functions. Examples of hardware features are

- Increased number of operations per instruction
- High degree of parallelism
- Use of mixed precision to reduce memory bandwidth and optimize on-chip memory capacity
- High computation throughputs commensurate with acceptable power/unit area, voltages, and clock frequency
- High interconnect to rapidly access data from memory
- Low overhead and low latency
- Low energy usage per instruction execution
- Efficient linear algebra kernels, well matched to ML software libraries, such as Google's TensorFlow or NVIDIA's CUDA-X

The first-generation TPU (TPUv1) had dedicated local memory that was sized to the level of the arithmetic precision being used [60]. Since the TPUv1 was designed for very fast inference operations, the computations were done in a large matrix multiply unit of size 256×256 and 8 bits multiply-accumulators. The multiply-add or multiply-accumulate (MAC) was implemented using a systolic array architecture. With a systolic array architecture, simultaneous multiply-add operations are executed in one clock cycle (to maximize throughput without additional overhead when accessing memory).

A systolic array architecture is very well matched to simple linear algebra operations—for example, in signal processing applications—where real-time performance is key [61, 62]. For instance, in adaptive array beamforming, a large number of multiply-add operations must be performed in real time with low latency. A systolic array architecture enables fast processing using simple hardware (to minimize power and cost). Using this case example as an analogy (i.e., needing high throughput at low power), similar specification requirements drive the need for TPUs.

In a TPU, the instructions were fetched from a host CPU via a PCIe interconnect bus to an instruction buffer. The implementation of the DNN algorithms was done using TensorFlow, as the domain-specific language, which enables portability.

As Dally explains about the benefits of design-specific accelerators, with general-purpose CPUs, significant energy is consumed in fetching instructions, decoding them, and reordering them. The actual energy used for arithmetic operations, in comparison,

Table 5.5 Energy usage comparison between nonarithmetic functions (fetching values from memory) and arithmetic operations (multiply and add) in a 45-nm CMOS process technology [64, 65]

Operation	Energy (pJ)	Relative Cost
32-bit integer ADD	0.1	1
32-bit float ADD	0.9	9
32-bit integer MULT	3.1	31
32-bit float MULT	3.7	37
32-bit 32 KBytes on-chip SRAM	5	50
32-bit DRAM	640	6400

Note: A picoJoule (pJ) is a unit of consumed energy = 10^{-12} J.

is relatively very small [63]. Thus, DSAs offer orders of magnitude improvements in throughput performance, lower latencies, lower power (or, equivalently, energy used), and reduced physical chip area compared to general-purpose CPUs [63].

Horowitz compared the energy consumed in accessing memory relative to energy used in performing an add or a multiply arithmetic operation. From Horowitz's presentation at the 2014 International Solid-State Circuits Conference [64], the approximate comparison of energy usage shown in table 5.5 makes this point very clearly. Also, Han et al. showed how the energy cost can be substantially reduced by compression (i.e., pruning redundant connections and sharing weights in a DNN model), such that the DNN weights can be accessed from faster on-chip static random access memory (SRAM) instead of slower dynamic random access memory (DRAM) [65].

The discussion so far has centered primarily on domain-specific hardware. Leiserson et al. foresees and recommends embracing a closer collaboration between software performance engineers (i.e., software developers focusing on performance gains) and hardware architects [66]. Leiserson et al. identify three important pillars as a way to increase computer system performance in a post–Moore's Law era for application-specific domains:

1. Software development: Emphasizing performance and exploiting hardware features such as parallelism, instead of focusing just on reducing software development time.

2. Algorithms: Particularly targeted at domain-specific applications, such as ML, signal processing, social networks, robotics, etc.

3. Domain-specific hardware architecture: Reinforcing the need to invest in hardware streamlining to get the most out of many parallel processing engines, consistent with the recommendations of Hennessy, Patterson, and Dally [27, 63]. The ability to reduce transistor feature size (i.e., Moore's Law; allowing more transistors to be packed into a given physical area), while at the same time operating at higher clock frequencies, has ended [67] (i.e., the end of Dennard Scaling).

Leiserson et al. showed the potential orders of magnitude improvement by employing performance gains in software coding and hardware instructions. Using the simple example of multiplying two 4096×4096 matrices, they demonstrated an absolute speed-up of $62,806x$ relative to software code written in Python, by exploiting several software- and hardware-related knobs, such as the C language, parallel loops, vectorization, and Intel's special Advanced Vector Extensions.

An absolute speed-up of $62,806x$, for that simple example of multiplying two matrices, translates into the following observations:

*   If Moore's Law had continued (at $2x$ every 1.5 years), it would have taken about 24 years to achieve that same absolute speed-up.
*   As shown by Hennessy and Patterson [27, 68], since about 2015, architectural hardware optimizations have reached a plateau, improving at about $2x$ every twenty years (or, about 3 percent every year). Therefore, at this rate, it would have taken about 318 years to achieve that same absolute speed-up!

We conclude this section by recalling one of the pitfalls highlighted by Dean et al., "designing ML hardware assuming the ML software is untouchable" [69]. Instead, as they pointed out, the goal should be a hardware and software system that more effectively solves the problem. ML models offer that opportunity for codesign by using domain-specific hardware together with software performance optimizations.

In the next section, we compare several modern computing microprocessors in terms of throughput and peak power for different levels of arithmetic precision. We also discuss classes of parallelism suitable for ML models. Since power usage is of paramount importance, we look briefly at energy consumption for chips targeted at embedded applications.

5.7 Contemporary Computing Engines and Integrated Systems

The background provided in sections 5.4–5.6 helps the reader better understand the capabilities available in contemporary microprocessors and cluster-level computing systems, specifically designed for high-performance applications (such as AI applications).

For example, continuing with the TPU as a use case, Google designers targeted no more than a 7-millisecond latency for delivering predictions from their most common ML MLP neural network model deployed in their data centers. For one of Google's MLP applications, they demonstrated an 83x improvement in throughput performance/watt and a 41x increase in the number of predictions per second relative to a contemporary CPU.

In this section, we illustrate different classes of chips, cards, and systems as a function of computation throughput and peak power. In figure 5.6, the neural network processing performance (peak) is shown as a function of the arithmetic precision for training, inference, or both, as reported by Reuther et al. [70]. The peak power metric, instead of actual consumed power, gives a useful relative comparison, since actual power consumption varies depending on the specifics of the application.

For the past several years, Reuther has done extensive analysis of chip architectures and performance throughputs per unit watt [70]. Figure 5.6 illustrates a comprehensive set of contemporary chips, computer cards, and systems developed for high-performance computing as required by, for example, ML models. The vertical axis is peak performance, as specified by the manufacturer, in billions of operations per second (GigaOps/sec). The horizontal axis is peak power in watts, as specified by the manufacturer. Several key insights can be summarized from this data—the data was aggregated for a presentation at the IEEE High-Performance Extreme Computing conference held in September 2021 [70].

Key insights include the following:

- Most hardware performance falls approximately within 100 GigaOps/W and 10 TeraOps/W, regardless if it is a chip, a computer card, or an integrated system.
- Several chips and computer cards have a peak power rating between 100W and 500W. This range is driven by the ability to dissipate the heat from the chip or the computer card containing the chip.
- Several of the low-power devices are also lower in throughput performance. However, they are well matched to either embedded (e.g., the IBM TrueNorth [71, 72]) or deployed at the edge (e.g., Google's TPUEdge [73]).
- The higher-power integrated systems (ranging between 1,000W and 10,000W) are rated at close to 10^6 and 10^7 in GigaOps/sec. For example, the NVIDIA DGX-station is rated at about 500 TeraFlops peak (mixed precision) and 1,500W (or, 333 GigaOps/W), integrating 4x NVIDIA Volta V100 GPUs.

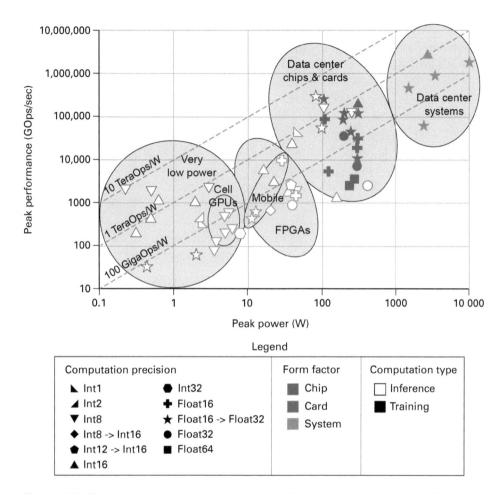

Figure 5.6 Peak performance versus peak power in contemporary chips, cards, and systems.

Although FPGAs are not as high-performance as DSA chips, there is a trend in the industry to customize them to neural network inference applications. For example, Intel acquired Altera, a manufacturer of FPGAs, in 2015. The Intel Arria-10 is customized to meet the needs of neural network inference algorithms (e.g., GoogLeNet) [74]. And in February 2022, AMD announced the acquisition of Xilinx, another FPGA manufacturer.

FPGAs have the benefit of ease of configurability for different types of applications, as an adjunct accelerator to a CPU. Another benefit is in the trade-off between nonrecurring engineering (NRE) cost and performance. Moeller and Martinez demonstrated the ability to achieve equal performance compared to a very-large-scale integration (VLSI) custom chip for a high-speed finite impulse filter (FIR) application—with a lower NRE cost only two years after the VLSI chip implementation was completed, using the Xilinx Virtex FPGA [75]. FPGAs provide higher throughput than CPUs. For verification of hardware design concepts or applications requiring reconfigurations, FPGAs can be a worthwhile approach to consider.

Modern computing architects must assess the computational requirements to meet performance throughputs, at a specified latency, and within the allowable SWaP and cost—for deployments in rugged environments—for airborne or undersea conditions. In addition, designers must address shock and vibration requirements when planning to deploy the AI capabilities in harsh environments.

Modern computing architects must also work closely with ML model designers. It is necessary to employ high-performance, domain-specific hardware, but not sufficient. Designers must also assess opportunities for parallelism to meet performance throughputs and latencies. For example, the ML model might be implemented in the hardware to exploit model parallelism (e.g., AlexNet). The application might also benefit from other types of parallelism, as we elaborate on later in this chapter.

Narayanan et al. described different types of parallelism to achieve an aggregate throughput of over 500 peta-FLOPS on a GPT model, requiring trillion model parameters in training using mixed precision. The different types of parallelism used were described under the rubric of pipeline, tensor, and data-parallelism [76]. Their integrated approach can be generalized to other ML models and computing engines besides GPUs.

Next, we describe three classes of parallelism. As pointed out earlier, there is a need for a close collaboration between hardware architects and software performance engineers. A close collaboration facilitates exploiting the best of hardware features coupled with the inherent structure of the ML algorithms, especially in this post-Moore's era that requires large data, exponential increase in model parameters, and exponential growth in required computation [66, 77].

Types of parallelism include the following:

- Data parallelism: Different training data are distributed across multiple compute engines. The ML model must be replicated across all compute engines. Therefore, the model must fit within the compute engine memory. There is also a need to coordinate weight updates.
- Model parallelism: The model parameters are distributed across processor engines. Data sets (mini-batches) must be copied to all processors. The weight application and weight updates (i.e., the forward dataflow and backpropagation) must be coordinated across layer boundaries among processors [53, 78].
- Pipeline parallelism: Layers and model parameters are distributed across processors. There is sparse communication among processors as the pipeline execution progresses. A mini-batch must be copied among all processors. Updated weights must be maintained across forward and backward passes.

Hybrid approaches, which incorporate different combinations of the above types of parallelism, are also used, balancing performance accuracy and the inherent parallelism available within the ML model structure [79]. Parallelism requires communication among processors. Synchronous and asynchronous communication techniques have been studied (for supervised learning ML models) for communicating among multiple physical processor engines employing established high-performance computing interconnect standards, such as the message passing interface [80].

Let us now conclude this section with a short discussion of energy consumption at the hardware level. This is important because any reduction in energy at the hardware level will be accompanied by a positive energy reduction in overall energy use at the system level. Energy usage assessment should be part of the systems engineering trade-offs prior to the development of AI capabilities, especially for ML model training. In section 5.10, on modern computing challenges, we discuss the need to reduce the equivalent CO_2 emissions from training ML models.

In addition to overall system energy use, we are interested in reducing the amount of energy consumed by embedded devices (i.e., at the edge), where power is at a premium. Canziani et al. did a comprehensive study of various DNN architectures and their implications for embedded systems in terms of accuracy, memory footprint, inference time, number of operations, and power consumption, implemented on an NVIDIA Jetson TX1 board [81]. They showed that accuracy (the Top-1 error rate, expressed as percent) improves as models require more operations (not surprisingly). However, an important finding was that a small increase in accuracy required a significant cost in terms of computational time. Therefore, the accuracy required of embedded systems must be a trade-off parameter relative to computational cost.

The peak power shown figure 5.6 is a function of several factors, as discussed earlier, such as arithmetic precision, internal and external memory (for instructions and data), power consumption by the interconnects, and operational intensity (Ops/Byte)—all factors affecting embedded systems. Sze et al. have researched and assessed ways to improve energy efficiency, particularly for embedded applications (e.g., robotics, drones, self-driving cars, IoT, etc.) [82]. They emphasized, as other scholars have done, the need for a joint algorithm and hardware design. As we discussed earlier, in addition to improvements in energy efficiencies by reducing the number arithmetic bits, they demonstrated an energy reduction of 45 percent in the Eyeriss chip, where the processing elements were designed to skip reads and multiply-accumulate operations when the inputs were zero [83]. Another improvement is achieved by performing compression to reduce the amount of data and weight storage or communication costs from external or internal memory. The costs of compression and decompression are much lower than the energy costs of communicating uncompressed values [55, 65].

In the next section, we elaborate further on another metric that helps when comparing compute engine architectures. The metric, roofline, provides a measure of attainable computation (GigaOps/sec) versus operational intensity (Ops/byte).

5.8   Roofline as a Metric

The roofline metric is a simple but powerful tool that AI architects and modern computing designers can use in assessing the architectural bounds of modern computing engines. The roofline can also show causes of bottlenecks, as a first-order metric [60].

First, let us define what the roofline metric illustrates. We adapt the discussion here from the original descriptions by researchers from Berkeley and Google [84, 85]. The roofline is an illustration of bounds in peak computation performance (e.g., as specified by the chip manufacturer), and operational intensity (e.g., as specified or measured in compute operations per byte accessed from DRAM). Peak computation performance is depicted, on the vertical axis, as operations per second. Operational intensity is shown on the horizontal axis as executed operations per byte of data accessed from DRAM. The ratio of computation throughput to operational intensity is a bound on memory bandwidth, in bytes/sec. Today, most processors are limited by memory access relative to computation throughputs. This limitation can result in significant latencies; which is a problem since many of the applications of interest demand low latencies in order to be useful [86]. Therefore, chip designers strive to fit as much as possible in fast memory

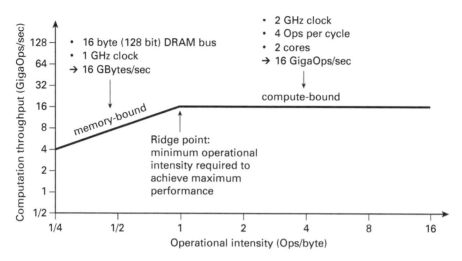

Figure 5.7   Roofline metric for a hypothetical processor and DRAM memory access.

(e.g., local SRAM), to keep the arithmetic processing units fed without waiting for a byte of data. However, in comparing compute engines, DRAM access drives the performance limitations. Thus, the roofline metric—quantifying computation throughput versus operational intensity—is a very useful metric to understand chip performance.

Figure 5.7 illustrates a roofline model for a hypothetical processor with a peak computation throughput of 16 GigaOps/sec and access from DRAM operating at a clock frequency of 1 GHz and a 128-bit-wide bus for a maximum capability of 16 GBytes/sec. The horizontal roofline sets the compute-bound of the processor. The slanted portion of the curve sets the memory-bound roofline with respect to data access from DRAM. A processor with increased computation throughput would move the horizontal line up. A processor with faster access from memory would move the slanted line to the top-left side of the chart. The ridge point (i.e., the knee on the curve) identifies the minimum operational intensity (in Ops/byte) to achieve maximum computation throughput performance (in GigaOps/sec).

Google's first-generation TPU (TPUv1) is a great use case example to analyze its roofline performance with respect to an MLP algorithm. Based on the analysis performed by Jouppi et al. the TPUv1 operated at 700 MHz, built in 28-nanometer process technology, with a maximum computation throughput of 92 TeraOps/sec on 8-bit integer (int8) inference operations. The DRAM used had a reported maximum bandwidth of 34 GBytes/sec. The TPUv1 had a horizontal roofline pegged at 92 TeraOps/sec. Since the TPU horizontal axis represented MAC, there are two additional operations/byte relative to the simple hypothetical illustration in figure 5.7. The ridge point for TPUv1 was about 1,350 MAC Ops/byte [59]. In section 5.12, we present an exercise to calculate the new ridge point, as if Google had used faster DRAM.

In Google's analysis, one of the dominant MLP applications (MLP0) required an operational intensity of 200 MAC Ops/byte. Therefore, with a memory bandwidth of 34 GBytes/sec, the MLP0 ran in the TPUv1 at about 13.6 TeraOps/sec, right against the memory-bound portion of the TPUv1 roofline.

The TPUv1 use case illustrates how to best use the roofline metric. For this purpose, there are a number of important takeaways (look at figure 5.7 for reference):

1. The ridge point is the point where for a minimum operational intensity (given a DRAM memory bandwidth), the maximum computation throughput is achieved. It helps identify how well the best processor performance might meet the algorithm requirements.

2. For a given algorithm, the designer can plot a point on the chart for its computation throughput requirement versus the required operational intensity.

3. If the algorithm point on the chart is to the left of the processor ridge point and at or near the memory-bound slanted line, then the algorithm is being executed consistent with the memory access capabilities of the processor.

4. If the algorithm point on the chart is significantly below either the memory-bound line or the compute-bound ceiling, as algorithm designers, we have enough room to optimize the algorithm executing in the same hardware.

5. If the algorithm point on the chart is above any of those two lines (memory-bound or compute-bound), the hardware is not able to efficiently execute the algorithm specifications (and still meet, for example, the latency requirement), because the algorithm exceeds either the memory access bound or the peak computation throughput.

6. In the case given in item 5, if memory-bound, we would need to get faster memory—to move the ridge line higher and to the left on the graph. And, if compute-bound, we would need to implement the algorithm in a processor with increased peak computation throughput; moving the horizontal line higher on the graph. Another alternative is to scale up the number of chips, assuming that the algorithm can be parallelized among a larger number of chips.

Norrie et al. detailed the design choices made by Google's designers in their evolution of the TPUs from TPUv1 to the TPUv2 and TPUv3 [87]. The TPUv2 and TPUv3 are able to perform both training and inference. The training was implemented using the bfloat16 arithmetic precision discussed earlier, reducing hardware and energy costs. The chip's performance also was assessed using its roofline characteristics.

As discussed earlier, AI organizations must include on their team algorithm designers working closely with hardware and software designers. An AI-driven organization must have, and/or must subcontract, a multidisciplinary team (from data conditioning through human-machine teaming/HMT) to best architect, design, develop, and deploy competitive AI products and services.

So far, the discussion in this chapter has focused on modern computing capabilities and how these hardware and software technologies are key enablers to making narrow AI feasible. Unfortunately, modern computing also presents potential threat surfaces available to attackers to exploit and make an AI system useless. In the next section, we discuss approaches to securing modern computing.

5.9 Securing Modern Computing

There are significant concerns about the vulnerability of AI systems—cyberattacks can make AI capabilities existential if not enough attention is paid to them and actions are not taken. In this section, we focus on making modern computing systems secure. In chapters 7 and 8, we focus on vulnerabilities affecting other parts of the end-to-end AI architecture shown in figure 5.1.

We divide the discussion into two subsections. One addresses AI as a technology to improve the cybersecurity posture, and the second discusses approaches for making AI more cybersecure. Both of these topics are directly relevant to making modern computing secure:

1. *AI for cybersecurity:* The application of AI for making systems more robust against cyberattacks is of great interest. Cyberthreats are dramatically increasing in frequency, are being executed by sophisticated threat actors, and are causing grave damage to businesses and their operations.

2. *Security for AI:* In the past decades, there has been a growth in cybersecurity research, leading to operational capabilities for making systems more secure. The good news is that the AI community can benefit from its many years of investments in cybersecurity.

For the ensuing discussion, it is useful to refer to figure 5.8 as an illustration of a compute environment. The compute environment is representative of a cloud-based system consisting of

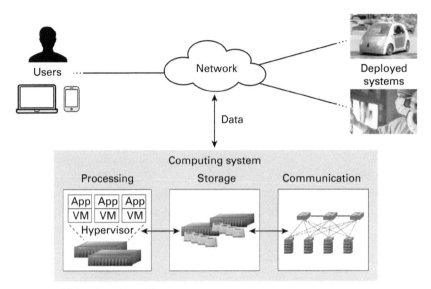

VM = virtual machine

Figure 5.8   Representative compute environment (cloud-based system).

- Users interfacing with deployed systems and a representative computing system via a network
- Deployed AI systems (e.g., driverless cars, medical devices)
- A computing system consisting of a processing subsystem (e.g., a subsystem composed of CPUs, GPUs, or TPUs), a storage subsystem (e.g., high-speed disks), and a computational network (e.g., PCI-Express [88])

As figure 5.8 shows, there is no loss of generality since the system illustrates potential threat surfaces available to an attacker—for example, phishing attacks on users, infiltrating deployed systems, or directly attacking the computing system. All or portions of these systems are present in small to large enterprises or at the edge.

Let us now discuss opportunities for AI to make a system more secure, like securing the system environment shown in figure 5.8.

Artificial Intelligence for Cybersecurity

One of the biggest issues confronted by cybersecurity practitioners is the need to make sure that compute system providers adhere to a constant and diligent attention to cyberhygiene. Cyberhygiene is the lowest bar that all information technology (IT) service providers must attend to and users must comply with (an example is in the use of multifactor authentication).

However, the complexity of these computing systems is so large that humans are overwhelmed. Therefore, AI is a technology that can help cyberanalysts in reducing their work loads. Martinez presented a framework for leveraging AI in *Emerging Areas at the Intersection of Artificial Intelligence and Cybersecurity* [89]. This framework addresses a systematic approach to reducing vulnerabilities across an end-to-end compute system as represented in figure 5.8.

The framework assumes a "cyber kill chain," in which an attacker progresses through a set of four offensive stages: (1) performing reconnaissance; (2) engaging in an attack; (3) maintaining presence; and (4) achieving effects and assessing damages. The defender prevents the attacker from succeeding by employing AI across five defensive stages:

1. Identify threats (perform reconnaissance): By implementing predictive analytics, the defender can monitor and anticipate the threat. This type of AI capability can be very complex, and it incorporates many implementation details that are beyond the scope of this discussion. Predictive analytics must be adapted to identify threats at the user level, within the computing system, or within the deployed system shown in figure 5.8.

2. Protect users, computing systems, and deployed systems: By implementing clustering and classification, the defender can use AI to infer and prioritize vulnerabilities. We show later a use case example of using principal component analysis (PCA), followed by a support vector machine (SVM) algorithm to classify potential counterfeit hardware parts.

3. Detect threats (identify new attacks): By implementing information extraction, the defender can detect and characterize the attacks. Information extraction algorithms must be adapted to work within a computing system or a deployed system.

4. Respond (i.e., stop attacks): By implementing automated tools for planning and prioritizing—depending on the severity of the threat—the defender can recommend courses of action (e.g., from the outputs of a recommender system).

5. Recover from an attack: By interfacing with the output of the ML system (i.e., at the HMT stage in the end-to-end AI architecture shown in figure 5.1), the defender can enact solutions—for example, by halting operations of a deployed system or disabling a computer system from continuing to access the organization's private network.

Let us look at an example of preventing a computing system—defensive stage 2—from operating with counterfeit parts. Koziel et al. automated the process of evaluating hardware parts by monitoring the power levels of hardware devices [90]. As shown in figure 5.9, their approach consisted of collecting unique signatures of power traces from an authentic part (the golden part), and correspondingly collecting power trace signatures from suspected parts. PCA was performed on the collected input data to reduce the dimensionality to the most dominant subspace components, followed by a tenfold cross-validation SVM to determine if the suspected part was close enough to the golden part. Their approach enabled rapid testing of devices under tests before integrating them into an embedded system. Zhu et al. also implemented a K–nearest neighbor algorithm to classify anomalies potentially present in a printed circuit board, indicating malicious attacks [91].

An example of capabilities, in support of defensive stage 3, was presented by Sven Kraser (CrowdStrike, Inc.) describing capabilities available in the Machine Learning–Enabled Security Platform, where typically 250 billion events are ingested from endpoint sensors and processed every day [89]. CrowdStrike minimizes false positives and negatives by having humans in the loop to ascertain the validity of the ML results.

In the next subsection, we discuss the complementary side: the use of many years of cybersecurity research and deployed operational capabilities to secure AI systems. We adapt the same terminology as reported by the US National Science and Technology Council (NSTC) subcommittee on Networking and Information Technology Research and Development (NITRD) and the subcommittee on ML and AI, who sponsored a three-day workshop to address AI for Security (AI4Sec) and Security for AI (Sec4AI) [92]. The workshop chairs' briefing also addressed AI for cybersecurity, requiring research in a number of areas including enhancing trustworthiness of systems.

Security for Artificial Intelligence

The attack surfaces across an AI system are numerous, both at the macro-level and at the micro-level. Referring back to figure 5.8, at the macro-level, users can be fooled by cyberattackers enabling access to deployed systems and/or the computing environment—for example, through credential stealing, spoofing, or spear phishing. There is also the potential for causing damage to organizations by insider threats. In addition, at the macro-level, attackers can infiltrate a deployed system—for example, by accessing medical devices through a wireless network.

At the micro-level, any of the subsystems illustrated in figure 5.8 are vulnerable to attacks. AI system vulnerabilities include data from sensors and sources while being

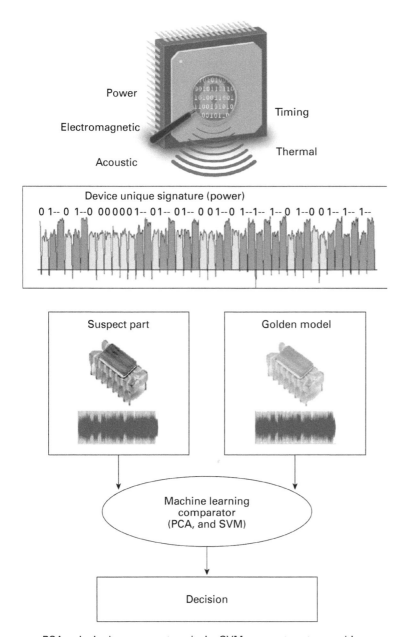

PCA: principal component analysis SVM: support vector machines

Figure 5.9 Automated detection of counterfeit parts using PCA followed by SVM.

processed (data in use), while in storage (data at rest), or while being communicated through the networks (data in transit).

An example of cyberattacks on a processing subsystem occurred in 2018, with massive implications to the hardware [93, 94]. The two types of cyberattacks were

- Meltdown: Designed to exploit side effects in the hardware's out-of-order execution, enabling access to memory and thus enabling the stealing of personal data and passwords
- Spectre: Exploited the processor's speculative execution

Both examples were attacks on CPUs, affecting millions of computers worldwide. It is worth reemphasizing that DSAs, as discussed earlier, do not employ these hardware features (e.g., out-of-order and speculative executions). However, AI systems use CPUs to control other parts of the hardware within a computing system. Therefore, AI systems can also be compromised.

Another area that continues to be very concerning to computing system manufacturers is the dependency on a trusted supply chain. The cyber-supply chain has been studied in detail by the US Defense Science Board [95]. The limited number of microprocessor foundries also present vulnerabilities, potentially disrupting the development and deployment of trusted computing systems.

Supply chain vulnerabilities are also concerning; they can affect both hardware and software. Software supply chain vulnerabilities can damage AI systems since much of the code comes from reusing open-source models. Enck and Williams reported a dramatic increase in malware, where attackers implanted malware directly into open-source software code affecting the commercial supply chain [96]. The Open-Source Security Foundation, with governing members from large commercial software providers, is addressing several of these key challenges. Our systems engineering approach, discussed in chapter 2, is very consistent with some of their recommendations of employing rigorous testing, verification, and validation to making AI systems more robust.

The report from the US NSTC subcommittees mentioned earlier recommended four research areas to advance the state of the art in securing AI systems (Sec4AI) [92]:

1. Develop techniques supporting AI system specifications and verification through standards, tools, and metrics to tractably and compliantly reason about the correctness of a system. The goal is to provide a level of security guarantees.

2. Ascertain trustworthy decision-making, especially in high-stakes applications, including the ability to provide explainability and accountability.

3. Make ML methods robust, for example, by limiting adversarial manipulations and ensuring data privacy and model fairness.

4. Establish new AI/ML engineering/design principles to ensure the secure integration of AI products and services.

The ultimate goal of fully securing AI systems will continue to present a challenge to AI system architects, designers, and implementers. A clean slate approach is needed across all seven layers of the Open Systems Interconnection model of a computing system. One such effort is being pioneered by Hamed Okravi (MIT Lincoln Laboratory) and Howie Shrobe (Massachusetts Institute of Technology) in the Resilient Mission Computer (RMC) project as a radical but feasible cybersecurity moonshot [97]. Even if only a subset of clean slate technologies are successfully developed initially, the capabilities will be revolutionary to AI computing systems, and, more generally, to general classes of mission computers. In this context, a mission computer is assumed to be performing a task only over a limited and finite duration.

Organizations and government funding agencies must make significant investments in the areas of AI4Sec and Sec4AI. There are many challenges. The well-being of secured societies—and their dependency on digital systems—are too important to ignore. Of course, AI is at the core of digital systems that we presently use and will continue to depend on in our daily lives. On May 3, 2022, Eric Horvitz, Microsoft's chief scientific officer, testified before the US Senate Armed Services Committee Subcommittee on Cybersecurity, offering several important and timely actions needed at the intersection of AI and cybersecurity [98].

5.10   Modern Computing Challenges

In a little over a decade, massive amounts of data, ML algorithms, and modern computing have converged very nicely, enabling AI researchers, developers, and providers of products and services to deliver unprecedented narrow AI capabilities. As expected, with new technologies, there are still significant modern computing challenges that must be addressed. In this section, we highlight some of these challenges.

Yann LeCun, renowned AI researcher and Turing Award winner, in his speech "Deep Learning Hardware: Past, Present, and Future," had a very apropos lesson to deliver: "The availability of suitable hardware and simple-to-use open-source software are critical to the wide adoption of a class of methods by the community. Good results are essential but not sufficient" [33]. In addition to continuing advancements in ML algorithms, there is a need to transition from the primary emphasis on algorithm performance to also emphasizing the operational environment, using computing technologies well matched to meeting the low latencies, and the optimized SWaP demanded of efficient AI deployments at an affordable cost.

In this section, we discuss a set of challenges as a function of time horizons, consistent with the format we have used in earlier chapters. However, prior to doing so, we want to direct the needed attention toward reducing the carbon footprint of AI applications. It is widely known that large-ML models can require many compute hours of training time, as well as in the inference stage while the deployed model is in use. The computing energy consumed amounts to a large equivalent CO_2 footprint, with potential deleterious impacts on the environment [99]. Addressing this important problem requires a multiprong approach, including a close collaboration between the AI community and policy makers [100].

Fortunately, the problem is recognized, and it is being addressed, by a number of AI researchers. For example, Patterson et al. have discussed a reduction in carbon footprint close to $100x$–$1,000x$ by a combination of sparse DNNs (without loss of accuracy), efficient infrastructure centers (including their physical location), and ML-oriented accelerators [101]. They make a recommendation that is paramount in addressing this problem: "energy usage and equivalent CO_2 emission should be part of the model evaluation metrics." They also recommend the use of a Machine Learning Emission Calculator, as an initial assessment, which takes into account the hardware used, the compute platform, hours of usage, and the region where the compute takes place [102]. Lannelongue et al. published a freely available tool for calculating the carbon footprint of computation [103].

Similarly, Schwartz et al. make recommendations for AI practitioners to publish their ML model energy usage when achieving a given algorithm performance [104]. They coined the term "Green AI." In addition to the importance of reporting novel results, Green AI is meant to emphasize the need to state the computation energy costs in achieving the algorithm results (i.e., algorithm performance relative to energy usage).

Before we leave the topic of energy consumption, AI can also be used for reducing data centers energy costs by using ML techniques to monitor computing usage and by further optimizing cooling. Google reported a reduction of 40 percent in the energy used for cooling by monitoring and training ML neural networks with data from multiple sensors collecting temperature, power use, pump speeds, and other statistics [105, 106].

In table 5.6, we enumerate a set of modern computing challenges. We categorize them as a function of one of three horizons: content-based insight, collaboration-based insight, and context-based insight.

For several of the challenges and recommendations outlined in table 5.6, we have offered techniques and approaches as discussed in earlier sections—for example, the need to accelerate model generation, minimize cybersecurity vulnerabilities, or reduce the computational carbon footprint. Next, we elaborate on other challenges and recommendations that we have not discussed yet:

Table 5.6 Modern computing challenges and recommendations

Challenge Area	Modern Computing
Horizon 1: Content-Based Insight (1–2 years)	• Accelerate model generation. • Deploy computing at the edge (i.e., the computing of many IoT inputs). • Advance serverless computing. • Minimize exposure to cybersecurity vulnerabilities. • Improve performance tools for AI hardware and software. • Reduce the carbon footprint.
Horizon 2: Collaboration-Based Insight (3–4 years)	• Compute across distributed platforms (e.g., robotics interfacing with humans). • Reduce SWaP and the cost of embedded systems as devices collaborate with humans to generate insight. • Improve computing performance to enable metaverse applications (i.e., minimize latencies).
Horizon 3: Context-Based Insight (5+ years)	• Develop efficient computing for neuro-symbolic reasoning. • Scale neuromorphic computing. • Incorporate privacy preserving computing techniques. • Employ quantum computing. • Scale computing to exploit multimodal data (i.e., text, speech, video, etc.).

- *Deploy computing at the edge:* The demands for AI at the edge are increasing significantly due to the exponential growth in IoT devices [107]. Unfortunately, computing at the edge presents significant challenges in SWaP, cost, and need for simplicity in the implementation of computing technologies for DNN. Several researchers have looked at techniques—for example, data compression, pruning of the neural network, variable precision (i.e., different levels of precision for the NN weights and the activation function) [108, 109]—to tune the hardware to edge computing. These approaches will enable DNN-processing hardware with more efficiency and a lower cost point than enterprise-level DNN hardware. There are many other challenges that need to be addressed in addition to improved performance, including privacy and security at the edge, latencies, and programmability [110].

- *Advance serverless computing:* A higher-level abstraction of the underlying computing system (e.g., cloud computing system) [111]. The developers need to know less about the details of where the application is running. The processor suite would be very heterogeneous (e.g., CPUs, GPUs, TPUs), matched to the application, and enabling scalability.

- *Improve performance tools for AI hardware and software:* There are performance benchmarks that are becoming popular in assessing ML algorithm performance (e.g., NAS-Bench suites and MLPerf) [112, 113]. More benchmarks are needed to address a full end-to-end AI system, from data sensors and sources through deployment, as emphasized in chapter 2. The *Artificial Intelligence Index Report* performs a comprehensive update every year of progress in AI, and the performance tools are critical for assessing progress and trends [114].

- *Compute across distributed platforms:* As DNN models become larger, data increases, and user and machines work collaboratively, there is a need to expand from centralized AI products and services to a constellation of users and machines. This topology presents a challenge since AI must employ distributed computing with a need for seamless operation.

- *Improve computing performance to enable metaverse applications:* AI has a central role to play in making the metaverse concept useful. Collaboration will improve as computing technologies advance in support of mixed-reality (e.g., virtual, augmented, extended reality). There are many applications where the metaverse can be extremely useful, such as healthcare, manufacturing, so-called digital twins for industry 4.0, and education in a hybrid mode (i.e., classroom and virtual) [115].

- *Develop efficient computing for neurosymbolic reasoning:* Advance computing technologies to more efficiently compute neural network algorithms, complemented with symbolic reasoning, to introduce more context into AI capabilities.

- *Scale neuromorphic computing:* Neuromorphic computing has the potential to deliver neural network computations at a small size and low power (well matched to edge computing). However, more advances are needed to scale neuromorphic computing to commercially viable and practical applications, including the supporting software ecosystem and benchmarks [72, 116, 117].

- *Incorporate privacy preserving computing techniques:* The need for security and privacy is becoming more important as ML applications expand. Further research is required to making techniques like differential privacy, secure multiparty computation, and homomorphic encryption practical in real-life applications [118]. At present, these techniques require substantial computation and communication overhead. However, the ability to share data, with security and privacy guarantees, can enable bringing more context into the application of AI. These techniques can also allow federated learning to improve model performance [119].

- *Leverage quantum computing:* There are two efforts (at the macro-level) that must be attended to now. First, in the near term, we must prepare for the postquantum computing era [120]. ML algorithms, and AI in general, which rely on encryption to protect private and valuable business content, must be made resilient to the postquantum computing era. The second emphasis is on continuing to advance quantum computing to improve the ability to model and reduce the computation time of advanced algorithms. One example of such an algorithm is the HHL algorithm (named after its authors, Aram Harrow, Avinatan Hassidim, and Seth Lloyd), which solves a set of linear equations commonly found in linear algebra computations using a quantum computer [121]. Once feasible, the solution will require log (N) computations, with a quantum computer, instead of order N required of a classical computer, where NxN is the size of a sparse matrix A, as part of $Ax = b$; where x and b are vectors of length N.

There are also strategic road maps incorporating the use of classical computing complemented with quantum computing as an approach to take advantage of the best of both and, when quantum computing is available, integrate quantum computing into the mainstream [122]. AI researchers at IBM are pursuing multiple hardware approaches, including specialized hardware based on a von Neumann architecture (as discussed earlier in this chapter), analog AI devices, and quantum computing for AI [123].

- *Scale computing to exploit multimodal data (i.e., text, speech, video):* By incorporating several modalities, AI systems can introduce more context, therefore improving reasoning about the environment and providing more insight into the AI decision support cycle (i.e., explainable AI/XAI). However, the computation throughputs to ingest multimodal data, perform data conditioning, and process multimodal data through ML algorithms—while maintaining spatial-time coherency—are daunting. There are high-stake applications where multimodality data is very beneficial, including driverless cars, medical diagnosis, autonomous drones, and robotics [57, 124, 125]. More research is needed to scale computing across multimodalities.

Foundation models will continue to demand more computational performance as model sizes increase. An adaptation of GPT-3 technologies was demonstrated to auto-generate code. OpenAI developed Codex, which it is trained with a 12-billion-parameter model employing over 150 gigabytes of an open-source corpus of code available in

GitHub [126]. The system can receive a sentence on what code is desired, and it can automatically generate the code in the desired programming language, such as Python. After the development of Codex, GitHub announced the availability of Copilot—an AI code generation assistant trained from billions of lines of open-source code and hosted in Microsoft cloud servers [127]—and its autocomplete functionality is similar to a human's experience when writing emails. DeepMind has also created an autocode generation capability named AlphaCode [128]. Autocoding, or assisting code programmers in code generation, can have many benefits. However, there are also challenges, such as the accuracy of the code, copyright issues, and ownership of intellectual property.

There is also concern about the levels of security vulnerabilities, either unintentional or intentional. Malicious actors could exploit these capabilities to create malware. Several researchers are addressing these issues. Pearce et al. have investigated the autocode vulnerabilities based on MITRE's "Top 25 Common Weakness Enumeration" of the most dangerous software weaknesses [129, 130]. They found that in eighty-nine different scenarios where Copilot completed the code, producing 1,689 programs, approximately 40 percent of the code was vulnerable. Again, autocoding assistance can be very helpful in improving programmer productivity, but these types of AI tools must be used diligently regarding code correctness and cybersecurity vulnerabilities.

We conclude this section by highlighting the Oak Ridge National Laboratory (ORNL) Frontier supercomputer—the most powerful exascale computer in the world (at the time of this book writing), and at the top of the 2022 Top 500 list. The system is rated at over 1.5 exaflops (1.5 x10^{18} floating-point operations per second) peak theoretical capacity [131]. The supercomputer system is based on the Hewlett Packard Enterprise Cray HW (HPE Cray EX235A), the HPE Cray OS, and an AI-optimized subsystem integrated with AMD EPYC processors and AMD Radeon Instinct GPU accelerators [132]. As pointed out by Jack Dongarra, a professor at the University of Tennessee and recipient of the 2021 Turing Award, the ORNL Frontier supercomputer will be "increasingly used to train deep neural networks, where 16-bit precision can suffice" [131]. However, Frontier's world-record performance comes at a significant power level, drawing over 20 megawatts (with a computation throughput per watt of about 50 Gigaflops/watt), but this is a good overall performance for a general-purpose supercomputer.

This exascale supercomputer system will be increasingly crucial to continuing advancement in ML modeling and training—among other modeling applications— to further advance the eras of ML [133, 134]. The challenge is in using such a powerful compute resource during the early stages of ML research, and ascertaining that the findings are also transitioned to commercial operations with a low carbon footprint.

5.11   Main Takeaways

Modern computing technologies, as an AI enabler (together with massive data and ML algorithms), have been the engine making feasible remarkable advances in narrow AI. As Hennessy and Patterson said, "we are in a new golden age," with joint collaboration among hardware and software architects, designers, and implementers working to invent and advance state-of-the-art AI.

AI, as a general-purpose technology, will depend on these modern computing advances to gain wide acceptance and provide business value to stakeholders. Modern computing is the bedrock needed across all the subcomponents shown in figure 5.1. However, as Moore's Law continues to slow, innovative approaches are needed, evolving from the original von Neumann architecture. Domain-specific hardware and software will continue to demonstrate unprecedented performance when supplemented with analog computing, and later with quantum computing.

The present inability to continue at the exponential improvement rate in Moore's Law has led designers to more optimally use the large increase in the number of transistors. As feature sizes decrease to below 7-nanometer process technology—currently, semiconductor foundries are instrumenting their facilities to support a 2-nanometer feature size—more cores will be available within a single device. However, the increase in number of compute cores requires hardware and software designers to address the ability to parallelize the ML algorithms (i.e., overcoming Amdahl's Law). Also, computing devices are limited by power dissipation, operating voltages, and clock frequencies (i.e., the end of Dennard Scaling).

Hardware and software must be optimized to deliver the requisite computing performance at the enterprise levels and at the edge. It is predicted that, globally, there will be in excess of 400 exabytes of data created every day. Some of this data growth is due to an exponential growth in IoT devices. Many AI applications, using IoT devices as sensors, demand that processing take place close to the IoT physical location, avoiding long latencies.

There are similarities and also significant differences in computing systems deployed at the enterprise versus at the edge. Similarities include the need for increasing data volume, variety, and velocity. In both environments, there is a high degree of concern about adversarial attacks. Furthermore, edge AI must be implemented in small SWaP form factors, and at low cost. For edge AI, latency requirements are stringent for achieving results in a timely manner. There is also little tolerance for errors.

ML algorithm computational kernels require large computational throughputs and low latencies. The GPUs and TPUs, as representative ML accelerators, are well matched to meeting these computational requirements because of the parallel structure inherent in the ML algorithms. The model parameters have been growing at approximately $2x$ every 2.3 months, and the corresponding computation throughput requirements have been increasing at $2x$ every 3.4 months. A large number of cores with high-speed memory close to or within the device enable both parallelism and low latencies. Many of the neural network key computational kernels (e.g., kernels in an MLP) are linear algebra operations: vector-vector adds, matrix-vector multiplies and adds, matrix-matrix multiplies and adds.

Because of these high-computation throughput demands, hardware designers have looked at ways to reduce arithmetic precision wherever possible. Most ML algorithms require higher precision during training than during the inference stage. Thus, today's AI processors use mixed-precision arithmetic. Microprocessors based on DSA also eliminate many of the unnecessary instructions typically found in CPUs (such as branch prediction, out-of-order execution, and different levels of cache hierarchies). A chip designed under the rubric of a DSA is a special-purpose device that performs a narrow range of tasks extremely well.

It is important to emphasize the points made by hardware and software pioneers that the future will involve exploiting hardware features with software—for example, simpler instructions, pruning, compression, parallel loops, vectorization, and linear algebra optimized instructions [27, 63, 66]. This architecture trends need to continue to meet the evolving demands of AI systems.

For the past decade, there has been a Cambrian Explosion in the number of computing engines, from specialized data centers to embedded devices. However, at present, most computing systems fall within 100 GigaOps/watt to 10 TeraOps/watt in power density. As the devices scale down to meet the form factors needed for edge AI, their computation throughputs per unit watt also scale correspondingly. Similarly, for enterprise computing, as the computation throughputs increase, the peak power also increases, resulting in about an equivalent range of computational power density.

In addition to hardware and software optimizations, another powerful lever in the implementation of ML algorithms is in exploiting parallelism. There are several types of parallelism that can be exploited individually or in a hybrid way: data parallelism, model parallelism, and pipeline parallelism. These techniques are used today in the implementation of the best-performing ML DNN algorithms.

Given all the different knobs that algorithm architects, together with hardware and software designers, can use, it is of paramount importance that metrics, standards, and tools be developed. A very useful metric for assessing chip performance is the roofline metric, commonly illustrated using a two-dimensional graph showing computation throughput versus operational intensity. A specific point in this relationship is the ridge point, which represents the minimum operational intensity required to achieve maximum computation throughput performance.

As discussed earlier, modern computing is the bedrock enabling the implementation of the critical subcomponents of our end-to-end AI system architecture. Unfortunately, these subsystems present multiple places of vulnerability as attack surfaces. Therefore, AI systems must be made secure. There are techniques for both using AI to improve the cybersecurity posture, as well as techniques for using cybersecurity technologies to make AI more secure.

AI techniques for improving the cybersecurity posture, AI4Sec, can be addressed by identifying defensive stages in a cyberkill chain, and implementing AI techniques to make digital systems more secure. Typical defensive stages include the following:

- Identifying threats
- Protecting users, computing systems, and deployed systems
- Detecting existing threats
- Responding against threats
- Recovering from attacks

The other side of the cybersecurity coin is employing cybersecurity techniques to make AI more resilient to adversarial attacks, also referred to as Sec4AI. Both of these areas require significant more attention to ascertaining that AI systems, and more generally digital systems, are secure.

As we conclude this chapter, another very important area needing serious effort across our triad of people-process-technology is reducing the carbon footprint of computing systems. We cannot afford to augment human capabilities by advancing narrow AI at the expense of deleterious impacts on our environment.

We bring up the people-process-technology triad because dealing with climate change requires a cultural shift by people to recognize and act to reduce (and hopefully someday also eliminate) the damage to the environment caused by the significant power usage needed by today's AI massive computations. We also need to establish guidelines in the form of policies and governance (i.e., processes) without atrophying AI innovations. Finally, AI technologies can help improve the use of natural and renewal sources of energy.

5.12   Exercises

1. A von Neumann computer architecture means that both the data and the program instructions can reside in the same computer memory.
   (a) True
   (b) False

2. In 1975, Gordon Moore updated his original prediction of 1965 to predict that the transistor density would double every year.
   (a) True
   (b) False

3. Amdahl's Law states that when executing algorithms in parallel, the maximum theoretical speed-up is limited by the sequential portion of the code.
   (a) True
   (b) False

4. Dennard Scaling is a power density rule stating that for metal-oxide transistors, the power dissipation stays constant regardless of the feature size if the current and voltage can scale correspondingly to the transistor area.
   (a) True
   (b) False

5. ML learning algorithms applied to data emanating from IoT devices demand:
   (a) High power efficiencies
   (b) Low cost
   (c) Low latencies
   (d) All of the above

6. Both enterprise systems and edge AI offer opportunities to minimize the need for monitoring data drifts and/or algorithm drifts.
   (a) True
   (b) False

7. The serial portion of ML algorithms, implemented in GPUs, is the main reason why GPUs have become popular.
   (a) True
   (b) False

8. Model parameters used for training DNNs have been growing at approximately 2x every 3.4 months and at a computation throughput rate of approximately 2x every 2.3 months.

   (a) True

   (a) False

9. In an MLP (such as the forward propagation fully connected neural network), the dominant computational kernel is:

   (a) The softmax processing to convert output numerical values into probabilistic values

   (b) The matrix-vector operations

   (c) The ReLU activation function

   (d) None of the above

10. ML algorithms require greater precision during the training of the model compared to the precision needed during the inference stage (i.e., the application of the model).

    (a) True

    (b) False

11. Google introduced a new type of precision referred to as "brain floating-point," shortened to bfloat. Bfloat16 requires 8 exponent bits plus 8 mantissa bits.

    (a) True

    (b) False

12. Derive and show that the fp16 representation (under the IEEE-754 floating-point standard) for the numerical value of -123 (base 10) is 1 10101 111011.

    Hint: $-2^{(21-15)} \times 1.921875 = -123$.

13. Describe what is meant by a DSA and explain why a chip based on a DSA design is well matched to ML arithmetic operations.

14. Explain the reasons for a computational power density in the range of approximately 100 GigaOps/W to 10 TeraOps/W, and how the limitation in Dennard Scaling affects the computation peak performance versus peak power.

15. Types of parallelism employed in the implementation of ML algorithms are:

    (a) Data parallelism

    (b) Model parallelism

    (c) Pipeline parallelism

(d) All of the above

(e) None of the above

16. Assume a processor with a peak computation throughput of 92 TeraOps/sec and faster DRAM, such as the GDDR5 memory used in the NVIDIA K80, which is capable of 180 GBytes/sec bandwidth. Based on that, derive what the operational intensity ridge point is in MAC Ops/byte? *Hint:* A MAC represents two operations (a multiply and an add).

17. Elaborate on what is meant by each of the following topics:

   • AI4Sec

   • Sec4AI

18. In addressing CO_2 emissions (during model training), it is most critical to take into account which of the folowing?

   (a) The ML algorithm used,

   (b) The number of hours spent in computing the model

   (c) The computing hardware used

   (d) The region of computation (i.e., where the computation takes place)

   (e) All of the above

19. Why is the location where the computation takes place so critically important in achieving high performance?

20. How can hardware and software designers use the technological advances demonstrated with the Top 500 supercomputers (e.g., the Frontier supercomputer) but apply them to building AI systems?

5.13   References

1. Brynjolfsson, E., D. Rock, and C. Syverson, The productivity J-curve: How intangibles complement general purpose technologies. *American Economic Journal: Macroeconomics*, 2021. 13(1): 333–372.

2. Crafts, N., Artificial intelligence as a general-purpose technology: An historical perspective. *Oxford Review of Economic Policy*, 2021. 37(3): 521–536.

3. Abadi, M., P. Barham, J. Chen, et al., TensorFlow: A System for Large-Scale Machine Learning, in *12th USENIX Symposium on Operating Systems Design and Implementation (OSDI 16)*, Savannah, GA. 2016.

4. Paszke, A., S. Gross, F. Massa, et al., Pytorch: An imperative style, high-performance deep learning library. *Advances in Neural Information Processing Systems*, 2019. 32.

5. Chollet, F., Keras documentation. https://*keras. io*, 2015. 33. https://github .com/fchollet/keras.

6. Jouppi, N. P., C. Young, N. Patil, et al., In-datacenter performance analysis of a tensor processing unit, in *Proceedings of the 44th Annual International Symposium on Computer Architecture,* Toronto. 2017.

7. Strubell, E., A. Ganesh, and A. McCallum, Energy and policy considerations for deep learning in NLP, 2019. arXiv preprint arXiv:1906.02243.

8. Nilsson, N. J., *The quest for artificial intelligence.* 2009, Cambridge University Press.

9. Turing, A., On compatible numbers. *Proceedings of the London Mathematical Society*, 1936: 230–265.

10. Turing, A. M., Lecture to the London Mathematical Society on 20 February 1947. *MD Computing.* n.d. 12: 390–390.

11. Haigh, T., M. Priestley, C. Rope, and W. Aspray, *ENIAC in action: Making and remaking the modern computer.* 2016, MIT Press.

12. Eckert, J. P. Jr., J. W. Mauchly, H. H. Goldstine, and J. G. Brainerd, *Description of the ENIAC and comments on electronic digital computing machines.* 1945, Moore School of Electrical Engineering.

13. Goldstine, H. H. and J. Von Neumann, *On the principles of large-scale computing machines.* 1946: ca.

14. Haigh, T. and M. Priestley, Von Neumann thought Turing's universal machine was "simple and neat." but that didn't tell him how to design a computer. *Communications of the ACM*, 2019. 63(1): 26–32.

15. Von Neumann, J., First draft of a report on the EDVAC. *IEEE Annals of the History of Computing*, 1993. 15(4): 27–75.

16. McCulloch, W. S. and W. Pitts, A logical calculus of the ideas immanent in nervous activity. *Bulletin of Mathematical Biophysics*, 1943. 5(4): 115–133.

17. Shannon, C. E., A symbolic analysis of relay and switching circuits. *Electrical Engineering*, 1938. 57(12): 713–723.

18. *MIT CSAIL News, Founding father of AI Oliver Selfridge dies at 82.* 2008.

19. Selfridge, O. G., Pattern recognition and modern computers, in *Proceedings of the March 1–3, 1955, Western Joint Computer Conference,* Los Angeles. 1955.

20. Selfridge, O., Pandemonium: A paradigm for learning, in *Proceedings of Symposium on the Mechanization of Thought Processes.* 1959.

21. Husbands, P., O. Holland, and M. Wheeler, *The mechanical mind in history.* 2008. MIT Press.

22. Chiou, S., C. Music, K. Sprague, and R. Wahba, A marriage of convenience: The founding of the MIT Artificial Intelligence Laboratory. *Structure of Engineering Revolutions*, 2001.

23. Buchanan, B. G., A (very) brief history of artificial intelligence. *AI Magazine*, 2005. 26(4): 53–53.

24. Stanford University, *The history of artificial intelligence*. https://exhibits.stanford.edu/ai.

25. Moore, G. E., *Cramming more components onto integrated circuits*. 1965, McGraw-Hill New York.

26. Moore, G., *The 40th anniversary of Moore's Law with Gordon Moore, co-founder and chairman emeritus, Intel, in conversation with Carver Mead, chairman and founder, Foveon*, C. Mead, Editor. 2005. *Computer History Museum Presents: The 40th Anniversary of Moore's Law with Gordon Moore and Carver Mead*. Mountain View, California.

27. Hennessy, J. and D. Patterson, A new golden age for computer architecture: Domain-specific hardware/software co-design, enhanced, in *ACM/IEEE 45th Annual International Symposium on Computer Architecture (ISCA)*, Los Angeles. 2018.

28. Hennessy, J. L. and D. A. Patterson, *Computer architecture: A quantitative approach*. 2011, Elsevier.

29. Fuller, S. H. and L. I. Millett, *The future of computing performance: Game over or next level?* 2011, National Academies Press.

30. Amdahl, G. M., Validity of the single processor approach to achieving large scale computing capabilities, in *Proceedings of the April 18–20, 1967, Spring Joint Computer Conference*, Atlantic City, NJ. 1967.

31. Dennard, R. H., F. H. Gaenssien, H.-N. Yu, et al., Design of ion-implanted MOSFET's with very small physical dimensions. *IEEE Journal of Solid-State Circuits*, 1974. 9(5): 256–268.

32. Dally, W. J., F. Labonte, A. Das, et al., Merrimac: Supercomputing with streams, in *SC'03: Proceedings of the 2003 ACM/IEEE Conference on Supercomputing*. 2003, IEEE.

33. LeCun, Y., Session 1.1, deep learning hardware: Past, present, and future. in *2019 IEEE International Solid-State Circuits Conference (ISSCC)*. 2019, IEEE.

34. Bonomi, F., R. Millito, J. Zhu, and S. Addepalli, Fog computing and its role in the Internet of Things, in *Proceedings of the first edition of the MCC Workshop on Mobile Cloud Computing*. 2012.

35. Nickolls, J. and W. J. Dally, The GPU computing era. *IEEE Micro*, 2010. 30(2): 56–69.

36. OpenAI, *AI and Compute*. November 7, 2019. https://openai.com/blog/ai-and -compute/.

37. Vaswani, A., N. Shazeer, N. Parmar, et al., Attention is all you need. *Advances in Neural Information Processing Systems*, 2017. 30.

38. Brown, T., B. Mann, N. Ryder, et al., Language models are few-shot learners. *Advances in Neural Information Processing Systems*, 2020. 33: 1877–1901.

39. Wolfram, S., What is ChatGPT doing . . . and why does it work? 2023. https:// writings.stephenwolfram.com/2023/02/what-is-chatgpt-doing-and-why-does-it -work/.

40. Zhang, H., X. Cheng, H. Zang, and D. H. Park, Compiler-level matrix multi- plication optimization for deep learning, 2019. arXiv preprint arXiv:1909.10616.

41. Fleischer, B., S. Shukla, M. Ziegler, et al., A scalable multi-TeraOPS deep learn- ing processor core for AI training and inference, in *2018 IEEE Symposium on VLSI Circuits*. 2018, IEEE.

42. Oser, P., *The IEEE-754 format*. http://mathcenter.oxford.emory.edu/site/cs170 /ieee754/.

43. Gupta, S., A. Agrawal, K. Gopalakrishnan, and P. Narayanan, Deep learning with limited numerical precision, in *International Conference on Machine Learn- ing*. 2015, PMLR.

44. Johnson, J., Rethinking floating point for deep learning, 2018. arXiv preprint arXiv:1811.01721.

45. Zhao, J., S. Dai, R. Venkatesan, et al., Low-precision training in logarithmic number system using multiplicative weight update, *IEEE Transactions on Com- puting*, 2022 71: 3179–3190. arXiv preprint arXiv:2106.13914.

46. Kurama, V., A review of popular deep learning architectures: AlexNet, VGG16, and GoogleNet. https://blog.paperspace.com/popular-deep-learning-architectures -alexnet-vgg-googlenet/.

47. Krizhevsky, A., I. Sutskever, and G. E. Hinton, ImageNet classification with deep convolutional neural networks. *Advances in Neural Information Processing Systems*, 2012. 25.

48. Simonyan, K. and A. Zisserman, Very deep convolutional networks for large- scale image recognition, 2014. arXiv preprint arXiv:1409.1556.

49. Szegedy, C., W. Liu, Y. Jia, et al., Going deeper with convolutions, in *Proceedings of the IEEE Conference on Computer Vision and Pattern Recognition*, Boston. 2015.

50. Szegedy, C., V. Vanhoucke, S. Ioffe, et al., Rethinking the inception architecture for computer vision, in *Proceedings of the IEEE Conference on Computer Vision and Pattern Recognition*, Las Vegas. 2016.

51. He, K., X. Zhang, S. Ren, and J. Sun, Deep residual learning for image recognition, in *Proceedings of the IEEE Conference on Computer Vision and Pattern Recognition*, Las Vegas. 2016.

52. Han, S., *Efficient methods and hardware for deep learning*. 2017, Stanford Course.

53. Dean, J., G. Corrado, R. Monga, et al., Large scale distributed deep networks. *Advances in Neural Information Processing Systems*, 2012. 25.

54. Sutter, H., The free lunch is over: A fundamental turn toward concurrency in software. *Dr. Dobb's Journal*, 2005. 30(3): 202–210.

55. Choquette, J., W. Gandhi, O. Giroux, et al., NVIDIA A100 tensor core GPU: Performance and innovation. *IEEE Micro*, 2021. 41(2): 29–35.

56. Silver, D., J. Schrittwieser, K. Simonyan, et al., Mastering the game of Go without human knowledge. *Nature*, 2017. 550(7676): 354–359.

57. Dean, J., 1.1 the deep learning revolution and its implications for computer architecture and chip design, in *2020 IEEE International Solid-State Circuits Conference-(ISSCC)*. 2020, IEEE.

58. Martinez, D., M. Arakawa, and R. Bond, Keynote speech, HPEC: Looking back and projecting forward, in *High Performance Embedded Computing Workshop*, Burlington, MA. 2006. https://archive.ll.mit.edu/HPEC/agendas/proc06/Day2/01_Martinez_Keynote.pdf.

59. Jouppi, N. P., C. Young, N. Patil, et al., A domain-specific architecture for deep neural networks. *Communications of the ACM*, 2018. 61(9): 50–59.

60. Patterson, D. 50 years of computer architecture: From the mainframe CPU to the domain-specific TPU and the open RISC-V instruction set, in *2018 IEEE International Solid-State Circuits Conference (ISSCC)*. 2018, IEEE. https://ieeexplore.ieee.org/xpl/conhome/8304413/proceeding.

61. Martinez, D. R., R. A. Bond, and M. M. Vai, *High performance embedded computing handbook: A systems perspective*. 2008, CRC Press.

62. Song, W. S., M. M. Vai, and H. T. Nguyen, High-performance low-power bit-level systolic array signal processor with low-threshold dynamic logic circuits, in *Conference Record of Thirty-Fifth Asilomar Conference on Signals, Systems and Computers (Cat. No. 01CH37256)*. 2001, IEEE.

63. Dally, W. J., Y. Turakhia, and S. Han, Domain-specific hardware accelerators. *Communications of the ACM*, 2020. 63(7): 48–57.

64. Horowitz, M., 1.1 computing's energy problem (and what we can do about it), in *2014 IEEE International Solid-State Circuits Conference Digest of Technical*

Papers (ISSCC). 2014, IEEE. https://ieeexplore.ieee.org/xpl/conhome/6747109/proceeding.

65. Han, S., X. Liu, H. Mao, et al., EIE: Efficient inference engine on compressed deep neural network. *ACM SIGARCH Computer Architecture News*, 2016. 44(3): 243–254.

66. Leiserson, C. E., N. C. Thompson, J. S. Emer, et al., There's plenty of room at the Top: What will drive computer performance after Moore's Law? *Science*, 2020. 368(6495): eaam9744.

67. Danowitz, A., K. Kelley, J. Mao, et al., CPU DB: Recording microprocessor history. *Communications of the ACM*, 2012. 55(4): 55–63.

68. Hennessy, J. L. and D. A. Patterson, *Computer architecture: A quantitative approach*. 6th ed. 2017, Morgan Kaufmann Series in Computer Architecture and Design. Elsevier. https://www.elsevier.com/books/computer-architecture/patterson/978-0-12-811905-1.

69. Dean, J., D. Patterson, and C. Young, A new golden age in computer architecture: Empowering the machine-learning revolution. *IEEE Micro*, 2018. 38(2): 21–29.

70. Reuther, A., P. Michaleas, M. Jones, et al., AI accelerator survey and trends, in *2021 IEEE High Performance Extreme Computing Conference (HPEC)*. 2021, IEEE. https://github.com/areuther/ai-accelerators

71. Feldman, M., *IBM finds killer app for TrueNorth neuromorphic chip*. 2016, September.

72. Merolla, P. A., J. V. Arthur, R. Alvarez-Icaza, et al., A million spiking-neuron integrated circuit with a scalable communication network and interface. *Science*, 2014. 345(6197): 668–673.

73. Yazdanbakhsh, A., B. Akin, J. Laudon, et al., An evaluation of edge TPU accelerators for convolutional neural networks, 2021. arXiv preprint arXiv:2102.10423.

74. Abdelfattah, M. S., D. Han, A. Bitar, et al., DLA: Compiler and FPGA overlay for neural network inference acceleration, in *2018 28th International Conference on Field Programmable Logic and Applications (FPL)*. 2018, IEEE.

75. Moeller, T. J. and D. R. Martinez, Field programmable gate array based radar front-end digital signal processing, in *Seventh Annual IEEE Symposium on Field-Programmable Custom Computing Machines (Cat. No. PR00375)*. 1999, IEEE.

76. Narayanan, D., M. Shoeybi, J. Casper, et al., Efficient large-scale language model training on GPU clusters using megatron-LM, in *Proceedings of the International Conference for High Performance Computing, Networking, Storage and Analysis*, St. Louis. 2021.

77. Sherry, Y. and N. Thompson, *How fast do algorithms improve?* 2021, IEEE. https://doi.org/10.1109/JPROC.2021.3107219

78. Krizhevsky, A., One weird trick for parallelizing convolutional neural networks, 2014. arXiv preprint arXiv:1404.5997.

79. Ben-Nun, T. and T. Hoefler, Demystifying parallel and distributed deep learning: An in-depth concurrency analysis. *ACM Computing Surveys (CSUR)*, 2019. 52(4): 1–43.

80. Jin, P. H., Q. Yuan, F. Iandola, and K. Keutzer, How to scale distributed deep learning? 2016 arXiv preprint arXiv:1611.04581.

81. Canziani, A., E. Culurciello, and A. Paszke. Evaluation of neural network architectures for embedded systems, in *2017 IEEE International Symposium on Circuits and Systems (ISCAS)*. 2017, IEEE.

82. Sze, V., Y.-H. Chen, J. Emer, et al., Hardware for machine learning: Challenges and opportunities, in *2017 IEEE Custom Integrated Circuits Conference (CICC)*. 2017, IEEE.

83. Chen, Y.-H., T. Krishna, J. S. Emer, and V. Sze, Eyeriss: An energy-efficient reconfigurable accelerator for deep convolutional neural networks. *IEEE Journal of Solid-State Circuits*, 2016. 52(1): 127–138.

84. Jouppi, N., C. Young, N. Patil, et al., Motivation for and evaluation of the first tensor processing unit. *IEEE Micro*, 2018. 38(3): 10–19.

85. Williams, S., A. Waterman, and D. Patterson, Roofline: An insightful visual performance model for multicore architectures. *Communications of the ACM*, 2009. 52(4): 65–76.

86. Patterson, D. A., Latency lags bandwith. *Communications of the ACM*, 2004. 47(10): 71–75.

87. Norrie, T., N. Patil, D.-H. Yoon, et al., The design process for Google's training chips: TPUv2 and TPUv3. *IEEE Micro*, 2021. 41(2): 56–63.

88. Tekin, A., A. Durak, C. Piechurski, et al., *State-of-the-art and trends for computing and interconnect network solutions for HPC and AI*. Partnership for Advanced Computing in Europe, Available online at www.praceri.eu, 2021.

89. National Academies of Sciences, E. and Medicine, *Implications of artificial intelligence for cybersecurity: Proceedings of a workshop*. 2020, National Academies Press. https://www.nationalacademies.org/our-work/implications-of-artificial-intelligence-for-cybersecurity-a-workshop.

90. Koziel, E., K. E. Thurmer, L. Milechin, et al., Side channel authenticity discriminant analysis for device class identification, in *Government Microciruit Applications & Critical Technology Conference*, Orlando. 2016.

91. Zhu, H., H. Shan, D. Sullivan, et al., *PDNPulse: Sensing PCB anomaly with the intrinsic power delivery network*, 2022. arXiv preprint arXiv:2204.02482.

92. McDaniel, P. and J. Launchbury, *AI-cybersecurity workshop briefing to the NITRD and MLAI subcommittees*. July 31, 2019. https://www.nitrd.gov/documents/AI -Cybersecurity-Workshop-Briefing-Patrick-McDaniel-2019-07-31.pdf.

93. Kiriansky, V., I. Lebedev, S. Amarasinghe, et al., DAWG: A defense against cache timing attacks in speculative execution processors, in *2018 51st Annual IEEE/ ACM International Symposium on Microarchitecture (MICRO)*. 2018, IEEE.

94. Lipp, M., M. Schwartz, D. Gruss, et al., Meltdown, 2018. arXiv preprint arXiv:1801.01207.

95. Hoeper, P. and J. Manferdelli, *DSB task force on cyber supply chain*. 2017, Defense Science Board, Washington, DC.

96. Enck, W. and L. Williams, Top five challenges in software supply chain security: Observations from 30 industry and government organizations. *IEEE Security & Privacy*, 2022. 20(2): 96–100.

97. Okhravi, H., A cybersecurity moonshot. *IEEE Security & Privacy*, 2021. 19(3): 8–16.

98. Horvitz, E. Applications for artificial intelligence in Department of Defense cyber missions. 2022, Microsoft Corporate Blogs, https://blogs.microsoft.com /on-the-issues/2022/05/03/artificial-intelligence-department-of-defense-cyber -missions/#_edn37.

99. Monserrate, S. G., *The cloud is material: On the environmental impacts of computation and data storage*. 2022, MIT Schwarzman College of Computing. https:// mit-serc.pubpub.org/pub/the-cloud-is-material/release/1.

100. Dhar, P., The carbon impact of artificial intelligence. *Nature Machine Intelligence*, 2020. 2(8): 423–425.

101. Patterson, D., J. Gonzalez, Q. Le, et al., Carbon emissions and large neural network training, 2021. arXiv preprint arXiv:2104.10350.

102. Lacoste, A., A. Luccioni, V. Schmidt, and T. Dandres, Quantifying the carbon emissions of machine learning. arXiv preprint arXiv:1910.09700, 2019. https:// mlco2.github.io/impact/.

103. Lannelongue, L., J. Grealey, and M. Inouye, Green algorithms: Quantifying the carbon footprint of computation. *Advanced Science*, 2021. 8(12): 2100707. http://www.green-algorithms.org/.

104. Schwartz, R., J. Dodge, N. A. Smith, and O. Etzioni, Green AI. *Communications of the ACM*, 2020. 63(12): 54–63.

105. Evans, R. and J. Gao, DeepMind AI reduces Google data centre cooling bill by 40%. *DeepMind Blog*, 2016. 20: 158. https://www.deepmind.com/blog/deepmind-ai-reduces-google-data-centre-cooling-bill-by-40.

106. Kissinger, H. A., E. Schmidt, and D. Huttenlocher, *The age of AI: And our human future.* 2021, Hachette UK.

107. Bouguettaya, A., Q. Z. Sheng, B. Benatallah, et al., An Internet of Things service roadmap. *Communications of the ACM*, 2021. 64(9): 86–95.

108. Han, S., H. Mao, and W. J. Dally, Deep compression: Compressing deep neural networks with pruning, trained quantization and Huffman coding, 2015. arXiv preprint arXiv:1510.00149.

109. Sze, V., Y.-H. Chen, T.-J. Yang, and J. Emer, Efficient processing of deep neural networks. *Synthesis Lectures on Computer Architecture*, 2020. 15(2): 1–341.

110. Shi, W., J. Cao, Q. Zhang, et al., Edge computing: Vision and challenges. *IEEE Internet of Things Journal*, 2016. 3(5): 637–646.

111. Schleier-Smith, J., V. Sreekanti, et al., What serverless computing is and should become: The next phase of cloud computing. *Communications of the ACM*, 2021. 64(5): 76–84.

112. Mehta, Y., C. White, A. Zela, et al., NAS-Bench-suite: NAS evaluation is (now) surprisingly easy, 2022. arXiv preprint arXiv:2201.13396. https://github.com/automl/naslib.

113. Reddi, V. J., C. Cheng, D. Kanter, et al., The vision behind MLPerf: Understanding AI inference performance. *IEEE Micro*, 2021. 41(3): 10–18.

114. Zhang, D., N. Maslej, E. Brynjolfsson, et al., *The AI Index 2022 annual report.* 2022, Stanford Institute for Human-Centered AI.

115. Huynh-The, T., Q.-V. Pham, X.-Q. Pham, et al., Artificial intelligence for the metaverse: A survey, 2022. arXiv preprint arXiv:2202.10336.

116. Davies, M., Benchmarks for progress in neuromorphic computing. *Nature Machine Intelligence*, 2019. 1(9): 386–388.

117. Davies, M., N. Srinivasa, T.-H. Lin, et al., Loihi: A neuromorphic manycore processor with on-chip learning. *IEEE Micro*, 2018. 38(1): 82–99.

118. Naehrig, M., K. Lauter, and V. Vaikuntanathan. Can homomorphic encryption be practical? in *Proceedings of the 3rd ACM Workshop on Cloud Computing Security Workshop,* Chicago. 2011.

119. Li, T., A. K. Sahu, A. Talwalkar, and V. Smith, Federated learning: Challenges, methods, and future directions. *IEEE Signal Processing Magazine*, 2020. 37(3): 50–60.

120. LaMacchia, B., The long road ahead to transition to postquantum cryptography. *Communications of the ACM*, 2021. 65(1): 28–30.

121. Harrow, A. W., A. Hassidim, and S. Lloyd, Quantum algorithm for linear systems of equations. *Physical Review Letters*, 2009. 103(15): 150502.

122. Menard, A., I. Ostojic, M. Patel, and D. Volz, *A game plan for quantum computing.* 2020. https://www.mckinsey.com/business-functions/mckinsey-digital/our-insights/a-game-plan-for-quantum-computing.

123. Welser, J., J. Pitera, and C. Goldberg. Future computing hardware for AI, in *2018 IEEE International Electron Devices Meeting (IEDM)*. 2018, IEEE.

124. Noda, K., H. Aire, Y. Suga, and T. Ogata, Multimodal integration learning of robot behavior using deep neural networks. *Robotics and Autonomous Systems*, 2014. 62(6): 721–736.

125. Zhou, T., S. Ruan, and S. Canu, A review: Deep learning for medical image segmentation using multi-modality fusion. *Array*, 2019. 3: 100004.

126. Chen, M., J. Tworek, H. Jun, et al., Evaluating large language models trained on code, 2021. arXiv preprint arXiv:2107.03374.

127. GitHub, Your AI pair programmer. 2021. https://copilot.github.com/.

128. Li, Y., D. Chou, J. Chung, et al., Competition-level code generation with AlphaCode, 2022. arXiv preprint arXiv:2203.07814.

129. MITRE. CWE top 25 most dangerous software weaknesses. 2020. https://cwe.mitre.org/top25/archive/2020/2020_cwe_top25.html.

130. Pearce, H., B. Ahmad, B. Tan, et al., An empirical cybersecurity evaluation of GitHub Copilot's code contributions. arXiv e-prints, 2021: p. arXiv: 2108.09293.

131. Schneider, D., The Exascale era is upon us: The Frontier supercomputer may be the first to reach 1,000,000,000,000,000,000 operations per second. *IEEE Spectrum*, 2022. 59(1): 34–35.

132. Leinhauser, M., R. Widera, S. Bastrakov, et al., Metrics and design of an instruction roofline model for AMD GPUs. *ACM Transactions on Parallel Computing*, 2022. 9(1): 1–14.

133. Sevilla, J., L. Heim, A. Ho, et al., *Compute trends across three eras of machine learning*, 2022. arXiv preprint arXiv:2202.05924.

134. Potok, T. E., C. Schuman, S. R. Young, et al., A study of complex deep learning networks on high-performance, neuromorphic, and quantum computers. *ACM Journal on Emerging Technologies in Computing Systems (JETC)*, 2018. 14(2): 1–21.

6

Human-Machine Teaming

Fundamental Rules of Robotics: 1. A robot may not injure a human being, or, through inaction, allow a human being to come to harm; 2. A robot must obey the orders given to it by human beings except where such orders would conflict with the First Law; 3. A robot must protect its own existence, as long as such protection does not conflict with the First or Second Law.
—Isaac Asimov (science fiction writer, 1941)

Ultimately the value of artificial intelligence (AI) is in providing increased capabilities to humans by augmenting human intelligence, making our day-to-day activities easier to accomplish, and relegating dull-dirty-dangerous tasks to intelligent machines. There are challenges but also opportunities for advancing human-machine teaming (HMT), which is the focus of this chapter.

There are many definitions of HMT, so it is important for us to clarify what we mean by the term. As shown in figure 6.1, a simplified illustration of the more detailed AI system architecture shown in figure 1.3 in chapter 1, HMT is the building block in the AI pipeline that transforms machine learning (ML) outputs (i.e., knowledge) into actionable insights delivered to stakeholders (e.g., users, consumers, or customers).

HMT, as one of the building blocks in the AI system architecture, spans a spectrum. It ranges from most or all insights being created by humans, after receiving output from the ML stage; to full autonomy, where the output of the ML stage is passed to an autonomous engine to perform a task. An example of the former is in cancer diagnosis, where the ML step (e.g., using computer vision) identifies potential cancerous cells, and the radiologist determines if that is indeed the case, creating insight to the attending physician. An example of the latter is in robotics, where the output of a reinforcement

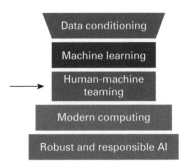

Figure 6.1 Simplified AI system architecture.

ML algorithm provides an input to a robot directing it to undertake a task (e.g., taking the role of a server by bringing the dishes with the meals to a table in a restaurant).

The three laws in the quote from Asimov were driven by the moral employment of robots, with a hope that robots will not overtake humans and potentially cause harm to them [1, 2]. These famous, fundamental Rules of Robotics imply that someday, robots could be existential to humanity. We want to emphasize that in our judgment, we are nowhere close to having robots with the ability to sense, plan, and act at the same levels as humans can. However, in the context of HMT, as discussed in this chapter, there are many important and productive roles that robots are performing and will continue to perform, providing value to humans.

In the following sections, we set the stage for the rest of the chapter by starting a discussion on augmenting human capabilities. We then elaborate on three levels of HMT:

1. AI as a search-and-discovery tool; an example is time-frequency/inverse document frequency (TF/IDF) or Latent Dirichlet Allocation algorithms for text analysis and retrieval
2. AI as a teammate (examples include semiautonomous robotics)
3. Full autonomy (examples include driverless vehicles)

These three levels represent a spectrum of HMT capabilities, relative to courses of action, as shown in figure 6.2, and operating under a trade-off between confidence in the machine making the decision versus the consequence of actions.

Another important topic discussed in this chapter is the use of quantitative and qualitative performance metrics. In figure 2.1 in chapter 2, we elaborated on the importance of performing integration-verification-validation at each of the building blocks of the end-to-end AI system architecture. There are many performance metrics for assessing ML

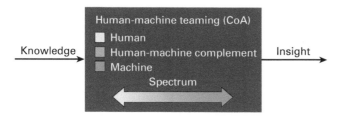

Figure 6.2   HMT spectrum.

algorithms. However, we need rigor not only in using performance metrics and benchmarks at individual stages of the AI system architecture, but also in assessing the AI system performance across all stages, from data conditioning through the HMT functional block.

We will also briefly discuss several recent analyses looking at the work of the future from the perspective of advances in AI. We conclude the chapter by addressing key HMT challenges, the main takeaways, and a set of exercises.

6.1   Augmenting Human Capabilities

The dramatic exponential growth in computing, as shown in figure 2.3, has led to a rapid acceleration in intelligent machines with a phenomenal ability to augment human capabilities. Today, it is common to see, especially in developed countries, what have become household names, such as an Amazon's Alexa or a Google Home pod. Although these devices have many limitations—for example, with automatic speech recognition for people with strong accents—they serve useful roles as users look for answers to general questions.

It is useful to assess HMT through the lens of confidence in the machine making the decision versus the courses of action, as shown in figure 6.3. If the confidence is high—for example, the ML output predicts an outcome or provides a recommendation with high accuracy—and the consequence of taking an action based on that prediction or recommendation is low (i.e., the cost of error writ large), then the user can opt to let the AI system act without human intervention. One example is when Alexa is asked about today's weather forecast. If Alexa returns the daily temperature but is off by a few degrees, it might be an inconvenience to the user, but it does not fall into the category of a high-stakes application.

On the other hand, as Marcus and Davis point out in their book *Rebooting AI* [3], there are many instances when Alexa and other recommender systems have made

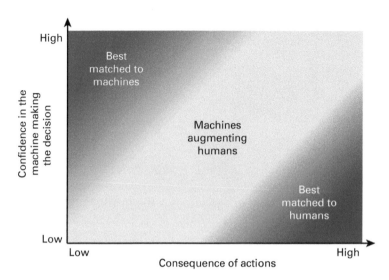

Figure 6.3   Confidence in the machine making the decision versus consequence of actions.

significant mistakes (see chapter 8 for additional examples). They point out the existence of a chasm between today's AI capabilities and full autonomy. There are many useful functions performed with AI today that are having and will continue to have major benefits to businesses and society in general. However, we need to be very vigilant about today's AI. Thus, we emphasize the need for an AI systems engineering approach to keep tight rails on narrow AI, but without unduly stifling innovation.

We also warn AI users and practitioners not to expect AI to replace a human brain in capabilities per unit of power for the foreseeable future. For instance, there have been comparisons made in terms of the number of neurons and synapses in a fly (with approximately 10^5 neurons and 10^7 synapses) or a mouse (with approximately 10^8 neurons and 10^{11} synapses) and a human (with approximately 10^{11} neurons and 10^{15} synapses). However, a human, on average, uses 100 watts of power on typical tasks, and 20 percent of that power is used by the brain. Even with exponential growth in computation (and growth in power consumption, as discussed in chapter 5) in the present or in the foreseeable future, it will be a long time before today's AI will evolve to meet the intelligent insight that a human can create per unit watt.

With all the present limitations of narrow AI, when seen through the trade-off lens shown in figure 6.3, several AI applications were identified in a comprehensive study undertaken by the MIT Lincoln Laboratory [4]. Figure 6.4 illustrates examples in two

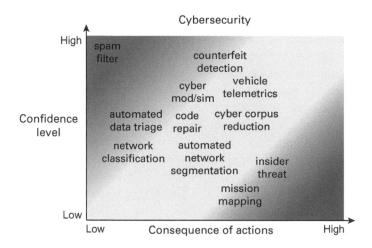

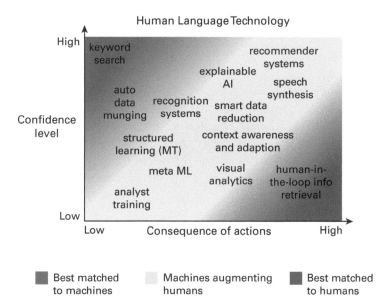

Figure 6.4   Examples of AI applications in cybersecurity and human-language technology.

domains, cybersecurity and human-language technology. Today, it is routine to have a spam filter (i.e., a recommender system) identify incoming spam emails. Another routine example, with little to no human intervention, is in keyword search (e.g., using TF/IDF to perform word searches with the objective of creating a word cloud). However, in personnel-critical applications, humans must make the final decision because of the high consequence of actions (i.e., the cost of making an error). Some examples include assessing potential insider threats (in the domain of cybersecurity) and human-in-the-loop information retrieval (in the human-language technology domain) with a low confidence level due to the proliferation of deepfakes.

These examples fall in the category of either AI as a search-and-discovery tool or AI as a teammate (discussed further in sections 6.2 and 6.3, respectively). Representative examples of AI as a search-and-discovery tool are virtual personal assistants [5]. Representative examples of AI as a teammate are decision support systems [6]. Similarly, there have been impressive robotics applications in manufacturing [7, 8], warehouse automation [9, 10], precision agriculture [11, 12], and robots disinfecting areas potentially contaminated with the COVID-19 virus [13, 14], to name a few. We address these latter classes of AI systems in section 6.4.

In the creation of insight through the HMT building block—from the knowledge created at the output of the ML stage—the role of the human versus the machine occupies a spectrum, as shown in figure 6.2. We use this spectrum topology in the following sections to parse the courses of action insight created between the human and/or the machine. In section 6.2, the HMT courses of action are principally decided by the human. In section 6.3, the discussion centers on the human and the machine operating closer to the middle of the spectrum. The human and machine complement each other in the process of reaching courses of action. In section 6.4, the machine is given autonomy to take a course of action but guided by the human if and when intervention is needed.

A study undertaken by Ransbotham et al. elaborates on the importance of integrating humans and machines within their organization's AI infrastructure, such that humans learn from AI and vice versa [15]. One of their key findings, from their research and global survey of more than 3,000 managers, is that only when organizations take this approach are they able to accrue significant financial business gains. They present five distinct modes of human-machine interaction that are complementary to the recommended spectrum shown in figure 6.2. Similarly, Davenport and Kirby formulated a very useful framework for assessing progress on this cognitive technologies spectrum, starting with support for humans, repetitive task automation, context awareness and learning, and progressing to machines with self-awareness, highlighting representative applications requiring higher degrees of complexity [16].

The field of HMT, sometimes also referred to as "human-machine interaction," is very broad. So, in this chapter, we take a high-level perspective to give the reader a working knowledge of how HMT fits within the construct of an end-to-end AI system architecture. We also present use cases in chapters 14–18, representing examples of the classes of HMT systems discussed in this chapter.

6.2   AI as a Search-and-Discovery Tool

The large increase in available data from both sensors and sources will continue to demand the use of AI tools to rapidly search and discover insights from the output of ML algorithms. As reported in the *Digital 2022—Global Overview Report* [17], as of January 2022, the worldwide population exceeded 7.9 billion. Of that population, 67.1 percent were unique mobile phone users (about 5.3 billion); 62.5 percent of the population were internet users (about 4.9 billion); and 58.4 percent were social media users (about 4.6 billion). These percentages are projected to increase with the advent of 5G/6G connectivity and the continuing proliferation of smart phones. Also, as reported by Juniper Research, industrial Internet of Things (IoT) connections will reach 37 billion globally by 2025 [18]. The World Economic Forum predicts that by the year 2025, 463 exabytes (1 exabyte $= 10^{18}$ bytes) will be created everyday [19].

As worldwide data become prolific, all of us, as consumers, will depend on our virtual personal assistants (VPAs), either in our homes, offices, smart phones, or in vehicles. Although today's virtual personal assistants (e.g., Alexa, Google Home, Microsoft Cortana, Apple Siri, etc.) depend on automatic speech recognition (ASR), there will likely be advancements that include other modalities besides speech as part of human-machine interaction, such as gestures, videos, images, and touch [20]. However, as we continue to employ these VPAs for searching and discovering insights, there will be more emphasis on further research regarding levels of trust in the output [21].

As discussed in section 6.1, levels of trust must be assessed through the lens of a trade-off between the confidence in the machine making the decision (in this case, a VPA) versus the consequence of actions, as illustrated in figure 6.3. Today, VPAs are somewhat reliable when, for example, searching for a weather report, general statistics, or a famous person's biography, all of which have a low consequence of action if an error is made by the VPA.

In today's information technology (IT) departments, many organizations are using online chatbots for automatic help. Initially, customers and clients don't speak to humans but instead interact with a machine. These algorithms use prior aggregated knowledge and natural-language processing (NLP), with the chatbots learning from records of past conversations to come up with appropriate responses [22]. There are different levels depending on how critical the issue is and difficulty in providing a recommendation. If the chatbot is unable to resolve the issue, the topic is escalated to a human.

On the other hand, searching and discovering insightful content, in high-stakes and critical applications, will require a very careful assessment of the accuracy or validity of the output from the ML stage. These applications will require humans to intervene in the final courses of action. If the recommendation time cycle permits a relatively long time constant (meaning if the HMT stage operates in hours or on a longer time scale), then the human analysts can make their own judgments on a valid course of action (or recommendation).

We want to illustrate one example of this from the cybersecurity domain. Often, cyberanalysts are inundated with documents (cyberreports), which contain valuable cybercontent in a sea of other documents that are of limited value (i.e., void of cyber-information). Figure 6.5 shows a prototype AI system developed at the MIT Lincoln Laboratory to reduce the time that normally takes for forensic cyberanalysts to search and discover malicious discussions in social media by leveraging natural-language ML techniques [23].

The input data on the left of the chart represented surrogate data from sources of social media, such as Twitter, Reddit, Stack Exchange, and others, consisting of tens of thousands of documents. After performing a simple feature generation as the first step, using a TF/IDF algorithm, through word stemming (e.g., "hacker," "hacking," "hacks"), the output resulted in documents containing approximately 1 percent cyberrelated content. This step was followed by a logistic regression classifier to separate documents with relevant cybercontent from those documents that were void of any cybercontent of interest. This two-step process (i.e., ML stages) resulted in more than 80 percent of documents containing cybermaterial of interest provided to the forensic cyberanalyst, with the ability to discover 90 percent of the cyberdiscussions with less than 1 percent false alarms. Thus, the prototype AI system reduced the workload of the cyberanalyst, who typically performs these functions manually. Although this protype was a proof-of-concept, a real-life cybersecurity application requires the human to make the final determination on courses of action due to the nature of this high-stakes application.

AI as a search-and-discovery tool will continue to be an important technique to augment human capabilities. The volume-velocity-variety of data are continuing to grow at an exponential rate, requiring the use of AI to make the search-and-discovery task manageable. However, the use of these techniques must be integrated into the

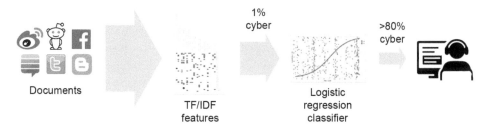

Figure 6.5 Searching and discovering malicious cyberdiscussions in social media.

human workflow, and there must be rigorous integration, verification, and validation of the end-to-end process, including feedback, as discussed in chapter 2.

AI aiding humans in search-and-discovery tasks will also have an impact at the enterprise level. For example, with advances in computer vision and cameras, managers will be able to predict potential problems in operations by monitoring the factory floor in real time [24].

Another application of search and discovery, at the enterprise level, is using software agents to parse a large volume of documents—for example, reducing the workload of humans performing structured and repetitive tasks as part of enterprise resource planning. This type of enterprise application is commonly referred to as "robotic process automation"—it is not, strictly speaking, a robotic function, but the community, advancing enterprise processing using AI, has adopted robotic process automation to refer to functions performed by software agents automating routine functions, thus enabling humans to dedicate more time to cognitive functions [25, 26].

In section 6.3, we address AI as a teammate. Remember that HMT fits on a spectrum. Although AI as a teammate falls between a human and a machine on the spectrum shown in figure 6.2, there are neither fixed nor hard boundaries. This loose separation is primarily driven by the level of complexity with machine intelligence, plus the dependency on human-machine interplay and the level of confidence versus consequence of actions at the output of the HMT stage. As the members of the human-machine team work closer together and complement each other, there is an implied higher degree of trust in the results of the ML stage.

6.3   AI as a Teammate

Undoubtedly, AI products and services will continue to advance to a point where humans will be able to accept and integrate machine intelligence into their day-to-day business operations. AI practitioners and researchers can assess where applications best fall on the HMT spectrum by understanding the level of judgment necessary to pursue courses of action, as described by Agrawal et al. [27, 28]. However, as discussed earlier and illustrated in figures 6.3 and 6.4, the human and the machine working as teammates must be carefully evaluated in terms of confidence level versus consequence of action. Dhar described a similar two-dimensional evaluation topology and gave several examples in terms of the cost of error versus a predictability axis [29].

Daugherty and Wilson offer a very clear delineation of activities well matched to humans, activities best matched to machines, and activities where a human and a machine can complement each other [30]. Although, as discussed in section 6.2, these roles or activities do not fall exclusively in one category or another, their structure can be interpreted as activities where, after the ML stage where the output represents knowledge, the human can determine the best courses of action. As Daugherty points out: "If you're going to look at where value is created, look at how the people in your organization can drive more value in collaboration with machines" [31].

The framework used in this section applies to any of the following scenarios:

- A single human working with a machine or multiple machines as teammates
- Multiple humans integrating a single machine as a member of their team
- Multiple humans leveraging multiple intelligent machines as teammates

As an example of multiple humans (cyberoperators) working with a single intelligent machine, we present an AI prototype concept demonstrated at the MIT Lincoln Laboratory and shown in figure 6.6. ML techniques are able to amplify the capabilities of cyberoperators. A common approach is to patch identified cybervulnerabilities, starting with the highest, and working down to the lowest vulnerability according to the Common Vulnerability Scoring Systems (CVSS) [32, 33]. Unfortunately, the cyberoperators can spend an immense amount of time trying to patch everything. Using NLP with a logistic regression classifier, the machine can recommend to the cyberoperators the vulnerabilities with the highest probability to be exploited, informed from a public database containing a set of exploitable vulnerabilities. Typically, only a small percentage of vulnerabilities are actually exploitable.

Figure 6.6 illustrates the input data, the ML techniques (i.e., NLP followed by a logistic regression classifier), and performance results. The performance results show the probability of exploited vulnerabilities detected (i.e., identified) via the ML techniques versus the probability of false alarm. There is a substantial improvement with an area under the curve (AUC) of 0.86 with the ML techniques, compared to the CVSS approach, with an AUC = 0.64 (the ideal maximum is AUC = 1.0) [4].

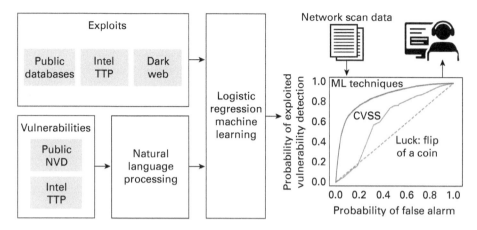

Figure 6.6   ML used to prioritize vulnerabilities, reducing cyberoperators' workload.

AI adoption will continue to vary across industry sectors, depending on company size, risks, and ability to generate revenue. In 2017, the McKinsey Global Institute performed a very comprehensive survey of AI adoption and use [34]. It surveyed AI-aware C-level executives at 3,073 companies. They covered fourteen industrial sectors, across ten countries in Europe, North America, and Asia, with a workforce ranging from 10 employees to over 10,000. One of the sectors, midway in terms of adoption and the future AI demand trajectory, was in healthcare. From their survey, the leading industry sectors were in the financial services, high-tech, and telecommunications. These findings were not surprising since the leading sectors, in general, do not have the same level of concern relative to the confidence level versus consequence of actions. Plus, these leading sectors can manage (for simple narrow AI applications) some of the issues relating to fairness, accountability, safety, transparency, ethics, privacy, and security (FASTEPS; discussed in more detail in chapter 8) easier than sectors such as healthcare to achieve faster adoption, acceptance, and regular use.

6.4   Autonomy and Near-Term Barriers

The progress toward more autonomy is a very exciting and active area of research and development (R&D). The machines are expected to operate farther to the right of the HMT spectrum shown in figure 6.2, converting knowledge from the ML stage into insight. One example mentioned earlier is in the field of precision agriculture [11], where, using sensor data and computer vision, an autonomous tractor can deliver nutrients to fertilize crops and pesticides to control infested plants. Similarly, Anastasia Volkova, the chief executive officer (CEO) of Regrow, and her team have developed a platform to analyze crop performance, soil health, and carbon outcomes using measurements from drones, satellites, and ground sensors applied to advanced agriculture practices [35, 36].

There are many roles that autonomous robots can perform quite effectively today. For example, Misty II is a robot developed by Misty Robotics (owned by the social robotics company Furhat Robotics). Tim Enwall, the former head of Misty Robotics, expressed his vision for this type of social robot as follows: "These robots will be seen and treated as our friends, our teammates, and a part of our families—performing helpful tasks, providing safety, and interacting with humans in entertaining and friendly ways that have only been seen before in science fiction" [37]. In chapter 14, we describe a concept developed by MIT graduate students using Misty II as a companion, while at the same time providing support to the elderly with an onset of Alzheimer's disease.

Pieter Abbeel, a professor at University of California at Berkeley and the recipient of the IEEE Kiyo Tomiyasu Award (the award is for outstanding early-career to midcareer contributions to technologies holding the promise of innovative applications), emphasizes that the next challenges for autonomous robots lie in performing tasks that people do with their hands: "The way automation is going to expand is going to be robots that are capable of seeing what's around them, adapting to what's around them, and learning things on the fly" [38].

Another area that continues to have significant growth is autonomous vehicles. The DARPA Urban Grand Challenge, held in November 2007, was the first time that autonomous vehicles interacted with both manned and unmanned vehicle traffic in an urban (and congested) environment. Of the eleven teams that entered the competition, six were able to finish the 85-kilometer course. Three of these teams finished the course without human intervention. Boss, from the Carnegie Mellon University (Tartan Racing Team), was the winner, a 2007 Chevrolet Tahoe modified for autonomous driving [39, 40]. Figure 6.7 illustrates its key system architecture, including sensing, AI modules for perception complemented by the world model of the urban environment, and a set of planning modules informing the actuators (i.e., steering, braking, and speed) [41].

From observations to actions, figure 6.8 shows a representative command-and-control architecture, including deep learning for the perception tasks and reinforcement learning for determining courses of action (based on goals and policies) [42]. This architecture is very consistent with the Boss architecture shown in figure 6.7, undertaking sense-plan-act tasks. However, the Boss architecture is fully autonomous. The architecture shown in figure 6.8 also includes an interface for communicating back to the human, who might be located in or connected to a cloud computing center.

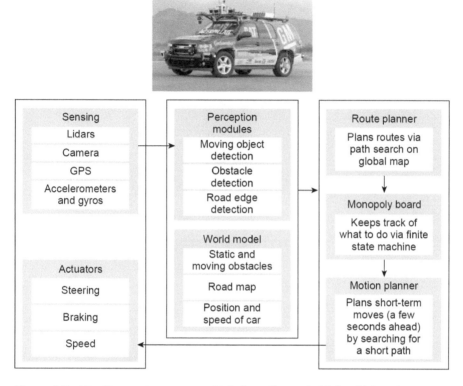

Figure 6.7 The Boss autonomous vehicle from Carnegie Mellon University, winner of the DARPA Urban Grand Challenge [41].

The DARPA Urban Challenge was a remarkable demonstration of autonomous capabilities achieved over a very short period of time from the initial event of March 2004, when no vehicle finished the difficult desert course. However, achieving full vehicle autonomy is still very hard today. As Karaman points out, full vehicle autonomy is difficult to accomplish in scenarios that are complex and where high speed is involved, as shown in figure 6.9 [43]. However, in scenarios where low speeds are acceptable and the scenario is simple, as with farming tractors, full autonomy is available.

Another challenge is the ability to navigate autonomously over difficult terrain (e.g., suburban towns). Researchers at Berkeley have devised an approach to employ auxiliary sensors and sources of data—like an inaccurate global positioning system, complemented with satellite images, contour maps, and prior experience learned from other environments—to build a learning-based method that it is more indicative of

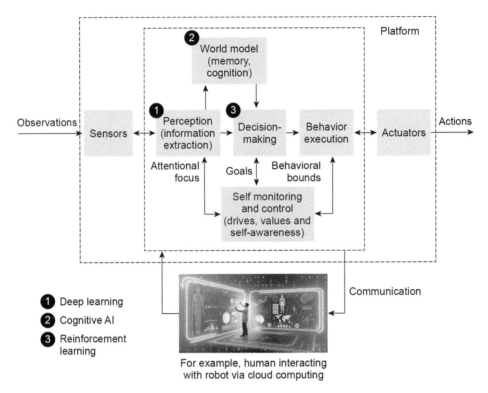

Figure 6.8   Representative example of a command-and-control architecture for a robotics system [42].

traversable patterns [44]. This innovative approach is consistent with building an informed world model, as shown in figure 6.8.

In typical DARPA fashion, based on its charter, their projects serve as initial pathfinders for others to follow. Several of the US tech giants are spending billions of R&D dollars on so-called frontier technologies, a category that autonomous vehicles fall into [45]. For example, Amazon invested in Zoox, Aurora, and Rivian; Apple acquired Drive.AI, a self-driving car start-up; Microsoft invested in a self-driving car firm named Wayve; and Google owns Waymo and Nuro. Similarly, Tesla is also allocating significant investment into making the electric vehicles more autonomous. All these efforts are very commendable and worthwhile because each year, a large number of people are killed in car accidents. According to the US Department of Transportation, in 2016 there were 37,461 people killed in car crashes, and 1.25 million worldwide (as reported by the World Health Organization). In addition to providing relief to human drivers (in terms of fatigue, boredom, loss of

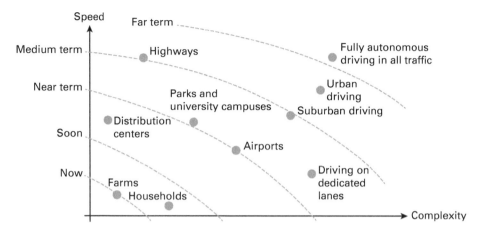

Figure 6.9 Autonomous vehicle capabilities relative to speed versus scenario complexity [43].

productivity, exposure to danger, and other negative elements), even a small reduction in car crashes caused by using driverless vehicles would save a large number of human lives.

6.5 Quantitative and Qualitative Performance Metrics

The HMT stage is one of the most critical building blocks of the end-to-end AI system architecture shown in figure 1.3 in chapter 1. We can have suboptimal ML algorithms and still achieve acceptable insights useful to the ultimate stakeholders, especially if humans are responsible for recommending the final courses of action. Conversely, we can have the best-curated data and the best-performing ML algorithms, but if the HMT function does not lead to useful (and practical) insights that are pertinent to the ultimate user, then the overall AI capability is useless.

However, much effort has been devoted to achieving the best ML performance without looking at the AI system holistically (i.e., from an end-to-end systems engineering viewpoint). In this section, we take a hierarchical approach to assessing HMT as an integral element of an end-to-end AI system. The hierarchical performance assessment is divided into four tiers. At each tier, the AI developers must ask the following questions:

1. *Tier 1—Confidence:* What is the confidence level in the knowledge created at the output of the ML stage?

2. *Tier 2—Consequence:* What is the consequence of actions relative to the confidence in the machine making the decision?

3. *Tier 3—Model representation:* Do the human-machine models accurately represent the real-world AI application for which tasks are undertaken as human-machine teammates?

4. *Tier 4—Responsible AI:* Are the insights created at the output of the HMT stage conforming with the FASTEPS principles?

This multitier approach forms a critical step as part of the integration-verification-validation in the overall systems engineering implementation framework discussed in chapter 2 and repeated in figure 6.10 (adapted from the Amazon's white paper on a well-architected framework [46]). HMT must be addressed as one important implementation building block of the overall AI development and deployment cycle. The quantitative and qualitative performance metrics also inform other critical system attributes, such as risk management, quality assurance, cost optimization, and reliability and security, all of which lead to achieving operational excellence.

It is for these reasons that this book's emphasis is on a systems engineering approach when developing and deploying AI products and services. In the authors' personal experience building complex engineering systems, there is always a trade-off among these system attributes and shown in the outer perimeter of figure 6.10. AI practitioners and researchers must keep this perspective in mind when the ultimate goal is to successfully deploy AI capabilities in operations.

Performance assessment at levels 1 and 2 (i.e., confidence versus consequence) are necessary to determine the proper operating point in the HMT spectrum shown in figure 6.2. As discussed earlier, if the confidence level is low and the consequence of action is high, the decision is best left to the human. In contrast, if the confidence is high and the consequence is low, the decision can be fully exercised by the machine (autonomously). Many of today's applications fall between these two ends.

Common quantitative performance metrics for assessing confidence levels (i.e., at the level 1 tier), used in practice, are measures of performance (MoPs) discussed in chapter 4, such as accuracy, precision-recall (and F1-score), confusion matrix, receiver operating characteristics (ROCs), and AUC. These metrics can then be aggregated into a measure of effectiveness (MoE) to make a determination of where the AI system should operate with respect to the HMT spectrum. In chapter 10, we elaborate in more detail on system performance tests, benchmarks, and metrics.

These performance metrics are implemented iteratively, as shown in figure 6.10. In this process, AI designers and developers must also take into account counterfactuals. Counterfactuals, in AI systems, are what-if assessments. If changes in the assumed scenarios or data inputs occur, how would the output of the system respond [47]? In

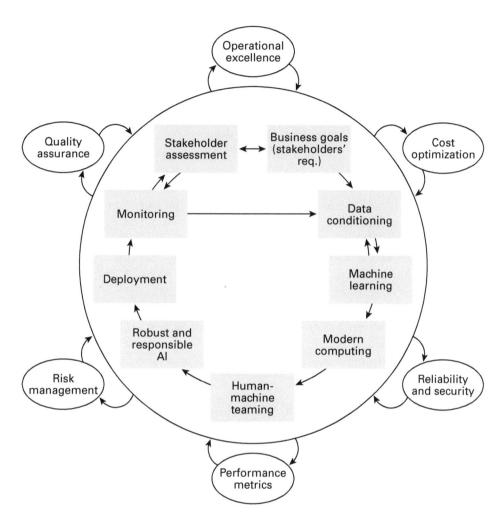

Figure 6.10 The AI system architecture implementation framework.

assessing AI decisions, another approach that is beginning to be applied is in the use of Shapley values to help in explaining the impact of input features on the output of an AI system. Shapley values had their origins in the game theory context, in terms of the players' influence on a game's outcome [48]. In the context of ML, the game is the output of the ML model relative to the features (i.e., the players) and their respective influence on the output [49–51]. These approaches serve to determine output sensitivity to data inputs and ML parameters, agnostic to ML models.

There are other more subjective metrics used in assessing HMT performance that are also useful for both tiers 1 and 2 (again, confidence versus consequence). One of these metrics is the Likert scale, adapted from a 1932 paper on measuring attitudes [52]. The Likert scale, as a metric, can be subjective since it is based on a set of questions requiring users to provide an assessment, and these questions are formulated from a set of AI system attributes (e.g., measured on a 5-point range). The Likert scale can give a quick visual of users' acceptance, satisfaction/dissatisfaction with the AI capability, and improvements in efficiency, usefulness, and system modularity. There are several visuals commonly used to depict the Likert scale results, offering a quick qualitative picture [53].

The next tier of the four-tier hierarchical performance assessment is model representation. The frame of reference that we use here is inspired by a very comprehensive report describing the challenges, opportunities, and future directions from a workshop called "Future Directions in Human Machine Teaming," led and attended by several distinguished scholars [54]. They formulated the HMT framework illustrated in figure 6.11. As pointed out in their report, the majority of today's intelligent machines are tools, not teammates. For intelligent machines to be teammates, there will need to be joint perceptions and effective communication, leading to joint actions. Advancements toward this goal will require human models and machine models containing accurate representations of the self, desired team intentions, and working in the role of teammates.

In the context of HMT performance assessments at tier 3, we need to evaluate the interplay of a human (or humans) and a machine (or machines), using the model-based framework shown in figure 6.11. In an excellent report titled, *A 20-Year Community Roadmap for Artificial Intelligence Research in the U.S.,* sponsored by the Computing Community Consortium and the Association for the Advancement of AI (AAAI), Gil and Selman recommended prioritizing research on meaningful interactions between humans and machines to achieve societal benefits [55]. Such interactions include techniques for productive collaboration in mixed teams of humans and machines (leveraging multimodalities). Today, most HMT exploits a single modality (e.g., images, speech, or text). However, achieving HMT as teammates will require detailed performance assessment as highlighted in tier 3.

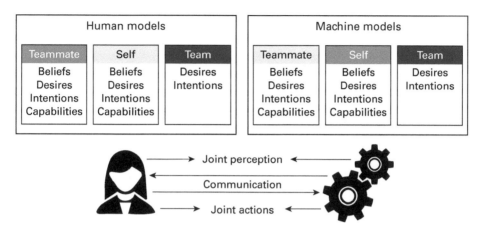

Figure 6.11   A model-based framework for HMT [54].

There is ongoing and very interesting human-machine (robot) research focusing on building mental models such that the human knows when and when not to rely on intelligent machine agents (i.e., a human finds an accurate mental model of the AI strengths and weaknesses) [56]. Another exciting area of research is in the human teaching the robot and vice versa, and the associated lessons, to strengthen the human-machine models, especially in real-world and messy, unstructured environments [57, 58].

However, performance metrics for these research areas are still evolving. One could, for example, assess the effectiveness in meeting the goals and objectives of a task or set of tasks. Some of this metrics can include:

- Quantitative assessment of confidence level versus consequence of action (as per tiers 1 and 2)
- Time to reach a trustworthy decision by the human operating alone versus the improvement (or lack of such) by humans-machines as teammates
- Adaptability to various real-world scenarios
- Counterfactual analysis of sensitivity of human-machine interactions to data inputs
- Compliance with the FASTEPS principles as the tier 4 metric under the rubric of responsible AI (RAI; discussed further in chapter 8)

Before we conclude this section, we want to emphasize that performance metrics should also be informed by HMT failures. Quoting Henry Petroski on famous historical failures, as described in his book *To Engineer Is Human: The Role of Failure in Successful Design* [59], "A good judgment is usually the result of experience. And experience is frequently the result of bad judgment. But to learn from the experience of others requires those who have the experience to share the knowledge with those who follow."

In the building of accurate and reliable human-machine models, we must also learn from successes as well as failures (i.e., learning from humans' and machines' experiences and judgments), preferably prior to the deployment of HMT as teammates in high-stake applications.

6.6   Human-Machine Teaming Challenges

Human-machine teaming (HMT) is one of the most crucial building blocks in our end-to-end AI system architecture. For many years, the HMT functional block had received little attention since at that stage, the decisions were primarily delegated to the human (see the HMT spectrum in figure 6.2). However, as AI became more capable of augmenting human intelligence, the role of the HMT function has increased in importance. There are many benefits that AI can provide with great societal impact, but these benefits must be assessed through the lens of the trade-off between confidence in the machine versus consequence of actions.

Since the roles between the human and the machine continue to evolve, there are many challenges. In this section, we limit our discussion to a smaller subset following the structure discussed here. For a more in-depth discussion of additional HMT challenges, we refer the reader to several reports [54, 55, 60].

We highlight some important HMT challenges by breaking the discussion into two categories:

1. *Micro-level challenges:* Here, we focus on AI as a search-and-discovery tool, AI as a teammate, and autonomy. The challenges are in addition to those discussed in the earlier sections of this chapter.

2. *Macro-level challenges:* We return to the framework established earlier in terms of the three horizons. We also briefly address a view into the future of work affected by the advances in HMT.

One of the most limiting factors experienced today with AI, either as a search-and-discovery tool or as a teammate, is in NLP, including speech recognition and understanding. VPAs rely on the ability to perform ASR and understanding in order to know how to respond to the human—either as an assistant or as a teammate. There have been tremendous improvements in generative pretrained language models (e.g., OpenAI

GPT-3). However, there are still many limitations and challenges with the ability to understand what humans are looking for. At times, these systems must also perform language translations. GPT-3 is capable of such translation, at the expense of significant computational time. In some instances, at least today, the computational latencies are unacceptable.

Another related set of challenges is in demonstrating quantitative and qualitative performance, as per the discussion in section 6.5. These performance metrics inform the operating point in the HMT spectrum explored earlier. The approach that we recommend is to assess performance—employing quantitative performance analysis—through the trade-off between the confidence in the machine making the decision and the course of action.

As an example in the healthcare field, Regina Barzilay and her students at MIT have demonstrated a major improvement in the early detection of breast cancer [61]. The onset of breast cancer is increasing. One in eight women in the US will have the disease in their lifetime. By using AI techniques, Barzilay et al. have been able to increase the true positive rate. They employ several of the metrics discussed in chapter 4, such as the confusion matrix and AUC. The HMT teaming takes place via their AI approach providing a risk score to a radiologist. The radiologist then screens the mammograms in more detail for those patients identified as high risk.

Another example, also in the healthcare domain, is the work by Li et al., where they have been able to develop a recurrent deep neural learning technique to help clinicians in determining alternative courses of action [62]. The approach uses counterfactuals, as discussed in section 6.5, applied to determining future treatment trajectories based on past patients observed history. Again, the HMT hinges on the performance of the neural network as a recommender system to the attending physician.

We like to highlight these AI applications in the healthcare domain as proxies to other high-stake applications. The role of the human and the role of the machine require very careful attention to the end-to-end system performance, including issues with data drifts (e.g., data set shifts due to changes in demographics) [63]. There are, of course, a number of other challenges related to the FASTEPS principles, within the rubric of responsible AI (RAI), that also must be addressed in addition to technical performance, including ethical and privacy considerations.

So far, we have addressed how tier 1 (confidence) and tier 2 (consequence) are important metrics in assessing HMT challenges. We briefly touched on tier 4 (RAI), which will be discussed further in chapter 8. Similarly, tier 3 (model representation) is a very active area of research focusing on improving the human-machine models, which is very pertinent to the model-based framework illustrated in figure 6.11. There are many challenges in this area that are being addressed both in academia and industry, but it is not feasible to address all these challenges in a single chapter. Instead, we give the reader some of recent research highlights and associated challenges and provide references for the readers with interest in exploring further:

- Neurosymbolic AI is a research area focusing on architectures that more closely resemble how humans employ intuitive actions (closely aligned with neural networks), plus more careful and deliberate reasoning (closely aligned with symbolic AI). Rossi and colleagues have developed a very innovative architecture framework [64] (which was also discussed during the AAAI 2022 conference). The architecture structure is inspired by the Nobel Prize winner Daniel Kahneman in *Thinking Fast and Slow* [65]. Rossi et al. introduced a metacognition engine to arbitrate between system 1 (thinking fast) and system 2 (thinking slow). There are many challenges, particularly with creating an accurate world model, having enough information to best determine when to employ system 1 or system 2 processing, and employing robust performance metrics.

  Several other researchers are emphasizing the importance of coupling neural networks and symbolic AI to achieve improved common-sense reasoning and, hopefully, also reduce the dependency on labeled data [66–68]. The challenge to bringing these techniques to practice is in forming an accurate representation of a world model. As Tennenbaum and colleagues, from the field of cognitive science, point out, children learn by forming world models through observing, experimenting, and confirming action-reactions (i.e., interactions in real-world environments) [69].

- Thomas Dieterich, a professor at Oregon State University, in his 2017 AAAI presidential address, outlined steps toward achieving robust AI even in the presence of known-unknowns and unknown-unknowns [70]. He discussed eight approaches pursued by the research community and their associated benefits and challenges. One approach that he suggested, which is very germane to the discussion in this section on challenges, is developing a portfolio of methods to include multifaceted learning. Among other suggestions, he particularly emphasizes this multifaceted approach, saying: "The greater the variety of tasks that the computer learns to perform, the larger the number of different facets it will acquire, and the more robust its knowledge will become."

- Another area of research with the potential to improve HMT is self-supervised learning. Yann LeCun has pioneered much of the progress in this area. He says: "The next revolution in AI will be machines that learn not through supervised learning but through observation" [71]. The challenge will be in the development of internal physical models (as babies and animals do) in a task-independent way and via observation of the physical environment.

Let us now turn our attention to more of a macro-level set of challenges and recommendations. Using the same format as in previous chapters, we anchor our discussion in three horizons, as shown in table 6.1.

We conclude our discussion by briefly addressing recent findings and recommendations on the broad topic of the work that will happen in the future. As AI capabilities improve, for example in HMT advancements, there is a general concern that humans are going to be displaced from their jobs and supplanted by machines. As we pointed out early in the chapter, machines today are best at augmenting human capabilities in jobs that are dull, dirty, and/or dangerous. MIT's president chartered a task force to undertake a two-year study on the impact of technologies and potential changes on the work of the future [73]. Among the many important findings and recommendations in their

Table 6.1 Additional HMT challenges and recommendations

Challenge Area	HMT
Horizon 1: Content-Based Insight (1–2 years)	• Update relationship graphs in real time to strengthen coupling among multiple HMTs. • Advance NLP and understanding to increase the level of insight. • Incorporate human-machine interactions in the AI operational workflows and monitor performance.
Horizon 2: Collaboration-Based Insight (3–4 years)	• Increase communication, reasoning, and joint actions among HMTs • Employ sensors and sources multi-modalities and social-cultural networks to enrich collaborations among human-machine teammates (e.g., employing the metaverse). • Integrate physical and virtual worlds via "digital twins" (discussed further in chapter 10) to improve manufacturing of products and deployment of services.
Horizon 3: Context-Based Insight (5+ years)	• Incorporate physics-based knowledge into neurosymbolic AI architectures to increase context. • Reconfigure human-machine teammates to meet the scale of the mission. • Incorporate goal reasoning techniques in the decision-making process of human-machine teams operating in complex environments [72].

final report, in the context of our discussion in this chapter, is that there is no indication that technological advances will displace employment—instead, new work and jobs will be created. However, robotics and automation will play an increasing role in closing the gap between new jobs and the people available to fill those jobs. Another important finding of their study is that it takes a long time—measured in decades—between the birth of an invention and its commercial use. One of the important challenges will be how to prepare the workforce of the future for this inevitable transition of more and more operational roles being performed by intelligent machines.

6.7 Main Takeaways

It is indeed very exciting to witness and be part of the HMT evolution, as well as the many applications where HMT is contributing to important business benefits and is having societal impact. In this chapter, we gave several examples used to ground this discussion. However, since HMT is such a broad area, we also stated the working HMT definition we use in the book: *HMT is the building block in the AI pipeline that transforms ML outputs (i.e., knowledge) into insights delivered to stakeholders (e.g., users, consumers, or customers).*

HMT is a functional block in the simplified end-to-end architecture shown in figure 6.1. It spans a spectrum depending on the confidence that the human (as a joint role between a human and a machine) has in the machine making the decision versus the consequence of actions. Of course, the operating point in this spectrum is heavily dependent on the application domain. Several examples were illustrated to show how an AI practitioner would determine where to operate across this spectrum.

We discussed AI as a technology that augments human capabilities—meaning intelligent machines helping with the dull-dirty-dangerous roles, which allows humans to devote themselves to more cognitive functions. To reach this goal, we broke the discussion into three topics:

1. AI as a search-and-discovery tool
2. AI as a teammate
3. Full autonomy

Today, it is not uncommon to hear the question, "When will intelligent machines achieve the same level of human intelligence?" There is a lot of debate within the AI community on the right answer. We believe that even if the computation resources were there, it would be many years before intelligent machines can operate at the same levels of power efficiency as a human. Although there are many unprecedented and remarkable advances to date, they require an immense amount of computation and electrical power, as discussed in chapter 5.

VPAs, because of the large advent of data, especially with the fast rate of increase in the proliferation of IoT devices, will continue to play an important role in our daily lives as AI search-and-discovery tools. Unfortunately, they remain mediocre, at best, in answering more than just routine questions. Their limitations are in their ability to perform automatic speech recognition (e.g., users with strong foreign accents), and the ability to achieve natural language understanding (e.g., in real-time applications).

Another role, occupying the middle of the HMT spectrum, is AI as a teammate. One example that we elaborated on was in the cybersecurity domain, where ML techniques were implemented as a recommender system. Again, to identify the best HMT operating point, one must view this through the lens of a trade-off between confidence level versus consequence of actions.

As we continue to advance AI systems to be able to make decisions autonomously, there is a need to ascertain whether these systems are operating under the rubric of RAI. There must be, at minimum, trust in the AI system. More generally, we emphasized the importance, regardless of where the HMT falls in the spectrum discussed earlier, of complying with FASTEPS principles. That topic will be discussed further in chapter 8.

Full autonomy, particularly with autonomous vehicles, continues to be difficult to achieve because these systems are not yet able to operate in environments like city traffic, which are complex and require decision making while operating at high speed. Nevertheless, there have been impressive achievements with autonomous systems either in low-complexity scenarios (e.g., in farming) or cases where the autonomous vehicles are limited by geofenced boundaries.

HMT is becoming the part of the AI pipeline where the "rubber meets the road." We can have superb data and well-performing ML algorithms, enabled by modern computing, but unless the HMT step is able to create insights that provide value to the stakeholders, the AI system will fail to find acceptance in operations. Therefore, it is imperative that the AI community pays attention to qualitative and quantitative performance metrics, which will help practitioners determine if AI products or services are suitable for deployment. We formulated a hierarchical, four-tier set of performance metrics:

1. Confidence
2. Consequence
3. Model representation
4. RAI

These performance metrics must be integrated into AI implementation and integration-verification-validation as part of the framework illustrated in figure 6.10. We will discuss in chapter 10 the process of assessing, in an iterative way, the overall AI system.

The AI research community is demonstrating very innovative architectures to have humans and machine work more effectively as teammates. These efforts emphasize the importance of building representative humans and machine models along with realistic world models.

We elaborated on a number of HMT challenges structured under the topology of micro-level and macro-level challenges. Admittedly, since HMT is such a broad topic, we were only able to highlight a subset of HMT challenges in this chapter. The reader is encouraged to follow up with the extensive references provided in this chapter. One very exciting area, but one that is also experiencing a number of research challenges, is the confluence of neural network and symbolic AI architectures. We also highlighted additional areas of research, such as multifaceted learning and self-supervised learning.

We concluded the chapter with a view of the future. All the indications are that intelligent machines will not displace the workforce. Instead, AI is likely to create new jobs, just as history has shown us that general-purpose technologies contributed to an economic landscape change with new and rewarding work. This point was made as follows by the recent MIT report alluded to in this chapter: "Most of today's jobs hadn't even been invented in 1940."

6.8   Exercises

1. What is the working definition of HMT used in this chapter? Elaborate on the important concepts captured in this definition.

2. Explain the HMT spectrum and associated roles between human and machine.

3. Within HMT functions, decisions are best left to only the machine with the confidence level is low and the consequence of action is low.
   a. True
   b. False

4. A human, on average, uses about 100 watts, and 80 percent of that (i.e., 80 watts) is consumed by the brain.
   a. True
   b. False

5. Provide real-world examples of VPAs being used today as AI search-and-discovery tools and discuss their limitations.

6. One of the biggest challenges with robots today is their ability to perform tasks that humans commonly do with their hands.
   a. True
   b. False

7. Explain why full vehicle autonomy is hard, and discuss examples with respect to complexity and speed.

8. Describe the four tiers of performance metrics and elaborate on their respective importance.

9. Provide examples of micro-level and macro-level HMT challenges.

10. Explain what is meant by neurosymbolic AI.

11. What is the meaning of self-supervised learning? Why is it an important next step in AI?

12. It is expected that as autonomy and robotics advance, humans will be left without jobs to perform.
    a. True
    b. False

13. How do you recommend that organizations prepare for a future where AI tools and techniques will do more of the routine tasks, thus permitting humans to undertake higher-level cognitive tasks?

14. With the advent of generative AI (e.g., GPT-4), what approaches do you recommend to verify the validity and veracity of the results?

6.9 References

1. Asimov, I., *Reason. I, Robot.* 1941, 59–77.

2. Asimov, I., Runaround. *Astounding Science Fiction*, 1942. 29(1): 94–103.

3. Marcus, G. and E. Davis, *Rebooting AI: Building artificial intelligence we can trust.* 2019, Vintage.

4. Martinez, D., N. Malyska, and B. Streilein. *Artificial intelligence: Short history, present developments, and future outlook.* MIT Lincoln Laboratory, 2019. https://www.ll.mit.edu/media/11876.

5. Dubiel, M., M. Halvey, and L. Azzopardi, *A survey investigating usage of virtual personal assistants*, 2018. arXiv preprint arXiv:1807.04606.

6. Martinez, D. Architecture for machine learning techniques to enable augmented cognition in the context of decision support systems, in *International Conference on Augmented Cognition,* Heraklon, Greece. 2014. Springer.

7. Matheson, E., R. Minto, E. G. G. Zampieri, et al., Human–robot collaboration in manufacturing applications: A review. *Robotics*, 2019. 8(4): 100.

8. Schneier, M. and R. Bostelman, *Literature review of mobile robots for manufacturing*. 2015, US Department of Commerce, National Institute of Standards and Technology.

9. Bogue, R., Growth in e-commerce boosts innovation in the warehouse robot market. *Industrial Robot: An International Journal*, 2016.

10. Laber, J., R. Thamma, and E. D. Kirby, The impact of warehouse automation in Amazon's success. *International Journal of Innovative Science, Engineering and Technology*, 2020. 7: 63–70.

11. Sanchez, J., *MIT EmTech digital*, J. Strong, Editor. 2021, MIT Technology Review.

12. Sharma, A., A. Jain, P. Gupta, et al., Machine learning applications for precision agriculture: A comprehensive review. *IEEE Access*, 2020. 9: 4843–4873.

13. Gosselin, K., Now boarding: Robots will sanitize Avelo Airlines aircraft for COVID-19, other pathogens, in *Hartford Courant*. TechXplore 2022. https://techxplore.com/news/2022-03-boarding-startup-airline-robots-sanitize.html.

14. Javaid, M., A. Haleem, A. Vaish, et al., Robotics applications in COVID-19: A review. *Journal of Industrial Integration and Management*, 2020. 5(04): 441–451.

15. Ransbotham, S., S. Khodabandeh, D. Kron, et al., *Expanding AI's impact with organizational learning*. 2020, MIT Sloan Management Review and Boston Consulting Group.

16. Davenport, T. H. and J. Kirby, *Just how smart are smart machines?* 2016.

17. Kemp, S., *Digital 2022: Global overview report.* 2022, Datareportal.

18. Juniper Research, *Industrial IoT connections to reach 37 billion globally by 2025.* 2020. https://www.juniperresearch.com/press/industrial-iot-iiot-connections-smart-factories#:~:text=Hampshire%2C%20UK%20%E2%80%93%202nd%20November%202020,overall%20growth%20rate%20of%20107%25.

19. Desjardins, J., *How much data is generated each day?* 2019. https://www.weforum.org/agenda/2019/04/how-much-data-is-generated-each-day-cf4bddf29f/.

20. Kepuska, V. and G. Bohouta, Next-generation of virtual personal assistants (Microsoft Cortana, Apple Siri, Amazon Alexa and Google Home), in *2018 IEEE 8th Annual Computing and Communication Workshop and Conference (CCWC)*. 2018, IEEE. https://ieeexplore.ieee.org/xpl/conhome/8293728/proceeding.

21. Tenhundfeld, N. L., H. M. Barr, E. H. O'Hear, et al., Is my Siri the same as your Siri? An exploration of users' mental model of virtual personal assistants, implications for trust. *IEEE Transactions on Human-Machine Systems*, 2021.

22. Brown, S., *Machine learning, explained.* 2021, Ideas Made to Matter. https://mitsloan.mit.edu/ideas-made-to-matter/machine-learning-explained.

23. Lippman, R. P., W. M. Campbell, D. J. Weller-Fahy, et al., Toward finding malicious cyber discussions in social media, in *AAAI Workshops.* 2017. AAAI.

24. Moore, A. W., Predicting a future where the future is routinely predicted. *MIT Sloan Management Review,* 2016. 58(1): 18.

25. González Enríquez, J., A. Jiménez-Ramírez, F. J. Dominguez-Mayo, et al., Robotic process automation: A scientific and industrial systematic mapping study. *IEEE Access,* 2020. 8: 39113–39129.

26. Gotthardt, M., D. Koivulaakso, O. Paksoy, et al., Current state and challenges in the implementation of smart robotic process automation in accounting and auditing. *ACRN Journal of Finance and Risk Perspectives,* 2020.

27. Agrawal, A., J. Gans, and A. Goldfarb, What to expect from artificial intelligence. April 2017, *MIT Sloan Management Review.*

28. Agrawal, A., J. Gans, and A. Goldfarb, *Prediction machines: The simple economics of artificial intelligence.* 2018, Harvard Business Press.

29. Dhar, V., When to trust robots with decisions, and when not to. *Harvard Business Review,* 2016. 17.

30. Daugherty, P. R. and H. J. Wilson, *Human + machine: Reimagining work in the age of AI.* 2018, Harvard Business Press.

31. Daugherty, P. and J. Euchner, Human + machine: Collaboration in the age of AI: Paul Daugherty talks with Jim Euchner about a new paradigm for collaborative work between people and intelligent systems. *Research-Technology Management,* 2020. 63(2): 12–17.

32. Mell, P. and K. Scarfone, Improving the common vulnerability scoring system. *IET Information Security,* 2007. 1(3): 119–127.

33. Mell, P., K. Scarfone, and S. Romanosky, Common vulnerability scoring system. *IEEE Security & Privacy,* 2006. 4(6): 85–89.

34. Bughin, J., E. Hazan, S. Ramaswamy, et al., *Artificial intelligence: The next digital frontier?* 2017.

35. Romeo, J., P. Filippi, J. Baird, et al., Identifying within-season cotton crop nitrogen status using multispectral imagery, in *Proceedings of the 19th Australian Society of Agronomy Conference. NSW,* Wagga Wagga. 2019. http://www.agronomyaustraliaproceedings.org.

36. Volkova, A., Invited AI practitioner to the *AI strategies and roadmap: Systems engineering approach to AI development and deployment class,* D. Martinez. 2021, MIT Course.

37. Enwall, T., Invited AI practitioner to the *AI strategies and roadmap: Systems engineering approach to AI development and deploytment class*, D. Martinez. 2021, MIT Course.

38. Ackerman, E., Covariant uses simple robot and gigantic neural net to automate warehouse picking. *IEEE Spectrum Automaton Article*, 2020.

39. Buehler, M., K. Iagnemma, and S. Singh, *The DARPA Urban Challenge: Autonomous vehicles in city traffic*. Vol. 56. 2009, Springer.

40. Urmson, C., J. Anhalt, D. Bagnell, et al., Autonomous driving in urban environments: Boss and the Urban Challenge. *Journal of Field Robotics*, 2008. 25(8): 425–466.

41. Gerrish, S., *How smart machines think*. 2018, MIT Press.

42. Bond, R. and S. Mohindra, *Command and control architecture*. 2019, MIT Lincoln Laboratory.

43. Karaman, S., Autonomous vehicles: Emerging technologies, new challenges and shifting opportunities. in *Autonomy*. 2020, MIT Industrial Liason Program. https://ilp.mit.edu/watch/2020-autonomy-day-1-sertac-karaman.

44. Shah, D. and S. Levine, ViKiNG: Vision-based kilometer-scale navigation with geographic hints, 2022. arXiv preprint arXiv:2202.11271.

45. *The Economist*, What America's largest technology firms are investing in. 2022. https://www.economist.com/briefing/2022/01/22/what-americas-largest -technology-firms-are-investing-in.

46. Amazon, Machine learning lens—AWS well-architected framework. December 20, 2021. https://docs.aws.amazon.com/wellarchitected/latest/machine-learning-lens /wellarchitected-machine-learning-lens.pdf#machine-learning-lens.

47. Fernandez, C., F. Provost, and X. Han, Explaining data-driven decisions made by AI systems: The counterfactual approach, 2020: 1–33. arXiv preprint arXiv:2001 .07417.

48. Shapley, L., A value for n-person games. *Annals of Mathematics Studies (Contributions to the Theory of Games*, ed. H. W. Kuhn and A. W. Tucker), 1953: 307–331.

49. Giudici, P. and E. Raffinetti, Shapley-Lorenz eXplainable artificial intelligence. *Expert Systems with Applications*, 2021. 167: 114104.

50. Lundberg, S. M. and S.-I. Lee, A unified approach to interpreting model predictions. *Advances in Neural Information Processing Systems*, 2017. 30.

51. Rozemberczki, B., L. Watson, P. Bayer, et al., *The Shapley value in machine learning*, 2022. arXiv preprint arXiv:2202.05594.

52. Likert, R., A technique for the measurement of attitudes. *Archives of Psychology* 22(140) (1932): 55.

53. Robbins, N. B. and R. M. Heiberger, Plotting Likert and other rating scales, in *Proceedings of the 2011 joint statistical meeting*. 2011, American Statistical Association.

54. Laird, J., C. Ranganath, and D. S. Gershman, *Future Directions in Human Machine Teaming Workshop*. Department of Defense, Office of Prepublication and Security Review, 2020.

55. Gil, Y. and B. Selman, A 20-year community roadmap for artificial intelligence research in the US, 2019. arXiv preprint arXiv:1908.02624.

56. Mozannar, H., A. Satyanarayan, and D. Sontag, *Teaching humans when to defer to a classifier via examplars*, 2021. arXiv preprint arXiv:2111.11297.

57. Booth, S., S. Sharma, S. Chung, et al., How to understand your robot: A design space informed by human concept learning, in *International Conference on Robotics and Automation,* Xi'an, China, 2021.

58. Booth, S., S. Sharma, S. Chung, et al., Revisiting human-robot teaching and learning through the lens of human concept learning, in *Proceedings of the 2022 ACM/IEEE International Conference on Human-Robot Interaction,* Sapporo, Japan. 2022.

59. Petroski, H., *To engineer is human: The role of failure in successful design*. 1985, St. Martin's Press.

60. Zhang, D., N. Meslej, A. Barbe, et al., *The AI index 2022 annual report*. 2022, Stanford Institute for Human-Centered AI.

61. Yala, A., C. Lehman, T. Schuster, et al., A deep learning mammography-based model for improved breast cancer risk prediction. *Radiology*, 2019. 292(1): 60–66. https://www.washingtonpost.com/technology/2021/12/21/mammogram-artificial-intelligence-cancer-prediction/.

62. Li, R., S. Hu, M. Lu, et al., G-Net: A recurrent network approach to G-computation for counterfactual prediction under a dynamic treatment regime, in *Machine learning for health*. 2021, PMLR. https://news.mit.edu/2022/deep-learning-technique-predicts-clinical-treatment-outcomes-0224.

63. Finlayson, S. G., A. Subbaswamy, K. Singh, et al., The clinician and dataset shift in artificial intelligence. *New England Journal of Medicine*, 2021. 385(3): 283.

64. Ganapini, M. B., M. Campbell, F. Fabiano, et al., *Thinking fast and slow in AI: The role of metacognition*, 2021. arXiv preprint arXiv:2110.01834.

65. Kahneman, D., *Thinking, fast and slow*. 2011, Macmillan.

66. Arabshahi, F., J. Lee, M. Gawarecki, et al., Conversational neuro-symbolic commonsense reasoning, 2020. arXiv preprint arXiv:2006.10022.

67. Mao, J., C. Gan, P. Kohli, et al., The neuro-symbolic concept learner: Interpreting scenes, words, and sentences from natural supervision, 2019. arXiv preprint arXiv:1904.12584.

68. Marcus, G., The next decade in AI: Four steps towards robust artificial intelligence, 2020. arXiv preprint arXiv:2002.06177.

69. Friston, K., R. J. Moran, Y. Nagai, et al., World model learning and inference. *Neural Networks*, 2021. 144: 573–590.

70. Dietterich, T. G., Steps toward robust artificial intelligence. *AI Magazine*, 2017. 38(3): 3–24.

71. LeCun, Y., The power and limits of deep learning: In his IRI Medal address, Yann LeCun maps the development of machine learning techniques and suggests what the future may hold. *Research-Technology Management*, 2018. 61(6): 22–27.

72. Aha, D. W., Goal reasoning: Foundations, emerging applications, and prospects. *AI Magazine*, 2018. 39(2): 3–24.

73. Autor, D., D. Mindell, and E. Reynolds, *The work of the future: Building better jobs in an age of intelligent machines*. 2020, MIT Task Force on the Work of the Future. https://workofthefuture.mit.edu/.

7

Robust AI Systems

> If we use, to achieve our purposes, a mechanical agency with whose operation we cannot efficiently interfere once we have started it, because the action is so fast and irrevocable that we have not the data to intervene before the action is complete, then we had better be quite sure that the purpose put into the machine is the purpose which we really desire and not merely a colorful imitation of it.
> —Norbert Wiener (MIT, 1960)

As we will discuss in this chapter, artificial intelligence (AI) outputs are very susceptible to minor changes to machine learning (ML) models. Unfortunately, what goes on inside these models, in terms of learning from data, is difficult to explain. There are significant research efforts dedicated to understanding AI system vulnerabilities caused by either intentional or unintentional means. There are also approaches proposed toward explainability of the ML results—explainable AI (XAI), as well as mitigation techniques against AI system vulnerabilities.

In this chapter, our focus is on understanding AI system brittleness and types of vulnerabilities across the AI system architecture discussed in chapter 1 and shown in figure 1.3. A simplified depiction of the AI system architecture is shown in figure 7.1, highlighting the robust AI building block. In chapter 8, we describe responsible AI (RAI), based on the principles of fairness, accountability, safety, transparency, ethics, privacy, and security (FASTEPS). As expected, there is a lot of synergy between robust AI and RAI. We opted to dedicate a separate chapter to each topic because of the high degree of brittleness and vulnerability that AI systems are experiencing today—and likely will continue to experience in the future.

AI systems cannot be robust unless these systems are designed, developed, and deployed consistent with the RAI principles discussed in chapter 8. However, while AI

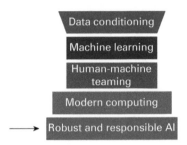

Figure 7.1   Simplified AI system architecture.

systems can comply with RAI principles, they can still suffer from not being robust to intentional or unintentional modifications to the input data, algorithms, or outputs.

Norbert Wiener stated the quote that started this chapter in his paper "Some Moral and Technical Consequences of Automation," published in Science in 1960. His statement, "Better be quite sure that the purpose put into the machine is the purpose which we really desire" [1], is as relevant today as it was over sixty years ago. As we discussed in earlier chapters of this book, Gary Marcus properly points out that today's AI capabilities are limited to narrow tasks that require an immense amount of training data and computation, show a high degree of brittleness, and fail in unpredictable ways (as we will discuss later in the chapter). His proposed hybrid approach can potentially lead to making AI systems more robust by incorporating both knowledge—synthesized from a variety of sources to better reason about its surrounding world—and cognitive models [2].

We are very optimistic that even today's narrow AI, which provides limited capabilities for a set of well-defined tasks, is adding important societal value and solving real-world problems by augmenting human capabilities. For example, researchers at Stanford, OpenAI, Google, and other organizations recently have shown the ability to use foundation models, trained on broad data at scale for one type of task but with the ability to generalize to other application domains. Any new technology will offer opportunities, but it also will present challenges; for instance, foundation models offer great potential to affect, in a positive way, the reusing and adapting of pretrained models [3]. As discussed by Ilya Sutskever (the cofounder of and chief scientist at OpenAI), commercial companies are starting to use GPT-3 in a number of real-world production-level applications, such as in customer service, education, and suicide prevention [4]. This remarkable rapid advancement in ML is very exciting; however, we as a community still must be very judicious and diligent about the use of pretrained models because there continue to be vulnerabilities that may alter the output of AI systems.

Since the topic of robustness has many different definitions and meanings, we provide a definition here that we will use throughout this chapter:

Robust AI system: The ability of a narrow AI system to deliver capabilities, assessed through the lens of confidence in the machine versus consequence of actions, constrained to providing value to stakeholders in a responsible manner.

The goal of this chapter is to introduce and to elaborate on some of the important issues afflicting AI when it comes to achieving full robustness. We also discuss mitigation approaches and provide an outlook for the future. The intent is not to imply that robustness of AI systems has been solved as a challenge—in fact, the opposite is true. We have a long way to go before we can be confident that our AI systems are fully robust.

It is useful to break this definition of a robust AI system into its key elements. First, the goal of narrow AI is to deliver capabilities with value to stakeholders, as discussed in earlier chapters. Second, an AI system must be evaluated, during integration-verification-validation, at each of the building blocks shown in the AI system architecture. The output of the AI pipeline must be assessed with respect to the confidence in the machine making a decision and the consequence of actions, as illustrated in figure 6.3 in chapter 6. As discussed in chapter 6, if the confidence is high and the consequence is low, the stakeholders can reliably accept the output of the AI system. Conversely, if the confidence is low and the consequence is high, humans must make the final decisions about recommended courses of action. Many of today's use cases operate between these two ends of the spectrum. As previously noted, Thomas Dietterich presented an excellent taxonomy on "Steps towards Robust Artificial Intelligence," including vulnerabilities due to known-unknowns and unknown-unknowns, that all AI designers and implementers should diligently attend to, with the objective of improving AI robustness [5].

In section 7.1, we elaborate on AI brittleness and associated vulnerabilities in terms of attack surfaces and effects across the end-to-end AI system architecture. Caceres's lecture material is also discussed and addressed later in this chapter [6].

In section 7.2, we discuss classes of adversarial AI in more detail. We also touch on deepfakes and provide some examples in section 7.3. In sections 7.4–7.6, the discussion is focused on XAI and mitigation approaches. We conclude the chapter by looking ahead at some robust AI system challenges, the main takeaways, and exercises.

7.1   Systems Perspective on AI Vulnerabilities

The focus of our discussion in this chapter is on robust AI systems, with emphasis on systems. The operative definition described earlier centers on the output of an AI system assessed, in terms of robustness, through a trade-off between confidence in the system versus consequence of actions. As discussed in preceding chapters and illustrated in figure 6.10, in the AI system architecture implementation framework, every building block must undergo verification and validation (V&V) during subsystem integration. However, prior to deployment, AI designers, developers, and implementers must determine if the AI capability is within the guardrails of a robust AI and RAI system. This process is part of a diligent systems engineering approach, as emphasized in chapter 2.

In the context of a systems engineering approach, it is critical that AI practitioners understand the points of vulnerability across the end-to-end architecture. Figure 7.2 illustrates the stages in the end-to-end AI system architecture that are vulnerable to attacks and the type of effects of attacks across the architecture.

Although the following discussion focuses on malicious attacks, it is important to clarify that some vulnerabilities of an AI system can be caused by unintentional errors during the design, development, or use of AI products and services. For example, at the human-machine teaming (HMT) stage, humans can make erroneous decisions after analyzing ML results. Therefore, it is necessary, *but not sufficient*, to evaluate the system output within the envelope of ML metrics (accuracy, precision, recall, and others). AI practitioners must also determine the level of confidence, or trust, that the stakeholders can reliably put in the system, especially in high-stakes decision support. Daniel Kahneman et al. point out that humans can introduce unreliable outputs due to errors in judgment resulting from, for example, preconceived notions or biases [7].

Let us return to figure 7.2 and explain the actions (attacks/vulnerabilities) and effects at each of the building block stages. Starting with inputs to the AI system, there are vulnerabilities caused by data poisoning attacks. Poisoning attacks are described in more detail in section 7.2. For now, it is sufficient to highlight a number of examples of poisoning attacks, such as physically altering a stop sign to cause the output to change to a speed limit sign; polluting Twitter and Facebook with unreliable data, causing misinformation to be released; or posting wrong content in Facebook posts. The effects at the data conditioning stage are to cause the designer to act on noisy, missing, or incomplete data, affecting the ML model during training or testing, and at the time of deployment.

Similarly, at the ML stage, attacks can be in the form of evasion attacks or black-box attacks. In the former, the malicious attacker generates inputs that cause model errors. In the latter, the adversary re-creates the algorithm model by probing the output stimulated from different inputs. The likely effects can be in the form of prediction or classification errors, reducing confidence in the output.

The AI hardware and software are also vulnerable to nefarious attacks. For example, the adversary could alter the model parameters, the stored data, or the software code.

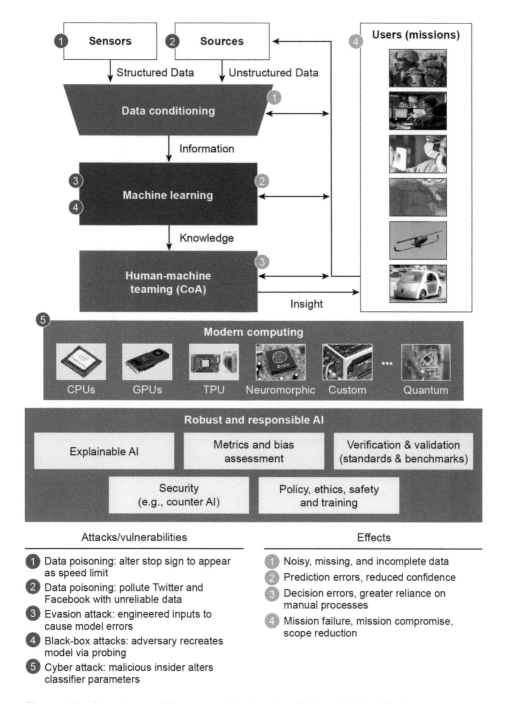

Figure 7.2 AI system architecture points of vulnerability and their effects.

These attacks can take the form of tampering with the computing system by physical or cyber means, resulting in hardware and software failures. The Carnegie Mellon University/ Software Engineering Institute has formulated a road map for addressing some of these issues under the rubric of robust and secure AI [8].

AI system vulnerabilities and failures propagate and become amplified throughout the system. For example, at the HMT stage, decisions with life-and-death consequences can be erroneously made. Or, in the process of creating insight into the knowledge generated at the output of the ML stage, the human might need to ignore the machine (low confidence in the decisions) and revert to manual processes.

Ultimately, the user, consumer, or customer must gain value from the system. Because of the wide range of vulnerabilities that an AI system can experience, if inflicted and affected, a mission might be compromised or even result in complete failure.

As discussed earlier, there is a lot of overlap in content between the robust AI and RAI topics. Robust AI is primarily focused on several of the boxes shown at the bottom of figure 7.2. For example, some important topics are XAI, metrics and bias assessment, V&V (including the need for standards and benchmarks), and physical and cybersecurity. The topics of policy, ethics, safety, and training in the context of RAI are discussed in more detail in chapter 8, but here, we touch on these latter topics briefly to provide some context.

In May 2017, the Defense Advanced Research Projects Agency (DARPA) launched a very timely program of XAI under the leadership of the program manager, David Gunning [9]. As Gunning emphasized, there is a tension between the accuracy of an ML model (learning performance) and the ability to explain how it reached its answers. For example, deep neural networks (DNNs) can have many training parameters, but it is difficult to explain to users how the learning performance was reached—this is in contrast to, for example, decision trees. Albert Einstein had a famous quote: "If you can't explain it simply, you don't understand it well enough." This quote is very apropos to today's complex neural network systems. There is a need for much additional research and development (R&D) investment in understanding AI systems before we can reliably accept these systems as robust. Thus, we must maintain a diligent balance between confidence in the machine making the decision versus the consequence of action. In the two-dimensional framework of the DARPA XAI program, in addition to learning performance versus explainability, the third axis should include consequence of actions. Fortunately, many of today's unprecedented successes with AI can be categorized as high (or acceptable) algorithm performance with low consequence of actions, even under limited explainability.

Metrics and bias assessments are also paramount to achieving robustness. Although bias assessment is addressed under the principles of RAI under fairness, we include it here as well, since bias can manifest itself starting with the input data shown in figure 7.2. The

US National Institute of Standards and Technology (NIST) has put forward an excellent AI Risk Management Framework with the goal of helping AI organizations assess risks throughout the AI system life cycle [10]. This framework is not a checklist for certifying an AI system. Instead, it provides a process for understanding risks relating to the development, deployment, and use of AI systems, consistent with our emphasis throughout this book under the foundational umbrella of systems engineering.

V&V of AI systems must be performed at all levels of the modified Vee-model shown in figure 2.1 in chapter 2. AI practitioners must start by fully understanding the stakeholders' required specifications (which help inform the measures of V&V) through system realization and at all stages of the system life cycle. A thorough V&V implementation calls for well-defined standards and benchmarks, which are lacking in most of today's evaluations of AI systems. The ML benchmarks (known as MLPerf) are excellent tools to use when formulating a set of accepted industry standards [11].

Another important component in making AI systems more robust is ascertaining the protection against physical access or cyber access to modern computing environments. There is a move in the cybersecurity community to take a zero-trust architectural approach to any local or global enterprise computing system (including commercial cloud computing platforms). In the context of a zero-trust architecture, Phil Leplante and Jeffrey Voas suggest that "any critical AI-based product or service should be continuously questioned and evaluated," even under increased overhead costs [12]. We definitely advocate for this approach, particularly in high-stakes applications with potential catastrophic consequences. We elaborated on the "how" approach to continuously evaluate AI systems by implementing the framework in figure 2.2 in chapter 2.

Before we conclude this section, it is worth pointing out several ongoing initiatives addressing policy, ethics, safety, training, and other vital topics, as shown in figure 7.2. These topics are elaborated in more detail in chapter 8. Pelillo and Scantamburlo, the editors of the book *Machines We Trust: Perspectives on Dependable AI*, provide several references to efforts undertaken in the US, the European Commission, China, the United Nations, and other entities to define a set of guidelines toward reaching the goal of reliable AI systems that can be lawful, ethical, and robust [13]. One of their recommendations is proposed by Rieder et al. as an "ethos of trustworthy AI that generates safe and reliable AI products." Trustworthy AI must be driven by the triad of people-process-technology discussed in this book, including AI business culture and beliefs.

7.2   Classes of Adversarial Artificial Intelligence

For the better part of the past decade, significant attention has been paid to the classes of adversarial attacks (e.g., against DNNs). Szegedy et al. showed how small perturbations

Pandas vs. gibbons

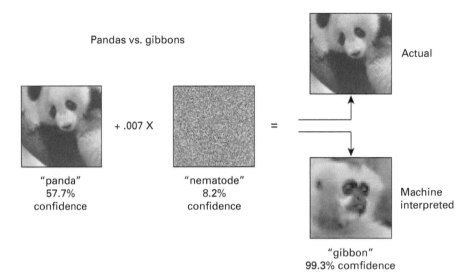

+ .007 X

=

Actual

"panda"
57.7%
confidence

"nematode"
8.2%
confidence

Machine
interpreted

"gibbon"
99.3% comfidence

Figure 7.3 A small perturbation of the original image of a panda, resulting in the wrong classification [15].

to the input data (indistinguishable to the human eye) can cause the AlexNet (DNN) algorithm to create the wrong classification [14]. Similarly, Goodfellow et al. explained that the misclassification by neural network models was not caused by model nonlinearity or regularization effects. Instead, they showed that a small linear perturbation resulted in the wrong classification, even among different neural network architectures operating on the same adversarial examples. One of the most famous examples is shown in figure 7.3 [15]. The GoogLeNet algorithm was given an image of a panda perturbed by small percentage representing a nematode. As a result, the GoogLeNet neural network misclassified the image as a gibbon with 99.3 percent confidence versus the indistinguishable actual panda picture.

In addition to this seminal work in understanding the vulnerabilities of DNNs to adversarial attacks, Papernot et al. formulated a very clear taxonomy of threat models affecting deep learning [16]. In the application of computer vision, they demonstrated misclassification by the DNN for specific target classes with a 97 percent success rate. This success rate required only that the adversary, on average, modified a little over 4 percent of the pixels; for example, in the MNIST data set with a digit consisting of 28×28 pixels (784 pixels), only 32 pixels (about 4 percent) had to be modified to achieve a successful misclassification.

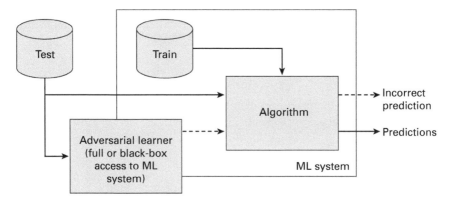

Figure 7.4   Simplified ML system.

The AI research community commonly refers to three topologies when identifying the amount of knowledge that an adversary has prior to formulating classes of adversarial attacks. Figure 7.4 illustrates a simplified ML system depicting training and test data input into an ML algorithm.

Although the earlier discussion centered on computer vision (image misclassification), adversarial attacks can occur in other application domains as well, such as natural-language processing (NLP), autonomous vehicles, and reinforcement learning (e.g., games) [17]. The three adversary topologies are white-box, black-box, and gray-box:

1. *White-box:* The adversary has full knowledge of the ML algorithm architecture, model parameters, and training data.

2. *Black-box:* Neither the model nor the training data is available to the attacker. However, the adversary can use the output of the ML algorithm to infer pieces of the training set or the ML model. For example, Papernot et al. showed the vulnerability of ML models—the adversary only needed to prove the output of the model to infer model vulnerabilities [18].

3. *Gray-box:* The adversary has limited knowledge of either the data, the algorithm, or both.

With this background given, we now introduce the reader to a simple set of classes of adversarial attacks. As discussed earlier and illustrated in figure 7.2, adversarial attacks can be categorized as follows:

• *Poisoning attacks:* An adversary pollutes training data to skew a decision boundary or model behavior. This type of attack can make the ML system completely unusable since the ML decisions would be wrong. An example can be determining whether an email is spam; with a poisoning attack on the training data, the decision boundary can be altered.

- *Evasion attacks:* The adversary designs inputs to misclassify the target class. The example shown in figure 7.3 falls in this category. Or it can make the ML system lack integrity, as in the case of altering a physical stop sign to make it look like a speed limit sign at the point of inputting test data illustrated in figure 7.4 [19]. Instead, if the training data is modified, then it is categorized as data poisoning (i.e., poisoning attack), as explained previously.

- *Model inversion:* The adversary reconstructs the ML model via probing or builds a proxy model to discover training data characteristics, which helps with designing poisoning attacks. This type of attack can lead, for example, to privacy breaches, or to stealing of intellectual property (as it relates to ML model intellectual property).

All of these classes of adversarial attacks can employ cyberapproaches to penetrate the training data, derive the model parameters, or infer the model architecture. The AI designers must be concerned with cyberattacks to the AI platform (hardware or software), as shown in figure 7.2. In addition to significant research ongoing in understanding and protecting AI systems against adversarial AI, there is some emphasis on understanding and defending AI systems by applying cybersecurity to AI robustness. However, more research is needed in this area [20–22].

Later in this chapter, we elaborate on some mitigation approaches. However, the adversarial AI area is still maturing. It is well known that researchers at times postulate mitigation approaches against adversarial AI at such events as AI conferences, and before or soon after such a conference is over, other researchers have already identified counterapproaches. A DARPA program called "Guaranteeing AI Robustness against Deception (GARD)" was started in order to establish theoretical ML system foundations to identify system vulnerabilities, characterize properties that will enhance system robustness, and encourage the creation of effective defenses [23].

We strongly advocate for the need to assess and evaluate AI systems throughout their life cycles—from design, to development, to the final deployment and use (including constant monitoring). An important aspect of this continual monitoring is to take a systems engineering approach following the framework illustrated in figure 2.2. Goodwin and Caceres stressed the importance of a system analysis approach for responsible design of AI/ML systems [24]. System analysis is a necessary component in the development of robust AI systems. In chapter 10, we discuss and recommend an AI ecosystem, where prototyping (including system analysis) is an important step toward successful deployment of AI systems.

7.3 Deepfakes and Examples

The many remarkable and unprecedented advances in ML with impressive value to society, and humanity in general, unfortunately also have led to nefarious uses of the technology. Deepfakes constitute such an example. The technology is only a handful of years old, but in such a short time, deepfakes has generated a lot of concern. In 2017, one of the first uses of deepfakes exploited material from social media to create false videos of celebrities in compromising positions. Since then, there have been many articles that mostly focused on the creation of deepfakes. Less attention has been devoted to detecting, defending, and preventing such deceptions. Hopefully, this R&D imbalance will change since deepfakes can cause major damages to society, political systems, business corporations, and national security.

In this chapter, we include a short discussion of deepfakes under the rubric of robust AI systems since it is a capability generated by the use of ML algorithms. In contrast to our previous discussion of adversarial AI and associated vulnerabilities—generated by attacking the machines at potential different attack surfaces shown in figure 7.2—the malicious use of deepfakes is created by humans to fool other humans. However, adversarial AI and deepfakes can cause existential damage to AI (i.e., lack of trust) at the point of providing insight, as depicted at the end of the pipeline in our AI system architecture. The nefarious use of deepfakes has a detrimental effect on the users and consumers of AI.

In the literature, there are several articles that do an excellent job in surveying and reviewing the state of the art of deepfakes, including enumerating the technologies used in the creation (i.e., algorithms), application, and recommended protection and prevention of deepfakes. Westerlund presented a review of eighty-four publicly available online articles published over the 2017–2019 period [25], and posits that solutions to addressing deepfakes requires a multiprong approach with stronger regulations, policies, education and training, and the development and deployment of robust technologies.

The nefarious uses of deepfake technology, as well as its penetration into the fabric of our society, have much synergy with the need for cybersecurity, such as the detection, prevention, intervention, attribution, and authentication of deepfakes. Corporations and government organizations should take a stand—similar to centers dedicated to exchanging cybersecurity threat information—to inform institutions and the general public about identified deepfakes. Perhaps, again similar to what is done in the cybersecurity domain, a database can be created containing detected deepfakes (and the associated responsible actors, if they have been identified). In the US, NIST has established an Open Media Forensics Challenge, a follow-up to an evaluation of DARPA's Media Forensics program (2017–2020) [26].

Before we delve deeper into malicious uses of deepfakes, there are also some beneficial uses of this technology. For example, in the entertainment realm, one can re-create instantiations of classic black-and-white movies in different languages; for education purposes, videos can be reanimated with historical figures adding additional instructional content; and the recent excitement about the metaverse can use deepfakes to build realistic avatars. A very impactful video was created with the help of the famous soccer player David Beckham as a worldwide call to end malaria [27]. Beckham's face was modeled capturing his key features, and then dialogue was superimposed from different speakers communicating the same message in nine languages. In the future, there will be realistic representations of people, scenarios, and speech articulated by attendees in different languages, all integrated into videoconferencing with the goal of breaking language barriers.

An area of great concern when it comes to deepfakes is in national security. Actors, such as political activists, foreign governments, fraudsters, and others, can influence the beliefs of the general public in very damaging ways. This topic is often referred to as "influence operations." A report commissioned by the US Congress stresses the importance of devoting more research efforts to address influence operations [28]. It also offers the following very clear and succinct deepfakes definition (which we have modified with the content shown in brackets):

Deepfakes: Computer-generated [images], video, or audio [e.g., speech] (particularly of humans) so sophisticated that it is difficult to distinguish from reality. Deepfakes have also been referred to as "synthetic media."

Deepfakes are very difficult to detect, mitigate, or attribute accurately. It involves more than just applying innovative technologies. A. Cybenko and G. Cybenko point out that a person possibly rejects information if any of the following four "cognitive safeguards" are present [29]:

1. The information is incompatible with the existing worldview.
2. It does not comprise a coherent story.
3. It does not come from what is considered a credible source.
4. It is perceived that others in the same community do not believe it.

Mirsky and Lee performed an in-depth technical survey of deepfakes approaches [30]. They analyzed the foundational neural network architectures used in the creation of deepfakes. At their core, deepfakes employ generative neural network technologies. Mirsky and Lee elaborated in detail on the following networks: Encoder-Decoder; convolutional neural networks (CNNs); generative adversarial networks (GANs), including two additional variants: Image-to-Image Translation and CycleGAN; and recursive neural networks (RNN). They also discussed approaches for more robust prevention and mitigation of deepfakes such as maintaining rigorous data provenance and employing techniques (similar to approaches used in adversarial AI) for altering images with a small amount of noise, such that the adversaries using deepfake technologies will encounter difficulties in finding relevant images.

A very modern approach in the ML community is to use GANs, which consist of two neural network models. A generator neural network model is used to create data (e.g., images videos, or speech), and the output is compared, in a probabilistic way, to known target data inside another network known as the "discriminator model." This process is repeated, wherein the goal is for both models to improve until the synthetic data are indistinguishable from the target data. Once that stage is reached, the discriminator network can be discarded and synthetic data created using the generator neural network model.

It is important to point out that GAN technology has advanced the ML field in very significant and revolutionary ways relevant to many valuable applications [31]. Unfortunately, malicious actors have found ways to adapt this technology to create nefarious deepfakes.

To bring out the danger and vulnerability that deepfakes present to society, researchers at the MIT Center for Advanced Virtuality re-created a realistic video of former president Richard Nixon (using his actual resignation speech) but altered with a speech draft of a failed moon landing [32, 33]. The reenactment won the Creative Media Award presented by the Mozilla Foundation, whose mission is "to realize more trustworthy AI in consumer technology, and to examine AI's effect on media and truth."

Another very interesting research study was undertaken by the MIT Media Lab and Johns Hopkins University to determine how well crowds of humans, a leading model, and a combination of both perform on deepfake video detection [34]. They determined that humans, when given the confidence of the leading ML model and an opportunity to update their choices, outperformed the human or the leading model alone. The leading model resulted from a 2019–2020 contest ran by the Partnership for AI, in collaboration with several commercial companies, including Facebook, Amazon, and Microsoft, to determine the most accurate deepfake video detection model. During this contest, the leading model had an accuracy of 65 percent on the holdout data, consisting of 2,000 deepfake and 2,000 real videos. The data set used in the competition, called Deepfake Detection Challenge (DFDC) was the largest open-source deepfakes data set to date [35].

In the next section, we return to looking at techniques for improving the AI system explainability (XAI). As pointed out earlier, the understanding of how deep learning algorithms reach their answers is still limited. However, improvements in XAI can lead to a more robust AI system.

7.4   Explainable Artificial Intelligence

In contrast to knowledge-based expert systems, where knowledge is programmed into the algorithm by someone like a subject matter expert (SME), with ML algorithms (especially DNNs), it is difficult to understand and explain their output. The fundamental reason is that these DNN models learn from data. During training, parameters are adjusted until an acceptable minimum cost function is reached. This cost function, as described in chapter 4, is based on the difference between the desired target output and the actual output.

The algebraic steps describing the mathematical operations inside DNNs are well understood, such as the formulas for calculating weights, biases, activation functions, regularization, backpropagation, and learning rate. These steps represent the functional mechanics of the model. However, what we lack is the ability to explain how the model arrived at its classification or prediction. For example, given a set of labeled training data in the instance of a supervised learning algorithm, it is hard to explain how the model reached, with high probability, one class versus another in the same training distribution. There is no equivalence to human-understandable rules, as there are in knowledge-based expert systems. This nascent R&D field is commonly known as XAI.

We pointed out earlier that DARPA recognized this limitation early on and started a research program under the same name [9]. This program consisted of three foundational thrusts:

- A deep explanation of how a model learned key features or representations
- Interpretable models with the goal of getting closer to formulating a set of rules describing interpretable model structure
- Model induction to infer explanations of how the model arrived at its output given a set of inputs (i.e., treating the ML model as a block-box)

These three program areas required modifications to the standard ML-processing pipeline, introducing a new learning process, an interpretable model, and explainable interfaces. We highlight these DARPA research areas because they were well thought out, formulated, and necessary to achieve explainability of AI systems. And the DARPA program is an important step to help the AI research community, as a starting point, further these concepts.

However, although necessary, these three focus areas are not sufficient given the recent AI regulations (at national and international levels), such as the Global Data Protection Regulation (GDPR) put forward by the European Union (EU) [36, 37], a likely future US AI Bill of Rights, and the White House guidance for the regulation of AI applications [38]. AI designers, developers, implementers, policy makers, and anyone else responsible for decisions made by AI systems also must clarify the level of risk, balancing confidence in the machine versus consequence of actions. The high-dimensional decision space and complexity of data used in training DNNs will continue to present a challenge to fully understanding a neural network system. AI organizations must be diligent about transparency and take responsibility for explaining the decisions reached by AI systems. There will be an increased call for the right to an explanation, especially in those circumstances when the AI system is used in high-stakes applications (e.g., medicine, transportation, national security, law enforcement, finance, and many others).

There is an urgency to attend to a more rigorous explainability of the technical underpinnings of an AI system. Also, there are moral and business continuity obligations by AI corporations to provide explanations of the risks inherent with AI systems. Thus, we elaborate on a recommended approach to do so. As we emphasized earlier, our recommendation involves a multiprong approach addressing the people-process-technology triad within the context of a systems engineering methodology.

By incorporating those three elements of the people-process-technology triad, we make sure that AI practitioners do not just focus on a technical solution. We now highlight each of these respective elements:

- *People:* AI businesses must educate and train those responsible for designing, developing, and deploying AI products and services. In high-risk applications, SMEs must be included in the decision cycle. For example, in medicine, doctors should be integral to the final decision (i.e., AI playing the role of augmenting human capabilities).

- *Process:* AI organizations must establish policies, safe guardrails, and governance consistent with its own internal practices and compliant with government regulations. Just as in the case of cybersecurity, where regulatory agencies and business boards (with fiduciary responsibilities) require that organizations be able to demonstrate its defensive measures, AI businesses must also demonstrate their ability to deploy trustworthy AI. Mitchell et al. have put forward the concept of reporting the ML model performance, as well as relevant characteristics, using *Model Cards for Model Reporting* [39]. These authors also suggested that the model cards concept can be extended to provide details on adversarial AI testing. Model cards are complementary to the *Datasheets for Datasets* referenced in chapter 3, with the goal of increasing AI system transparency.

- *Technology:* AI systems must undergo a systems engineering integration-verification-validation using modern V&V technologies, employing risk assessments as discussed in chapter 2, starting with stakeholder requirements/specifications, architecture subcomponent definition, end-to-end integration, and continual AI system monitoring.

This guidance is broader than only focusing on XAI. It is relevant to making AI systems robust. The guidance must be reinforced with RAI under the umbrella of FASTEPS principles. Synergistic to our recommendation, Arrieta et al. provide an outstanding treatise on XAI (including an in-depth discussion on technical concepts, taxonomies, opportunities, and challenges) [40]. We were delighted to see the synergy between our framework and the topology of Arrieta et al. for XAI challenges and its impact on the principles of RAI.

In addition to these guidelines, as we look ahead, there is a large body of knowledge and approaches employed by the computing community that are relevant to making AI systems more explainable and robust. Within the rubric of trustworthy computing, there have been advances in computing properties such as reliability, safety, security, privacy, availability, usability, and correctness-by-construction (e.g., formal methods to ensure trustworthiness), which can be employed for trustworthy AI as discussed by Wing [41].

In the following sections, we give the reader a brief introduction to mitigation techniques. We also recommend a testing methodology to improve the robustness of AI systems.

7.5   Mitigation Techniques

In the following discussion, we elaborate on mitigation techniques applicable to protecting systems against adversarial AI and making AI systems more explainable. An AI system must be robust and compliant with the RAI principles, and it also must achieve the desired functionality as an AI product or service. Furthermore, it is not sufficient to attend to only a single subsystem block—what stakeholders care about is the end-to-end AI system delivering the desired capabilities.

In general, there is a set of desired properties that we want robust AI systems to demonstrate:

- Resiliency:
  - Is the ML model stable to small perturbations?
  - Can the AI system continue to operate even in the presence of attacks (i.e., graceful degradation)?
  - Can transfer learning succeed using pretrained models if faced with adversarial threats?
- Explainability:
  - Can the AI system be explained in enough detail to understand the decision-making process?
  - Are there rules interpreting the model structure?
- Reliability:
  - Does the AI system do what it is expected to do?
  - Are there any potential failure modes (including analysis of counterfactuals)?
- Verifiability:
  - Can the end-to-end AI system guarantee protection against data leakage, stealing of model intellectual property, privacy, and other breaches?
  - Are there guardrails to identify drifts in data, algorithm, and/or overall performance?

Mitigation techniques for addressing the resiliency property have been proposed and extensively evaluated in the literature—a small subset is discussed here. These representative techniques have limitations—for details, see the discussion by Carlini and other studies [17, 42–44]. Our objective is to expose the reader to a few of these techniques, not to give an exhaustive list, since this defense-attack (analogous to a mouse-cat game) evolves very rapidly:

- *Adversarial training:* The training set includes representative adversarial samples.
- *Data sanitization:* Cleaning of data from adversarial poisoning attacks.
- *Defensive distillation:* True class labels are converted into soft labels. Soft labels are approximate probabilistic outputs representing the correct classification, providing more flexibility to the classifier.
- *Feature squeezing:* Typical training samples have a wide space where adversaries can add perturbations. This technique reduces that space—for example, by special

smoothing. If the output with feature squeezing is different than the output from the original undisturbed (known) input, then it is likely that the squeezed input is from an adversary.

- *Defensive-GAN:* This technique employs a GAN, but the generator input comes from minimizing the reconstruction error between the generator samples and the legitimate sample, making it harder for adversaries to generate output containing attacks.

- *Projected Gradient Descent (PGD):* Several adversarial techniques exploit the error norm used in the gradient descent steps in a neural network. This approach, suggested by Madry et al. formulates a PGD based on an outer minimization (i.e., minimizing a risk function) and an inner maximization to constraint the gradient descent [45]. As a result, this generalization allows DNN designers to build more robust DNNs.

Ilyas et al. provide a theoretical explanation for the reason why data sets are nonrobust to small feature perturbations indistinguishable to humans, such as images [46].

The goal of being able to explain AI, or more specifically how the AI system arrives at its decision (also referred to as "interpretable models"), is still evolving (i.e., significant attention is being devoted at the research levels). In terms of helping answer the first question under explainability—can the AI system be explained in enough detail to understand the decision process?—approaches are often divided into white-box and black-box scenarios.

For white-box scenarios, where details of the model are known, the techniques mentioned earlier for protecting against adversarial AI are also used to gain insight into the workings of an ML model. In addition, simpler ML models (often called "shallow AI") are also easier to explain. Examples of these shallow ML models, in contrast to DNN models, are support vector machines (SVMs), decision trees, and logistic regression. These models also provide insights into the model structure to help with the second question—are there rules interpreting the model structure? There are several tools available for addressing model explainability, one of which is AI Explainability 360, an open-source tool developed by IBM [47].

More effort, in comparison to white-box scenarios, has been devoted to explaining AI for DNNs in black-box scenarios. Two common approaches are known as Local Interpretable Model-Agnostic Explanation (LIME) [48] and SHapley Additive exPlanation (SHAP), inspired by the Shapley value in game theory with a player-game construct) [49]. In LIME, a new data set is formed, with samples close to the input that we wish to explain. With this data set, a simpler and more interpretable model is created.

In SHAP, with a set of features (representing "players") and a model output (representing the "game"), data inputs (features) and ML parameters are assessed to understand model sensitivity and what features have the most influence on ML outputs. These representative techniques are considered most useful for local explainability, meaning ML model outputs for specific inputs. These approaches, and additional techniques, are compared and discussed in great detail by Bhatt et al. [50], Mittelstadt et al. [51], and Linardatos [52], including some of their limitations.

Bhatt et al. [50], emphasized that these approaches are focused on local explainability (i.e., relevant to specific model inputs). We lack techniques that are more focused on global explainability. This observation reinforces our approach in this book of looking at AI as a system, from its end-to-end architecture design through its deployment.

The steps discussed in chapter 2, following the AI system architecture implementation framework illustrated in figure 2.2, delineate a more global approach by focusing on stakeholder requirements, subsystem assessments, interplay among architecture building blocks, risk management, and end-to-end system performance. This more global approach also helps with addressing two of the desired technical properties outlined here—namely, having a reliable and verifiable robust AI system. Ultimately, people with domain knowledge must take part in helping explain the AI system outputs—especially in high-stakes use cases—including an assessment of the confidence level versus the consequence of actions, as discussed in chapter 6. For example, a radiologist must help a physician understand ML output to determine if the results of X-rays show the presence of a malignant tumor.

Another useful mitigation technique is using counterfactuals. The employment of counterfactuals can be motivated by the work of Byrne [53], since humans often think about what would happen if different inputs or scenario are taken into account. Counterfactuals are inputs to a system or model motivated by asking what-if questions. They can be used either at the end-to-end level (global explainability) or at the specific target input to a model (local explainability). However, one limitation in the use of counterfactuals is how to select those system inputs that best represent and that can lead to a more thorough explanation, with the goal of achieving a robust AI system [54].

We conclude this section on mitigation techniques by briefly pointing out some techniques for protecting against data leakage, stealing of model intellectual property, individual privacy, and other issues. These are very important protection objectives necessary to make an AI system robust. Some of the seminal work by Dwork and Roth, in protecting privacy, involves a technique known as "differential privacy" [55]. Differential privacy is a very powerful technique based on adding a small amount of predetermined noise (known only by the algorithm designer), which makes it very difficult to determine if an individual's information exists in a data set (protecting personal identifiable information, or PII). Song, with her students, proposed practical applications of differential privacy [56]. Researchers have also proposed differential privacy as a way to balance privacy versus explainability and accuracy [57], which is relevant to maintaining compliance with government regulations like GDPR. Another application is in crowdsourcing data sets used in the development of DNNs, with limited privacy loss [58].

Another technique, offering tremendous promise, is federated learning. One application of this is in the development of more robust models without necessarily disclosing the data being used. Separate organizations can develop a first-pass model with their own training data. Then, these models, not the data, are shared to derive a new model from the individual models. The final model is shared with the individual organizations' servers. One application of federated learning is in medicine, where separate hospitals have patients' data (localized in their own private servers), but they wish to develop a robust model, where the characteristics of a given disease can be better diagnosed from a larger data set without ever disclosing patients with similar diseases. Exciting work in federated learning is ongoing at Carnegie Mellon University by Smith et al. [59].

7.6   Methodology for Testing against Adversarial Attacks

The discussion in the previous section gives a foundation for examining a methodology for testing against adversarial attacks. We propose the methodology shown in figure 7.5. The methodology incorporates several of the techniques discussed here to protect systems against adversarial attacks. For simplicity, in illustrating the methodology, not all known adversarial AI defenses are included.

At the core of our approach is the use of a cybersecurity framework to make AI systems more robust against attacker offenses. Typical offense stages that an attacker might undertake are

- *Prepare for an attack:* Know the targeted system (e.g., users, communication networks, computing platform, white-box exploitation, black-box exploitation)
- *Engage:* Execute the attack
- *Maintain presence:* Persist with additional adversarial threats
- *Achieve effects*: Assess damage to targeted system

In response to these potential stages, the defender of an AI system can perform the defense stages and measure the defense posture using the metrics shown in table 7.1.

A technique that is useful in understanding and interpreting what goes on at each of the layers of a DNN is known as "saliency maps" [61]. Saliency maps work on images and videos to provide a view of what parts of the image features most contribute to

Figure 7.5 Methodology for testing against adversarial attacks.

activating specific neurons (i.e., triggering the activation function) in a neural network layer. For example, what portions of an image are most dominant in a given neural network layer. One observed limitation is that saliency maps offer more of a qualitative assessment of the internals of a neural network. However, most recently, Chakraborty et al. proposed a technique that modifies the gradient descent process based on a Bayesian approach in order to quantify and augment saliency maps with uncertainty scores [62]. Saliency maps can also be used to identify vulnerable attack surfaces by interpreting

Table 7.1   Defense stages, impact, and metrics

Defense Stages	Impact	Metrics
Identify	Understand likely attack surfaces [60]	• Loss with respect to attack magnitude. • Neuron/decision coverage.
Protect	Prevent and deflect attacks	• Change in loss from a known baseline.
Detect	Identify, characterize, and quantify attacks	• If an attack occurs, quantify the level of vulnerability and loss via measures of performance (e.g., probability of detection versus PFA; ROC; AUC).
Respond	Stop additional attacks	• Update decision boundaries and accuracy loss. • Limit or stop use of AI system. Inform stakeholders.
Recover	Assess ability to "fight" through attack and retest	• Derive confidence levels versus consequence of actions.

regions on the maps where an adversary can inject content in a way that is imperceptible by humans. Yosinski and collaborators have built an open-source tool to help in interpreting saliency maps [63, 64].

As shown in figure 7.2, another important step in developing and deploying robust AI systems is to protect the AI platforms from cybersecurity attacks. We elaborated on example approaches in chapter 5. Protecting the AI computing infrastructure, against adversarial attacks, is critically important. Advancements are ongoing under two broad categories:

1. Cybersecurity for protecting AI (security of AI) [65]
2. AI applied to making systems more secure (AI for cybersecurity; AICS) [66–68]

Robust AI systems require diligent attention to both the individual building blocks and the assessment of an end-to-end AI system, as we have emphasized throughout this book. Several researchers (from industry, government, and academia) have proposed the use of a Machine Learning Technology Readiness Levels (MLTRL) framework to ensure robust, reliable, and responsible systems [69]. MLTRL is adapted from the engineering of systems in mission-critical applications, such as spacecraft systems, that follow well-defined processes and testing standards. We briefly discuss MLTRL in chapter 10, since the topic is very relevant to ultimately having the rigorous process ingrained in the development and deployment of high-risk AI systems.

In the next three sections, we provide a discussion of robust AI system challenges, a summary of the main takeaways, and a set of exercises.

7.7   Robust AI System Challenges

In this section, we describe the robust AI system challenges with a focus on three areas: (1) adversarial AI, (2) deepfakes, and (3) XAI. This is consistent with our earlier discussion and provides local and global coverage of many of the issues requiring attention by academia, government, and industry.

As we discussed in section 7.2, and also illustrated in figures 7.2 and 7.4, there are many points of potential adversarial attacks on an AI system. Adversaries can use such tactics as poisoning attacks, evasion attacks, model inversion attacks, physical attacks, and cyberattacks. These all require that the adversary have either a white-box or a black-box or gray-box knowledge of the ML models.

An example of a very concerning adversarial attack is against autonomous or computer-assisted driving systems. This application is mission critical, and human lives are at stake. Shafiee et al. reviewed a number of common adversarial attacks and provided a clear summary of the applicable defense mechanisms [70].

In general, one of the biggest challenges in making AI systems robust is the limited shelf-life of defenses against adversarial attacks. In addition to continuing R&D investment by academia, government, and industry, there must be a unified approach following a disciplined process of constant system testing and monitoring, as described in chapter 2 and elaborated further in chapter 10.

Similarly, deepfakes can have a detrimental impact on trusting AI. There are moral and core values (i.e., an integrity compass) that societies operate under to make the world a better and safer place. A very disturbing use of deepfakes, presenting a significant challenge to protecting and responding against it, is the malicious application of deepfakes to influence or alter the beliefs of the public. The solution space demands a multipronged construct, and AI technologists, government entities, and industry must be part of the solution. Because of the damage that deepfakes can have on national security, Jack Clark, former policy director of OpenAI, in his testimony to the US House Permanent Select Committee on Intelligence in 2019, pointed out the need for three types of interventions: technical, institutional, and political [71].

Since there is a lot of synergy between securing cybersystems from attacks and making AI systems robust, we should stand up AI emergency response teams against deepfakes and adversarial AI—analogous to Computer Emergency Response Teams (CERTs) established to respond rapidly against cyberthreats. In the cybersecurity realm, Landwehr proposed the idea of a "building code for building code" [72]. In addition to model cards for model reporting [39], which it is an excellent idea, we should consider more the requirement that AI organizations in academia, government, and industry comply with a building code equivalent for AI systems from the architecture design through system deployment. Key to this approach is the continual inspection, testing, and monitoring of AI systems. Shneiderman offered an excellent governance structure that complements our recommendation for more robust and responsible AI [73]. Many of these ideas need to be codified to achieve long-lasting traction among academia, government, and industry while not atrophying research progress and innovation.

There are recognition and concerted efforts to advance the state of the art in XAI. Prominent AI conferences, including conferences and workshops as part of the Neural Information Processing Systems (NeurIPS) conference, Association for the Advancement of AI (AAAI) conference, and International Joint Conference on Artificial Intelligence (IJCAI), which hold sessions, tutorials, and workshops focused on XAI [74]. To date, one of the major challenges, as expressed by researchers and practitioners, is the lack of XAI tools for model monitoring, debugging, and interfacing with AI developers and users [50]. Verma et al. formulated a set of eight challenges, from the perspective of industry, requiring additional XAI research to improve the adoption of XAI (specifically explainable ML) [75].

In sections 7.2–7.4, we described and recommended an approach that encompasses the people-process-technology triad (at the AI system level) which addresses adversarial AI, deepfakes, and XAI. AI organizations have an obligation to characterize and explain risks, especially in high-stakes applications, similar to today's businesses being required by their boards to explain the level of cybersecurity risks.

In table 7.2, based on the earlier discussion, we outline challenges and recommendations, using the same horizon structure as in previous chapters.

These horizon time lines, challenges, and recommendations provide opportunities for academia, government, and industry for improving the present limitations of AI systems and helping with making them more robust.

7.8 Main Takeaways

There are many classes of vulnerabilities affecting today's AI systems. Some of these vulnerabilities can take the form of intentional or unintentional changes to the data, ML algorithms, or derived outputs. The AI community is devoting significant efforts to making systems more robust. Here, we repeat the operative definition employed throughout the chapter:

Table 7.2   Robust AI Systems challenges and recommendations

Challenge Areas	Recommendations for Robust AI Systems
Horizon 1: Content-Based Insight (1–2 years)	• Develop benchmarks and metrics to assess robustness. • Improve XAI tools for debugging, monitoring, and explaining AI model performance. • Design adversarial AI countermeasures. • Create an infrastructure to facilitate a methodology for testing against adversarial AI (for more, see section 7.6). • Institute a process following the MLTRL framework [69]. • Formulate a building code for AI systems, including enforcement.
Horizon 2: Collaboration-Based Insight (3–4 years)	• Assess the vulnerability of foundation models and use in human-machine applications [76]. • Define evaluation guidelines for computer-assisted driving systems. • Establish AI emergency response teams to defend against adversarial AI and deepfakes. • Learn system vulnerabilities through AI hackathons, with domain experts in the loop. • Take action on the three interventions: technical, institutional, and political [71], consistent with a people-process-technology triad.
Horizon 3: Context-Based Insight (5+ years)	• Bring context (expert knowledge and stakeholders) into the evaluation of AI systems. • Develop AI systems using a correctness-by-construction technique (e.g., formal methods to ensure trustworthiness). • Recommend actionable insights in context, based on understanding the AI system's decision process.

Robust AI system: The ability of a narrow AI system to deliver capabilities, assessed through the lens of confidence in the machine versus consequence of actions, constrained to providing value to stakeholders in a responsible manner.

One stumbling block to achieve robustness is caused by the high degree of brittleness that AI systems have to adversarial AI. There are four types of adversarial attacks;

- Poisoning attacks
- Evasion attacks
- Model inversion
- Physical attacks and cyberattacks

We also discussed the significant overlap between making AI systems robust and the need to comply with RAI principles (see chapter 8 for a more detailed description). It is important to emphasize that AI systems can comply with RAI principles but still not be robust. We need to achieve both robustness and compliance with RAI before AI systems can be deployed in operations, and once deployed, AI systems must undergo continual monitoring.

Adversaries can gain knowledge about the workings of ML models in the following three scenarios:

- White-box: Full knowledge of the ML model
- Black-box: No knowledge of the ML model internals, but able to derive its functionality by proving the model outputs given a set of inputs
- Gray-box: Limited knowledge of the ML model

In addition to adversarial AI, another topic that has received attention recently is the ease of creating deepfakes, computer-generated images, videos, or speeches that are very difficult to distinguish from reality. There are useful applications of deepfakes—for example, in entertainment, education, and videoconferencing. However, there are also nefarious uses of deepfakes that have very damaging impacts on individuals, businesses, societies, national security, and other areas. Both of these system vulnerabilities—adversarial AI and deepfakes—are existential threats to AI, making the technology useless. Therefore, R&D investments are crtical to stay ahead of malicious adversaries.

One approach to improve the trustworthiness of AI products and services is XAI, the ability to explain how these systems arrive at their decision outputs. Explainability requires more than technical solutions. We recommended approaches that address solutions under a people-process-technology triad.

We also elaborated on mitigation techniques with the goal of achieving four desired properties that robust AI systems must demonstrate:

- Resiliency
- Explainability
- Reliability
- Verifiability

In addition, we outlined several approaches to achieve resiliency in terms of sensitivity to small model perturbations. We pointed out that these techniques were useful for understanding the model internals, particularly in white-box scenarios. Two common techniques used for explaining ML model behavior, particularly in black-box scenarios, are LIME and SHAP. Both of these techniques have limitations against adversarial AI, a rapidly evolving area between the defender capabilities and the attackers' abilities to improve their adversarial techniques.

We proposed a methodology for testing against adversarial AI. The recommendations centered on a framework adapted from cybersecurity defense stages. The methodology consists of five defense stages:

1. Identify: Understand likely attack surfaces.
2. Protect: Prevent and deflect attacks.
3. Detect: Identify, characterize, and quantify attacks.
4. Respond: Stop additional attacks.
5. Recover: Assess the ability to fight off an attack, complete the mission, and then retest.

We summarized a set of challenges, offered a set of recommendations, and provided opportunities for academia, government, and industry for improving the present limitations of AI systems and helping to make them more robust.

In chapter 8, we will discuss many of the issues under the rubric of RAI. AI researchers, designers, developers, integrators, and anyone in the business of providing AI products and services must ascertain robustness and compliance with RAI principles. Both of these goals go hand in hand.

7.9 Exercises

1. Describe and explain the key attributes of the robust AI systems definition used in this chapter.

2. Types of adversarial attacks include which of the following?
   a. Poisoning
   b. Evasion
   c. Model inversion
   d. Physical attacks and cyberattacks
   e. All of the above
   f. None of the above

3. For an AI system to be robust, all building blocks in the AI system architecture must be robust against adversarial attacks.
   a. True
   b. False

4. Scenarios describing the knowledge attained by the adversary can be described as white-box, black-box, and gray-box.
   a. True
   b. False

5. If an adversary alters a stop sign, and the ML model interprets it as a speed limit sign at testing time, this is considered a type of poisoning attack.
   a. True
   b. False

6. Some examples of beneficial uses of deepfakes include entertainment, education, and avatars embedded in video conferencing.
   a. True
   b. False

7. Explain key attributes of the deepfakes definition used in the chapter.

8. GANs are not useful for generating deepfakes.
   a. True
   b. False

9. Elaborate on why it is difficult to explain how AI systems, specifically DNN models, arrived at the model outputs.

10. Robust AI systems must demonstrate four desirable properties: resiliency, explainability, reliability, and verifiability.
    a. True
    b. False

11. PGD is an example of a mitigation approach requiring only a minimization of a risk function.
    a. True
    b. False

12. Two common approaches for explaining the behavior of ML models, in a black-box scenario include which of the following?
    a. LIME (Local Interpretable Model-Agnostic Explanation)
    b. SHAP (SHapley Additive exPlanation)
    c. Both LIME and SHAP
    d. None of the above

13. Differential privacy is a technique that helps in protecting PII.
    a. True
    b. False

14. Federated learning allows multiple parties to share their respective ML models and disclose their data to achieve a more robust aggregated model.
    a. True
    b. False

15. Enumerate and describe the five defense stages outlined in the methodology for testing against adversarial AI.

16. Saliency maps can be used to identify vulnerable attack surfaces by interpreting regions on the maps where an adversary can inject content in a way that is imperceptible to humans.
    a. True
    b. False

17. Why is addressing robust AI in the context of the people-process-technology triad so important?

18. Assume that you have access to social network analytical tools. Can you formulate an innovative process to use such tools to making AI systems more robust?

7.10   References

1. Wiener, N., Some moral and technical consequences of automation: As machines learn they may develop unforeseen strategies at rates that baffle their programmers. *Science*, 1960. 131(3410): 1355–1358.

2. Marcus, G., The next decade in AI: four steps towards robust artificial intelligence, 2020. arXiv preprint arXiv:2002.06177.

3. Bommasani, R., D. A. Hudson, E. Adeli, et al., On the opportunities and risks of foundation models, 2021. arXiv preprint arXiv:2108.07258.

4. Sutskever, I. GPT-3 use cases, in *2022 HAI Spring Conference*. Stanford University. https://hai.stanford.edu/2022-hai-spring-conference-agenda.

5. Dietterich, T. G., Steps toward robust artificial intelligence. *AI Magazine*, 2017. 38(3): 3–24.

6. Caceres, R., *Robust AI; Artificial intelligence: A systems engineering approach course*. 2020, MIT Lincoln Laboratory.

7. Kahneman, D., O. Sibony, and C. R. Sunstein, *Noise: A flaw in human judgment*. 2021, Little, Brown.

8. Barmer, H., R. Dzombak, M. Gaston, et al., *Robust and secure AI*. 2021, National AI Engineering Initiative.

9. Gunning, D. and D. Aha, DARPA's explainable artificial intelligence (XAI) program. *AI Magazine*, 2019. 40(2): 44–58.

10. Chambers, A., *Re: Artificial intelligence risk management framework*. 2021.

11. Mattson, P., V. J. Reddi, C. Cheng, et al., MLPerf: An industry standard benchmark suite for machine learning performance. *IEEE Micro*, 2020. 40(2): 8–16.

12. Laplante, P. and J. Voas, Zero-trust artificial intelligence? *Computer*, 2022. 55(02): 10–12.

13. Pelillo, M. and T. Scantamburlo, *Machines we trust: Perspectives on dependable AI*. 2021, MIT Press.

14. Szegedy, C., W. Zaremba, I. Sultskever, et al., Intriguing properties of neural networks, 2013. arXiv preprint arXiv:1312.6199.

15. Goodfellow, I. J., J. Shlens, and C. Szegedy, Explaining and harnessing adversarial examples, 2014. arXiv preprint arXiv:1412.6572.

16. Papernot, N., P. McDaniel, S. Jha, et al., The limitations of deep learning in adversarial settings, in *2016 IEEE European Symposium on Security and Privacy (EuroS&P)*. 2016, IEEE.

17. Carlini, N. and D. Wagner, Towards evaluating the robustness of neural networks, in *2017 IEEE Symposium on Security and Privacy*. 2017, IEEE.

18. Papernot, N., P. McDaniel, I. Goodfellow, et al., Practical black-box attacks against machine learning, in *Proceedings of the 2017 ACM on Asia Conference on Computer and Communications Security*, Abu Dhabi. 2017.

19. Eykholt, K., I. Evtimov, E. Fernandes, et al., Robust physical-world attacks on deep learning visual classification, in *Proceedings of the IEEE Conference on Computer Vision and Pattern Recognition*. 2018. https://ieeexplore.ieee.org/xpl/conhome /1000147/all-proceedings.

20. Banerjee, S., C.-Y. Chen, J. Takudar, et al., Towards functionally robust AI accelerators, in *2021 IEEE Microelectronics Design & Test Symposium (MDTS)*. 2021, IEEE.

21. Martin, P., J. Fan, T. Kim, et al., Toward effective moving target defense against adversarial AI, in *MILCOM 2021–2021 IEEE Military Communications Conference (MILCOM)*. IEEE.\

22. Sadi, M., B. M. S. Bahar Tukuder, K. Mishty, and M. T. Rahman, Attacking deep learning AI hardware with universal adversarial perturbation, 2021. arXiv preprint arXiv:2111.09488.

23. Draper, B. *Guaranteeing AI robustness against deception (GARD)*. https://www .darpa.mil/program/guaranteeing-ai-robustness-against-deception.

24. Goodwin, V. H. and R. S. Caceres, *System analysis for responsible design of modern AI/ML systems*, 2022. arXiv preprint arXiv:2204.08836.

25. Westerlund, M., The emergence of deepfake technology: A review. *Technology Innovation Management Review*, 2019. 9(11).

26. Guan, H., Y. Lee, and L. Diduch, *Open Media Forensics Challenge 2022 Evaluation Plan*. 2022.

27. *David Beckham launches the world's first voice petition to end malaria*. 2021. Malaria No More. https://malariamustdie.com/news/david-beckham-launches -worlds-first-voice-petition-end-malaria.

28. Schmidt, E., et al., *National Security Commission on Artificial Intelligence (NSCAI final report)*. 2021, National Security Commission on Artificial Intelligence. https://www.nscai.gov/2021-final-report/.

29. Cybenko, A. K. and G. Cybenko, AI and fake news. *IEEE Intelligent Systems*, 2018. 33(5): 1–5.

30. Mirsky, Y. and W. Lee, The creation and detection of deepfakes: A survey. *ACM Computing Surveys (CSUR)*, 2021. 54(1): 1–41.

31. Goodfellow, I., J. Pouget-Abadie, M. Mirza, et al., Generative adversarial networks. *Advances in Neural Information Processing Systems*, 2014. 27.

32. Panetta, F., H. Burgund, P. Amer, et al., In event of moon disaster, in *SIGGRAPH Asia 2020 Computer Animation Festival*. 2019. 1–1.

33. Parnell, M. *Misinformation epidemic with "In Event of Moon Disaster."* 2020. https://virtuality.mit.edu/tackling-the-misinformation-epidemic-with-in-event-of-moon-disaster/.

34. Groh, M., Z. Epstein, C. Firestone, and R. Picard, Deepfake detection by human crowds, machines, and machine-informed crowds. *Proceedings of the National Academy of Sciences*, 2022. 119(1).

35. Dolhansky, B., J. Bitton, B. Pfaum, et al., The deepfake detection challenge (DFDC) dataset, 2020. arXiv preprint arXiv:2006.07397.

36. Hamon, R., H. Junklewitz, I. Sanchez, et al., Bridging the gap between AI and explainability in the GDPR: Towards trustworthiness-by-design in automated decision-making. *IEEE Computational Intelligence Magazine*, 2022. 17(1): 72–85.

37. Regulation (EU) 2016/679 of the European Parliament and of the Council of 27 April 2016 on the protection of natural persons with regard to the processing of personal data and on the free movement of such data, and repealing Directive 95/46/EC (General Data Protection Regulation), in *Official Journal of the European Union L.* 2016. https://eur-lex.europa.eu/legal-content/EN/TXT/PDF/?uri=CELEX:32016R0679.

38. Guidance for regulation of artificial intelligence applications, in *Memorandum for the heads of executive departments and agencies.* 2020.

39. Mitchell, M., S. Wu, A. Zaldivar, et al., Model cards for model reporting, in *Proceedings of the Conference on Fairness, Accountability, and Transparency,* Atlanta. 2019. https://dl.acm.org/doi/proceedings/10.1145/3287560.

40. Arrieta, A. B., N. Díaz-Rodríguez, J. Del Ser, et al., Explainable artificial intelligence (XAI): Concepts, taxonomies, opportunities and challenges toward responsible AI. *Information Fusion*, 2020. 58: 82–115.

41. Wing, J. M., Trustworthy AI. *Communications of the ACM*, 2021. 64(10): 64–71.

42. Asha, S. and P. Vinod, Evaluation of adversarial machine learning tools for securing AI systems. *Cluster Computing*, 2022. 25(1): 503–522.

43. Carlini, N., A. Athalye, N. Papernot, et al., On evaluating adversarial robustness, 2019. arXiv preprint arXiv:1902.06705.

44. Chakraborty, A., M. Alam, V. Dey, et al., Adversarial attacks and defences: A survey, 2018. arXiv preprint arXiv:1810.00069.

45. Madry, A., A. Makelov, L. Schmidt, et al., Towards deep learning models resistant to adversarial attacks, 2017. arXiv preprint arXiv:1706.06083.

46. Ilyas, A., S. Santurkar, D. Tsipras, et al., Adversarial examples are not bugs, they are features. *Advances in Neural Information Processing Systems*, 2019. 32.

47. Arya, V., R. K. E. Bellamy, P.-Y. Chen, et al., One explanation does not fit all: A toolkit and taxonomy of AI explainability techniques, 2019. arXiv preprint arXiv:1909.03012.

48. Ribeiro, M. T., S. Singh, and C. Guestrin. "Why should I trust you?" Explaining the predictions of any classifier, in *Proceedings of the 22nd ACM SIGKDD International Conference on Knowledge Discovery and Data Mining*. 2016. ACM.

49. Lundberg, S. M. and S.-I. Lee, A unified approach to interpreting model predictions. *Advances in Neural Information Processing Systems*, 2017. 30.

50. Bhatt, U., A. Xiang, S. Sharma, et al., Explainable machine learning in deployment, in *Proceedings of the 2020 Conference on Fairness, Accountability, and Transparency*, Barcelona. 2020.

51. Mittelstadt, B., C. Russell, and S. Wachter. Explaining explanations in AI, in *Proceedings of the Conference on Fairness, Accountability, and Transparency*, Atlanta. 2019. https://dl.acm.org/doi/proceedings/10.1145/3287560.

52. Linardatos, P., V. Papastefanopoulos, and S. Kotsiantis, Explainable AI: A review of machine learning interpretability methods. *Entropy*, 2020. 23(1): 18.

53. Byrne, R. M., Counterfactual thought. *Annual Review of Psychology*, 2016. 67: 135–157.

54. Byrne, R. M. Counterfactuals in explainable artificial intelligence (XAI): Evidence from human reasoning, in *IJCAI*. 2019. https://www.ijcai.org/proceedings/2019/876.

55. Dwork, C. and A. Roth, The algorithmic foundations of differential privacy. *Foundations and Trends in Theoretical Computer Science*, 2014. 9(3–4): 211–407.

56. Johnson, N., J. P. Near, and D. Song, Towards practical differential privacy for SQL queries. *Proceedings of the VLDB Endowment*, 2018. 11(5): 526–539.

57. Harder, F., M. Bauer, and M. Park. Interpretable and differentially private predictions, in *Proceedings of the AAAI Conference on Artificial Intelligence*, New York. 2020.

58. Abadi, M., A. Chu, I. Goodfellow, et al., Deep learning with differential privacy, in *Proceedings of the 2016 ACM SIGSAC Conference on Computer and Communications Security*, Vienna. 2016.

59. Li, T., A. Kumar Sahu, A. Talwalkar, and V. Smith, Federated learning: Challenges, methods, and future directions. *IEEE Signal Processing Magazine*, 2020. 37(3): 50–60.

60. Fazelnia, M., I. Khokhlov, and M. Mirakhorli, Attacks, defenses, and tools: A framework to facilitate robust AI/ML systems, 2022. arXiv preprint arXiv:2202.09465.

61. Simonyan, K., A. Vedaldi, and A. Zisserman, Deep inside convolutional networks: Visualising image classification models and saliency maps, in *Workshop at International Conference on Learning Representations*. 2014, Citeseer. https://iclr.cc/archive/2014/workshop-proceedings/.

62. Chakraborty, S., P. Gurram, F. Le, et al., Augmenting saliency maps with uncertainty, in *Artificial intelligence and machine learning for multi-domain operations applications III*. 2021, International Society for Optics and Photonics.

63. Yosinski, J., J. Clune, A. Nguyen, et al., Understanding neural networks through deep visualization, 2015. arXiv preprint arXiv:1506.06579.

64. Yosinski, J., J. Clune, A. Nguyen, et al., Deep visualization toolbox. Yosinski.com. https://yosinski.com/deepvis.

65. McDaniel, P., J. Launchbury, B. Martin, et al., *Artificial intelligence and cyber security: Opportunities and Challenges Technical Workshop summary report*. Networking & Information Technology Research and Development Subcommittee and the Machine Learning & Artificial Intelligence Subcommittee of the National Science & Technology Council, 2020.

66. Barash, G., M. Castillo-Effen, N. Chhaya, et al., Reports of the workshops held at the 2019 AAAI Conference on Artificial Intelligence. *AI Magazine*, 2019. 40(3): 67–78.

67. Holt, J., E. Raff, A. Ridley, et al., *Proceedings of the Artificial Intelligence for Cyber Security (AICS) Workshop at AAAI 2022*. arXiv e-prints, 2022: p. arXiv: 2202.14010.

68. Ross, D., A. Sinha, D. Staheli, and B. Streillein, *Proceedings of the Artificial Intelligence for Cyber Security (AICS) Workshop 2020*, 2020. arXiv preprint arXiv:2002.08320.

69. Lavin, A., C. M. Gilligan-Lee, A. Visnjic, et al., *Technology readiness levels for machine learning systems*, 2021. arXiv preprint arXiv:2101.03989.

70. Shafiee, M. J., A. Jeddi, A. Nazemi, et al., Deep neural network perception models and robust autonomous driving systems: Practical solutions for mitigation and improvement. *IEEE Signal Processing Magazine*, 2020. 38(1): 22–30.

71. Watts, C., *The national security challenges of artificial intelligence, manipulated media, and deepfakes*. 2019. Foreign Policy Research Institute.

72. Landwehr, C., We need a building code for building code. *Communications of the ACM*, 2015. 58(2): 24–26.

73. Shneiderman, B., Responsible AI: Bridging from ethics to practice. *Communications of the ACM*, 2021. 64(8): 32–35.

74. Miller, T., R. Weber, and D. Magazenni, Report on the 2019 IJCAI Explainable Artificial Intelligence Workshop. *AI Magazine*, 2020. 41(1): 103–105.

75. Verma, S., A. Lahiri, J. P. Dickerson, and S.-I. Lee, Pitfalls of explainable ML: An industry perspective, 2021. arXiv preprint arXiv:2106.07758.

76. Johnson, S., A.I. is mastering language. Should we trust what it says? *New York Times Magazine*. 2022. https://www.nytimes.com/2022/04/15/magazine/ai-language.html.

8

Responsible Artificial Intelligence

Those that fail to learn from history are doomed to repeat it.
—British statesman Winston Churchill

Artificial intelligence (AI) is a wide-ranging tool that is revolutionizing technology across industries. Advancements in computing and the proliferation of data have enabled significant improvements to the quality of products and services through performance gains and more tailored user experiences. From personalized Netflix and YouTube recommendations to self-driving cars, conversational virtual assistants like Siri and Alexa, and AI-powered medical diagnoses that are capable of outperforming doctors, AI is transforming the way that we live. However, these benefits have come paired with increasing concerns about the unintended consequences of AI on people and society [1]. For example, recent studies have uncovered increased rates of anxiety and depression among teens as a result of the increasing role that AI-powered social media platforms like Instagram play in worsening body image issues for teenage girls and mental health among young adults [2]. Moreover, the widespread adoption and application of AI-powered decision-making systems in highly consequential domains such as law enforcement, financial services, and medicine have raised concerns around the adherence of AI to key value-based principles such as fairness, accountability, safety, transparency, ethics, privacy, and security (FASTEPS). These concerns have become a pervasive topic of recent public discussion as the stakes of AI grow higher. In this chapter, we emphasize the importance of responsible AI (RAI) to ensure that the benefits of AI are realized while minimizing unintended harm.

RAI describes the practice of designing, developing, and deploying AI that is safe, ethical, trustworthy, accountable, and used with the good intention of benefiting all people and society at large [3]. Increasing public concern and mistrust of AI in recent years have contributed to the rapidly growing interest and investment in RAI across academia, industry, and government. As AI continues to revolutionize industries, RAI practices are important to ensuring that its benefits are maximized without compromising public trust or creating unintended harm.

We chose to highlight the quote from the former British prime minister Sir Winston Churchill at the start of this chapter because it emphasizes the importance of learning from history to avoid repeating past mistakes. Churchill's words are especially important and relevant to the development of AI, as attending to RAI from the start of a system design can help minimize the likelihood of repeating prior errors made in the past with the deleterious use of powerful technology. As discussed by Kissinger et al. in the book *The Age of AI and Our Human Future* [4], "societies could permit an AI to be employed only after its creators demonstrate its reliability through testing processes." The authors' argument highlights the importance of RAI practices and rigorous testing to build trust in AI and prevent potential harm to individuals and society.

Since the third Industrial Revolution (beginning around the mid-1940s), the engineering profession has made significant progress in defining frameworks and processes for putting users at the center of product design and development [5]. However, the rise of RAI requires going beyond traditional user-centered design principles to ensure that AI systems are designed to uphold important, value-based FASTEPS principles. These principles must be integrated into the entire AI development life cycle, from the conceptualization stage to development, deployment, and beyond. It's important to note that the development of AI has led to a renewed focus on ethical principles such as fairness and transparency. These principles were not always as important or relevant in engineering-based systems, but they are essential for ensuring that AI systems are used in a responsible and beneficial manner. Ultimately, progress toward RAI requires the consideration of all AI-powered technologies as sociotechnical systems (STSs). An STS is one that inherently requires an understanding of the complex interaction between technology, people, and society and how such interactions influence the functionality and usage of the technology [6]. Failure to consider both technical and social factors is likely to result in not only degraded system performance and utility, but serious harm to individuals and society at large [7].

In this chapter, we start by describing the role of AI in society and provide a short overview of efforts made across academic, industrial, and governmental circles to promote the design, development, and use of RAI. We also describe the progress that has been made internationally to track AI principles and shape AI policy. To further illustrate the importance of RAI, we present a review of nine case studies showcasing examples of

AI causing harm at the individual, collective, and societal levels. These case studies demonstrate instances where AI technology has been subject to various types of biases, resulting in unintended consequences and harms.

In section 8.3, we describe the important considerations for STSs. In doing so, we provide a vocabulary for classifying different types of biases in AI systems and AI-induced harms. In this context, we define "harms" as the real-life risks, impacts, and implications of AI-powered systems.

With an understanding of the biases in AI systems and their real-world harms, we describe a set of seven value-based AI principles with respect to FASTEPS guardrails. We then review important considerations that should be made throughout the stages of the AI development life cycle to operationalize these principles, from problem identification and solution design to data collection and processing, model development, and finally deployment.

The later sections build on these important concepts to outline a set of technical and tactical steps that practitioners can take to work toward RAI in practice. We finally conclude with a set of RAI challenges, the main takeaways, and a set of exercises for the reader.

8.1 AI and Society

The concept of "technology" is defined as the application of scientific knowledge to practical, societal needs [8]. However, this narrow definition fails to capture the ethical duty that comes with the development and implementation of new technologies. As a result, considerations regarding the ethics and moral principles that govern how technology should be developed and used have long been considered tangential to the primary functions of practitioners in the design and development of technological solutions [9]. However, recent cases of AI harm have motivated concerted efforts across government, industry, and academia to emphasize and enforce the design, development, and deployment of AI systems that comply with FASTEPS principles.

Within computing professional societies, longstanding efforts have been made to prevent the misuse of technology by more concretely defining the role of engineers and scientists in society. Acknowledgment of such responsibilities is seen in ethics codes defined, for example, by the Association for Computing Machinery (ACM), Institute of Electrical and Electronics Engineers (IEEE), and National Society of Professional Engineers (NSPE) [10–12]. Civil forces are also advocating for RAI. A great overview of prominent voices from civil society organizations in the push toward AI regulation and RAI is presented by de Laat [13].

In academia, leading institutions have begun making significant investments to revamp research and education efforts in RAI. Stanford University launched the Institute for Human-Centered AI with a commitment to "studying, guiding, and developing human-centered AI technologies and applications" [14]. Harvard University introduced its Embedded EthiCS initiative to train its computer scientists to effectively "think through the ethical and social implications of their work" [15]. The Massachusetts Institute of Technology (MIT) established the Schwarzman College of Computing, a $1.1 billion commitment to addressing the global opportunities and challenges presented by computing across industries and academic disciplines. One of the college's major emphases is in research focusing on the social and ethical responsibilities of computing (SERC) to further enable MIT to emerge as a leader in the "responsible and ethical evolution of technologies that are poised to transform society" [16]. Academic conferences and workshops have also emerged, bringing together students, researchers, and practitioners to advance the field of RAI and discuss challenges and solutions.

Similar investments have also been made in industry circles as well. The Partnership on AI represents a coalition of diverse voices from across the AI community to ultimately advance positive outcomes for people and society at large [17]. Since the early coalition of large AI companies such as Amazon, Apple, Facebook, Google, IBM, and Microsoft in 2016, efforts have been made across industry to develop benchmarks and best practices for AI [13]. Beyond public commitments to AI principles, some of these organizations have made strides toward self-regulation, establishing hubs for ethical AI research, defining internal RAI standards and principles, and even contributing to the development and open sourcing of new RAI software tools [18–20].

However, there are many who argue that self-regulation alone is not enough. Public demand for government regulation of AI has motivated growing national and international efforts in the establishment of AI regulation and legislation [21]. In the US, while many efforts have been made to increasingly support and conduct AI research and development (R&D), invest in AI technologies, establish initiatives, form committees, and even define principles for AI regulation [22], there currently is no comprehensive federal legislation specifically designed to regulate the development and use of AI. However, AI systems are partly regulated through other existing laws and regulations that apply to certain aspects of AI development and use, such as privacy and data protection laws, antidiscrimination laws, and regulations related to specific industries such as healthcare and finance.

For example, to date, three states have developed comprehensive consumer data privacy laws (the California Consumer Privacy Act, Virginia Consumer Data Protection Act, and Colorado Privacy Act) that define user rights and protections and establish standards governing the collection and use of user data [23]. Efforts are also emerging federally, with the White House's Office of Science and Technology Policy (OSTP) collaborating with academia, civil society, the private sector, and the public to develop a so-called AI Bill of Rights outlining a set of guidelines for the responsible design and use of AI and guarding against data-driven technologies [24]. Internationally, the European Union (EU) has served as a leader in effective regulatory governance. For example, the General Data Protection Regulation (GDPR) is a regulation on data protection and privacy in the European Union. Implemented in 2018, GDPR sets strict policies and regulations for businesses operating on individuals in the European Union and has served as an important component of its privacy law [25]. More recently, the European Commission's ambitious proposal for a new AI act sets out core rules for the development and use of AI-driven products, services, and systems within the European Union [26].

Beyond the efforts of individual governments, international bodies such as the World Economic Forum have also developed ethical frameworks around the responsible development and use of AI [27]. The AI policy observatory of the Organisation for Economic Cooperation and Development (OECD) serves as a crucial source of real-time information on AI principles and policy initiatives in over sixty countries and territories [28].

The need for RAI investment and AI regulation is clear, and efforts are being made across industry, academia, and government to invest in RAI and regulate AI development and its use. However, one of the biggest challenges faced by organizations and governments is in finding ways to achieve regulatory governance without stifling innovation [29].

8.2   Case Studies: Harms from Artificial Intelligence

From the watershed moment of AlexNet in 2012 [30] to the major milestone in AI research with AlphaGo's 2016 victory over a human world champion at the game of Go [31], the AI industry and community have celebrated many successes. However, these achievements have been accompanied by an equally noticeable number of blunders and AI harms. From self-driving car errors resulting in fatal car accidents to the use of biased face recognition technologies, real-world AI harms have fueled public mistrust and concern around the adoption of AI. In this section, we will examine nine case studies that illustrate a range of AI-induced harms in the real world. Our goal here is not to undermine the advancements made in AI or to scrutinize the actions of particular organizations, but rather to identify lessons from these past incidents, with the goal of preventing similar mishaps in the future.

Bias in Commercial Face Recognition Technologies: A 2018 study by Buolamwini and Gebru uncovered racial and gender bias in commercial face recognition technologies. The study evaluated three commercial gender classification systems and found that darker-skinned females were the most misclassified group, with error rates of up to 34.7 percent, while the maximum error rate for lighter-skinned males was only 0.8 percent. Furthermore, the study found gender and skin-type skews in existing data sets, with benchmark data sets overrepresenting lighter-skinned individuals, particularly men [32]. The implications of such findings are significant, considering the potential downstream application of such technologies in law enforcement and healthcare. The disparities in performance across different subpopulations raise concern about the fairness, inclusiveness, and overall transparency of AI systems broadly.

Emergent Bias in Microsoft Conversational Chatbot: In 2016, Microsoft's AI conversational chatbot, Tay, revealed the dangers of concept drift [33] as the bot began spewing abusive and offensive content within twenty-four hours of being released. Tay was implemented with extensive user studies, stress testing, and filtering in preparation for many types of potential abuses of the system. Tay was released to a broad group of users on Twitter, where increased interaction was anticipated to provide additional data that would enable the AI to learn and improve. However, the company cited that a coordinated attack exploited a vulnerability in Tay, leading the chatbot to generate inappropriate and harmful content [34]. This incident highlights that the task of designing and developing such a system is not solely a technical problem, but a social one as well. The design of Tay failed to consider the context in which it would be deployed and edge cases around potential system vulnerabilities, attacks, and subsequent harms. Such vulnerabilities raise concern around the robustness, reliability, and safety of AI systems broadly.

Public Safety of Self-Driving Vehicles: The vast amounts of data generated by sensors, paired with breakthroughs in deep learning, have enabled autonomous driving systems capable of simulating human perception and decision-making processes. However, the emergence of self-driving vehicles on today's roads has raised concerns around machine ethics and public safety. Several accidents resulting from human and AI error, such as the Uber self-driving car accident in 2018 in Arizona that resulted in a pedestrian fatality, have further intensified these concerns about the safety of self-driving cars. These incidents highlight the real-world ethical dilemmas that driverless cars must be designed to confront, such as having to choose between saving the life of a passenger or the life of a pedestrian. These dilemmas were once the subject of thought experiments in ethics, such as the "trolley problem," but they are now a reality that we must grapple with as self-driving cars become more widespread [35]. At the core of this ethics experiment is the choice between one harmful (even deadly) option versus another, as we elaborate on next. The Moral Machine experiment at MIT, which crowdsourced millions of human perspectives on moral decisions made by machine intelligence, showed significant variations in ethical preferences in different countries. Such findings reflect the role of many factors such as geography, religion, and culture in informing one's ethical principles, as well as the challenge of aligning on a standard around machine ethics [36]. Nonetheless, companies are beginning to program ethical decisions into self-driving vehicles and AI systems broadly. The question then becomes: Whose ethical and moral principles are reflected in the consequential decisions that AI systems are making? Furthermore, beyond ethics, who is being held accountable for the decisions and subsequent consequences of AI systems such as self-driving vehicles? Such open questions raise concern around AI systems' adherence and commitment to FASTEPS principles.

Facebook, Cambridge Analytica, and Data Privacy: In 2018, Facebook gave unauthorized access of personally identifiable information (PII) of more than 87 million Facebook users to the data firm Cambridge Analytica. The company, hired by Donald Trump's 2016 presidential campaign, used this data to identify the personalities of American voters and microtarget US voters with tailored messages across digital channels to influence their behavior at the ballot box [37]. The 2018 Facebook–Cambridge Analytica data scandal sparked public outcry around user data privacy, security, and the ethics of political microtargeting.

Bias in AI-Powered Decision-Making Tools: AI has enabled performance and efficiency gains in sectors such as medicine, where studies have identified models capable of outperforming doctors on tasks like breast cancer identification [38]. However, such systems also have been proven to encode and perpetuate social bias. In medicine, a 2019 study found racial bias in an Optum algorithm widely used in US hospitals for allocating specialized healthcare to nearly 200 million patients each year [39]. In criminal justice, predictive algorithms have been widely used to predict where crimes will occur, who is likely to commit crimes, and who is likely to reoffend in the future. A 2016 analysis of the Correctional Offender Management Profiling for Alternative Sanctions (COMPAS) system, a widely used criminal risk assessment tool developed in 1998 and used to asses more than 1 million offenders, found that the predictions were unreliable and racially biased, favoring white defendants by underpredicting recidivism (with lower false positive rates), and overpredicting for black defendants (with higher false positive rates) [40]. Such instances are a reminder that while AI can be a powerful tool, it can also play a harmful role in encoding and perpetuating social injustices and inequities.

Bias in Word Embeddings and Language Models: Natural language processing (NLP) is a subfield of AI concerned with how computers process and understand human language. Advancements in NLP have had practical applications, from chatbots and predictive text features like autocorrect and autocomplete to semantic search and language translation systems. Word embeddings, foundational to NLP, are learned vector representations of words that are used to encode semantic meaning. However, research has found that word embeddings quantify and encode real-world stereotypes and biases [41]. Subsequently, cases of racial or gender bias emerge in downstream applications of word embeddings, such as language translation systems. For example, research has shown that when translating from gender-neutral languages, Google Translate exhibits a strong tendency toward male defaults for science, technology, engineering, and math (STEM) occupations like "engineer" or "doctor" [42]. These biases are concerning, given the foundational role of word embeddings in enabling NLP and the many downstream applications, including everyday tools such as chatbots. The harmful stereotypes and biases encoded and reinforced by AI could further perpetuate systematic inequality and discrimination, making it imperative to address these issues.

Public Health versus Privacy: While the COVID-19 pandemic took the world by storm, the virus was also a breakthrough for scientific innovation. Many governments turned to AI tools to manage public health. For example, the city of Seoul collected extensive personal data including credit card transactions, geolocation data, and closed-circuit television footage to monitor citizens and contain the spread of the virus [43]. However, extensive surveillance has raised some serious questions and concerns around privacy [44]. Should human rights and data privacy be violated for the greater good of fighting the immediate threat of COVID-19 and protecting citizens from future public health emergencies? What kind of dangerous precedents are being set by the hasty introduction of technologies that pose grave privacy and security concerns for citizens? For now, widespread surveillance and AI tools have proven invaluable in fighting the pandemic and managing public health, but the implications of such widespread surveillance are deeply concerning to the core AI principles of privacy and security.

AI and the Labor Market Robotics: AI continues to transform a range of sectors, including education, medicine, transportation, and financial services. However, such transformations come with large implications for the labor market, particularly AI's role in job displacement and skill shifts. While researchers are divided on the exact number of jobs that will be lost, gained, or changed, the impact of AI on the job market is inevitable. Such ongoing and future AI impacts will undoubtedly result in a shift in the workforce and economic landscape [45].

AI and Recruiting: While the impact of AI on the labor market is clear, the concern is not only around the jobs that AI will automate away, but also those that AI may gate-keep. A 2014 AI recruiting tool, rolled out by Amazon to screen résumés and identify potential candidates, was scrapped after it was shown to discriminate against women [46]. This is because Amazon's tool was trained to vet new applications by observing and identifying patterns in résumés submitted to the company over a ten-year period. Since most of them came from men, the recruiting AI engine taught itself to prefer male candidates over female candidates. Such a system highlights the potential role of AI in reinforcing historical biases and inequities, as well as the need for identifying and mitigating representation bias in data sets.

The case studies presented in this section serve to illustrate the various types of harms caused by the misuse of AI, ranging from decision-making systems to language and vision models and their impacts at the individual, collective, and societal levels. Reviewing and analyzing these case studies can help establish overarching frameworks to proactively identify and prevent similar harms in the future.

8.3   Considerations for Sociotechnical Systems

In the previous section, we examined a series of case studies to illustrate various AI blunders that have occurred over the past decade. The chosen examples are representative of real incidents, and the companies in question are working diligently to avoid future incidents and ensure that their AI systems operate within the principles of RAI. However, there still remains a lack of clarity on how to classify the different types of biases that emerge in AI systems and proactively identify the types of harms that AI can pose. In this section, we will define a vocabulary for identifying the types of biases that can enter an AI system and describe the classes of AI-induced harms.

Bias in AI systems can take many forms, but the ultimate result is the same—an unfair system that disproportionately affects one or more subpopulations. In the following discussion, we will present a vocabulary for identifying three classes of biases that can occur in AI systems: preexisting bias, technical bias, and emergent bias [47]. It is important to point out that fairness (the first element of FASTEPS) is exacerbated by bias issues.

Preexisting Bias: Preexisting bias is rooted in social or collective practices, attitudes, and stereotypes. Such biases can enter a system consciously or unconsciously [47]. For example, individuals or institutions responsible for designing the system can make decisions that explicitly encode personal bias (e.g., designing a credit allocation system to negatively weight individuals in specific ZIP codes because they are deemed to be a higher financial risk). Such biases can also enter a system implicitly because of preexisting cultural norms and biases. In the case of the Amazon recruitment tool, data was measured and sampled to be representative of past applicants, but the system failed to consider preexisting biases that existed in the data itself, ultimately resulting in inequities being perpetuated by the model used for résumé screening. Recognizing preexisting bias requires a careful, retrospective understanding of how biases, stereotypes, and structural discrimination have manifested over time in a given domain [48].

Technical Bias: Technical bias is rooted in the technical constraints and decisions made in the design of an AI system [47]. This could include technical considerations in the collection and preparation of data, selection training of the machine learning model, and the overall user interface/user experience design of the system. For example, representation bias could result in data sets with sampling methods that underrepresent specific parts of the population or simply are not representative of the target population (e.g., data representative of one city is not representative if used to analyze the population of another city) [48]. From a modeling perspective, learning bias resulting from modeling choices could amplify performance disparities across different subpopulations in the data (e.g., selection of a specific objective function can result in cases where prioritizing one objective such as overall accuracy damages another objective, such as disparate impact) [48, 49]. Recognizing technical bias requires a thorough review and validation of decisions made in data collection and processing, as well as model development and deployment.

Emergent Bias: Emergent bias arises as AI systems learn through interaction with real-world users [47]. It's important to note that the addition of new users and new types or sources of data can result in bias being introduced into a system in ways that would have been difficult, if not impossible, to foresee when the system was being designed and developed. Such bias can be a result of the lack of quantitative metrics during AI system evaluation prior to deployment. For example, evaluation bias occurs when the benchmark data used for a task does not represent the user population, resulting in unforeseen degradation in quality and harms to users (e.g., biased commercial facial analysis tools that failed to evaluate the model performance on a diverse benchmark representative of the global population) [48]. With deployment bias, a mismatch between the primary functionality and intended use of an AI system and the ways in which users engage or use the system could result in unintended harm (e.g., self-driving cars that are intended to operate at level 2 driving automation with alert and active human supervision, but are actually operated as fully autonomous vehicles) [48]. Recognizing emergent bias requires a robust design and development process for AI systems, along with rigorous evaluation metrics that are resilient to attacks and manipulation. In addition, continuous monitoring of deployed systems is necessary to ensure upkeep of performance, identify potential biases that may emerge over time, and ensure alignment of the intended and the actual uses.

What are the consequences of such biases entering AI systems? We define these consequences as harms, the risks, impacts, and unintended outcomes from AI-powered technology. It's important to distinguish among three types of harms that can arise in the context of AI—individual harm, collective harm, and societal harm. In some cases, these types of harms do overlap (collective harms that are reducible to instances of individual harm), but it's also important to note that not all of them are reducible to each other (e.g., societal harms are not always reducible to demonstrable individual harms) [7]. In the following discussion, we will present a vocabulary for these three types of harms.

Individual Harms: Harms at the individual level are meant to capture the impact and risks of AI-powered decisions or recommendations that affect the rights and interests of an individual [7]. Such harms could take both physical forms (e.g., injury or death due to self-driving vehicle malfunctions or medical misdiagnoses due to AI diagnosis tool errors) and nonphysical forms (e.g., denial of consequential services such as loans, job/ educational opportunities as a result of AI-powered recommendations or decision-making systems, or private data being used without consent and potential data breaches resulting in downstream individual implications).

Collective Harms: Collective harms are meant to capture the impacts and risks of AI on the rights and interests of a collective or group of individuals, brought together through membership of an informal or formal group or subpopulation (e.g., race, gender, socioeconomic status, or employment). Collective harms can be thought of as the sum of harms suffered by individual members of the collective [7]. As such, many individual harms can give rise to collective harms if they pose similar risks and affect a specific group or groups of people. For example, bias in face recognition systems give rise to collective harm as it disproportionately affects darker-skinned females subjected to the AI system. Another example is bias in an AI recruiting tool that gives rise to collective harm as it disproportionally discriminates against female applicants. Ultimately, collective harm arises from the sum of similar risks and adverse impacts on the rights and interests of individuals, regardless of what connects those individuals (e.g., skin color, gender, employment, age, or other characteristics).

Societal Harms: Societal harms are meant to capture implications of AI that go beyond impacts that are reducible to individuals or a collective, but rather present risks to the interests of society at large [7]. At the societal level, AI can be a matter of grave social (e.g., AI perpetuation of structural oppression and discrimination), political (e.g., manipulation of national institutional process through misinformation and microtargeting), and economic concern (e.g., black-box financial instruments creating increased heightened volatility and instability in financial markets).

8.4   Responsible AI Principles

Governments around the world have made efforts to define principles and guidelines for responsible AI (RAI), and much has been done internationally to track these shared principles and shape policy around the responsible development of AI. Namely, the OECD and its AI policy observatory have served as a source of real-time tracking and information about global efforts in this space [28]. These principles help in guiding and ascertaining organizations meet the FASTEPS principles that need to be addressed early in the design and development of AI systems [50]. Building on concerted global efforts to define and track AI principles, we briefly describe here this set of seven FASTEPS principles, in support of RAI, that all AI STSs should uphold.

Fairness (Inclusiveness): Are we ensuring fair performance and treatment across all subpopulations and users? Fairness seeks to mitigate unwanted bias and provide AI-powered decisions, recommendations, and services that are fair to all subpopulations and users under a specific and clearly communicated definition. It's important to note that there are many definitions of fairness in the context of AI [51]. Is a fair system one that achieves demographic parity, in which the proportion of each subpopulation of a protected class receives positive outcomes at equal rates? Or is a fair system one that equalizes false negatives and odds across subpopulations of a protected class? Clearly, there is no one right definition, and there are strong contextual considerations that should be made when defining fairness and deciding how to minimize disparate impacts. However, consideration of fairness ensures that systems are prioritizing a predefined definition of fair and equal treatment across subpopulations.

Inclusiveness is part and parcel of fairness. How are we ensuring that AI systems work well for all users? First, organizations must ensure inclusive practices in the teams responsible for designing AI systems. This requires that diverse backgrounds and perspectives are solicited (e.g., race, gender, educational background), and that the input of subject matter experts (SMEs) are considered (i.e., doctors in the development of a medical AI tool or legal experts in the development of a legal AI tool). In addition to inclusive practices in the design and development of AI systems, inclusiveness requires that products are made to cater to and perform well for all members and subpopulations of the target populations (e.g., face recognition should work for all, regardless of race or gender).

Accountability: Who is being held accountable for the consequential actions and decisions made by AI systems? Accountability requires clear understanding of the concept for AI-powered decisions and their subsequent harms. The growing widespread adoption of AI will continue to come with high consequences of action (e.g., self-driving cars, medical diagnoses). Thus, it's important for organizations to establish clear accountability for the proper functioning and potential physical and nonphysical harms resulting from AI systems.

Safety (Reliability): How are we ensuring that our AI systems are robust, reliable, and prioritize the safety of users? Safety and reliability require that a system performs and is used in such a way that is consistent with the original design ideas. This requires AI systems to be robust in order to respond safely to edge cases and resist unintended attacks or manipulation. Reliability of AI systems is key to mitigating concept drift, minimizing physical and nonphysical harm, and ensuring the safety of users.

Transparency (Explainability): How are we ensuring that our AI-powered recommendations and decisions are understandable and explainable? Transparency and explainability require that AI-powered systems are designed to provide users insight into how their data is being used and how AI models arrive at their decisions and recommendations. Far too often, AI systems are thought of as black boxes. However, as the widespread adoption of AI continues to drive more consequential AI-powered decisions and recommendations in consequential domains such as medicine, financial services, and law enforcement, it's imperative that AI models are transparent and the resulting decisions and recommendations are explainable.

Ethics: Are we ensuring that AI systems are making decisions that uphold important moral values? The term "ethics" refers to the moral principles and techniques that inform the design, development, and use of AI systems. In the case of self-driving vehicles, how are we choosing the lesser of two values? Whose personal, religious, and geopolitical considerations are being considered in determining "right" or "ethical" behavior? The core goal is to develop AI that is not only compliant with regulations, but also is ethical and upholds moral values.

Privacy: How are we ensuring the privacy of user data? Privacy requires that organizations avoid the use of sensitive user data without consent and user data is protected to prevent data breaches and subsequent downstream misuse, with individual implications.

Security: How are we ensuring that our data is secure? Any of the guidelines espoused in these FASTEPS principles can be affected by unintentional or intentional lack of security. For example, as discussed in chapter 7, AI systems are very sensitive to attacks on data at rest, data in motion, or data in use, leading to incorrect results.

Ultimately, beyond frameworks and principles, progress toward RAI is founded on an ecosystem of trust. Trustworthy AI requires ethics of data, ethics of algorithms, and ethics of practice and use. However, trust in technology is only one piece of the puzzle. Successful RAI development and adoption require trust in AI producers, policy makers, and organizations, countries, and legal systems. Successful RAI development requires a multicultural and a multistakeholder ecosystem of trust (illustrated in figure 8.1),

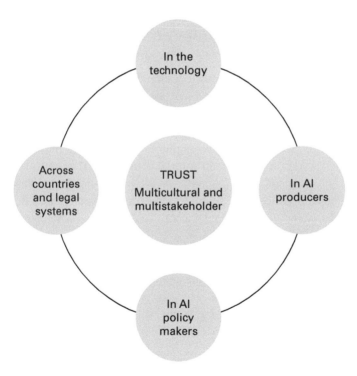

Figure 8.1   An ecosystem of trust.

as clearly formulated by Francesca Rossi, a former president of the Association for the Advancement of AI (AAAI) and International Joint Conference on Artificial Intelligence (IJCAI) [52].

The principles described here are certainly not exhaustive, but they do seek to capture and synthesize important RAI principles tracked across organizations, governments, and civil societies around the world. In the following section, we describe how users can incorporate and operationalize these principles throughout the stages of the AI development life cycle.

8.5   RAI Considerations in the AI Development Life Cycle

With a foundational understanding of different types of biases in AI systems, potential AI-induced harms, and key FASTEPS principles, we describe in this section a set of considerations throughout the AI development life cycle that seek to operationalize the FASTEPS principles explored in the previous section. The simplified AI development life cycle, presented in figure 8.2, is informed by the widely adopted cross-industry standard process for data mining (CRISP-DM) and is an iterative process comprised of solution design, data understanding and preparation, model development and evaluation, and deployment [53]. This life-cycle model and sequence presented are not strict; rather, they are flexible and can be customized, with variations in the dependencies that can exist between the stages and the back-and-forth iterations between specific stages. However, throughout each stage of this development life cycle, there are multiple decisions and considerations that must be made to ensure the development and deployment of RAI. The considerations described here, patterned after a set of well-thought-out questions and guidelines defined by the Defense Innovation Unit of the US Department of Defense, highlight important questions that should be raised and addressed throughout each stage of the AI life cycle [54].

Design refers to the process of planning and defining an AI system to solve a given problem. In this phase, important decisions are made around the prospective functionality, required resources, success metrics, and context in which the system will be deployed. In this phase, stakeholders responsible for the design and development of AI solutions should consider the following:

- Justification for AI solutions: Is there a clear need and value-add to using AI to solve this problem?
- Data sources and model selection: What candidate data sources and models should be used to create a solution for the given problem?

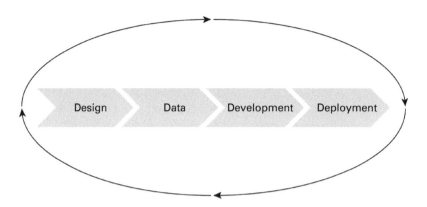

Figure 8.2 AI development life cycle (simplified).

- End user and stakeholder analysis: Who are the end users and relevant stakeholders, and what is the context in which the system will be deployed?
- Sociotechnical implications and potential harms: What are the sociotechnical implications and likelihood and magnitude of harms of the proposed AI solution?
- System reliability and safety measures: How are we ensuring system reliability and safety, and what is the process for identifying and correcting system errors?

Data involves the collection, cleaning, and processing of data that will be used to enable the underlying AI technology. The quality of data is foundational to the overall success of an AI product or service. In this phase, important decisions must be made about how data is sampled and collected, how it is then cleaned and processed, and how validation sets are created and developed to evaluate prospective real-world performance. Stakeholders responsible for the design and development of AI solutions should consider the following:

- Representativeness of the data set: How well does my data set reflect the target audience in the real world?
- Bias in data: Are there inherent biases in my data set because of historical and societal bias?
- Limitations of data: What are the limitations of my data, and how can that influence the model's behavior?
- Mitigating biases: How do we handle sensitive features such as race, gender, and socioeconomic status to mitigate potential biases? How do such features present themselves through other proxy features (e.g., socioeconomic status and ZIP code)?

- Regulatory considerations: Are we adhering to any regulations and laws that govern the use of data and ensuring compliance?

Development involves the iterative process of building out the planned AI system, featuring important decisions around model selection and training, but also the overall user interface/user experience and engineering needed to package and deliver core AI technology as an end product. In the development phase, the following important considerations must be made:

- Success metrics and trade-offs: How do we measure and monitor success, and what trade-offs are we making among the various metrics (e.g., fairness and accuracy)?
- System reliability and safety: Is the system designed with the appropriate guardrails and reliability to safely respond to edge cases and resist unintended attacks or manipulations?
- Explainability and transparency: Is the AI system transparent, and are the decisions and recommendations being made explainable?
- Postdeployment performance monitoring and alerting: What is the process for monitoring postdeployment system performance, and who is accountable for implementing these processes?

Deployment refers to the process of making the developed AI system available to real-world users in a production environment. In this stage, continuous testing and evaluation efforts must be made to ensure minimizing the likelihood and magnitude of harms and ensure the upkeep of the system's performance and reliability. In chapter 10, we discuss in detail ten guidelines in the deployment of AI, as well as the overall role of test harness (i.e., the test infrastructure needed to evaluate an AI system), benchmarks, and metrics in AI system deployment. In this stage, a number of important considerations must be made, such as the following:

- Data input validation and monitoring: How are data inputs being qualitatively and quantitatively audited to protect against manipulation and subsequent concept drift?
- Performance and functionality evaluation: Is the capability still meeting the desired functional goals, and is production use aligned with intended use?
- Postdeployment monitoring and incident management: How are postdeployment monitoring signals being used to identify and rectify potential harms or errors?

8.6   RAI Challenges

In the earlier sections, we addressed many of the challenges with RAI that require cooperative efforts from government, academia, and industry. In recent years, regulatory efforts, such as the establishment of the National Artificial Intelligence Initiative Office and the National AI Initiative Act of 2020 [55, 56], have been made, and more AI-focused bills are being introduced in the US Congress to call for federal AI regulation. The National Institute for Standards and Technology (NIST) has also released the AI Risk Management Framework [57], a guidance document for voluntary use by organizations designing, developing, deploying, or using AI in order to help them "enhance AI trustworthiness while managing risks" [58].

Broderick et al. provide very useful guidance using a taxonomy of trust for probabilistic ML [59]. The authors addressed the full cycle, from real-world goals (desired at deployment) through algorithm implementation (i.e., development) and culminating with future real-world output. As they emphasized, trust can break down at any of these stages, including training. They also provide methods for increasing trust at different stages within their taxonomy.

The recent rapid progress in the AI field, such as OpenAI's ChatGPT, has motivated accelerated attention being paid to policy and regulation. A nonprofit organization called the Future of Life Institute recently issued "Pause Giant AI Experiments: An Open Letter," calling for a six-month moratorium on the training of AI systems that were any more powerful than the recently released GPT-4 in order to accelerate the "development of robust AI governance systems." [60]. The call has sparked debate and criticism within the AI community and the public, with some arguing that it could provide time for reflection, evaluation, and planning around the potential risks and benefits of advanced AI systems, while others claim that it will stifle innovation, progress, and economic growth in the field.

As the field of AI continues to advance rapidly, it is crucial that regulatory efforts keep pace with the technological progress to ensure safe and responsible development and deployment of AI systems, again without stifling innovation. Further research is needed to provide technical approaches to help in the design and development of AI systems that use generative AI technologies, including the designs to avoid potential harm [61].

In table 8.1, we formulate a set of challenges as a function of the horizons described in chapter 1, but we propose this set as a function of near-term, mid-term, and far-term (i.e., horizons 1, 2, and 3) challenges and associated opportunities.

The challenges and opportunities discussed in this chapter are crucial for the successful deployment and adoption of RAI. While there is much to be done, we are seeing the

Table 8.1 RAI challenges and opportunities

Horizon	Opportunities
Horizon 1 (1–2 years): Content-Based Insight	*Apply AI to gain insight on the content of interest in disparate types of data.* • Design, develop, and deploy AI with attention to system testing-verification-validation, consistent with the FASTEPS principles. • Address the three types of biases: preexisting, technical, and emergent. • Formulate AI system capabilities with business value, but within the guardrails, in order to avoid individual harm, collective harm, and societal harm.
Horizon 2 (3–4 years): Collaboration-Based Insight	*Extend AI roles to include multiple human-machine teams (HMTs) working together.* • Develop AI capabilities while ascertaining ethics of data, ethics of algorithms, and ethics of practice and use are met. • Test-verify-validate output at the human-machine teaming (HMT) stage. • Build a multicultural and multistakeholder ecosystem of trust. • Adopt, within the culture of the organization, an iterative AI life cycle from early design, data preparation, development, and deployment.
Horizon 3 (5+ years): Context-Based Insight	*Incorporate semantic context in the application of AI.* • Confidence in the machine versus consequence of actions also should incorporate FASTEPS principles. • Employ counterfactuals to ascertain that the context is consistent with societal moral and core values. • Avoid company embarrassments by employing an AI ecosystem to perform early prototyping to obtain user feedback (i.e., lessons learned in context) prior to full deployment (see chapter 10).

emergence of regulatory efforts and standardization frameworks to address AI risks. The rapid progress of AI, especially with recent release of powerful language models, calls for more proactive measures, while the call for a moratorium on AI development highlights the need for continued dialogue and collaboration among stakeholders. Looking forward, it is imperative that AI practitioners, researchers, policy makers, and international organizations work together to ensure RAI for the benefit of all.

8.7 Main Takeaways

With great technological AI power undoubtedly comes great responsibility. RAI is an emerging field that is rapidly and continuously evolving. While we can't cover the fundamentals of all aspects of this technology, this chapter provided readers with a foundation to approach AI systems as STSs. This involves considering potential harms, identifying biases, and upholding important FASTEPS principles throughout the various stages of the AI development life cycle to meet the ultimate objective of deploying RAI systems. In this chapter, we

- Provided a short history on ethics in computing and described the current landscape of RAI initiatives across academia, industry, professional organizations, and civil society
- Explored nine case studies that highlighted different AI harms over the past decade
- Outlined important considerations for the design and development of STSs, classifying different types of biases and harms at the individual, collective, and societal levels
- Described a set of seven FASTEPS principles in support of RAI that all AI-based STSs should uphold
- Operationalized such principles through important RAI considerations that should be made in the various stages of the AI development process across the full AI life cycle (design, data collection, development, and deployment)

We concluded the chapter by formulating a key set of challenges and associated opportunities, categorized into three horizons (i.e., in the near term, midterm, and far term). Given the lack of attention to RAI in today's design, data collection, development, and deployment workflow of AI capabilities, we felt that it was important to progress through the three horizons demonstrating progress without attempting to solve all the challenges at once. Looking forward, it is imperative for AI researchers, designers, developers, and organizations to commit to developing AI technologies that adhere to the FASTEPS principles and are beneficial to all individuals and society.

8.8 Exercises

1. What is meant by RAI?
2. What are the FASTEPS principles, and what is the importance of each one?
3. Provide examples of potential harms that arise if each FASTEPS principle is not addressed.
4. Which of the following stages of the AI development life cycle is most important to consider when working to address the FASTEPS principles?
    a. Design
    b. Data
    c. Development and deployment
    d. All of the above
5. What are some examples of important RAI considerations that should be made throughout each stage of the AI development life cycle?
6. All societal harms are reducible to demonstrable individual harms.
    a. True
    b. False
7. Review each of the nine case studies in this chapter to classify the types of bias (preexisting, technical, and emergent), and the harms (individual, collective, and societal) presented by the AI system.
8. What are the key elements of an ecosystem of trust?
9. Review the OECD's AI principles and elaborate on the main takeaway from the principle focusing on "Human-Centered Values and Fairness" (https:// oecd.ai/en/dashboards/ai-principles/P6).
10. Describe how counterfactuals can help to bring more context to AI systems in addressing the FASTEPS principles.

    Hint: Review the paper by Byrne on counterfactuals in XAI [62].

11. There have been significant recent advancements in the development of large language models, such as GPT-4. What are some potential RAI concerns associated with the development of these technologies, and how can we ensure that they are developed and used in a responsible and ethical way?

12. You are responsible for designing an AI system for determining the eligibility of a loan applicant. To ensure that your model does not exhibit bias against specific races, your boss suggests removing the race feature from your training data set. However, it is important to understand that this step alone does not guarantee that your model will be completely free of bias. Explain why.

13. Imagine you are a policy maker tasked with regulating the development and use of AI. Discuss some key principles that should guide regulatory efforts and analyze the potential challenges of regulating AI, as well as some ways to overcome these challenges.

14. Use an open-source AI fairness toolkit (e.g., Microsoft Fairlearn or IBM AI Fairness 360) to examine, report, and mitigate discrimination and bias for a simple classification model.

8.9   References

1. Zhang, B. and A. Dafoe, US public opinion on the governance of artificial intelligence, in *Proceedings of the AAAI/ACM Conference on AI, Ethics, and Society,* Oxford, UK. 2020.

2. Gayle, D. Facebook aware of Instagram's harmful effect on teenage girls, leak reveals. *The Guardian.* September 14, 2021. https://www.theguardian.com/tech nology/2021/sep/14/facebook-aware-instagram-harmful-effect-teenage-girls-leak -reveals.

3. Cheng, L., K. R. Varshney, and H. Liu, Socially responsible AI algorithms: Issues, purposes, and challenges, 2021. arXiv preprint arXiv:2101.02032.

4. Kissinger, H. A., E. Schmidt, and D. Huttenlocher, *The age of AI: And our human future.* 2021, Hachette UK.

5. Mao, J.-Y., K. Vredenburg, P. W. Smith, and T. Carey, The state of user-centered design practice. *Communications of the ACM,* 2005. 48(3): 105–109.

6. Baxter, G. and I. Sommerville, Socio-technical systems: From design methods to systems engineering. *Interacting with Computers,* 2011. 23(1): 4–17.

7. Smuha, N. A., Beyond the individual: Governing AI's societal harm. *Internet Policy Review,* 2021. 10(3).

8. Nichols, S. P., Professional responsibility: The role of the engineer in society. *Science and Engineering Ethics,* 1997. 3(3): 327–337.

9. Cech, E. A., Culture of disengagement in engineering education? *Science, Technology, & Human Values,* 2014. 39(1): 42–72.

10. *ACM Code of Ethics and Professional Conduct*, A.f.C. Machinery, Editor.

11. *IEEE Code of Ethics*, I.o.E.a.E. Engineers, Editor.

12. *NSPE Code of Ethics for Engineers*, N.S.o.P. Engineers, Editor.

13. de Laat, P. B., Companies committed to responsible AI: From principles towards implementation and regulation? *Philosophy & Technology*, 2021: 1–59.

14. *Stanford University Human-Centered Artificial Intelligence*. https://hai.stanford.edu/.

15. *Embedded EthiCS @ Harvard*. https://embeddedethics.seas.harvard.edu/.

16. MIT Schwarzman College of Computing. https://computing.mit.edu/.

17. Heer, J., The partnership on AI. *AI Matters*, 2018. 4(3): 25–26.

18. Bellamy, R. K., K. Dey, M. Hind, et al., AI Fairness 360: An extensible toolkit for detecting and mitigating algorithmic bias. *IBM Journal of Research and Development*, 2019. 63(4/5): 4, 1–4, 15.

19. Bird, S., M. Dudík, R. Edgar, et al., *Fairlearn: A toolkit for assessing and improving fairness in AI*. Microsoft Technical Report MSR-TR-2020-32, 2020.

20. Wexler, J., M. Pushkarma, T. Bolukbasi, et al., The what-if tool: Interactive probing of machine learning models. *IEEE Transactions on Visualization and Computer Graphics*, 2019. 26(1): 56–65.

21. Howard, A., The regulation of AI—should organizations be worried? *MIT Sloan Management Review*, 2019. 60(4): 1–3.

22. Vought, R. T., *Guidance for regulation of artificial intelligence applications*, Office of Science and Technology Policy (OSTP). January 7, 2020.

23. Lively, T. K., *US State Privacy Legislation Tracker*. 2021. https://iapp.org/resources/article/us-state-privacy-legislation-tracker/.

24. Lander, E. and A Nelson., *ICYMI: WIRED (Opinion): Americans need a bill of rights for an AI-Powered World*. 2021, The White House.

25. Union, E., *The history of the General Data Protection Regulation*. n.d., European Union.

26. Kop, M. *EU Artificial Intelligence Act: The European approach to AI*. 2021, Stanford-Vienna Transatlantic Technology Law Forum, Transatlantic Antitrust.

27. Ratte, E. *World Economic Forum: Responsible use of technology*. World Economic Forum's Shaping the Future of Technology Governance: Artificial Intelligence and Machine Learning Platform. https://www.weforum.org/projects/responsible-use-of-technology.

28. Organisation for Economic Co-operation and Development (OECD). Artificial intelligence. https://www.oecd.org/digital/artificial-intelligence/.

29. Organisation for Economic Co-operation and Development (OECD) Directorate for Science, Technology and Industry, *Regulatory reform and innovation*. https://www.oecd.org/sti/inno/2102514.pdf.

30. Krizhevsky, A., I. Sutskever, and G. E. Hinton, ImageNet classification with deep convolutional neural networks. *Advances in Neural Information Processing Systems*, 2012. 25: 1097–1105.

31. Silver, D., A. Huang, C. J. Maddison, et al., Mastering the game of Go with deep neural networks and tree search. *Nature*, 2016. 529(7587): 484–489.

32. Buolamwini, J. and T. Gebru. Gender shades: Intersectional accuracy disparities in commercial gender classification. *Proceedings of Machine Learning Research,* 2018. 81: 1–15.

33. Tsymbal, A., The problem of concept drift: Definitions and related work. *Computer Science Department, Trinity College Dublin*, 2004. 106(2): 58.

34. Lee, P., *Learning from Tay's introduction*, Microsoft, Editor. 2016, Microsoft.

35. Nyholm, S. and J. Smids, The ethics of accident-algorithms for self-driving cars: An applied trolley problem? *Ethical Theory and Moral Practice*, 2016. 19(5): 1275–1289.

36. Awad, E., S. Dsouza, R. Kim, et al., The moral machine experiment. *Nature*, 2018. 563(7729): 59–64.

37. Isaak, J. and M. J. Hanna, User data privacy: Facebook, Cambridge Analytica, and privacy protection. *Computer*, 2018. 51(8): 56–59.

38. McKinney, S. M., M. Sieniek, V. Godbole, et al., International evaluation of an AI system for breast cancer screening. *Nature*, 2020. 577(7788): 89–94.

39. Ledford, H., Millions of black people affected by racial bias in health-care algorithms. *Nature*, 2019. 574(7780): 608–610.

40. Dressel, J. and H. Farid, The accuracy, fairness, and limits of predicting recidivism. *Science Advances*, 2018. 4(1): eaao5580.

41. Garg, N., L. Schiebinger, D. Jurafsky, and J. Zou, Word embeddings quantify 100 years of gender and ethnic stereotypes. *Proceedings of the National Academy of Sciences*, 2018. 115(16): E3635–E3644.

42. Prates, M. O., P. H. Avelar, and L. C. Lamb, Assessing gender bias in machine translation: A case study with Google Translate. *Neural Computing and Applications*, 2020. 32(10): 6363–6381.

43. Sonn, J. W., Coronavirus: South Korea's success in controlling disease is due to its acceptance of surveillance, in *The Conversation*. 2020. https://theconversation.com /coronavirus-south-koreas-success-in-controlling-disease-is-due-to-its-acceptance

-of-surveillance-134068#:~:text=The%20focus%20has%20largely%20
been,test%20in%20the%20first%20place..

44. Naudé, W., Artificial intelligence vs COVID-19: Limitations, constraints and pitfalls. *AI & Society*, 2020. 35(3): 761–765.

45. Harris, L. A., *Artificial intelligence: Background, selected issues, and policy considerations*, C.R. Service, Editor. 2021.

46. Dastin, J., Amazon scraps secret AI recruiting tool that showed bias against women, in *Reuters*. 2018.

47. Friedman, B. and H. Nissenbaum, Bias in computer systems. *ACM Transactions on Information Systems (TOIS)*, 1996. 14(3): 330–347.

48. Suresh, H. and J. Guttag, A framework for understanding sources of harm throughout the machine learning life cycle, in *Equity and Access in Algorithms, Mechanisms, and Optimization* (EAAMO '21), New York, October 5–9, 2021, ACM.

49. Hooker, S., Moving beyond "algorithmic bias is a data problem." *Patterns*, 2021. 2(4): 100241.

50. Fjeld, J., N. Achten, H. Hilligoss, et al., *Principled artificial intelligence: Mapping consensus in ethical and rights-based approaches to principles for AI*. 2020, Berkman Klein Center Research Publication, (2020–1).

51. Mehrabi, N., F. Morstatter, N. Saxena, et al., A survey on bias and fairness in machine learning. *ACM Computing Surveys (CSUR)*, 2021. 54(6): 1–35.

52. Rossi, F., *Fostering an environment of trust*. 2020, MIT Technology Review EmTech Digital.

53. Wirth, R. and J. Hipp, CRISP-DM: Towards a standard process model for data mining, in *Proceedings of the 4th International Conference on the Practical Applications of Knowledge Discovery and Data Mining*. 2000, Springer-Verlag London.

54. Dunnmon, J., B. Goodman, P. Kirechu, C. Smith, and A. Van Deusen, *Responsible AI Guidelines in Practice: Lessons Learned from the DIU AI Portfolio. Defense Innovation Unit (DIU)*. 2021. https://www.diu.mil/responsible-ai-guidelines.

55. *National Artificial Intelligence Initiative*. https://www.ai.gov/.

56. Thornberry, W. M. M., *National AI Initiative Act of 2020*. 2020.

57. Technology, N.I.o.S.a., *Artificial Intelligence Risk Management Framework (AI RMF 1.0)*. 2023.

58. Boutin, C. *NIST Risk Management Framework aims to improve trustworthiness of artificial intelligence*. 2023. https://www.nist.gov/news-events/news/2023/01/nist-risk-management-framework-aims-improve-trustworthiness-artificial.

59. Broderick, T., A. Gelman, R. Meager, et al., Toward a taxonomy of trust for probabilistic machine learning. *Science Advances*, 2023. 9(7): eabn3999.

60. *Pause giant AI experiments: An open letter*. https://futureoflife.org/open-letter/pause-giant-ai-experiments/.

61. Weisz, J. D., M. Muller, J. He, and S. Houde, *Toward general design principles for generative AI applications*, 2023. arXiv preprint arXiv:2301.05578.

62. Byrne, R. M. Counterfactuals in explainable artificial intelligence (XAI): Evidence from human reasoning, in *IJCAI*. 2019: 6276–6282.

II
Strategic Principles

9

AI Strategy and Road Map

Without strategy, execution is aimless. Without execution, strategy is useless.
—Morris Chang, founder and former CEO of Taiwan Semiconductor Manufacturing Company (TSMC)

In part I of the book, as illustrated in figure 1.1 in chapter 1, we presented an overview of the artificial intelligence (AI) system architecture, elaborated on the fundamentals of systems engineering, and described in detail the key architecture subcomponents. These chapters helped the reader gain a deeper understanding of the building blocks used in the development of AI products and services.

These building blocks, across the AI pipeline, are necessary to transform data input into insights. The created insights are delivered as AI value in the form of predictions or classifications to the ultimate stakeholders. However, to successfully deliver business value, it is imperative that AI practitioners institute a systematic approach within their respective organizations that conforms to a well-defined strategy and associated road map.

Part II begins with chapter 9, which discusses the steps involved in the formulation of an AI strategy, culminating in a strategic road map. The strategic road map serves as the blueprint for upper management, technical leaders, AI architects, designers, and implementers to progress from architecture principles to deployment. The strategic road map, also referred to as the "strategic blueprint," must be an integral part of any AI-based organization to successfully develop and deploy AI capabilities.

During his illustrious career as an engineer, manager, and business leader, Morris Chang, the founder and former chief executive officer (CEO) of the Taiwan Semiconductor Manufacturing Company (TSMC), emphasized the importance of formulating a strategy and executing that strategy—meaning to "plan the work and work the plan."

Table 9.1 AI transformation playbook—how to lead your company into the AI era

Playbook Steps
1.   Execute pilot projects to gain momentum.
2.   Build an in-house AI team.
3.   Provide broad AI training.
4.   Develop an AI strategy.
5.   Develop internal and external communications.

Source: Ng (2021).

The strategy is the cornerstone for leaders and team members to identify their collective efforts under a unified AI strategic direction, to work toward a set of well-defined goals and establish actionable milestones.

One of this book's authors, David Martinez, with his colleagues Stephen Rejto and Marc Zissman, instituted a process for formulating a strategic direction and road map, in their roles as coleaders of the Cyber Security and Information Sciences Division at MIT Lincoln Laboratory. Some of the techniques described in this chapter are drawn from many years of practical experience successfully developing and deploying complex prototype systems. The formulated strategic road map served as a first step toward embarking on a prototype development and ultimately in its successful deployment.

In this chapter, we provide the reader a step-by-step procedure to create an AI strategic road map. The creation of a strategy is not a final destination, but a journey to get the whole AI team marching in the same direction. A strategy is not a static deliverable; instead, it is a living blueprint that must be revisited and, if necessary, updated throughout the year while the AI system is undergoing development leading to the final deployment.

Andrew Ng, a faculty member at Stanford University and the CEO and founder of Landing.ai, outlined the five-step playbook shown in table 9.1 during a fireside chat at the MIT Technology Review conference on AI-focused EmTech Digital [1].

Ng's five steps are very consistent with several of the topics discussed in part II of this book. The need to execute pilot projects to gain buy-in and momentum—we use pilot projects and prototypes interchangeably—is necessary to demonstrate initial AI capabilities in the form of a minimum viable product (MVP). Step 2, building an in-house AI team, allows AI organizations to grow and preserve organic capabilities, since AI technologies evolve very rapidly. Also, as discussed in earlier chapters, AI excellence requires teams to be multidisciplinary, so there is a need for continued training to broaden the AI team's levels of expertise. The fourth recommendation is to develop an AI strategy, which is the focus of this chapter. Finally, the last recommendation—step 5—is to develop internal and external communications. We would augment that recommendation to include relationships as well. It is important to communicate the AI value proposition internally, to your team and other internal business units, and externally, to stakeholders. But to be successful in the development and deployment of AI capabilities, an AI organization must also use its internal and external expertise by cementing enduring relationships.

Part I of the book focused on the AI system architecture, representing the "what" that forms the end-to-end building blocks. Part II focuses primarily on the "how" of implementing the AI system architecture. After the AI strategy discussion in this chapter, we devote chapters 10 and 11 to AI deployment, using techniques initially formulated under the field of development, security, and operations (DevSecOps), and applied to the nascent discipline of machine learning operations (MLOps). Chapter 12 delves more deeply into the fostering of an innovative team environment, addressing the important topics of AI leadership and technical talent. Chapter 13 provides tools and techniques to communicate technical topics clearly and focuses on communicating effectively, internally, and externally to the AI organization—a topic that is very important for technology-based businesses, their leadership, and their technical employees.

In section 9.1, we begin by setting up the stage and defining the process of strategic thinking. We then present an AI strategic development model (AISDM) that, in addition to being used in our class projects, has been used during the creation of strategic development road maps for technical divisions. The remaining sections in this chapter describe in detail the key parts of the AISDM framework, employing a selective class project example from one of our MIT classes. We conclude the chapter with the main takeaways and a set of exercises.

9.1 Introduction to Strategic Thinking

In this section, we explicate the importance of developing a strategic thinking competency. Initially, this competency might appear relevant to only the C-suite and upper management. However, strategic thinking, to work effectively, must include members of the AI team who are responsible for different aspects of AI products and services throughout the AI system life cycle. A superior and well-defined strategic road map is not executable unless there is buy-in from the top management and the staff.

Let us start by addressing a simple but powerful perspective on strategic thinking. David Briggs, emeritus director of the MIT Lincoln Laboratory, stressed the importance

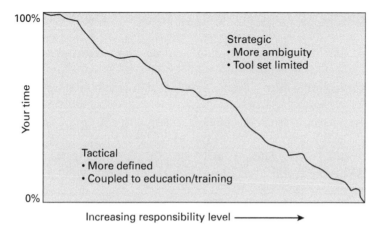

Figure 9.1   Strategic and tactical divide.

of technical personnel developing strategic thinking skills as they advance in their careers. His perspective, and what he referred to as the "strategic and tactical divide," is shown in figure 9.1. There are two important messages captured in the graph here:

- As the AI leadership and project managers increase in responsibilities, more and more of their time must be spent on strategic thinking.
- In contrast to tactical competencies, where the work is more defined and typically coupled to the individual's education and training, strategic thinking is more ambiguous and the tool set is limited.

Patrick McGovern, an American businessman who was the chairman and founder of the International Data Group, built a multibillion-dollar corporation by focusing on a clear mission and vision while embracing emerging technologies. Glenn Rifkin, in his book *Future Forward: Leadership Lessons from Patrick McGovern* [2], highlights one of these lessons, exemplified in the following quote from McGovern: "In a fast-changing business, long-term vision plus short-term operational excellence will outperform any other strategy." This lesson is also consistent with the important message mentioned earlier: to plan the work and work the plan. Plan the work—meaning formulating a strategic plan—should encompass multiple years, as shown in the horizons framework discussed in part I. The execution approach should be short-term and focused on operational excellence.

Strategic thinking puts a framework around determining the overall direction of an AI organization. More specifically, strategic thinking helps answer critical management questions, such as the following:

- What should the AI organization look like in the near term (horizon 1; 1–2 years), midterm (horizon 2; 3–4 years), and far term (horizon 3; ≥ 5 years)?
- Who benefits from strategic thinking and the formulation of a strategic plan?
- Who should we be hiring (i.e., talent)?
- What should we be learning?
- What technologies do we need to enter or exit?
- What contracts and channels should we grow or shrink?
- Who will be willing and able to pay for it?
- What processes are or are not needed?
- How federated should the portfolio be with respect to internal and external partners?
- What should be the mix of direct and contracted employees?

It is also important to emphasize that the process of strategic thinking and the strategic plan that results from this process help in the following ways:

Individual leaders:

- The planning process causes you to stop and think about what's important.
- The plan itself helps guide decision making going forward.
  - It is a "compass" of sorts that can help remind you what your desired destination is when the seas get rough.

AI staff:

- The planning process is an opportunity to promote and get buy-in.
- The plan can be used as a performance goal guide in the evaluation of AI staff.

Upper management:

- The plan explains where your AI team is going in the development and deployment of AI products and services.
- The plan can be used to justify decisions and show progress.

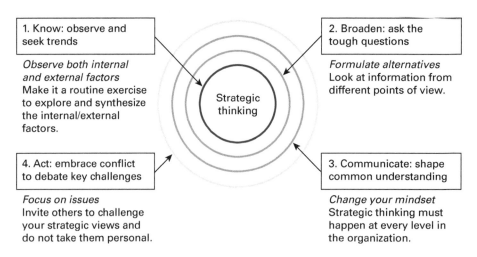

Figure 9.2   Four ways to improve your strategic thinking skills. Adapted from [4].

There are also important questions that must be asked at the start of the process. The *Harvard Business Review* (HBR) guide *Thinking Strategically* presents a set of questions to inspire strategic thinking, with dozens of prompts to get started [3]. As pointed out in appendix A of the *HBR* guide, answers to these questions drive clarity, alignment, and strategic insight.

Nina Bowman, in her *HBR* article "4 Ways to Improve Your Strategic Thinking Skills" [4], presents a clear approach to reduce the ambiguity that it is often in the strategic thinking process. We have adapted her four-way approach with some modifications to maintain the same terminology that we have used in this chapter. Figure 9.2 illustrates the four areas of focus during the process of strategic thinking. First, *Know* emphasizes the need to observe and seek trends to identify both internal and external factors likely to affect your strategic direction.

As pointed out earlier, a strategic planning process should begin by asking key questions to *Broaden* the AI value proposition understanding and how it affects the business. Answers to key and tough questions must be addressed from different points of view. Third, strategic planners must *Communicate* the strategic plan at every level within the AI organization that is responsible for delivering value. Communicating effectively (see chapter 13) shapes a common understanding. Finally, the fourth component of the approach is to *Act* by identifying goals and actions that should be debated. The strategic views should be challenged to ascertain that the strategic direction is consistent with the originally defined objectives while formulating answers to key questions.

A 2018 survey by McKinsey & Company, performed across over 2,000 participants representing a range of regions, industries, and company sizes, identified a lack of clear AI strategy as the most frequently cited barrier to AI adoption [5]. This barrier was identified by 43 percent of the respondents. The next-largest barrier, identified by 42 percent of the respondents, was the lack of personnel with appropriate skills to be effective at their assigned AI duties.

In the next section, we introduce the reader to the AISDM. We have used the AISDM framework for our own real-world applications. We have also taught, as part of our MIT classes, the AISDM framework as a step-by-step guide for creating a strategic plan used in the team class projects. The use cases discussed in chapters 14–18, illustrate the application of the AISDM framework to a broad range of AI applications.

The AISDM approach helps in addressing the barriers identified in the McKinsey & Company report. The output of the AISDM framework is a strategic development road map that formulates the direction for the successful development and deployment of AI capabilities.

Before we describe in detail the AISDM framework, readers interested in additional tools to assess emerging trends that could have an impact on business opportunities are referred to Christensen et al. [6], who have devised a theory to understand why companies have difficulty responding to disruptive innovations. Their resource, process, value (RPV) theory is very useful and well aligned, as a tool, with the first strategic thinking step, *Know*: observe and seek trends. The RPV theory incorporates resources (what an organization has in the form of people, technology, and equipment); processes (how an organization does its work); and values (what the organization wants to achieve). Again, the strategic road map is the blueprint that codifies the near-term, midterm, and long-term strategic directions of an AI organization informed by its resources, processes, and values.

Another useful resource, on how other companies approach their long-term assessment of opportunities, is by Schwartz, who describes techniques for scenario planning in his book *The Art of the Long View* [7]. Scenario planning is not about trying to predict the future. Instead, it is a technique to help in formulating alternatives—*Broaden* the AI value proposition options—as discussed in the four-way approach illustrated in figure 9.2.

The ability to communicate effectively is so important, from technical staff to upper management, that we dedicate chapter 13 to this topic. As shown in figure 9.2, the third step in improving strategic thinking skills is to *Communicate* at every level in the organization to shape a common understanding.

In the subsequent sections, we describe each of the components that form the AISDM framework.

9.2   AI Strategic Development Model

One of the most successful visionaries of the modern information age was Steve Jobs, who, under his leadership, transformed Apple multiple times. One of his famous quotes well reflects his emphasis incorporated into Apple's strategic planning: "I'm as proud of what we don't do as I am of what we do." A strategic development road map helps in defining the business direction, priorities, and areas to emphasize and deemphasize.

The earlier discussion, in part I of the book, centered on the AI system architecture shown in figure 1.3. The development of a strategy that is executable requires the contributions of diverse participants with AI expertise across the full end-to-end architecture. AI demands a multidisciplinary approach with responsibility for each of the subcomponents in the AI architecture. Stadler et al. discuss the need to "open up your strategy" [8]. As the authors pointed out, making a company strategy behind closed doors is a prescription for failure. Our experience at MIT Lincoln Laboratory is to include all levels, with lead responsibilities within a technical group, to help in defining a group's strategy. The group's strategy then feeds into the overall division-level strategy, consistent with the long-term direction of the organization.

Michael Tushman led the Program for Leadership Development (PLD) within the Harvard Business School for many years [9]. The depiction shown in figure 9.3 is adapted from Harvard's PLD discussion of strategic planning. This high-level description incorporates simple but meaningful explanations of some of the components that form part of a strategic road map. Many organizations struggle with the mission and vision statements. As shown in figure 9.3, the mission should be a short statement on why the organization, the business unit, a group, or other collective exists. Sinek, in his book titled *Start with Why*, discusses the importance of clearly defining the *why* [10]—it must be an inspirational statement to take action. The vision should also be a short statement, focusing on what the organization wants to be and complementing the mission statement.

Let us present a real-life example that in practice describes very inspirational mission and vision statements. Daniela Rus, director of the MIT CSAIL Laboratory, describes its mission and vision as follows [11]:

- *Mission:* MIT CSAIL pioneers research in computing that improves how people work, play, and learn.
- *Vision:* A world where computing empowers all people and enhances all human experiences.

The mission statement has a clear focus on action within the laboratory. MIT CSAIL's mission is also very well aligned with the overall mission of the Massachusetts Institute of

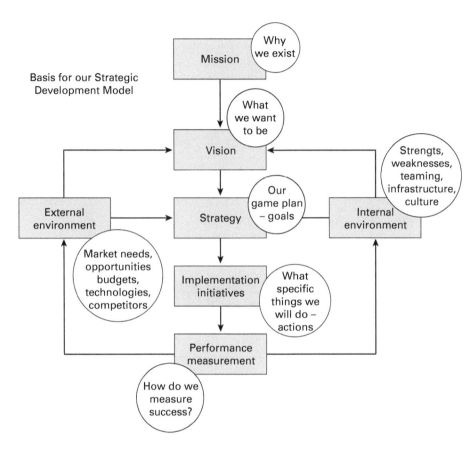

Figure 9.3   High-level description of a strategic plan [9].

Technology (MIT), which is to advance knowledge and educate students in science, technology, and other areas of scholarship that will best serve the nation and the world in the twenty-first century. MIT CSAIL's vision statement complements its mission statement, describing a purpose for the laboratory's research: namely, to have a worldwide impact.

Other factors influencing a strategic road map are the internal and external environments. We discuss the organization's strengths and weaknesses (internal environment), and the opportunities and threats (external environment) later in this section, when we discuss the strengths, weaknesses, opportunities, and threats (SWOT) analysis. For now, it is sufficient to point out that successful organizations must formulate an AI strategy that is executable.

Organizations must focus on the present or near term (relative to execution), midterm, and the future (long term), as we will address during the envisioned future discussion in section 9.3. Smith et al. referred to this duality (i.e., focusing on the near term through the long term) as not an either/or choice but instead as "Both/and Leadership" [12]. Similarly, O'Reilly and Tushman called for an "Ambidextrous Organization" [13], where there is a need for a balance of the paradox between the exploitation of existing capabilities (near term) and looking forward (long term) by exploring opportunities (i.e., staying competitive through constant innovation) [14]. They emphasized the importance of a vision, the associated strategy, and formulating clear objectives as the bedrocks for managing innovation and change. These lessons are crucial to AI organizations since the technological underpinnings evolve very rapidly. We elaborate further on these topics as part of our upcoming discussion of the AISDM framework, starting with the organization's mission and vision, core values, and strategic direction.

We integrate the last two elements of the high-level strategic plan shown in figure 9.3—implementation initiatives (i.e., goals and actions) and performance measurement (i.e., performance metrics)—as part of the strategic road map deliverables.

The AISDM, shown in figure 9.4, building on the description of the high-level strategic plan shown in figure 9.3, provides more context and details relevant to formulating an AI strategic road map. The AISDM framework is applicable to formulating an AI strategic road map at the organization level, at the business-unit level, at the group-unit level, and at the AI-project level. However, since we want to cement many of these concepts by highlighting representative examples, we use MIT team class projects; which were focused on formulating a strategic road map for a proposed AI organization product or service. Additional use cases are discussed in chapters 14–18, and the reader is encouraged to follow up with these use cases to gain further understanding on how to use the framework.

It is useful to start by formulating the long-term mission and the vision for the AI product or service—after traversing the AISDM framework, culminating with a strategic blueprint, we can then return and refine the mission and vision statements. The mission and vision statements, plus the envisioned future, are discussed in more detail in section 9.3.

Once the envisioned future is formulated, we can then revisit the organization core values feeding into the strategic direction for an AI product or service, based on the organization's competencies, capabilities desired, and existing and future system applications. We discuss these components of the AISDM framework in section 9.4.

The next step in the building of the strategic road map is to identify the AI value proposition (shown at the center of figure 9.4). The AI value proposition is formulated

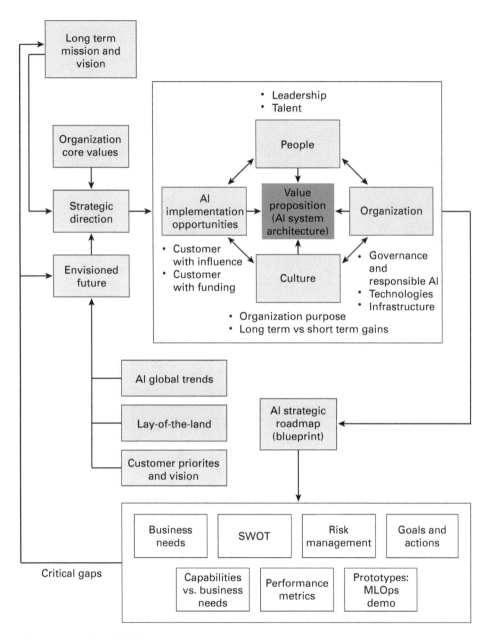

Figure 9.4   The AISDM.

based on the AI system architecture discussed in part I. The value proposition is informed by four key components, which are discussed in detail in section 9.5:

- AI implementation opportunities
- People (i.e., leadership and talent)
- Organization AI governance, responsible AI (RAI), technologies, and infrastructure
- Culture

The ultimate deliverable, from the AISDM framework, is the AI strategic road map (i.e., a strategic blueprint). The AI value proposition must be aligned with the value capture seen through the lens of the stakeholders. The AI strategic road map consists of seven components:

- The stakeholder's business needs
- SWOT analysis
- Risk management (including cost)
- Goals and actions
- Capabilities provided versus business needs
- Performance metrics (i.e., quantitative and qualitative assessment of progress)
- Prototypes (i.e., demonstrating an initial MVP or service through the implementation of an MLOps demo)

Since many trade-offs must be made in the formulation of a strategy that is executable, an important by-product of executing these seven components is enumerating the critical gaps found. Such gaps can result from any or several of the components. For example, the SWOT analysis can inform on limitations on what can or cannot be implemented due to internal weaknesses or external threats. Another example, informing identified critical gaps, can result from prototyping an initial MVP through a prototyping demo. The critical gaps can then feed back to the envisioned future (i.e., the three horizons discussed in earlier chapters). At this point, there is also an opportunity to refine the mission and vision statements.

It is important to stress that the development of a strategic road map is not an end point. It is a journey with a need to revisit, engage in dialogue, and debate the planned work and its execution (i.e., work the plan) on a regular basis (e.g., every three months, depending on the complexity of the AI system).

In the next sections, we address each of the components of the AISDM framework in greater detail.

9.3   Mission/Vision and Envisioned Future

In the previous section, we introduced the AISDM framework. In this section, we present more details on the mission and vision statements and the envisioned future. The mission and vision statements should be short, capturing the broad purpose of an AI product or service. Again, these statements should be inspirational and able to draw in prospective stakeholders.

Let us look at an example. During our MIT graduate engineering class, we invited the head of the Misty Robotics company, Tim Enwall, to present a challenge problem to the students [15]. The posed challenge required that the students, working as a team, develop a strategic plan following the AISDM framework. The students were given access to a Misty robot demo unit (https://www.mistyrobotics.com/). The team conceptualized an AI product using the Misty robot, where the robot would serve as an elderly person's companion, with the task of detecting the onset of a stroke and proceeding with an intervention. The MIT team consisted of graduate students Hannah Varner, Jeg Sithamparathas, Nicolas Zhang, and Zubin Wadia. In chapter 14, we describe another application, formulated by a different MIT student team, for the Misty companion robot to help elderly people suffering from Alzheimer's disease.

Stroke is a cardiovascular disease responsible for about 6.7 million deaths worldwide, second only to coronary heart disease in terms of the number of deaths annually [16]. Because of its severity, stroke is the leading cause of disability in the US, with a very low percentage of stroke patients ever recovering completely (about 10 percent). Early stroke detection can have a significant reduction on the afflicted damage caused by a stroke. There are early signs of a likely onset of a stroke, classed under the acronym BEFAST (which stands for "Balance, Eyes, Face, Arms, Speech, and Time") [17]. This set of early indicators and warnings are very well matched to the capabilities of a personal robot like Misty, when supplied with a range of sensors such as cameras, microphone, and natural-language processing (NLP), and equipped with an AI capability.

The MIT team formulated a mission and vision for its AI early detection and intervention product, as follows:

Mission (Misty robot; the "why"): To build a programmable robotics platform and marketplace that accelerates societal progress.

- **Programmable:** Application programming interfaces (APIs) and software development kits will use the Misty sensor suite fully, delivering high accessibility and productivity to science, technology, engineering, the arts, and math (STEAM) practitioners.

- **Marketplace:** Misty skills and labeled data sets for a variety of use cases. Both of these assets can be monetized with network effects: More skills + data lead to faster learning, faster iteration, and better solutions, plus more value delivered to society.

Vision (Misty robot; the "want to be"): To be the world's most accessible and useful robotics platform for hobbyists, educators, and professionals.

The Misty robotics mission statement is short but inspirational, and it captures its purpose. The additional supporting text (i.e., programmable and marketplace) only serves to clarify the context. The vision is a clear description of what the Misty robot's value-add and usefulness are for developers and educators.

As illustrated in figure 9.4, the envisioned future is another important part of the AISDM framework. It helps crystalize the near-term, midterm, and far-term prospects via the horizon structure discussed in part I. In the context of AI future directions, it is formulated and informed by the following:

- AI global trends: These trends can be synthesized based on latest study reports, surveys, and global projections [18–22].
- Lay of the land: This input informs content for the horizons but focused on the specific AI product or service. Here, we must analyze the competition, the market segmentation, channels, pricing, and other elements. Michael Porter, a renowned scholar on the proper way to formulate a strategy, emphasizes that a company can outperform rivals if it can identify a company differentiator that can be preserved [23]. Another useful reference, on how to identify a unique company differentiator in a competitive market, is by Kim and Mauborgne [24].
- Customer priorities and vision: It is not uncommon to have either existing or potential customers ask for help in formulating their own strategic road map. This presents an opportunity for you to align your envisioned future with the needs of your customers. As we have done in other parts of the book, we use the term "stakeholders" to mean customers or users (i.e., those people or organizations with the ability to influence or fund the development of AI capabilities) interchangeably.

These inputs give the AI team tangible knowledge that it can use to proceed with defining the company's envisioned future for an AI product or service.

Returning to our MIT Misty robot team, it formulated the three horizons illustrated in figure 9.5.

Horizon 1 (1–2 years) focuses on building the infrastructure and Misty skills employing existing ML tools. Horizon 2 (years 3–4) would incorporate multiple Misty robots

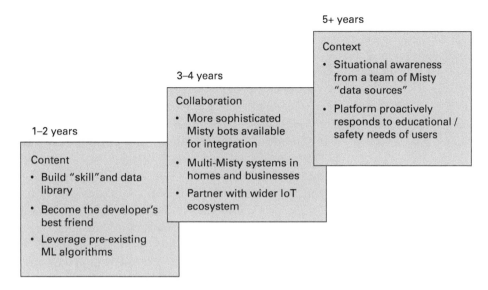

5+ years

Context
- Situational awareness from a team of Misty "data sources"
- Platform proactively responds to educational / safety needs of users

3–4 years

Collaboration
- More sophisticated Misty bots available for integration
- Multi-Misty systems in homes and businesses
- Partner with wider IoT ecosystem

1–2 years

Content
- Build "skill"and data library
- Become the developer's best friend
- Leverage pre-existing ML algorithms

Figure 9.5 Envisioned future (across three horizons) for the Misty robot applied to early detection and intervention of stroke.

collaborating in homes or businesses. For example, machine learning (ML) algorithms learn from patterns seen in other stroke patients as robots share information and use an Internet of Things (IoT) ecosystem. Horizon 3 (5+ years) integrates context in the robot decision support system. An example of context might be observing that an elderly person with the propensity to have a stroke is not going to bed at a normal time and appears to unable to speak (one of the BEFAST indicators). The robot can then respond by calling a nearby family member or neighbor. This example is drawn from a family member of one of the authors (David Martinez). A close relative was nonresponsive, sitting on a couch until a relative found her the next day. The person had suffered a massive stroke that could have been prevented if there was an immediate action taken; like calling a nearby relative and responding in a more timely manner to minimize the stroke severity.

The envisioned future helps in clarifying milestones and showing progress in the development of AI capabilities. It is also very well aligned with executing pilot projects to gain momentum, as outlined in table 9.1.

In the next section, we formulate a strategic direction by incorporating the mission, vision, and envisioned future, all while staying consistent with an organization's core values.

9.4   Organization Core Values and Strategic Direction

The organization mission and vision statements, plus the envisioned future, are the starting steps to help us formulate an organization's strategic direction for an AI product or service. However, the strategic direction must align with the organization core values.

It is important for us to have a common understanding of what organization core values are and are not. Lancioni articulates very clearly that "core values are the deeply ingrained principles that guide all of the company's actions" [25]. The core values are sacrosanct to the foundational bedrock of a well-functioning organization. The culture in the organization must breathe and live the company's core values. You can think of core values as the guideposts and the culture (discussed further in section 9.5) as the instantiation of the core values in the organization's day-to-day operations (i.e., walk the talk); and they must also be aligned with the mission and vision statements [26].

Effective core values should resonate with all business units within an organization. A set of core values that have served the book authors well in the development and successful demonstration of complex systems can be summarized as follows:

- Innovation: Build advanced systems that exemplify innovative differentiators.
- Technical excellence: In technology-based organizations, leaders and staff must maintain the highest technical standards.
- Learning-by-doing: *Mens et manus* (mind and hand) is the MIT motto, and it means that it is not sufficient to conceptualize an AI capability. It must be demonstrated in real-world applications.
- Meritocracy: Reward successes based on well-defined expectations (see chapter 12 for further details).
- Mentoring: Strengthen staff effectiveness by establishing different mentorship approaches across the organization. This core value implies reverse mentoring as well, meaning a mentee providing feedback to a mentor.
- Diversity, equity, and inclusion (DEI): Diversity of thought results from a well-established multidisciplinary team with diverse backgrounds and experiences.
- Integrity and openness: In the day-to-day work environment, we must celebrate and accept input from all ranks within the organization to do the right thing.

In addition to these broad core values, in the rapidly evolving field of AI, we must operate with purpose while adhering to a set of RAI principles. In chapter 8, we introduced and discussed the fairness, accountability, safety, transparency, ethics, privacy, and security (FASTEPS) principles. These principles must form part of the core values of an

AI organization to behave responsibly and be governed by moral values. Nonaka and Takeuchi highlighted this point by stressing that "businesses must root strategy in moral purpose to thrive in a complex, rapidly changing world" [27]. The strategic direction, for an AI product or service, must incorporate, either implicitly or explicitly, the organization core values.

The strategic direction element of the AISDM framework shown in figure 9.4 consists of three parts:

- *Competencies:* The organization's areas of existing or required expertise.
- *Core capabilities:* AI products or services that are built from the organization's competencies.
- *System applications:* Areas to which the AI products or services are applied. These application areas can be identified from a stakeholder's pull (i.e., business needs) or from an AI developer's innovation push.

We adopted these three parts of the strategic direction from the seminal research of Prahalad and Hamel [28], which pointed out: "The corporation, like a tree, grows from its roots. Core products are nourished by competencies and engender business units, whose fruit are end products." We adopted the same three-part structure since it applies well to AI products and services (meaning AI core capabilities).

Let us now crystalize this structure, applied to the AI product formulated by the MIT Misty robot team in enabling the robot to detect and intervene at the onset of a stroke. Figure 9.6 shows the three parts of the MIT team's strategic direction. The competencies represent the required technical skills needed to nourish the desired core capabilities. These capabilities in turn engender a set of system applications. In particular, the AI system application of early stroke detection and intervention falls under the responsibilities of the caretaker. The MIT team also formulated other AI system applications that are well matched to the set of core competencies, enabled by a set of competencies.

The organization core values are implicit in this AI application example, since the AI designers, developers, and implementers must attend to such issues as patient privacy. The ML algorithms must also comply with fairness across a broad range and diverse population, and the AI techniques must be safe and secure. Broadly speaking, this type of healthcare application must be deployed in a responsible way for the betterment of the aging population (accelerating societal progress and employing an accessible and useful robot, as described in the mission and vision statements).

Now that we have formulated what goes into a strategic direction, we expand on the elements of the AI value proposition in the next section. The AI value proposition is a key component of the overall AISDM framework.

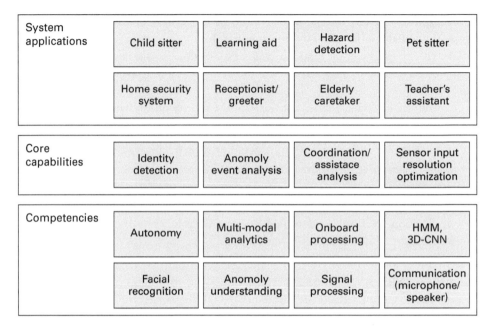

Figure 9.6   Strategic direction for the AI product using the Misty robot to enable the timely detection and intervention of a stroke.

9.5   AI Value Proposition

The AI value proposition forms the cornerstone of the AISDM framework, gluing together preceding elements of the model (i.e., mission and vision statements, envisioned future, organization core values, and strategic direction) with the AI strategic road map. As shown in figure 9.4, the AI value proposition is substantiated and reinforced by four key subelements:

- AI implementation opportunities
- People
- Organization
- Culture

The reader will notice the explicit emphasis on the people-process-technology triad that we have discussed throughout the earlier chapters of this book in the context of emphasizing the AI systems engineering approach across all aspects of an AI-driven organization. The first subelement—AI implementation opportunities—reflects the need for financial support. An AI value proposition is not realistic if there is no source of financial backing from either internal (e.g., internal research and development) or external support (e.g., investors, customers, or business clients). Typically, there are implementation opportunities from customers with influence and/or customers with funding resources. This distinction is made because, for example, a CEO of a corporation (with influence) can direct the development of a proposed AI capability, but an executor like a chief technology officer must allocate the financial resources.

Next in the subelements outlined here is people; we devote chapter 12 to the topic of leadership and talent. Both are needed to successfully execute an AI strategic road map. Furthermore, since AI requires a multidisciplinary team with broad expertise, leaders must be AI-aware, with sufficient competence (with emphasis on systems-level knowledge) to lead teams. Conversely, an AI team with limited skills—meaning lack of talent—will struggle in executing the AI value proposition.

Next, an AI organization must support the development and implementation of AI products or services by establishing a governance structure, facilitating access to AI technologies, and building the requisite infrastructure. The governance structure must ascertain compliance with the FASTEPS elements discussed in chapter 8. The AI governance, within the organization, should be light enough to avoid atrophying innovation, but robust enough to assert correction if the AI organization deviates from its moral compass, and to ensure operating within the guardrails complying with the FASTEPS principles.

Access to AI technologies is needed to enable the AI value proposition. The AI technologies are the innovative differentiators for an organization across the end-to-end AI system architecture shown in figure 1.3 in chapter 1. We include this component within the organization subelement because an AI organization must approach the design, development, and integration of their products and services, emphasizing every element of the system architecture. However, we do not elaborate further in this chapter since this topic has been covered in detail in part I of the book.

An AI-based organization must also work with internal and external partners by cementing relationships and using an AI ecosystem, which strengthens the prowess of an AI team to successfully develop and deploy AI capabilities. We elaborate on this topic in much greater detail in chapter 10.

As we traverse the inputs that influence the AI value proposition, culture is a subelement that makes or breaks the ability to effectively and efficiently execute the design, development, and implementation of AI capabilities. It is very well documented that without a supporting culture, a strategy will fail. Edgar Schein, an MIT professor emeritus and renowned scholar of organizational culture, clearly expressed what culture means for an organization in his seminal paper "Coming to a New Awareness of Organizational Culture" [29]. He defined organizational culture as "a pattern of basic assumptions that a given group has invented, discovered, or developed in learning to cope with its *problems of external adaptation and internal integration*, that has worked well enough to be considered valid and, therefore, to be taught to new members as the correct way to perceive, think, and feel in relation to those problems."

We emphasized the phrase "problems of external adaptation and internal integration" since developing AI capabilities depends on the successful internal integration, verification, and validation of the systems engineering Vee-model shown in figure 2.1. The Vee-model starts with understanding the stakeholders' specifications through system realization, while at the same time meeting the needs of external customers.

Van Maanen explains succinctly what culture is: "The organizational culture is not as visible as, for example, the organization's management structure, but it is the glue that drives the operating environment of an organization" [30]. As we pointed out earlier, organization core values can be thought as the guideposts. The organizational culture is how leaders and staff effect or espouse those core values in practice (i.e., walk the talk [31]). Organizational culture encapsulates its purpose while always balancing long-term versus short-term gains.

Westerman et al., in their article "Building Digital-Ready Culture in Traditional Organizations" [32], discussed areas to advance the state of traditional organizations as they go through a digital transformation by adhering to the following steps:

- **Build** an environment for rapidly experimenting, self-organizing, and driving decisions with data.
- **Preserve** practices of acting with integrity and seeking stability by attracting and retaining talent, fostering customer loyalty, and ensuing stakeholder confidence.
- **Reorient** the business direction, anticipating customers' needs, driving accountability for results, and setting rules that prevent abuses.

These steps are excellent recommendations for building a digital-ready culture. They provide guidelines very consistent with our emphasis in this chapter.

It is also very refreshing to see and to recognize that AI can also help in unifying the culture within an organization. Ransbotham et al. performed a comprehensive global survey across over 2,000 managers and interviews with eighteen executives. A common message that they heard was: "Business culture affects AI deployments, and AI deployments affect business culture." [33].

Another relevant survey was performed by Deloitte Global Boardroom to identify how well aligned the technology was with the organization's strategy. Touche et al.

summarized the global survey performed across fifty-five countries, consisting of more than 500 board directors, C-suite executives, and subject matter experts (SMEs) [34]. One of their findings motivates the clear need for formulating an AI strategic plan as described in this chapter. They found that about 50 percent of the respondents said that they: "didn't think—or didn't know if—technology aligned with the organization's strategy." This finding was common among a large range of industries, including financial services, manufacturing, healthcare, and retail.

Returning to the Misty robot application from one of the MIT teams working on early detection and intervention of the onset of stroke, they formulated their AI value proposition as shown in figure 9.7. The AI system architecture illustrates the labeled data feeding into the desired Misty apps and skills. The apps and skills, designed from the incoming labeled data, form the core ML techniques to enable BEFAST stroke detection and intervention. The architecture uses a multicloud computing environment, such as with Microsoft Azure, Amazon Web Services (AWS), or Google Cloud Platform. The team also included the use of multiple Misty robots and available open-source tools.

The prior descriptions, on the elements of the AISDM framework shown in figure 9.4, lead to coupling the near-term and long-term strategic directions with the AI technology underpinning the AI value proposition. We are now equipped to formulate the AI strategic road map (i.e., the strategic blueprint), aligning the value proposition with the customer's value capture.

9.6   AI Strategic Road Map: A Blueprint

The AI strategic road map bridges the five early subcomponents of the AISDM framework—long-term mission and vision and envisioned future (section 9.3), the organization core values and strategic direction (section 9.4), and the AI value proposition (section 9.5)—with the implementation blueprint. As we have emphasized throughout this chapter, a strategy without an implementation blueprint is not going to bring business value to an AI-driven organization. The seven elements of an AI strategic road map are:

- Business needs: A clear delineation of the stakeholder's requirements to ascertain that the AI value proposition is aligned with the value capture from the viewpoint of the customers.
- SWOT: An analysis of organization's strengths and weaknesses (internal factors), and opportunities and threats (external factors).
- Risk management: Risk categories and risk levels.

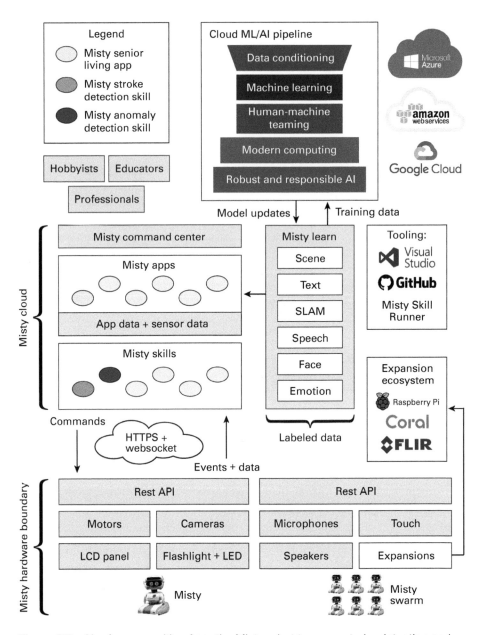

Figure 9.7   AI value proposition from the Misty robot team on stroke detection and intervention.

- Goals and actions: Roles and responsibilities within the leadership and management team to achieve a defined set of objectives.
- Capabilities versus business needs: Mapping of AI capabilities to the business needs of the stakeholders and determining if there are critical gaps.
- Performance metrics: Tools and techniques for evaluating execution progress when implementing the AI strategic road map.
- Prototypes (MLOps demo): Demonstration of the MVP to gain momentum (see table 9.1).

Let us now describe in more detail each of the seven elements of the AI strategic road map, starting with business needs. There have been a number of academic and how-to articles on the effective mapping of a business digital transformation [35–37]. An emphasis in these recommended approaches is to understand the pain points of the stakeholder (e.g., the customer), and how the value proposition addresses them. They point out the importance of focusing not just on technology, but instead on the business value (i.e., the customer's value captured from the AI value proposition).

As discussed in chapter 2, the strategic road map implementation—specifically, the AI value proposition—hinges on the effective execution of the AI system architecture implementation framework shown in figure 2.2. We will discuss the AI strategic implementation in more detail in chapter 10. In chapter 3, the questions to answer were focused on the data, ML algorithms, and computing requirements (as shown in table 3.1). Business understanding for creating value out of AI capabilities is intrinsic to the key phases of the cross-industry standard process for data mining (CRISP-DM) [38].

In formulating the AI strategic road map, we need to ask higher-level questions focusing on the expected value delivered to the customer. Table 9.2 addresses the key questions. Again, the AI team must be customer focused. Answers to these questions, during the formulation of the AI strategic development road map and prior to the implementation phase, are critical to the successful design, development, and deployment of AI capabilities.

The questions in the table are an adaptation of the well-known questions known as the "Heilmeier Catechism," or sometimes called the "Heilmeier Criteria," originally posed by George H. Heilmeier, in deciding whether to approve funding research projects.

The Misty robot with the MIT team formulated a set of answers to the questions posed here, which are given in table 9.3.

Next, we turn our attention to a SWOT analysis of the organization with respect to the offered AI product or service. The proper method to develop a SWOT analysis is to identify the internal and external factors that most critically affect the ability of

Table 9.2   Key questions to ask addressing business needs, adapted from the Heilmeier Criteria

Key Questions	Comments
What AI capability is the customer looking for from this AI product or service?	Work from the result expected back to the product or services needed.
How is it done today? And what are the limits of the current practice?	Understand the competition landscape.
What is new in your approach? And why do you think it will be successful?	Highlight the unique differentiator.
Who cares?	Identify the customers with influence and/or funding resources.
If you're successful, what difference will it make?	Emphasize the customer value aligned to the AI value proposition (using the AI system architecture).
What are the risks and the payoffs?	Assess the capabilities versus the needs through the lens of a risk management approach (i.e., risk categories and levels).
How much will it cost? How long will it take?	Time is money. Balance requirements against time and cost.
What are the performance metrics?	Define clear metrics to assess progress.
What are the milestones to check for success?	Formulate the success criteria as defined in the envisioned future.

the organization to design, develop, integrate, and deploy the AI capability in production. One can formulate an effective SWOT analysis as follows [39]:

- Start with either of the following:
  - Strengths and weaknesses (internal)
  - Then, opportunities and threats (external)
  - Or strengths and opportunities, and then weaknesses and threats
- Look at alternatives:
  - Debate alternatives viewed through the lens of making your product or service executable, and in line with the 1-to-5-year envisioned future (i.e., the three horizons: near term, midterm, and far term).

Table 9.3   Misty robot answers to questions addressing business needs

Key Questions	Answers
What AI capability is the customer looking for from this AI product or service?	Helping caretakers and the elderly as an early stroke detection and intervention solution for residential environments.
How is it done today? And what are the limits of the current practice?	Stroke victims go too long before treatment because recognition of symptoms by the patient or caretaker is slow. Average time to reach the emergency room is 5.5 hours.
What is new in your approach? And why do you think it will be successful?	Misty with ML capabilities will be a companion and vigilant attendant to notice the first physical signs or behavior changes caused by a stroke. Sensors and ML algorithms are available now
Who cares?	Elderly patients and their caretakers deserve peace of mind and a reliable backup (i.e., nonthreatening appearance)
If you're successful, what difference will it make?	Millions of people have heightened risk factors for stroke, and billions of dollars in treatment cost are on the line. Only 10 percent of today's stroke patients make a full recovery
What are the risks and the payoffs?	Risk is in the rejection of Misty in the target use case. The payoffs will be commensurate with the envisioned future time line.
How much will it cost? How long will it take?	Cost is in primarily software development. Time to completion is <6 months for proof of concept and <9 months to a prototype deployment.
What are the performance metrics?	High accuracy in early stroke detection; very low false negatives.
What are the milestones to check for success?	Checkpoints are as follows: 1. Proof of concept: Ability to recognize time series invariance in normal behavior patterns 2. Transition to Misty platform 3. Training for physical symptoms with compiled data

- Summarize these alternatives.
- Vote on top choices.
- Do a quick revisit again after completing all four categories.
- SWOT analysis output:
  - Highlight the most important drivers per quadrant (typically limited to about five)

The MIT team also performed a SWOT analysis for their Misty robot offering, as shown in table 9.4.

The third component of the AI strategic road map, risk management, deserves a detailed explanation since there are well-established tools within the systems engineering discipline [40], but that have not yet been adapted to the development of AI strategic plans. In this section, we bridge that gap by introducing the reader to these risk management tools. It is important to identify the risk categories and levels within the AI strategic road map since these are crucially important to the successful implementation of the AI system architecture framework illustrated in figure 2.2.

The *INCOSE: Systems Engineering Handbook* [40] addresses several types of risk categories and levels, as shown in figure 9.8. To be consistent with the International Council on Systems Engineering (INCOSE) standards and guidelines, we highlight the meaning of each of the following risk categories:

- Technical risk: The possibility that a technical requirement of the system may not be achieved.
- Cost risk: The possibility that the available budget will be exceeded.
- Schedule risk: The possibility that the project will fail to meet the scheduled milestones.
- Programmatic risk: Risk produced by events that are beyond the control of the project manager and can affect the risk in any of the other three risk categories

Note that these risk categories are exacerbated by a number of factors, such as requirement drift (i.e., scope creep), technical problems, compressed schedules, and limited funds. Therefore, there needs to be alignment among the envisioned future, the strategic direction—discussed early in the AISDM framework—and the AI value proposition with the execution approach defined in the AI strategic plan.

These risk categories have another important and complementary dimension when assessing the impact on the overall project. As illustrated in figure 9.8, the risk levels determine the severity of the risk categories on AI development. It is an assessment of the

Table 9.4 Misty robot SWOT analysis

Internal Factors	Strengths	Weaknesses
	• Versatile platform, navigation of a home is built into the design	• The short height of Misty robot means that some events may be out of view.
	• High-resolution cameras integrated on board	• The marketing platform as a tool for hobbyists might reduce trust in the life-or-death BEFAST application.
	• Proximity with the developer community	• Privacy violations have the potential to damage company image and reduce customer trust.
External factors	**Opportunities**	**Threats**
	• No existing commercial platform for AI detection of stroke on the market	• Ability to make 911 calls / proper emergency alerts
	• Misty robot is likely to be accepted given its form factor	• Possible regulatory prohibitions for use in nursing homes. (however, there is a large market potential).
	• Caretakers are increasingly trusting of and savvy with technology	• Personal robotics is a very active growth sector.

likelihood that an event will occur and the undesirable consequence of that event on the AI project [40, 41]. The likelihood is often presented as a probability of the event occurring (from low to high). The consequence, or impact of an event, depends on the type of event at any of the subcomponents in our AI system architecture causing a risk (e.g., lack of data, low-performing ML algorithms, limited computing resources, adversarial attacks, and many other technical risks). The same is true for programmatic, cost, and schedule risks. The risk assessments have a direct implication on the Vee-model subcomponents [42] in figure 2.1 and the AI system architecture implementation framework illustrated in figure 2.2.

The fourth element of the AI strategic road map is goals and actions. Goals enumerate objectives that must be met to successfully develop, integrate, and deploy an AI product or service. The actions are formulated with respect to the set of goals. It is also important to allocate a member of the AI team with the authority and responsibility to

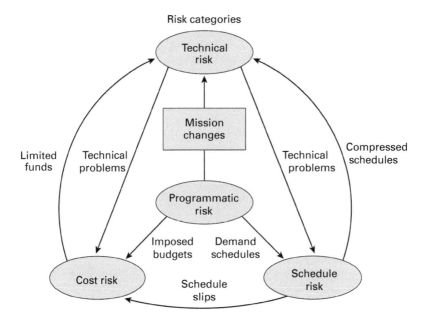

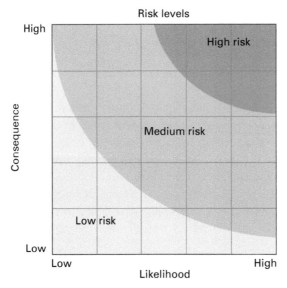

Figure 9.8   INCOSE risk management categories and levels [40]. © 2015 John Wiley & Sons, Inc.

meet a subset of the defined actions. Otherwise, the AI project can fail because of lack of ownership by the team members. It is useful to define the goals and actions relative to addressing the business needs, SWOT analysis, and risk management. These goals and actions must also have a deliverable time line to make them tangible. An example of a goal for the Misty robot application might be to receive approval from the US Food and Drug Administration (FDA) to operate in nursing homes. The action can be to fill out a formal request with the FDA office in question.

There are also practical tools to address capabilities versus business needs as part of a comprehensive AI strategic road map. A tool used by one of the authors, in the actual development of a strategic road map, is shown in figure 9.9. The chart depicts capabilities (existing or new) and stakeholder business needs (existing or new). This topology is very useful because an existing customer can be very receptive to advancing capabilities by evolving from existing to new. Or an existing capability, in development or delivered to an existing customer, might be very appealing to a new customer if it underwent some adaptations.

The "capabilities versus business needs" context is most useful when revenues are identified for each of the respective AI programs or projects. The arrows represent the use of an earlier program in a new program or programs. A well-functioning AI organization is likely to have most of its revenues (and backlog) in the Existing-Existing quadrant. There are likely less AI programs in the New-New quadrant since new capabilities for a new stakeholder require significant marketing and business development efforts. In the process of identifying different AI programs, the AI strategic road map must be directly coupled with the system applications of the AI strategic direction shown in figure 9.6. For example, for the Misty robot that we have discussed throughout this chapter, the focus has been primarily on the elderly caretaker AI application. However, the same AI capability (existing), with some modifications, can be well matched to a home security system, hazard detection, or teacher's assistant (representing either new capabilities for existing stakeholders or similar capabilities for new stakeholders).

The sixth element of the AI strategic road map must incorporate both quantitative and qualitative performance metrics to determine how well the execution of the AI strategic blueprint is progressing. An example of a quantitative metric is an evaluation of the accuracy and level of false negatives (i.e., meaning the robot indicated no stroke when the elderly person did have a stroke) in the early stroke detection.

A more macro-level metric, as a qualitative metric, is the level of maturity of the proposed AI product or service. All products or services go through a life cycle, depending on how mission critical the offering is to the AI organization versus the level of company differentiator, as illustrated in figure 9.10.

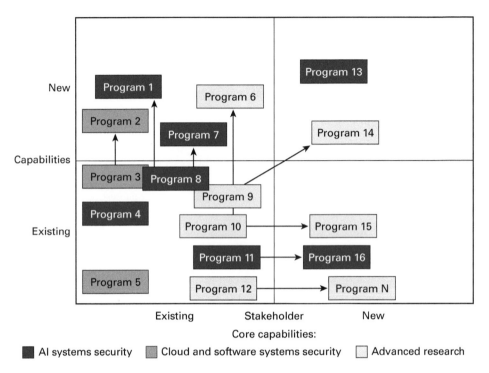

Figure 9.9   Capabilities versus stakeholder business needs.

A thriving technology-based organization must be constantly innovating. This represents the lower-right quadrant in figure 9.10. Once products or services demonstrate useful AI capabilities, the organization can then exploit this advancement to gain broader market share with existing customers or penetrate new sources of revenue as high company differentiators—this is represented by the top-right quadrant. Once the AI product or service reaches maturity, it is best to commoditize the AI capability to reap the benefits of the revenue stream, and some percentage of the revenue can also be used to infuse sources of funding into new, innovative AI products or services. Finally, the last quadrant of figure 9.10 represents the outsourcing of routine jobs (e.g., product support).

The last element of the AI strategic road map requires a description of the type of prototype under the category of MLOps. This component of the road map is so important to the ultimate success of an AI capability that we devote chapters 10 and 11 to AI deployment and operations. As emphasized earlier in table 9.1, executing pilot projects will help with gaining momentum. These pilot demonstrations serve to assess the

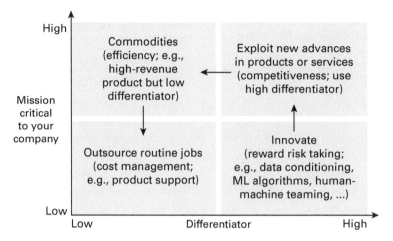

Figure 9.10   Macro-level assessment of an AI product or service.

progress in formulating new innovative AI concepts, as shown in the lower-right quadrant in figure 9.10.

The MIT team did an in-class demonstration to a panel of academia and AI practitioners, of an initial pilot prototype using the Raspberry Pi (representing the single-board computer that could house the Misty robot's "AI brains"). The team was able to successfully demonstrate the ability to learn normal patterns, spot movement pattern changes, and notify responsible caretakers if an abnormal behavior was detected.

As shown in figure 9.4, the AISDM framework must be used as an ongoing and formal document that encapsulates the overall focus, justification, and plans for the design, development, integration, and deployment of the organization's AI products or services. An important part of revisiting the strategic plan is to incorporate critical gaps. The critical gaps result from an in-depth analysis of the subelements of the AI strategic road map (i.e., its seven elements), with respect to what desired business needs have not yet been met.

9.7   Strategy and Execution: A Complementary Duo

We began this chapter with a quote from Morris Chang: "Without strategy, execution is aimless. Without execution, strategy is useless." Strategy depends on successful execution to provide meaning and benefits to an organization. Similarly, successful execution requires a directional vector—codified in the strategic development road map—for the organization and the AI team to have a clear understanding of the AI value proposition, and its intended business value to stakeholders, as a function of time (i.e., the envisioned future).

There have been many articles and studies showing lack of success in transitioning AI concepts into operation. Some of these barriers are particularly problematic for organizations that were not originally digital. Today, AI organizations must make timely decisions based on data-driven approaches while attending to a rapidly evolving AI market. As discussed by Fountaine et al., in order to scale up, AI organizations must have a clear understanding of what is feasible, the business value, and time horizons [43].

It is also critical to incorporate the RAI guidelines discussed in chapter 8. AI capabilities might show the right levels of performance according to the metrics described in the strategic road map but can result in complete failure and not be accepted by the stakeholders, if these capabilities do not adhere to the FASTEPS principles.

To increase the likelihood of success, in this chapter, we have described the AISDM framework for leaders, managers, and technical staff who are responsible for developing and deploying AI capabilities. The tools described offer more clarity and specifics of the strategic focus (near term, midterm, and far term) and road map. In the next several chapters, we turn our attention to AI execution and deployment.

9.8 Main Takeaways

In this chapter, we set the stage for part II of the book by discussing the steps involved in the formulation of an AI strategy that culminates with a strategic road map. It serves as a blueprint for upper management, technical leaders, AI architects, designers, and implementers to progress from architecture principles to deployment. A strategy is not a static deliverable, but rather a living blueprint that must be revisited and, if necessary, updated over time while the AI system is undergoing development and, until the final deployment, careful attention must be paid to its use and monitoring. A key part of a successful strategy is to include all ranks, from AI management to AI staff, in the process of strategic thinking and in the formulation of the AI strategic road map.

Strategic thinking puts a framework around determining the overall direction of an AI organization. More specifically, strategic thinking helps answer critical management questions, such as the following:

- What should the AI organization look like in the near term (horizon 1; 1–2 years), midterm (horizon 2; 3–4 years), and far term (horizon 3; ≥ 5 years)?
- Who benefits from strategic thinking and the formulation of a strategic plan?
- Who should we be hiring?

- What should we be learning?
- What technologies do we need to enter and exit?
- What contracts and channels should we grow or shrink?
- Who will be willing and able to pay for it?
- What processes are or are not needed?
- How federated should the portfolio be with respect to internal and external partners?
- What should be the mix of direct and contracted employees?

The mastery of strategic thinking helps staff and entry-level managers advance in a competency that has long-lasting value, especially in highly competitive, technology-based organizations. Four of the ways to improve your strategic thinking are: *Know* (observe and seek trends), *Broaden* (ask the tough questions), *Communicate* (shape common understanding), and *Act* (embrace conflict to debate key challenges).

The AISDM, shown in figure 9.4, provided context and details relevant to formulating an AI strategic road map. The AISDM framework is applicable to formulating an AI strategic road map at the organization level, the business-unit level, the group-unit level, or the AI project level.

The AISDM framework calls for first formulating a draft of the mission and vision statements. These statements can be refined after completing a full pass through the AISDM framework and after completing a draft of the AI strategic road map. The other key elements of the AISDM framework are the envisioned future and organization core values. These two elements serve to formulate the company's strategic direction for an AI product or service. The mission and vision statements, the envisioned future, the organization core values, and the AI strategic direction lead to the central part of the AISDM framework: the *AI value proposition*.

The AI value proposition is formulated based on the subsystems of the AI system architecture discussed in part I. The value proposition is informed by four key components:

- AI implementation opportunities
- People (i.e., leadership and talent)
- Organization AI governance, RAI principles, technologies, and infrastructure
- Culture

The prior key elements of the AISDM framework are necessary to understand how to properly create an AI strategic road map, as the blueprint for the near term, midterm, and far term. The AI strategic road map also serves to build clarity in transitioning from the company's strategy to an executable blueprint. The seven elements of an AI strategic road map are:

- Business needs: A clear delineation of the stakeholder's requirements to ascertain that the AI value proposition is aligned with the value capture from the perspective of the customers
- SWOT: An analysis of organization's strengths and weaknesses (internal factors), and opportunities and threats (external factors)
- Risk management: Risk categories and risk levels
- Goals and actions: Roles and responsibilities within the leadership and management team to achieve a defined set of objectives
- Capabilities versus business needs: Mapping of AI capabilities to the business needs of the stakeholders and determining if there are critical gaps
- Performance metrics: Tools and techniques for evaluating execution progress when implementing the AI strategic road map
- Prototypes (MLOps demo): Demonstration of the MVP to gain momentum.

No one should start formulating an AI strategic road map without well-defined answers to the stakeholders' business needs outlined in table 9.2. Often AI strategies fail because they focus primarily on the technologies; organizations must drive their AI products or services based on the business needs.

Once an AI-based company understands the driving business needs, the next task is formulating the organization's SWOT with respect to the offered AI product or service. The proper method to develop SWOT analysis is to identify the internal and external factors that most critically affect the ability of the organization to design, develop, integrate, and deploy AI capabilities in production.

An AI strategic road map must also balance the organization's desires against the achievable capabilities, as a function of time. Therefore, a crucial step in the formulation of the AI strategic road map is to assess the risk categories and levels.

The *INCOSE: Systems Engineering Handbook* [40] addresses several types of risk categories and levels, as shown in figure 9.8. We highlighted the meaning of each of the risk categories as follows:

- Technical risk: The possibility that a technical requirement of the system may not be achieved
- Cost risk: The possibility that the budget will be exceeded
- Schedule risk: The possibility that the project will fail to meet the scheduled milestones

- Programmatic risk: Risk produced by events that are beyond the control of the project manager; and can affect the risk in any of the other three risk categories

Once we have determined the business needs, performed a SWOT analysis, and assessed the categories and levels of risk, we then can proceed to identify a set of goals and actions. Members of the AI team must be assigned this responsibility and given the authority to attend to a subset of the goals and actions; this helps the team have ownership of the AI strategic road map.

Again, an important reason for developing the AI strategic road map is to clearly determine the achievable AI capabilities (existing and new) relative to the stakeholders' business needs (existing and new). From this analysis, one can begin to generate a set of critical gaps that can be delayed until future iterations of AI products or services.

One of the most important components of a comprehensive strategic plan is the performance metrics. We discussed quantitative and qualitative metrics. Performance metrics help in tracking and measuring progress. These metrics also help in enumerating critical gaps.

Ultimately, the AI team must develop the MVP, an early prototype to gain credibility and provide initial value to the stakeholders. An early prototype also leads to finding additional critical gaps. This effort falls under the rubric of ML operations, which will be discussed in the next two chapters.

9.9 Exercises

1. Is it correct to say that without strategy, execution is useless, and with a strategy, execution is guaranteed?
   a. True
   b. False
2. A strategic road map is addressed once and should never be revisited throughout the year in order to avoid misunderstandings.
   a. True
   b. False
3. Strategic thinking, as well as the strategic plan that results from the process, help for example AI staff, project leaders, and upper management.
   a. True
   b. False

4. Choose two of the strategic thinking questions that help answer critical management questions, outlined in section 9.1, and elaborate on what they are addressing and why they are important.

5. Pick one out of the four ways to improve your strategic thinking and discuss its meaning and benefit.

6. A 2018 McKinsey & Company report identified a lack of a clear strategy for AI as the most frequently cited barrier to AI adoption [5].
    a. True
    b. False

7. The organization's mission is about "what" the organization wants to be. The vision is about "why" the organization exists.
    a. True
    b. False

8. In a short set of paragraphs, give succinct definitions for the envisioned future, the organization core values, and the AI strategic direction.

9. The strategic road map is the blueprint that codifies the near-term, midterm, and long-term strategic directions for an AI organization, informed by its resources, processes, and values.
    a. True
    b. False

10. What are the seven elements of an AI strategic road map?

11. Pick one of the key questions to ask addressing business needs, and describe why that question is important in formulating an AI strategic road map.

12. Describe the four risk management categories.

13. Elaborate on the risk levels and their significance.

14. Highlight the four quadrants of the macro-level assessment of an AI product or service, in terms of mission critical to your company versus company differentiator, and provide a short description of each.

15. An early pilot project is useful because of which of the following:
    a. It helps in gaining momentum with demonstrating AI capabilities
    b. The AI team can identify additional critical gaps
    c. The hardware designers can perform a preliminary design review before starting the software coding or hardware fabrication

d. a. and b

e. None of the above

16. What performance metrics would you recommend, as part of the strategic development road map, to ascertain compliance with the RAI principles (FASTEPS) discussed in chapter 8?

9.10 References

1. Ng, A., *MIT technology review: AI-focused EmTech digital.* 2021.
2. Rifkin, G., *Future forward: Leadership lessons from Patrick McGovern, the visionary who circled the globe and built a technology media empire.* 2018, McGraw-Hill. https://www.futureforwardbook.com/.
3. *Harvard Business Review, HBR guide to thinking strategically.* 2019, Harvard Business School Publishing Corporation.
4. Bowman, N. 4 ways to improve your strategic thinking skills. *Harvard Business Review*, December 27, 2016. https://hbr.org/2016/12/4-ways-to-improve-your-strategic-thinking-skills.
5. Webb, N., *Notes from the AI frontier: AI adoption advances, but foundational barriers remain.* 2018, McKinsey & Company Report.
6. Christensen, C. M., S. D. Anthony, and E. A. Roth, *Seeing what's next: Using the theories of innovation to predict industry change.* 2004, Harvard Business Press.
7. Schwartz, P., *The art of the long view: Planning for the future in an uncertain world.* 2012, Currency.
8. Stadler, C., J. Hautz, K. Matzler, and S. F. von den Eichen, Open up your strategy. *MIT Sloan Management Review*, 2022. 63(2): 1–6.
9. Harvard Business School, Program for leadership development: Accelerating the careers of high-potential leaders. https://www.exed.hbs.edu/comprehensive-leadership-programs.
10. Sinek, S., *Start with why: How great leaders inspire everyone to take action.* 2009, Penguin.
11. *MIT computer science and artificial intelligence laboratory.* https://www.csail.mit.edu/about/mission-history.
12. Smith, W. K., M. W. Lewis, and M. L. Tushman, Both/and leadership. *Harvard Business Review*, 2016. 94(5): 62–70.
13. O Reilly, C. A. and M. L. Tushman, The ambidextrous organization. *Harvard Business Review*, 2004. 82(4): 74–83.

14. Tushman, M., M. L. Tushman, and C. A. O'Reilly, *Winning through innovation: A practical guide to leading organizational change and renewal.* 2002, Harvard Business Press.

15. Enwall, T. *Challenge: Formulate a strategic development roadmap helping the elderly using the misty robot,* 2019. https://www.youtube.com/watch?v=FDukxWJk6iU.

16. Chandra, A., C. R. Stone, X. Du, et al., The cerebral circulation and cerebrovascular disease III: Stroke. *Brain Circulation,* 2017. 3(2): 66. https://www.ncbi.nlm.nih.gov/pmc/articles/PMC6126259/.

17. Stull, M. *When it comes to recognizing a stroke, B.E.FA.S.T.* 2018. https://www.mymarinhealth.org/blog/2018/may/when-it-comes-to-recognizing-a-stroke-b-e-f-a-s-/.

18. Dutton, T., *Overview of national AI strategies. Medium,* June 28, 2018.

19. Gil, Y. and B. Selman, A 20-year community roadmap for artificial intelligence research in the US, 2019. arXiv preprint arXiv:1908.02624.

20. Parker, L. E., Creation of the National Artificial Intelligence Research and Development Strategic Plan. *AI Magazine,* 2018. 39(2): 25–32.

21. Schmidt, E., et al., *NSCAI final report.* 2021, National Security Commission on Artificial Intelligence. https://www.nscai.gov/2021-final-report/

22. Zhang, D., N. Maslej, E. Brynjolfsson, et al., *The AI Index 2022 Annual Report.* 2022, Stanford Institute for Human-Centered AI.

23. Porter, M. E., The five competitive forces that shape strategy. *Harvard Business Review,* 2008. 86(1): 25–40.

24. Kim, W. C. and R. Mauborgne, *Blue ocean strategy, expanded edition: How to create uncontested market space and make the competition irrelevant.* 2014, Harvard Business Review Press.

25. Lencioni, P. M., Make your values mean something. *Harvard Business Review,* 2002. 80(7): 113–117.

26. Collins, J. C. and J. I. Porras, Building your company's vision. *Harvard Business Review,* 1996. 74(5): 65.

27. Nonaka, I. and H. Takeuchi, *Strategy as a way of life.* 2021, *MIT Sloan Management Review.*

28. Prahalad, C. and G. Hamel, The core competence of the corporation. *International Library of Critical Writings in Economics,* 2003. 163: 210–222.

29. Schein, E. H., Coming to a new awareness of organizational culture. *Sloan Management Review,* 1984. 25(2): 3–16.

30. Van Maanen, J., *Early career leaders—Circle mentoring program*. 2015, MIT Lincoln Laboratory.

31. Sull, D., S. Turconi, and C. Sull, When it comes to culture, does your company walk the talk? 2020, *MIT Sloan Management Review*.

32. Westerman, G., D. L. Soule, and A. Eswaran, Building digital-ready culture in traditional organizations. *MIT Sloan Management Review*, 2019. 60(4): 59–68.

33. Ransbotham, S., F. Candelon, D. Kiron, et al., *The cultural benefits of artificial intelligence in the enterprise*. 2021, MIT Sloan Management Review and Boston Consulting Group.

34. Touche, W., D. Konigsburg, and J. Iwasaki, *Digital frontier: A technology deficit in the boardroom*, D. G. B. Program, Editor. 2022. Deloitte. https://www2 .deloitte.com/ca/en/pages/audit/articles/digital-frontier-a-technology-deficit-in -the-boardroom.html.

35. Fritscher, B. and Y. Pigneur, Visualizing business model evolution with the business model canvas: Concept and tool, in *2014 IEEE 16th Conference on Business Informatics*. 2014. IEEE.

36. Osterwalder, A., Y. Pigneur, G. Berrardo, and A. Smith, *Value proposition design: How to create products and services customers want*. 2015, John Wiley & Sons.

37. Mueller, B. How to map out your digital transformation. *Harvard Business Review*, 2022. https://hbr.org/2022/04/how-to-map-out-your-digital-transformation.

38. Kelleher, J. D., B. Mac Namee, and A. D'arcy, *Fundamentals of machine learning for predictive data analytics: Algorithms, worked examples, and case studies*. 2020, MIT Press.

39. Leigh, D., SWOT analysis. *Handbook of improving performance in the workplace*, Volumes 1–3, 2009. 115–140. International Society for Performance Improvement.

40. *INCOSE: Systems engineering handbook: A guide for system life cycle processes and activities*. 4th ed., edited by D. D. Walden, G. J. Roedler, K. J. Forsberg, et al. 2015, John Wiley & Sons.

41. Eisner, H., Systems engineering: Building successful systems. *Synthesis lectures on engineering, science, and technology*. Vol. 14. 2011, Morgan & Claypool Publishers. 1–139.

42. Forsberg, K. and H. Mooz, The relationship of system engineering to the project cycle, in *INCOSE International Symposium*. 1991, Wiley Online Library.

43. Fountaine, T., B. McCarthy, and T. Saleh, Building the AI-powered organization. *Harvard Business Review*, 2019. 97(4): 2–73.

10

AI Deployment Guidelines

> AI is a tool. The choice about how it gets deployed is ours.
> —Oren Etzioni, founding CEO of the Allen Institute of Artificial Intelligence (AI2)

In chapter 9, we described all the elements that go into the development of an artificial intelligence (AI) strategy and road map, as part of our AI strategic development model (AISDM). The AISDM framework serves as a blueprint for AI architects, developers, and implementers to follow a rigorous approach to successfully develop and deploy AI capabilities with value to the stakeholders. An important component of the strategic road map is the demonstration of an AI product or service via early prototypes—which, after multiple iterations, would mature into a final deliverable product or service. In chapters 10 and 11, we elaborate on AI deployment, and we discuss approaches, techniques, and representative tools.

Chapter 10 is a macro-view where we lay out the foundational structure needed for a successful AI deployment. It also helps the reader understand how systems engineering is applied to the development and deployment of AI systems. With this background, in chapter 11, we go deeper into approaches, techniques, and representative tools to effectively transition AI capabilities from deployment into operations. This latter topic is discussed within the rubric of a fast-evolving field known as "machine learning operations (MLOps)."

Before going further into the discussion on transitioning AI capabilities from development to deployment, it is worth noting that MLOps could be construed and interpreted as only addressing ML (meaning transitioning algorithms into operations). MLOps is much more encompassing than algorithms transitioned into operations. As noted by Kreuzberger et al., accepted MLOPs definitions, as a minimum, span development, integration, deployment, and continuous integration/continuous delivery

(CI/CD) [1]. Strictly speaking, our discussion and the theme of our book are better called "AI systems operations (AISysOps)." However, we opted to adapt MLOps to stay consistent with the AI development and deployment community and avoid yet another acronym including "Ops."

Many organizations fail at successfully transitioning AI prototypes into operations because they do not know how to establish a foundational infrastructure, and they also lack the knowledge of what is required to achieve a successful deployment. As properly pointed out by Etzioni in the quote at the start of this chapter, AI is a revolutionary tool (in this context, "tool" is meant to be interpreted broadly), but how we deploy it to provide beneficial value is totally up to the cognizant AI-driven organizations.

Neroda et al. emphasized the importance of carefully evaluating AI projects, measuring success, embedding responsible AI (RAI), monitoring AI model performance, pursuing continuous improvement, and checking other criteria to increase the chances of successfully adopting AI capabilities [2]. These are key tasks that are part of what we are going to discuss in this chapter and the next. Another very valuable reference is by Stoica et al., "A Berkeley View of Systems Challenges for AI," in which they identify a number of research areas that must be addressed to reap the benefits of AI [3]. They identified, very accurately, a key challenge with AI systems: the need to "design AI systems that learn continually by interacting with a dynamic environment, while making decisions that are timely, robust, and secure."

At the macro-level, meeting these challenges demands a careful analysis of stakeholder needs, a well-defined system architecture, a methodical and rigorous development approach, and quantitative and qualitative evaluations at each of the system integration steps in the discipline of AI systems engineering, as discussed in chapter 2. It is useful to repeat the definition used throughout our book, informed by the INCOSE's definition:

> *AI systems engineering:* "An integrated set of AI architecture elements, subsystems, or assemblies that accomplish a defined objective. These elements include *technologies* as enablers of AI products (hardware, software, firmware) or AI services, adhering to a set of *processes*, and using *people*, information, techniques, infrastructure, and other support elements. The integrated AI system must meet a minimum set of threshold requirements, while attending to FASTEPS issues."

We will refer to this definition throughout this chapter while addressing the AI deployment guidelines. As we presented in chapter 2, the Vee-model shown in figure 10.1 is a very useful construct in decomposing the end-to-end AI system architecture (illustrated in figure 1.3 in chapter 1). The left side of the chart depicts the architecture decomposition. The right side of the chart is focused on the system architecture realization.

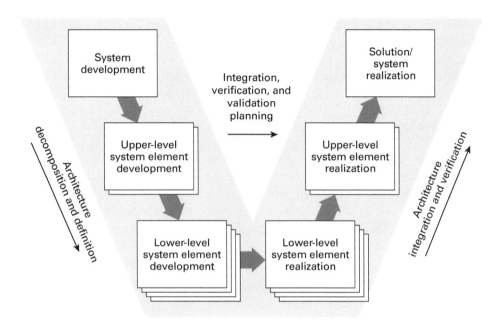

Figure 10.1   Modified Vee-model starting with stakeholders' specifications (scoping of the AI project) culminating with the system realization. Adapted from Forsberg and Mooz [4].

As we have brought up throughout the earlier chapters in the book, a critical first step is identifying the business needs by understanding the stakeholder requirements. In the next two sections, we discuss further the need to traverse the Vee-model, but view it through the lens of the triad of people-process-technology, as captured within our AI systems engineering definition. Chapters 9–11 are focused primarily on process, as part of this triad. The last chapters in part II, chapters 12 and 13, emphasize the people aspect of successfully deploying AI systems.

We begin in section 10.1 by addressing the challenges of deploying AI systems and the importance of focusing beyond technology to include processes and people. In section 10.2, we list ten guidelines to successfully deploy AI capabilities. In section 10.3, we introduce a set of system readiness levels (SRLs) to help AI practitioners in assessing progress in the development and deployment of AI systems. In section 10.4, we present four organizational clusters, depending on the level of AI maturity within the organization. This background leads to a discussion, in section 10.5, of an AI ecosystem structure to help AI designers and developers in transitioning from prototypes to production. A fundamental function, within the AI ecosystem, is to perform testing, verification, and validation at each of the building blocks of the AI system architecture. This topic is discussed in section 10.6.

We highlight, in section 10.7, a set of AI platform characteristics and benefits. Finally, we conclude the chapter with the main takeaways and a set of exercises.

10.1   Challenges in Deploying Artificial Intelligence

There are significant challenges in deploying AI, which we will discuss in this section. A recommended path forward to alleviate some of these challenges is to follow a structured approach to the development and deployment of AI capabilities. Without being too prescriptive—since every AI project is different—there are techniques that readers will find useful as they begin a new AI project or carry out product or service improvements to existing AI capabilities.

In reference to figure 10.2, a recommended approach is as follows:

- *Align:* To align the AI business value in the project's development by attending not just to the technology itself, but also to all elements encapsulated under the triad of people-process-technology.

- *Assess:* To assess the leadership and talent within the AI team. This assessment is the "people" part of the triad, which we will discuss in more detail in chapters 12 and 13. For now, it is sufficient to emphasize the need for a multidisciplinary team and knowledgeable AI leaders with strong support from the upper management and the C-suite.

- *Adhere:* To adhere to logical and methodical processes, including the development of an AI strategic plan (as discussed in chapter 9 in the context of the structure of the AISDM framework) and AI deployment approaches discussed in this chapter and the next.

- *Formulate:* To formulate the AI system architecture (and requisite technologies)—as discussed in chapters 1–8—to clearly identify the AI value proposition that meets the stakeholders' requirements.

- *Commit:* To commit to a disciplined systems engineering implementation approach.

We now highlight for the reader some of the challenges often experienced during the deployment of AI capabilities. We present these challenges in table 10.1. Since there are challenges that are germane to the specifics of each AI project, we focus our attention in particular on macro-level challenges.

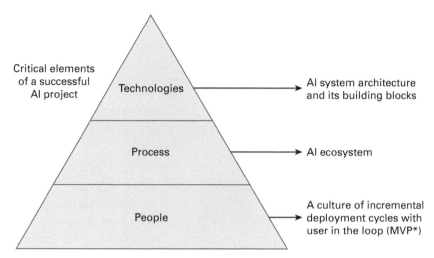

Critical elements of a successful AI project

Technologies → AI system architecture and its building blocks

Process → AI ecosystem

People → A culture of incremental deployment cycles with user in the loop (MVP*)

*MVP: Minimum Viable Product is a product with just enough features to be usable by early customers who can then provide feedback for future product development

Figure 10.2   Triad of people-process-technology as critical elements of a successful AI project. MVP = minimum viable product.

There are a number of topologies proposed for identifying levels of AI development and deployment maturity. Sadiq et al. looked at several maturity models and their definitions [5].

One maturity-level framework that we have found useful was published by Lavin et al. in *Technology Readiness Levels for Machine Learning Systems (MLTRL)* [6]. It has a similar nature as the famous National Aeronautics and Space Administration (NASA) TRL maturity levels [7]. MLTRL is intended on guiding and communicating AI and ML development and deployment, rather than space systems as developed and deployed by NASA. The MLTRL tends to emphasize maturity levels as an AI project transitions from research and development (R&D) into a real-world application. For example, TRL 9, in the MLTRL framework, is the first time that the AI system reaches deployment: "Monitoring current version and improving the next version."

In the next two sections, we focus more on the deployment stage. We begin by enumerating ten guidelines to successfully deploy AI capabilities. We then proceed to applying a systems engineering approach to AI deployment by using the AI system architecture implementation framework illustrated in figure 2.2 in chapter 2.

Table 10.1   Macro-level challenges in the deployment of AI

Critical Elements of a Successful AI Project	Areas of Emphasis	Challenges
Technology	AI system architecture and its building components (see figure 1.3)	• Seamless interoperability among tools across the full deployment pipeline (i.e., a modern pipeline stack)
Process	• AI ecosystem (see section 10.5 for more details)	• Well-defined governance and scalable infrastructure
		• Maintaining a robust and secure environment throughout the development and deployment life cycle
		• Managing risk (discussed in chapter 9)
		• Assuring quality and maintaining operational excellence
		• Meeting schedule and cost requirements
People	• A culture of incremental deployment cycles, with the user in the loop	• Fostering an innovative and productive team environment
		• Working in unison across all of the areas of responsibility (i.e., system architects, designers, developers, and implementers)

10.2   Ten Guidelines for Successfully Deploying AI Capabilities

Careful attention to deployment approaches and techniques will help in minimizing AI project risks and overcoming key challenges [8]. It is well accepted that AI solutions are set to upend a wide variety of industries. The future will bring humans working collaboratively with AI to enable the best capabilities from human-machine teams (HMTs). However, a systematic approach is needed to deploy those AI capabilities.

In table 10.2, we enumerate ten guidelines to successfully deploy AI capabilities, which were introduced in chapter 1.

Table 10.2   Ten guidelines for deploying AI capabilities and discussion pointers

Ten Guidelines	Description	Discussion Pointers
1.	Develop a clear strategic vision and project road map for the AI system.	AI teams must understand and implement their AI strategic blueprint (see chapter 9).
2.	Understand the stakeholder AI needs.	This is a critical element of the strategic road map discussed in section 9.6.
3.	Strengthen the AI team by fostering internal and external relationships.	AI offerings require leveraging both in-house and subcontract work (see chapter 12).
4.	Build a multidisciplinary and diverse team with complementary skills.	An end-to-end AI system architecture demands a broad range of skills (see chapter 12).
5.	Provide measurable objectives while mentoring AI talent.	Peter Drucker, a renowned modern management scholar, had a famous quote: "What gets measured gets managed." See chapter 12.
6.	Continue to expand AI team skills as the future of work evolves.	AI as a technical field is evolving very rapidly. Talent must stay up to date. See chapter 12.
7.	Demonstrate an initial AI capability via a prototype, then iterate.	Gain momentum by demonstrating an initial minimum viable product (MVP). See chapter 11.
8.	Verify individual subcomponents and validate the end-to-end AI system.	These are fundamental steps of a systems engineering approach (see section 10.6).
9.	Secure AI systems against both physical threats and against cyberthreats.	AI systems must be robust to adversarial attacks (see chapters 7 and 11).
10.	Attend to RAI principles.	AI organizations must comply with the FASTEPS principles discussed in chapter 8.

In chapter 12, we address guidelines 3–6 in more detail. However, in the context of successfully deploying AI capabilities, the present workforce is changing their skill sets and their need for advancement in technical areas. Historically, technical professionals could expect to be principal contributors within their organizations for seven to ten years (or even longer) before being promoted to leadership roles. That is no longer the case.

Early career professionals entering the workforce are not content with sitting on the sidelines. If they feel as though they have more skills or knowledge than the people in higher positions, they are going to demand more responsibility—and they'll often leave if they do not get it. But perhaps even more important, these younger workers are often correct in their understanding that they bring a high level of skill and talent with them that can be a major boon to their companies.

The tech sector has the highest employee turnover rate of any industry. And now there are even more opportunities available for workers with very specialized and in-demand skills, including the skills necessary to contribute effectively to AI teams. Therefore, AI leaders must pay careful attention to guidelines 3–6.

Guideline 7 is critically important to achieving success. In addition to starting small with a demonstration of an initial MVP, it is important to develop these AI systems based on a set of composable tools. Stoica et al. make the point that a "composable approach will allow developers to rapidly build and evolve new systems from existing components [3]." Davenport makes a similar recommendation to more effectively introduce AI into an organization by building systems that are modular and based on open-source tools [9].

Guidelines 8 and 9 are part of the systems engineering approach that we have emphasized throughout the book. AI systems must undergo testing, verification, and validation of each of the subcomponents of the AI system architecture. Kellogg and Sendak point out that AI projects can stall when end users resist adoption [10]. They recommend that end users be part of the AI system evaluation in order to ascertain the AI solution meets their needs.

Guideline 10 is on RAI, a topic that we addressed under the rubric of FASTEPS in chapter 8. Soklaski et al. have developed a set of tools that integrate well with common ML environments, such as PyTorch application programming interfaces (APIs) [11]. Their RAI toolbox (rAI-toolbox) enables the developer to assess the robustness of their AI system and the requirement for explainability. This rAI-toolbox and their hydra-zen tool, used for traceability and scalability of ML workflows, are very valuable for addressing aspects of guideline 10.

The AI industry at large needs further research into tools and techniques that are holistic across the full scope of robust AI and RAI. Most tools address portions of a full end-to-end AI system architecture.

In the next section, we go further into the process of applying a systems engineering discipline to AI deployment based on the AI system architecture implementation framework discussed in chapter 2.

10.3   A Process for Applying a Systems Engineering Discipline to AI Deployment

In this section, we show how the AI systems engineering discipline applies to improving the likelihood of success in transitioning from AI development to deployment. Our systems engineering methodology centers on successfully traversing through the Vee-model illustrated in figure 10.1. Key to this process is integrating a set of AI architecture elements, subsystems, or assemblies that accomplish a defined objective while focusing on people, process, and technology. This process must also comply with the RAI principles (i.e., FASTEPS principles).

Let us start by describing how the pieces discussed so far in the book fit together:

- The AI system architecture is a functional description enabling AI practitioners to address "what" building blocks (presented in part I of the book) form an AI product or service.

- The AISDM framework serves as the structure that lays out the AI strategic direction in the form of a road map, as discussed in chapter 9.

- The AI strategy must be implemented. The AI system architecture implementation framework, shown in figure 10.3, serves as "how" the AI system architecture is implemented. It is the scaffold bringing all the architecture building blocks together, driven by a set of attributes (i.e., operational excellence, cost optimization, reliability and security, performance metrics, risk management, and quality assurance). As discussed in chapter 2, we adapted a subset of the attributes from the Amazon Web Services (AWS) well-architected framework [12].

In this section, we describe the use of this implementation framework within a set of recommended steps when undertaking the development and deployment of AI capabilities. In the process of unpacking the components of the AI system architecture implementation framework, we outline a set of SRLs that help AI practitioners identify progress in their implementation and deployment stages. The format that we are following is to define the SRLs and include a short explanation of their meaning, similar to the structure described by Mankins to apply to the technology readiness levels used by NASA [7]. But instead of assessing technologies, we identify readiness levels at the system level.

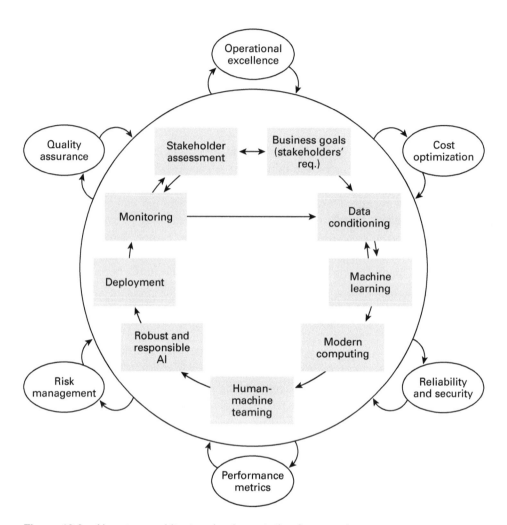

Figure 10.3   AI system architecture implementation framework.

Table 10.3 SRL definitions

SRL	Definition
SRL 1	Business goals and stakeholders' requirements defined
SRL 2	AI strategic development road map completed
SRL 3	AI system architecture formulated
SRL 4	AI ecosystem established and infrastructure, governance, and security instantiated
SRL 5	Initial integration, verification, and validation of AI architecture subsystems demonstrated
SRL 6	Minimum viable product or service demonstrated in a user's operational environment (on-premise and/or in an enterprise/edge environment)
SRL 7	Attributes of AI system architecture implementation framework assessed
SRL 8	Full end-to-end AI system subcomponents integrated, tested, verified, and validated against stakeholder requirements
SRL 9	AI system capability released into operations, and continuous integration/continuous delivery (CI/CD) and monitoring established

Here are short explanations of each SRL:

SRL 1 *Business goals and stakeholders' requirements defined*

Before we can start the system development and the architecture decomposition and definition shown in the Vee-model, we must first define and agree on the business goals and stakeholders' requirements. We formulate these requirements by answering the questions posed earlier in table 9.2 under business needs.

SRL 2 *AI strategic development road map completed*

Once the business needs are clearly formulated, with user inputs from the beginning of the AI project, the AI team must develop a strategic development plan as described in chapter 9. That plan should include the AI value proposition based on an initial AI system architecture, which it is further refined in SRL 3.

SRL 3 *AI system architecture formulated*

The end-to-end AI system architecture illustrated in figure 1.3, and as discussed in part I of the book, must be formulated starting from the data sensors and sources to the human-machine teaming (HMT) subsystem. It is expected that this architecture will undergo refinement as the initial prototype is evaluated and performance gaps are identified.

SRL 4 AI ecosystem established and infrastructure, governance, and security instantiated

The AI ecosystem, further elaborated on in more detail in section 10.5, includes the infrastructure where a modern AI stack is implemented. The AI ecosystem must also incorporate the governance, serving as the oversight body to ascertain that the FASTEPS guidelines are adhered to, and security measures are in place (for both physical security and cybersecurity). This ecosystem must also include the prototype environment (including computational resources), simulating or emulating the actual user operational environment.

SRL 5 *Initial integration, verification, and validation of AI architecture subsystems demonstrated*

As illustrated in the Vee-model, each of the subsystems must be integrated, verified, and validated during integration. An initial prototype is demonstrated using the AI ecosystem formulated in SRL 4. This initial prototype can be demonstrated on premise, in a commercial cloud, or in a hybrid of the two.

SRL 6 *Minimum viable product or service demonstrated in a user's operational environment (on-premise and/or in an enterprise/edge environment)*

The AI strategic development blueprint, as discussed in chapter 9, calls for an MVP or service demonstrated in a development environment, as explained in SRL 5. In SRL 6, this development protype must be transitioned into a user's operational environment. A full assessment of its performance must be undertaken, with the user in the loop. Typically, this assessment is performed in an *online* operational configuration to avoid disrupting the main functions of the operational environment.

SRL 7 *Attributes of AI system architecture implementation framework assessed*

The attributes illustrated in the perimeter of figure 10.3 (i.e., operational excellence, cost optimization, reliability and security, performance metrics, risk management, and quality assurance) are very important parts of the AI system implementation framework and form part of a rigorous systems engineering approach. It is necessary to ascertain that all the AI architecture subsystems are tested, validated, and verified during integration, but that alone is not sufficient. These attributes must also be part of the end-to-end evaluation process. Many AI organizations fail to transition prototypes into production because they assess only the performance accuracy of the ML algorithm, or at most the performance of the data conditioning with the ML subsystem. A successful AI deployment must also include an assessment of the AI value proposition with respect to these attributes.

SRL 8 *Full end-to-end AI system subcomponents integrated, tested, verified, and validated against stakeholder requirements*

After reaching SRLs 6 and 7, the AI team is now ready to perform a full end-to-end system testing, verification, and validation against the agreed-upon stakeholder goals and business needs. A full end-to-end AI system, in this SRL context, is the exercising of the full AI pipeline stack, including the interfaces to all the AI ecosystem components. Today, AI platforms depend on many open-source tools. These tools and APIs must work in unison to avoid down time in an operational environment. That is why SRLs 6, 7, and 8 are completed in an online operational configuration (i.e., often referred to as a "sidecar").

SRL 9 *AI system capability released into operations, and continuous integration/continuous delivery (CI/CD) and monitoring established*

The AI system can now be released into operations. At this SRL, the AI system capability is transitioned from an online configuration to an inline configuration in an operational environment. The system must also undergo CI/CD, and monitoring, as additional capabilities are added to improve on identified performance gaps.

The AI system architecture implementation framework shown in figure 10.3, together with the step-by-step process described under each of the SRLs, form part of the AI systems engineering discipline. However, for this rigorous process to be adopted and successfully executed, the right organizational structure must exist within an AI-driven organization. In the next section, we discuss four distinct organizational clusters.

10.4   AI Adoption: Four Distinct Organizational Maturity Clusters

Most of the discussion in this chapter has centered on process as one critical element of the people-process-technology triad illustrated in figure 10.2. In this section, we highlight four distinct organizational maturity clusters. This discussion is inspired by the work of Ransbotham et al. [13].

In chapter 12, we elaborate further on another aspect of people as a necessary and critical component for fostering an innovative team environment. Specifically, we delve into organizational culture, measuring progress and people success, and executive and technical leadership.

In this section, we wanted to bring up a comprehensive survey undertaken by Ransbotham et al., which provides several important insights directly relevant to AI deployment. Their research included 3,000 business executives, managers, and analysts. These businesses and personnel were located across 112 countries and twenty-one industries.

Table 10.4   Four distinct organizational clusters [13]

Organizational Clusters	Characteristics
Pioneers (19% of respondents)	• Organizations that both understand and have adopted AI • These organizations on the leading edge of incorporating AI into both these organizations' offerings and internal processes
Investigators (32% of respondents)	• Organizations that understand AI but are not deploying it beyond the pilot stage • Their investigation into what AI may offer, emphasizing looking before leaping
Experimenters (13% of respondents)	• Organizations that are piloting or adopting AI without deep understanding • Organizations that are learning by doing
Passives (36% of respondents)	• Organizations with no adoption or much understanding of AI

As shown in table 10.4, most of the respondents either fell into the category of Investigators (32 percent) or Passives (36 percent). Another 19 percent fell into the category of Pioneers. The remaining 13 percent were Experimenters.

Several insights were drawn from the respondents' survey. We only highlight and contrast the Pioneers compared to the Passives since they show very different emphases in terms of the people (i.e., organization) dimension. Investigators and Experimenters fit between the Pioneers and Passives. The Investigators were closer to the same dimensions achieved by the Pioneers, and the Experimenters were closely aligned with the emphasis of Passives.

Pioneers, as well as Investigators, showed the following similarities in incorporating AI within their respective organizations. They demonstrated strengths in the following capabilities:

- The organization collaborates effectively.
- The organization is open to change and receptive to new ideas.
- Executives have the vision and leadership required to navigate the coming changes (a capability directly coupled to the AISDM framework discussed in chapter 9).
- Their analytics capabilities are better than those of their competitors.
- The organization plans for the long term in terms of return on investment (a capability directly coupled to the AISDM framework).

- The overall business strategy is closely linked to their technology road map (a capability directly coupled to the AISDM framework and the AI system architecture implementation framework, shown in figure 10.3).
- The organization is able to change their existing products and services to take advantage of changing technology (a capability emphasized in SRL 9 in the context of CI/CD and monitoring).
- The organization governs data well (a capability encapsulated as part of SRL 4).

On the last capability in this list, we must emphasize that it is more than data governance. Organizations must also attend to standing up a governance structure with enough oversight to ascertain that the FASTEPS principles are being followed.

Passives, and to some degree Experimenters as shown in table 10.4, are slower at introducing AI into their organizations. They are also characterized by having limited knowledge and understanding of AI.

Organizations, regardless of where they belong in the organizational clusters, must formulate a data strategy [14] and show how the data strategy fits across the end-to-end AI architecture pipeline. Organizations must also maintain on-the-job training and keep leveling up employees' skills to stay abreast of the rapidly changing field of AI, as we indicated in guideline 6 for successfully deploying AI capabilities, as shown in table 10.2.

The discussion in this section provides the reader and AI-driven organizations insights into what organization strengths are necessary to successfully adopt and deploy AI capabilities. In the next section, we expand further on the need for a robust AI ecosystem and associated infrastructure.

10.5   The AI Ecosystem

Throughout the earlier sections, we have brought up the importance of having an infrastructure to serve as the foundation (i.e., the bedrock) necessary to implement AI capabilities. We refer to this infrastructure as the "AI ecosystem."

The AI ecosystem (shown in figure 10.4) enables the rapid transition from development (i.e., an AI capability prototype system) into operations. The AI ecosystem facilitates bringing together:

- The AI system architecture shown in figure 1.3 in chapter 1
- The AISDM shown in figure 9.4 in chapter 9
- The AI system architecture implementation framework illustrated in figure 10.3, while following the SRLs described in table 10.3

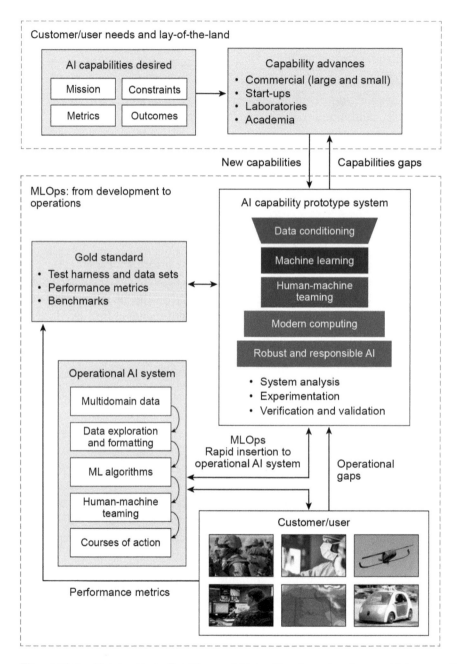

Figure 10.4   AI ecosystem: Enabling rapid transition from development into operations.

We now describe the main constituents of the AI ecosystem, starting from the top of figure 10.4 to the bottom and culminating with the AI capability delivered to customers and users.

SRL 1 calls for defining the business goals and stakeholder requirements. That is represented in the AI capabilities shown at the top of figure 10.4. The AI capabilities desired should also include the mission objective, any constraints, overall metrics to assess if the system meets the stakeholders' requirements, and the ultimate outcomes. For example, as discussed in chapter 9, the pieces of a desired AI capability in a medical application could be:

- Mission: Early detection of stroke onset
- Constraints: Balance, Eyes, Face, Arms, Speech, and Time (BEFAST) indicators
- Metrics: Achieve high probability of detection, while maintaining a very low level of false negatives
- Outcomes: Timely decision support system to inform elderly caretakers

SRL 2 requires the completion of the strategic road map guided by the AISDM framework. An important step within the AISDM step-by-step flow is to identify the lay of the land. One output from the lay of the land is to identify AI capability providers, such as large commercial companies, small start-ups, laboratories, and academia.

SRL 3 is achieved when the AI system architecture (discussed in part I of the book) is formulated. The architecture identifies the level of complexity of an AI system prototype shown in the middle of the AI ecosystem in figure 10.4. The complexity of the AI architecture is informed by rigorous system analysis, experimentation, and verification and validation (V&V).

AI system analysis is defined as "a methodical approach to assess the system requirements, the desired outcomes, quantitative performance measures, and responsible deployment of AI capabilities, including the AI system performance under counterfactual scenarios." Goodwin and Caceres described a methodical approach to performing system analysis, including use case examples [15]. The assessment steps involve a quantitative evaluation that uses a set of gold standards (discussed in more detail in section 10.6).

Once we have reached SRLs 1 through 3, we have the first AI ecosystem constituents. However, as called out in SRL 4, the AI ecosystem must include within its infrastructure (e.g., a computing center) the ability to transition from a prototype environment into an operational environment. The AI ecosystem must also have instantiated a governance body and security measures, as described earlier. The prototype environment preferably should be a high-fidelity representation of the ultimate customer/user operational environment.

Once the AI ecosystem is in place as described in SRL 4, the AI developers and users can proceed to SRL 5, where the initial integration, verification, and validation of the AI architecture subsystems are demonstrated in a protype system.

Once the protype system is successfully integrated and evaluated, for SRL 6 to be completed, an MVP or AI service must transition into an online demonstration in the users' operational environment (i.e., sidecar). This can be done on-premise, in an enterprise/edge environment, or both. The testing, verification, and validation should be assessed according to the set of attributes illustrated in the outer perimeter of the AI system architecture implementation framework shown in figure 10.3. This assessment, performed in close collaboration between the AI developers (and, of course, the AI integration team) and users, helps in reaching SRL 7.

SRL 8 involves an in-depth integration, verification, and validation of the full end-to-end system architecture against the stakeholder requirements. This evaluation must also include adhering to the FASTEPS principles. SRL 8, in an online configuration (often referred to as a "sidecar"), also helps in identifying operational gaps that can be integrated into future revisions. Completion of SRL 8 permits the AI team and the user to proceed to SRL 9.

For SRL 9, the online (i.e., sidecar) demonstrated in SRLs 5 through 8 can be released into operations. This step means transitioning from an online configuration to an inline configuration, where the AI system is part of the day-to-day operational environment. SRL 9 also requires CI/CD and monitoring to identify necessary upgrades resulting from operational gaps, or gaps identified after an initial capability is demonstrated (as defined in the envisioned future within the AISDM framework), or motivated by the desire to incorporate new technology.

The AI ecosystem consists of the foundational constituents necessary to implement MLOps practices. We discuss MLOps in great depth in chapter 11. However, it is useful to provide readers with our working MLOps definition.

Historically, MLOps evolved as a discipline within the AI and ML community, patterned after the well-established development, security, and operations (DevSecOps) methodology for developing software systems. DevSecOps is a repeatable and somewhat automated discipline that has been adapted as an approach to migrate from development into operations. A common DevSecOps definition—among other similar and useful definitions [16]—is:

> *DevSecOps:* A set of practices that works to automate and integrate the processes between software development and information technology (IT) teams so they can build, test, and release software faster and more reliably in a secure manner. The Dev part of the approach includes steps to plan, code, build, test, release,

and deploy a software system. The Ops component includes the tasks of operating and monitoring the deployed software system.

MLOps adopted similar practices but applied to the larger context of ML operations. The definition that we adopted is:

MLOps: A discipline that evolved from the practices employed in DevSecOps. It is a combination of philosophies and practices designed to enable AI teams to rapidly develop, deploy, maintain, and scale out ML models. MLOps incorporates data and models into DevOps cycles to improve the lead time and frequency of deliveries and to enable optimization of the entire ML project life cycle. Security is also a crucial component of MLOps.

DevOps, as well as its inclusion of security, began to be accepted by many software-driven organizations by about 2010, as hardware, software, and cloud computing environments became commoditized. Kim et al. discussed the growth of DevOps as a discipline, bringing the development teams closer to the IT teams responsible for the final operations of complex software systems [17]. IT teams demanded that software developers continuing improvements and accelerated deployments "to respond to the rapidly changing competitive landscape, as well as to provide stable, reliable, and secure service to the customer."

Kim et al. described how DevOps adapted techniques from the 1980s that were put into practice by the Lean Product Development and the Lean Manufacturing principles for efficient testing, evaluation, and operations, resulting in reduced time to market. Sundar et al. provided a review of the Lean Manufacturing implementation, originally performed by the Toyota production system [18]. The relevance of the Lean methodology to our discussion in this chapter on AI deployment guidelines is that the Lean Product Development and Lean Manufacturing principles define the value of a product or service as perceived by the customer. Thus, the development and production of an AI product or service must be driven by the stakeholders' (i.e., the customers') requirements, creating more value for customers with fewer resources.

Similarly, Galup et al. elaborated further on DevOps as a confluence of three practices [19]:

- Agile: Continually delivering product or service features
- Lean: Providing capabilities with minimum wasted efforts while meeting stakeholder requirements
- Information Technology Service Management (ITSM): Overseen by good governance controls (a topic that we have already highlighted as a critical element of SRL 4)

As a point of clarification, Galup et al. [19] emphasized that there are various ITSM frameworks, but "the most widely known framework is the Information Technology Infrastructure Library (ITIL)." The authors propose using ITIL knowledge to represent ITSM knowledge.

Since a rapid cycle of product or service development and deployment is central to the DevSecOps process, and its adaptation into the MLOps process, we conclude this section with an Agile primer for the reader. We also include additional references if the reader desires a more in-depth discussion.

A Primer on the Agile Development Process

The Agile software methodology evolved as an alternative to the classical waterfall software development methodology. Complex software is best developed in incremental versions (i.e., shorter cycles) to get customer feedback and to improve on the ultimate value to the customer instead of a long period of development before you see the system instantiated (as occurs in the case of a classical waterfall methodology). The Agile process methodology had its origins in a "Manifesto for Agile Software Development" [20, 21]. This manifesto is supported by twelve principles outlined in table 10.5.

The Agile development process is very well matched to the MLOps methodology for developing and deploying AI capabilities. The Agile implementation approach typically consists of small multidisciplinary teams, charged to develop a complex capability where there is a need to break up the problem into modules [22]. The solution is achieved from demonstrating component prototypes with a tight feedback loop collaborating with the users.

There are several important steps in properly implementing the Agile process. For example, the Scrum methodology, within the Agile approach, enables rapid and iterative demonstrations with incremental features as the team develops the overall system. For readers interested in getting a deeper understanding of the Agile process, there are many excellent references [22–24].

The Agile development approach enables us to implement the Vee-model consistent with the rapid demonstrations of SRLs 5 through 8, followed by an inline deployment into operations including CI/CD and monitoring as discussed in SRL 9. The Agile process is also very useful in adapting the systems engineering practice, as espoused by the INCOSE association (see chapter 2), to today's demands of complex engineering systems [25, 26].

The Vee-model has been accepted in the systems engineering community for many years—it has stood the test of time. The agile development methodology complements

Table 10.5   Principles behind the Agile Manifesto [21]

Agile Principle	Description
1	Our highest priority is to satisfy the customer through early and continuous delivery of valuable software.
2	Welcome changing requirements, even late in development. Agile processes harness change for the customer's competitive advantage.
3	Deliver working software frequently, from a couple of weeks to a couple of months, with a preference to the shorter timescale.
4	Business people and developers must work together daily throughout the project.
5	Build projects around motivated individuals. Give them the environment and support they need, and trust them to get the job done.
6	The most efficient and effective method of conveying information to and within a development team is face-to-face conversation.
7	Working software is the primary measure of progress.
8	Agile processes promote sustainable development. The sponsors, developers, and users should be able to maintain a constant pace indefinitely.
9	Continuous attention to technical excellence and good design enhances agility.
10	Simplicity—the art of maximizing the amount of work not done—is essential.
11	The best architectures, requirements, and designs emerge from self-organizing teams.
12	At regular intervals, the team reflects on how to become more effective, then tunes and adjusts its behavior accordingly.

the Vee-model in several important ways. For example, the agile approach, at the lowest level, can generate insights that should flow back up the planning process and affect the large-scale business decisions, as described under SRL 1. There is also an opportunity to modify the original strategic development road map, discussed in chapter 9, by revising the strategic decisions at the business level, when AI designers, developers, and implementers encounter difficulties while performing rapid iterations described in the agile implementation process.

The agile approach also provides a methodology throughout the interaction between the left and right sides of the Vee-model during the stages of integration, verification, and validation. Each of the building blocks of the AI system architecture, shown in figure 1.3 in chapter 1, must undergo this methodical process during the design, development, and integration to achieve a successful deployment.

The AI ecosystem also serves as the foundational environment where an Agile process can be executed, while effectively and efficiently traversing all the elements of the AI system architecture framework shown in figure 10.3.

In the next section, we focus on the evaluation stage of the AI ecosystem. The gold standard, consisting of the test harness (i.e., the infrastructure necessary to perform the verification and validation tests), performance metrics, and benchmarks, constitutes critical parts of an overall AI infrastructure.

10.6   Gold Standard: Test Harness, Performance Metrics, and Benchmarks

The AI ecosystem not only serves as the foundational structure to successfully executing the AI strategy, but it also helps in transitioning from development into production. Many organizations fail in this transition because the data science projects are decoupled from the business operations. Joshi et al. stressed the importance of coupling the end user throughout the development of data science projects [27]. The AI ecosystem facilitates this coupling.

However, the business goals and stakeholders' requirements are the start of the AI development and deployment journey, as discussed in SRL 1. As shown in the AI ecosystem in figure 10.4, there is also a need to assess the prototype system against a gold standard. The gold standard is integral to successfully completing SRLs 5–9.

In this section, we address how to properly use a gold standard. Unfortunately, this topic is not very well ingrained among the AI community in the practice of developing and deploying AI products or services. The proper use of a gold standard is at the core of the systems engineering discipline emphasized throughout this book and illustrated in figure 10.1, as part of the subsystem integration, verification, and validation of each of the AI system architecture building blocks, and the end-to-end AI capability.

One of the book authors (D. Martinez) has decades of firsthand experience in the development and deployment of complex, real-time signal processor systems [28, 29]. In addition to following a systems engineering methodology during unit-level testing, subsystem integration, and single-thread integration, a key enabler in successfully deploying a system was the V&V based on high-fidelity simulations used in modeling the user's environment. In the implementation of these signal processors, the data simulations were done using fixed-point arithmetic and input into the real-time system during different stages of development. The results were then compared to a gold standard, simulated with full precision, to assess the accuracy and measure quantitative losses of the signal processing algorithms when operating in real time. These complex real-time systems operated successfully during their first operational demonstrations, when real sensor data was input into the systems instead of simulated data. The rigor spent in creating high-fidelity simulations was instrumental in the successful system deployments.

Fortunately, there is an emerging concept referred to as "digital twins," which can now be employed in the development and deployment of complex AI systems. Digital twins are close to (or in some cases indistinguishable from) digital counterparts of physical systems or operational systems. A digital twin infrastructure can facilitate a test harness for implementing performance metrics and benchmarks—in some instances, the digital twin infrastructure might already have those tools available.

One of the best definitions we have seen about digital twins is from c3.ai [30], named a leader among AI and ML platforms by Forrester Wave Leader 2022 [31].

Digital Twins: Digital twins are virtual replicas of physical devices—assets, systems, or processes—that data scientists and IT personnel can use to run simulations. Digital twins are designed to detect and prevent problems, predict performance, and optimize processes through real-time analytics to deliver business value. C3 AI Digital Twin, in a unified object virtual mode, allows customers to gain critical business insights through simulations of their processes, assets, or systems.

Digital twins are used for much more than just performing AI system assessments. An AI platform, with a digital twin, can look at how an AI capability will function using a virtual environment with simulated sensors, physical sensors, or a hybrid of the two across the full AI pipeline. The advent of IoT (making many physical and inexpensive sensors available), high-speed interconnects like 5G/6G, advances in virtual/augmented reality, high-speed computation, and other technologies allow AI practitioners to validate the performance of an overall AI system [32]. Upper management and AI leaders, using a digital twin infrastructure, can also address opportunities and business value as part of their strategic plan.

However, an AI platform, with a digital twin infrastructure, still requires AI talent and expertise to assess the system's end-to-end capability. AI developers and implementers, working closely with the ultimate stakeholders, can do an in-depth analysis if problems and errors are identified. As we discussed in chapter 3, there are benchmark tools, such as Dawnbench, Dynabench, and MLPerf, that are useful in benchmarking ML algorithms using representative data sets [33–35].

The need for ML rigor continues to be an area of emphasis among researchers in the commercial sector and academia. At the International Conference on Learning Representations 2022, a workshop was held to address evaluation and benchmark standards [36, 37].

Marie et al., from the National Institute of Information and Communications Technology (NICT) in Japan, performed a large scale meta-evaluation of a total of 769 machine translation research papers, published between 2010 and 2020 [38]. They found that many publications lack rigor in their evaluation of machine translation techniques. They proposed guidelines for a more consistent and better automatic machine translation evaluations. Several other conference venues are also addressing the importance of ML reproducibility regarding research findings. Pineau et al. published the lessons learned from the NeurIPS 2019 reproducibility challenge program and formulated an ML reproducibility checklist to be included in research paper submissions [39].

AI development must also be evaluated more broadly than just assessing performance. All AI-driven organizations must attend to the impact of AI on environmental-social-governance. Couillet et al. argued for a new metric called "technical conviviality," to be used in place of the unsustainable metric of absolute performance. The authors emphasize a need to address, for example, resource availability including energy, environment preservation, and accessibility to society (in a socially responsible manner) [40].

The US National Institute of Standards and Technology (NIST) conducts R&D into metrics, measurements, and evaluation methods in emerging and existing areas of AI; contributes to the development of standards; and promotes the adoption of standards, guides, and best practices for measuring and evaluating AI technologies as they mature and find new applications. The emerging NIST AI guidelines (https://www .nist.gov/artificial-intelligence) can also be integrated within an AI platform infrastructure as part of the test harness.

AI leaders and developers can use the wide range of performance metrics discussed as part of each of the chapters in part I. We also strongly recommend addressing the outer perimeter of the AI system architecture implementation framework illustrated in figure 10.3. This step must be completed to reach SRL 7. In selecting the most appropriate AI platform, AI leaders and practitioners must determine if tools exist that facilitate the assessment of these attributes as part of SRL 7, as discussed earlier in the chapter.

Let us take one of the AI system architecture building blocks shown in figure 10.3—the modern computing subcomponent. The AI system architecture implementation requires computational resources, discussed in chapter 5. Today's ML algorithms require a significant amount of computing power, making them expensive and requiring attention to the carbon footprint. Thus, the AI developers and implementers must look at operational excellence and cost optimization as an example of addressing the SRL 7 attributes. A number of researchers at the Massachusetts Institute of Technology (MIT), Northeastern University, and the US Air Force are addressing this issue with a set of comprehensive challenges (based on the MIT Supercloud Dataset), metrics, and evaluation techniques [41, 42].

In the next section, consistent with the macro-level discussion in this chapter, we present AI platform characteristics and benefits that facilitate the choice of a commercial infrastructure. Of course, as time goes on, additional companies and tools come into the market to facilitate the development and deployment of AI products and services.

10.7   AI Platform Characteristics and Benefits

In prior sections, we have presented guidelines to deploy AI capabilities effectively. We emphasized the importance of following a systems engineering structure as shown in figure 10.1 (illustrating the Vee-model). The AI system architecture implementation framework, displayed in figure 10.3, provided the foundational framework to methodically implement all the subcomponents of the AI system architecture discussed in part I of this book.

The AI ecosystem, together with the SRLs, gave AI leaders and practitioners a process for transitioning from development prototyping into the production stage. The production environment represented the user's day-to-day operational infrastructure. The challenge is in having an AI platform that facilitates this methodical process for transitioning and increasing the likelihood of successfully deploying AI capabilities.

Sculley et al. described the notion of "hidden technical debt" [43]. They pointed out, as illustrated in figure 10.5, that the "ML code" (i.e., the algorithm software) is a very small piece of the ML implementation infrastructure required to successfully integrate complex AI systems. Technical debt is a metaphor for the potential large cost that can be incurred if other system components and infrastructure are not integrated together.

Figure 10.5 depicts the pieces of the infrastructure that are most relevant to the ML algorithm implementation. AI practitioners still need to use a comprehensive platform to integrate all the components of the AI system architecture implementation framework shown in figure 10.3.

There is no one solution that meets all the AI developers' requirements. AI organizations have different needs, ranging from on-premise processing, enterprise-level processing, processing at the edge, or hybrid configurations of all these. We highly advise that AI organizations, responsible for providing AI products or services, do not build their own custom AI platform solutions. This effort can be very frustrating and costly, requiring constant upgrades and maintenance. Instead, we strongly recommend acquiring commercial capabilities. However, developers should look for AI platforms with certain macro-level characteristics.

Next, we describe a set of characteristics that AI practitioners must address in their selection of an AI platform. In table 10.6, we outline the AI platform characteristics and the accrued benefits. When done successfully, the acquisition of an AI platform is

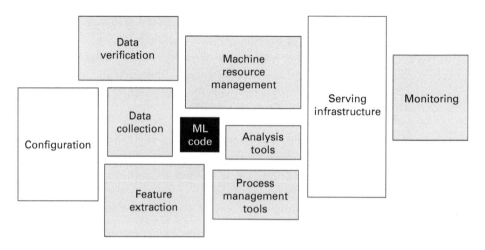

Figure 10.5 Hidden technical debt in ML systems [43].

an important step on the digital transformation journey discussed in chapter 3 and illustrated in figure 3.3. A digital transformation is of great importance for organizations to stay competitive and create value [44]. Kraus et al. did an in-depth review, in the form of a bibliometric map, of the rising interest in digital transformation, and its relevance to big data, AI, and data analytics [45].

To date, there have been a number of AI platforms and development environments meeting several of the desired characteristics outlined here. We mentioned earlier that Forrester performed a comprehensive assessment of several contemporary AI/ML platforms, which was published in 2022 [31]. Another very useful resource for assessing trends in the IT and enterprise landscape is regularly published by the Gartner consulting company under the name of "Magic Quadrant." Gartner's methodology and evaluation criteria offer insights into the market's direction, maturity, and company leaders in, for example, data science and ML platforms [46].

In chapter 11, we address the technologies that are commonly used today in making AI platforms more flexible and scalable within the rubric of the nascent field of MLOps. Together with the AI characteristics and benefits outlined in this chapter, we hope to help the reader in understanding the important technologies used in many of these AI platforms.

Next, we summarize the main takeaways from this chapter and conclude with a set of exercises.

Table 10.6 AI platform characteristics and benefits

AI Platform Characteristic	Benefits
Ability to overcome data silos	Permeable interfaces within the organization and to external partners.
Supportive of simulations, emulations, physical, and/or hybrid interfaces to sensors and sources	Capability to incorporate a digital twin and augmented/virtual/mixed realities.
Different types of data formats	Capable of processing structured, semistructured, and unstructured data.
Open access (facilitating adoption and swapping in and out of technologies as the organization needs evolve)	Ease in integrating a wide range of tools from different providers (i.e., technologies from large and small companies, laboratories, and academia).
Cloud agnostic	Interfaces to multiple elastic cloud computing providers.
Scalable architecture	AI developers can start with a small AI infrastructure with support in order to grow at the level of a large enterprise.
Broad range of resources to assess performance across the full AI pipeline	Availability of quantitative tools to address performance of all building blocks in the AI architecture.
Extensive "knobs" to evaluate and iterate meeting the FASTEPS principles	Compliant with RAI, including state, national, and international regulations.
Simple user interface and visualization dashboard	Minimum ramp-up required to gain, synthesize, and communicate AI insights; supportive of team collaboration.
Recognized leaders in the AI community	Established large customer base.
Mature training and customer support	Well-trained and educated staff to help customers reduce time to achieving proficiency with the AI platform.

10.8   Main Takeaways

Since every AI project has very specific requirements and challenges, in this chapter we have presented a macro-view, rather than a prescriptive description, of the foundational structure needed for a successful AI deployment. Hopefully, this has helped the reader understand "how" systems engineering is applied to the development and deployment of AI systems.

There are AI deployment guidelines that AI practitioners can follow—at the macro-level—that improve the likelihood of a successful AI deployment. At the macro-level, meeting these challenges demands a careful analysis of stakeholder needs, a well-defined system architecture, a methodical and rigorous development approach, and quantitative and qualitative evaluations at each of the system integration steps, as we discussed in chapter 2, under the discipline of AI systems engineering. Therefore, it is useful to repeat the definition of "AI systems engineering" used throughout our book, informed by INCOSE's definition, as follows:

> *AI systems engineering:* "An integrated set of AI architecture elements, subsystems, or assemblies that accomplish a defined objective. These elements include *technologies* as enablers of AI products (hardware, software, firmware) or AI services, adhering to a set of *processes*, and using *people*, information, techniques, infrastructure, and other support elements. The integrated AI system must meet a minimum set of threshold requirements, while attending to FASTEPS issues."

AI developers can follow a methodical systems engineering structure as illustrated in the Vee-model. The first step begins by decomposing the architecture subcomponents applicable to a given AI project and incorporating each of the AI system architecture building blocks shown in figure 1.3 in chapter 1. These building blocks (i.e., subsystem components) must then be integrated, verified, and validated, culminating with a full system realization.

As part of the recommended methodical approach discussed in this chapter, AI developers can adhere to the triad of people-process-technology shown in figure 10.2, as critical elements of a successful AI project development and deployment. It is imperative to pay careful attention to the following macro issues:

- *Align:* To align the AI business value by attending, in the project development, not just to the technology but to all elements encapsulated under the triad of people-process-technology.

- *Assess:* To assess the leadership and talent in the AI team. This assessment is the "people" part of the triad, which we will discuss in more detail in chapters 12 and 13. For now, it is sufficient to emphasize the need for a multidisciplinary team and knowledgeable AI leaders, but also strong support from the upper management and the C-suite.

- *Adhere:* To adhere to logical and methodical processes, including the development of an AI strategic plan (as discussed in chapter 9 under the structure of the AISDM framework), and the AI deployment approaches discussed in this chapter and the next.

- *Formulate:* To formulate the AI system architecture (and requisite technologies)—as discussed in chapters 1–8—to clearly identify the AI value proposition that meets the stakeholder's requirements.
- *Commit:* To commit to a disciplined systems engineering implementation approach.

In the interest of going a bit deeper, but still staying at the macro-level, we formulated ten guidelines for successfully deploying AI capabilities as shown in table 10.2. We stressed the importance of not ignoring any of these guidelines, since the biggest impediment in the adoption of AI capabilities today is transitioning from an AI development stage into the deployment stage.

In addition to these ten deployment guidelines, AI practitioners must execute the AISDM strategic road map by adhering to the AI system architecture implementation framework illustrated in figure 10.3. The inside of the framework incorporates the AI system architecture for a given AI product or service. The attributes in the outer perimeter must be exercised during integration of each of the system architecture building blocks and after all the subcomponents are fully realized and integrated together.

Since AI developers and implementers need a way to assess progress, we have formulated a set of nine SRLs, defined in table 10.3. Again, these help evaluate progress as the AI system is developed and deployed. It is very important to emphasize that the process implies an iterative implementation approach, so each SRL must be assessed at the subcomponent level—during integration, verification, and validation—and at the point of formulating an end-to-end AI capability.

Ransbotham et al. undertook a large survey of AI organizations, and identified a set of typical characteristics [13]. We discussed their four distinct organizational maturity clusters—in the context of this chapter on AI deployment guidelines—because the organizational tolerance to AI development and deployment has a direct bearing on successfully delivering AI capabilities. Over 50 percent of the survey's respondents were, in aggregate, Pioneers and Investigators, with a willingness to introduce AI practices within their respective organizations. The executives and upper management were visionaries with a willingness to adapt to change. They also demonstrated good alignment between their business strategy and their technology road map.

The AI organization must instantiate an AI ecosystem, in addition to adhering to the deployment guidelines described in the early sections of the chapter, the AI system architecture implementation framework, and the SRLs for assessing progress. The AI ecosystem, illustrated in figure 10.4, facilitates bringing together the following:

- The AI system architecture shown in figure 1.3 of chapter 1
- The AISDM shown in figure 9.4 of chapter 9

- The AI system architecture implementation framework illustrated in figure 10.3, while following the SRLs described in table 10.3

The rapid evolution of AI and the need for rapid and iterative demonstrations have led to adopting an Agile software methodology, which evolved as an alternative to the classical waterfall software development methodology. Complex hardware and software are best developed in incremental versions (i.e., shorter cycles) to get customer feedback and to improve the ultimate value to the customer instead of a long period of development before you see the system instantiated (as occurs with the classical waterfall methodology). The Agile process methodology had its origin in a "Manifesto for Agile Software Development" [20, 21], supported by twelve principles as outlined in table 10.5.

As shown in the AI ecosystem in figure 10.4, there is also a need to assess the prototype system against a gold standard. The gold standard is an integral part to successfully completing SRLs 5–9.

Unfortunately, this topic is not very well ingrained in the AI community when it comes to the practice of developing and deploying AI products or services. The proper use of a gold standard is at the core of the systems engineering discipline emphasized throughout this book and illustrated in figure 10.1 as part of the subsystem integration, verification, and validation of each of the AI system architecture building blocks, as well as the end-to-end AI capability.

Fortunately, there is an emerging concept referred to as "digital twins," which can now be used in the development and deployment of complex AI systems. A digital twin infrastructure can facilitate a test harness for implementing performance metrics and benchmarks—in some instances, that infrastructure might have those tools already available.

Digital twins are used for much more than just performing AI system assessments. An AI platform, with a digital twin, can look at how an AI capability will function using a virtual environment with simulated sensors, physical sensors, or a hybrid of the two across the full AI pipeline.

There are also more specific tools and benchmarks used to assess data quality, ML algorithms, and computing technologies, as discussed in this chapter and in earlier chapters of this book. However, as a community, we still lack rigorous end-to-end system benchmarks and standards.

We concluded this chapter with a discussion of the desired characteristics and the accrued benefits in the selection of an AI platform. An AI platform is necessary because of the likelihood of incurring "hidden technical debt," as described by researchers at Google [43]. The ML code is a very small component of what is needed to successfully implement and deployed ML algorithms. Furthermore, as described in section 10.7, an AI platform should facilitate more than just the ML algorithm implementation. The AI platform must also enable the integration, verification, and validation of all the elements described under the AI system architecture implementation framework illustrated in figure 10.3, including the attributes shown in the outer perimeter of the figure.

As described in table 10.6, a scalable and sustainable platform must also address the FASTEPS principles described in chapter 8.

10.9 Exercises

1. Provide a short description of the people-process-technology triad illustrated in figure 10.2.
2. Enumerate the five macro-level challenges often encountered in the process component of the triad.
3. Demonstrating an initial AI capability and iterating on it to gain momentum lead to schedule delays; therefore, it is not advisable.
   a. True
   b. False
4. What is meant by an MVP or service, and what is the importance of such a demonstration?
5. Provide a short description of each of the six attributes on the outer perimeter of the AI system architecture implementation framework shown in figure 10.3.
6. Describe SRL 9 and the role of CI/CD and monitoring of an AI capability deployment.
7. Pioneers and Investigators, as described in table 10.4, often demonstrate well-established data governance.
   a. True
   b. False
8. An AI ecosystem avoids bringing the user into the loop because in the process of doing a capability prototype demonstration, the AI capabilities can be derived.
   a. True
   b. False

9. An AI ecosystem facilitates online (prototyping) and inline (full deployment into a production environment) system demonstrations.
    a. True
    b. False

10. Pick the answer that incorporates a subset of the twelve principles behind the Agile Manifesto.
    a. The highest priority is to satisfy the customer through early and continuous delivery of valuable software.
    b. Freezing the system requirements after completing the first demonstration
    c. Businesspeople and developers need to work together daily throughout the project
    d. Continuous attention to technical excellence and good design enhances agility
    e. a. c., and d
    f. None of the above

11. Provide a definition of a digital twin (DT) and elaborate briefly on its value in the development and deployment of AI products and services.

12. Explain what is meant by "hidden technical debt."

13. A cloud-agnostic AI platform includes only interfaces to inelastic cloud computing providers.
    a. True
    b. False

14. The role of a digital twin, within an AI platform, is to incorporate capabilities in support of simulations, emulations, physical, and hybrid interfaces to sensors and sources of data.
    a. True
    b. False

15. Referring to the AI ecosystem illustrated in figure 10.4, explain how this ecosystem can help in the execution of the AI strategic road map (as described in the AISDM framework).

16. The Pareto principle (known as the 80/20 rule) can be interpreted to imply that completing the last 20 percent of a task takes 80 percent of the effort. How would you employ this principle to assess schedule risk in the AISDM road map and in progressing through SRLs 1–9?

10.10   References

1. Kreuzberger, D., N. Kühl, and S. Hirschl, *Machine learning operations (MLOps): Overview, definition, and architecture*, 2022. arXiv preprint arXiv:2205.02302.

2. Neroda, J., S. Escaravage, and A. Peters, *Enterprise AIOps*. 2021. O'Reilly Media.

3. Stoica, I., D. Song, R. A. Popa, et al., *A Berkeley view of systems challenges for AI*, 2017. arXiv preprint arXiv:1712.05855.

4. Forsberg, K. and H. Mooz, The relationship of system engineering to the project cycle, in *INCOSE International Symposium*. 1991, Wiley Online Library.

5. Sadiq, R. B., N. Safie, A. H. A. Rahman, and S. Goudarzi, Artificial intelligence maturity model: A systematic literature review. *PeerJ Computer Science*, 2021. 7: e661.

6. Lavin, A., C. M. Gilligan-Lee, A. Visnjic, et al., *Technology readiness levels for machine learning systems*, 2021. arXiv preprint arXiv:2101.03989.

7. Mankins, J. C., Technology readiness levels. White paper, Office of Space Access and Technology, National Aeronautics and Space Administration (NASA), April 1995. 6(1995): 1995.

8. Martinez, D. R. *Four challenges facing AI leaders*. 2020, MIT Professional Education.

9. Davenport, T. H., 7 ways to introduce AI into your organization. *Harvard Business Review Digital Articles*, 2016.

10. Kellogg, K. C., M. Sendak, and S. Balu, AI on the front lines. *MIT Sloan Management Review*, 2022. 63(4): 44–50.

11. Soklaski, R., J. Goodwin, O. Brown, et al., *Tools and practices for responsible AI engineering*, 2022. arXiv preprint arXiv:2201.05647.

12. Amazon. *Machine learning lens—AWS well-architected framework*. December 20, 2021, https://docs.aws.amazon.com/wellarchitected/latest/machine-learning-lens/wellarchitected-machine-learning-lens.pdf#machine-learning-lens.

13. Ransbotham, S., D. Kiron, P. Gerbert, and M. Reeves, Reshaping business with artificial intelligence: Closing the gap between ambition and action. *MIT Sloan Management Review*, 2017. 59(1). https://sloanreview.mit.edu/projects/reshaping-business-with-artificial-intelligence/.

14. *Harvard Business Review*, Data strategy: The missing link in artificial intelligence–enabled transformation, Cloudera, Editor. 2021, *Harvard Business Review Analytic Services: Pulse Survey*.

15. Goodwin, V. H. and R. S. Caceres, *System analysis for responsible design of modern AI/ML systems*, 2022. arXiv preprint arXiv:2204.08836.

16. Sen, A., DevOps, DevSecOps, AIOPS-paradigms to IT operations, in *Evolving technologies for computing, communication and smart world*. 2021, Springer. 211–221.

17. Kim, G., J. Humble, P. Debois, and J. Willis, *The DevOps handbook: How to create world-class agility, reliability, & security in technology organizations*. 2021, IT Revolution.

18. Sundar, R., A. Balaji, and R. S. Kumar, A review on lean manufacturing implementation techniques. *Procedia Engineering*, 2014. 97: 1875–1885.

19. Galup, S., R. Dattero, and J. Quan, What do agile, lean, and ITIL mean to DevOps? *Communications of the ACM*, 2020. 63(10): 48–53.

20. *Manifesto for agile software development*. 2001. https://agilemanifesto.org/.

21. Beck, K., J. Grenning, R. C. Martin, et al., *Manifesto for agile software development*. 2001.

22. Rigby, D. K., J. Sutherland, and A. Noble, Agile at scale. *Harvard Business Review*, 2018. 96(3): 88–96.

23. Rigby, D., S. Elk, and S. Berez, *Doing agile right: Transformation without chaos*. 2020, Harvard Business Press.

24. Schwaber, K. and J. Sutherland, The scrum guide. *Scrum Alliance*, 2011. 21(19): 1.

25. Darrin, M. A. G. and W. S. Devereux. The Agile Manifesto, design thinking and systems engineering, in *2017 Annual IEEE International Systems Conference (SysCon)*. 2017, IEEE.

26. Haberfellner, R. and O. De Weck. 10.1. 3 Agile systems engineering versus agile systems engineering, in *INCOSE International Symposium*. 2005, Wiley Online Library. https://doi.org/10.1002/j.2334-5837.2005.tb00762.x.

27. Joshi, M. P., N. Su, R. D. Austin, and A. K. Sundaram, Why so many data science projects fail to deliver. 2021, *MIT Sloan Management Review*.

28. Martinez, D., R. Bond, and M. Vai. Embedded digital signal processing for radar applications, in *2008 IEEE Radar Conference*. 2008, IEEE.

29. Martinez, D. Keynote: Future challenges in the development of real-time high performance embedded systems, in *IEEE 19th Real-Time Systems Symposium*. 1998, Madrid. https://doi.ieeecomputersociety.org/10.1109/REAL.1998.739725.

30. Siebel, T. M., *Digital transformation: Survive and thrive in an era of mass extinction*. 2019, RosettaBooks. https://c3.ai/glossary/artificial-intelligence/digital-twin/.

31. Gualtieri, M. and R. Curran. *AI/ML Platforms, Q3 2022*. 2022. https://reprints2.forrester.com/#/assets/2/1438/RES176365/report.

32. Apte, P. P. and C. J. Spanos, The digital twin opportunity. *MIT Sloan Management Review*, 2021. 63(1): 15–17.

33. Coleman, C., D. Narayanan, D. Kang, et al., Dawnbench: An end-to-end deep learning benchmark and competition. *Training*, 2017. 100(101): 102.

34. Kiela, D., M. Bartolo, Y. Nie, et al., *Dynabench: Rethinking benchmarking in NLP*, 2021. arXiv preprint arXiv:2104.14337.

35. Mattson, P., V. J. Reddi, C. Cheng, et al., MLPerf: An industry standard benchmark suite for machine learning performance. *IEEE Micro*, 2020. 40(2): 8–16.

36. *ML evaluation standards*. 2022. https://ml-eval.github.io/.

37. Agarwal, R., S. Chan, X. Bouthillier, et al., *ICLR Workshop on Setting up ML Evaluation Standards to Accelerate Progress*. 2022. https://iclr.cc/virtual/2022/workshop/4559.

38. Marie, B., A. Fujita, and R. Rubino, *Scientific credibility of machine translation research: A meta-evaluation of 769 papers*, 2021. arXiv preprint arXiv:2106.15195.

39. Pineau, J., P. Vincent-Lamarre, K. Sinha, et al., Improving reproducibility in machine learning research: A report from the NeurIPS 2019 reproducibility program. *Journal of Machine Learning Research*, 2021. 22.

40. Couillet, R., D. Trystram, and T. Ménissier, The submerged part of the AI-ceberg [Perspectives]. *IEEE Signal Processing Magazine*, 2022. 39(5): 10–17.

41. Tang, B. J., Q. Chen, M. L. Weiss, et al., *The MIT Supercloud Workload Classification Challenge*, in *2022 IEEE International Parallel and Distributed Processing Symposium Workshops (IPDPSW)*. 2022, IEEE.

42. Zhao, D., N. C. Frey, J. McDonald, et al., A green(er) world for AI, in *2022 IEEE International Parallel and Distributed Processing Symposium Workshops (IPDPSW)*. 2022, IEEE.

43. Sculley, D., G. Holt, D. Golovin, et al., Hidden technical debt in machine learning systems. *Advances in Neural Information Processing Systems*, 2015. 28: 2503–2511.

44. Verhoef, P. C., T. Broekhuizen, Y. Bart, et al., Digital transformation: A multidisciplinary reflection and research agenda. *Journal of Business Research*, 2021. 122: 889–901.

45. Kraus, S., P. Jones, N. Kailer, et al., Digital transformation: An overview of the current state of the art of research. *Sage Open*, 2021. 11(3): 21582440211047576.

46. Krensky, P., C. Idoine, and E. Brethenoux, Gartner magic quadrant for data science and machine learning platforms. *Gartner*, 2020. https://www.gartner.com/en/documents/3998753.

11

MLOps: Transitioning from Development to Deployment

For one person who is blessed with the power of invention, many will always be found who have the capacity of applying principles.

—Charles Babbage, pioneer English mathematician (1791–1871)

In this chapter, we focus further on the implementation of artificial intelligence (AI) products or services. We elaborate on the transition from a development stage to a deployment stage. In chapter 10, we presented macro-level AI deployment guidelines. We encourage the reader to first review chapter 10 before reading this chapter, as they complement each other.

Our discussion centers on machine learning operations (MLOps) as the overarching rubric encompassing all aspects of AI development and deployment. It is important to remember that MLOps is a nascent field that has begun to gain significant attention in both academia and industry. There have been several variants of this, referred to as "data operations (DataOps)," "AI operations (AIOps)," and "systems operations (SystemsOps)." Fundamentally, all these variants focus on the issues of transitioning from development to deployment. Khalajzadeh et al. published a survey of definitions and tools for DataOps, AIOps, and development operations (DevOps) [1] and separated each of these disciplines by type of activity. For example, the Data Conditioning processing, described in chapter 3, fits best with the DataOps part of their taxonomy. Our ML design discussed in chapter 4, from model creation to inference, fits best with the AIOps part of their taxonomy. We take a more holistic approach under the MLOps umbrella aligned with the data-centric AI initiatives [2, 3] across the end-to-end AI system architecture shown in figure 1.3 in chapter 1.

As defined by Andrew Ng [2], data-centric AI is "the discipline of systematically engineering the data used to build an AI system." This definition is well aligned with the working definition of MLOps presented in chapter 10 and discussed in this chapter:

MLOps: A discipline that evolved from the practices employed in development, security, and operations (DevSecOps). It is a combination of philosophies and practices designed to enable AI teams to rapidly develop, deploy, maintain, and scale out ML models. MLOps incorporates data and models into DevOps cycles to improve the lead time and frequency of deliveries and to enable optimization of the entire ML project life cycle. Security is also a crucial component of MLOps.

There are other MLOps definitions with similar emphases of transitioning from development to deployment, including tools for streamlining the ML life cycle [4, 5]. As pointed out by Gift and Deza, "the reason models are not moving into production [i.e., deployment] is the impetus for the emergence of MLOps" [4]. However, it is imperative that we emphasize that MLOps involves the entire life cycle, from data conditioning, machine learning (ML) model development and testing, continuous integration/continuous delivery (CI/CD), monitoring, performance and risk assessments, and ultimately meeting the stakeholders' business needs.

The MLOps discipline, within the context of data-centric AI, promotes a set of principles to deploy AI capabilities successfully. The quote at the start of this chapter, from the renowned mathematician Charles Babbage, reminds us that in practice, applying principles is of great importance, and the successful application of principles does not require unique inventions. Babbage's quote reinforces the need to innovate by assembling building blocks into a system. Walker's article in *Wired* magazine put it very succinctly: "An invention is usually a 'thing' while an innovation is usually [incorporating] inventions that cause change in behavior or interaction" [6]. A successful orchestration of MLOps requires following a process to instantiate, via innovation, philosophies and practices that have evolved from the well-established field of DevOps.

Chapters 10 and 11 help the reader to bring together the information in several of the prior chapters to transition AI capabilities into production. As a short recap, in part I of the book, we addressed the AI system architecture building blocks. In chapter 9, and illustrated in figure 9.4, we presented the AI strategic development model (AISDM), which enables the AI team to formulate a strategic development road map. In chapter 10, we introduced the system readiness levels (SRLs), and explained how SRLs can be used to assess progress in the implementation of the AI system architecture implementation framework illustrated in figure 10.3 in chapter 10.

An important deliverable, included in the AI strategic development road map, is the initial demonstration of an MLOps prototype. As Andrew Ng pointed out during the 2021 AI-focused MIT EmTech conference: "Pilot projects help in creating momentum" within the AI organization and customers [7]. These pilot projects can then continue to evolve—as part of the CI/CD and monitoring cycles of the MLOps structure—into developments that are intended to transition into operations. The AI system architecture implementation framework shown in figure 10.3 enables the rigorous execution of the AISDM, while adhering to the SRL steps.

In this chapter, we start in section 11.1 with an introduction to the MLOps fundamentals. We then explore the MLOps flow in the instantiation of the AI system architecture implementation framework, discussed in section 11.2. Since an important objective, intrinsic to the MLOps flow, is to automate parts of the AI/ML development life cycle, we present representative MLOps enabling techniques and contemporary tools in section 11.3.

We also discuss MLOps platforms, automated machine learning (AutoML), and low-code/no-code (LCNC) application development in section 11.4.

In section 11.5, we discuss a set of common pitfalls in AI development and deployment. We conclude the chapter with the main takeaways and exercises.

11.1   Introduction to MLOps Fundamentals

As we have described, within the MLOps definition, there are a set of fundamental elements that encompass the MLOps process. In this section, we introduce these key fundamentals, steps, and tasks that must be instantiated while following the MLOps process.

At a high level, and consistent with the cross-industry standard process for data mining (CRISP-DM) [8], the planning stage of an MLOps instantiation begins with a very good understanding of the business needs and stakeholder requirements of the organization, as discussed in chapter 3. Similarly, Baier et al. described the MLOps flow as follows: business understanding, data understanding, data preparation, modeling, evaluation, and deployment [9]. These early stages of the MLOps process are key to the successful development and deployment of AI capabilities.

MLOps has evolved from well-understood fundamental steps—such as the DevOps methodology—adapted to the development and deployment of AI systems. We highlight the common tasks found in the DevOps of software systems. The key steps are:

DevOps: Plan followed by Code, Build, Test, Release,
Deploy, Operate, and Monitor

Each of these steps or tasks must be implemented with security support, commonly referred to as "development, security, and operations (DevSecOps)."

For MLOps, we must modify these steps to include the full AI development and deployment life cycle:

> *MLOps:* Plan (i.e., business understanding), followed by Create
> (i.e., AI system architecture), Condition (i.e., data conditioning), Model
> (i.e., ML Modeling), Test (verification and validation), Package, Release,
> Configure, Deploy (i.e., CI/CD), and Monitor

All these tasks must be implemented with a governance infrastructure to ascertain compliance with the fairness, accountability, safety, transparency, ethics, privacy, and security (FASTEPS) principles discussed in chapter 8. The global goal in implementing the MLOps process is to automate some of the laborious and routine steps that are necessary to develop and deploy AI capabilities successfully.

Notice that the MLOps process described here maintains a very similar structure to the AI system architecture implementation framework shown in figure 10.3. The MLOps steps are necessary, but more needs to be incorporated, such as the attributes on the outer perimeter of the framework shown in figure 10.3 (i.e., operational excellence, cost optimization, reliability and security, performance metrics, risk management, and quality assurance), discussed as part of achieving SRL 7. These tasks need human control; thus, the goal is to establish a balance between MLOps automation and human control. Shneiderman calls for the need for both automation and maintaining human control—and some of these goals can be accomplished via a robust governance structure [10].

Since humans should exert control over the developed and deployed AI capabilities to meet the stakeholder's needs—compliant with the FASTEPS principles—we briefly address various AI team member roles here. We discuss the people aspect of the triad people-process-technology in much more detail in chapter 12.

The most effective AI teams incorporate a multidisciplinary skill set, discussed in part I of this book and chapters 9 and 10 and emphasized as part of the AISDM model regarding leadership and technical talent. For example, the proper implementation of MLOps requires a number of skills, including a subset of the following jobs:

AI systems engineer/architects: Formulate the end-to-end AI system architecture informed from detailed analysis of the business needs and stakeholder requirements.

Data scientists: Responsible for the tasks defined under the data conditioning subsystem discussed in chapter 3, and shown as one of the inner subcomponents in figure 10.3.

Data engineers: Facilitate and oversee the AI platform infrastructure to ascertain all the data formats, application programming interfaces (APIs), microservices, containers, and overall orchestration (discussed later in this chapter) are in place.

Machine learning (ML) scientists: Typically work on state-of-the-art ML models at the prototype and development stages.

ML Engineers: Responsible for implementing ML models and seeing them progress from development through deployment into operations (i.e., the production environment).

These roles are not mutually exclusive, nor does this list imply that all AI organizations must have all these disciplines to succeed. A lot will depend on the complexity of the AI system. In some instances, a data scientist might also share or own the responsibility of designing, developing, and deploying ML algorithms (i.e., data preparation and ML model creation) [11, 12].

These technical roles are likely to change with time as MLOps become adopted by many AI organizations and the development of AI products or services reaches a much wider community than those with specialized skills. Molino and Ré described an approach in which ML system developers do not need to know about the low-level details of the ML code while using declarative ML systems (leading to LCNC implementations, as discussed later in the chapter) [13]. They pointed out: "Technologies change the world when they can be harnessed by more people than those who can build them."

In the next section, we discuss how to instantiate the AI system architecture implementation framework by using MLOps and elaborate further on additional implementation details.

11.2   AI System Architecture Implementation Using MLOps

Each of the prior chapters of this book helps us in assembling a path forward for AI practitioners to succeed in the transitioning of AI capabilities from development to deployment. In this section, we walk the reader through a step-by-step approach to bringing together the AI system architecture implementation framework and the MLOps fundamentals. The AI system architecture implementation framework, shown in figure 10.3 of chapter 10, is the foundational construct that enables a holistic approach to instantiating the AI system architecture and executing the AISDM strategic road map.

As discussed in chapter 3, AI organizations must undergo a digital transformation to permit the use of the requisite platform infrastructure. To succeed in the digital transformation journey, we stress the importance of considering the triad of people, process, and technology to avoid making the mistake that many AI organizations make in only focusing on the AI technology. Baculard et al. found that out of hundreds of companies around the world, many had a dismal record in achieving the proper level of digital readiness [14].

Our AI systems engineering approach, complemented by the nascent field of MLOps, will improve the likelihood that a successful AI digital transformation will enable the transition from AI development to operations.

Before we delve further into how the AI system architecture implementation framework is complemented by the MLOps fundamentals (as discussed in the previous section), we give in table 11.1 the AI systems engineering key characteristics addressed in chapter 2. These key characteristics of the people-process-technology triad serve as the guiding principles that we espoused for bringing together the AI system architecture into a full realization, shown on the right side of the Vee-model (figure 10.1 in chapter 10).

The AI system architecture implementation framework integrates the AI system architecture building blocks outlined in table 11.1 and shown within its inner circle (refer to figure 10.3). We highlighted a short description of each of these AI pipeline building blocks and the associated risk analysis. An important subcomponent is the need to adhere to the FASTEPS principles under the rubric of responsible AI (RAI). Lack of attention to RAI is likely to lead to an unsuccessful deployment of AI products or services; or, just as important, a concern about the societal implications if AI is misused.

Next, we focus on the process set of key characteristics. As shown in table 11.1, we must start with a clear understanding of the stakeholder needs (i.e., the business goals). This process step, together with the AI technology building blocks, is part of the MLOps steps described in section 11.1.

The use of MLOps reinforces our approach in this chapter, where AI practitioners can put into practice a set of tools and techniques to instantiate the AI system architecture implementation framework. The available tools and techniques, growing out of the MLOps community, can be employed across the full AI development and deployment life cycle. We discuss MLOps enabling techniques and contemporary tools in section 11.3.

Although the MLOps implies a sequential process, we must not think of each step as complete until we undertake rigorous integration-verification-validation (I-V&V) of *each* of the AI building blocks and complete a performance assessment of the end-to-end AI system capability, as shown on the right side of the Vee-model. Furthermore, as outlined in table 11.1, the successful I-V&V must incorporate a deployment readiness review.

The deployment readiness review must follow the step-by-step assessment of the SRLs discussed in chapter 10. The AI architects, system developers, and integrators must show completion of SRLs 1–4 before taking the I-V&V step. Again, the MLOps fundamental steps discussed earlier are necessary, but unless there is rigor in the implementation, the AI development and deployment are most likely to fail.

Martinez et al. demonstrated the successful development through deployment of a complex signal processor system by following a methodical approach [15], which

Table 11.1 AI systems engineering key characteristics

Key Characteristic	Description	Risk Analysis
Technology		
System analysis trade-offs	Trade-offs must be performed incorporating the stakeholder's business goals and the AI capabilities provided. Clear articulation of the user/consumer/customer (stakeholder) AI application occurs at this early step. Perform trade-offs between threshold and target requirements.	Can the AI capability provided be scaled? Are the stakeholder requirements realistic? Risks in achieving target requirements versus threshold requirements.
Functional architecture	System architecture building blocks.	End-to-end measures of performance.
Sensors and Sources	Identify the data needed (structured and/or unstructured data).	Are the required data available?
Data conditioning, ML, modern computing, HMT	Formulate the data preparation needed; select the ML algorithms; identify the computing infrastructure (enterprise and edge computing); address HMT levels of collaboration to achieve augmented intelligence.	Technology management and levels of maturity
Robust AI	Vulnerability to adversarial AI.	What is the likelihood versus the consequence of unintentional or intentional adversaries?
RAI (FASTEPS)	Use tools and techniques to incorporate FASTEPS principles.	To what degree are the harms mitigated and the FASTEPS principles implemented?
Process		
Stakeholder needs (business goals)	The business goals drive the AI capabilities requirements.	Are needs and goals realistic?

(continued)

Table 11.1 (continued)

Key Characteristic	Description	Risk Analysis
I-V&V	I-V&V work in concert at each stage of the Vee-model. Verification is most often referred to as building things right, and validation is about testing the full end-to-end system. Is the AI system enabling the right capability?	I-V&V must be exercised at all levels of the Vee-model, from system definition to system deployment, including a deployment readiness review.
Development, deployment, and monitoring	The characteristics that follow are part of the development, deployment, and monitoring, plus the required risk analysis	
Operational excellence	Operational excellence is the ability to deliver business value (value capture) based on the AI value proposition, and to continually improve supporting processes and procedures through system monitoring.	Cuts across all the characteristics identified here under technology; must be assessed for each of the building blocks.
Cost optimization	Starts at the stage of system development definition (see the Vee-model) but it continues throughout the life cycle of the system	Minimize costs while still achieving the stakeholder's needs and goals.
Reliability and security	Each of the building blocks in the AI system architecture must be assessed in terms of their ability to meet the threshold and target requirements in a secured way	What is the likelihood of a failure and its associated consequences?
Operational/performance metrics	These metrics include measures of performance (MoPs) for each of the subcomponents and the overall AI architecture. MoPs lead to measures of effectiveness (MoEs). Metrics should be quantitative based on well-established system benchmarks (i.e., gold standards).	Accuracy, precision, recall, F-scores, and other metrics are assessed relative to the model performance and impact on the overall system.

Risk management	Includes technical risks (e.g., performance risk), management risks (e.g., cost and schedule risks), and organizational risks (e.g., societal risks)	Must address these broad range of risks in the context of likelihood versus consequence (impact)
Quality assurance	A broad term asserting that a product or service meets a set of specifications. Specs can include size, weight, power, shock, vibration, and humidity. In addition, ISO 9000 addresses manufacturing standards.	Should be addressed from the design trade-offs through system deployment and monitoring.
Cross-cutting systems engineering	In this category, there are tools and approaches to assess the performance of an AI system prior to deployment, such as modeling and simulation, emulation, and tabletop exercises simulating realistic environments.	Employing these tools and techniques should also include the use of operational metrics.
People		
Leadership and management	Leaders define and drive the AI strategic direction and road map. Management must be focused on execution.	Leaders are responsible for assessing the strategic risks. Managers are responsible for addressing execution risks. For example, is the necessary talent available to execute the plan?
Systems thinking and systems engineering skills	Every AI team must include a person with responsibility, and authority, for ensuring that the overall AI system meets a set of system requirements.	In concert with the AI leadership and management, the systems engineer must assess overall AI system risks.
Staff performance	Management must set clear staff performance goals and assess progress several times during the AI project.	There are tools to measure what matters, as discussed later in the book.

(continued)

Table 11.1 (continued)

Key Characteristic	Description	Risk Analysis
Multicultural environment	AI projects require a very multidisciplinary team such as systems engineers, data scientists, ML experts, computing technologists, human-machine social engineers, and other staff.	Assess the organization's mentorship and coaching approaches.
Effective communications	All members of an AI team must have the ability to communicate effectively at the subcomponent level up to the system level. Leadership and management also have the obligation to communicate effectively to their system stakeholders.	Communication is paramount to articulate well the subsystems through the final AI system. This attribute includes ethics in systems engineering.

equally applies to the transitioning of AI capabilities from the development stage to the deployment stage. We describe next this methodical approach.

SRLs 5 through 9 call for I-V&V of architecture subsystems, a single-thread, and the end-to-end AI system. The methodical approach described by Martinez et al. [15] consists of five steps:

1. *Unit-Test:* Integrate and verify (i.e., make sure that hardware and software operate correctly) that the AI capability provided by the subcomponent is operating properly. Validate hardware and software functionality. Let us take data conditioning as an example. Unit-testing for the data conditioning building block would involve verifying and validating the proper functioning of the software (for consuming input data) in a development environment (i.e., online within the AI platform).

2. *Subsystem:* Once the unit-test is completed (i.e., hardware and software are functioning properly), the subsystem (i.e., an AI building block) must verify that all the hardware and software steps are producing the correct output at the levels of expected performance. Let us take the ML building block as an example. The output from the ML stage (after data conditioning) must be evaluated to verify that the hardware, such as a graphics processing unit (GPU), is correctly implementing the ML algorithm (i.e., the software) producing the acceptable level of algorithm performance (i.e., transforming information into knowledge). Upon completing I-V &V of individual AI architecture subsystems, we can declare SRL 5 completed.

3. *Single-Thread:* Several subsystems are integrated, verified, and validated together as part of the AI pipeline. For example, in the AI application of computer vision for diagnosing lung cancer, we input X-ray data and then condition it to prepare it for the ML supervised learning algorithm. Results are delivered to a human-machine team where the AI team, working together with the subject-matter-expert (SME) radiologist, can make an assessment on the accuracy of the diagnosis. What differentiates a single-thread I-V&V from a full end-to-end AI system I-V&V is that not all permutations of data and counterfactuals need to be exercised. Therefore, we do not need to have all the computing fully in place. We only need to ascertain a smaller AI system showing a single-thread—from input data to insights at the output of the HMT are delivering the desired results. This smaller AI system can also serve as a minimum viable product (MVP) as part of meeting SRL 6. This MVP system validates the functionality and the AI product's value proposition as described in the AISDM strategic development plan.

   It is very important to perform the I-V&V at the unit-test, subsystem, and single-thread levels, while at the same time assessing each of the attributes shown on the outer perimeter of the AI system architecture implementation framework shown in figure 10.3, and described in table 11.1 (i.e., operational excellence, cost optimization, reliability and security, operational/performance metrics, risk management, and quality assurance). Upon successfully completing this step, the AI team would have reached SRL 7 and be ready to proceed to the next step.

4. *Online end-to-end AI system:* The single thread can now be expanded to include all aspects of the AI capability desired. In all likelihood, the AI system requires a greater number of AI platform elements. A full end-to-end AI system, in this context, is exercising the full AI pipeline stack, including the interfaces to all the AI ecosystem components. Today, AI platforms depend on many open-source tools. These tools and APIs must work in unison to avoid downtime in the operational environment. The successful completion of this I-V&V meets the objectives of SRL 8. SRLs 6, 7, and 8 are completed in an online operational configuration (i.e., often referred to as a "sidecar"). And again, the full complement of the attributes discussed earlier must be exercised in this online operational configuration.

5. *Inline end-to-end AI system:* The end-to-end AI system, demonstrated in step 4, is transitioned into an inline operational configuration. At this step, the AI system must be validated against the full suite of attributes. It meets the requirements of SRL 9, and the full system can be given full deployment readiness by the stakeholders. This step also incorporates CI/CD and monitoring as part of the iterative process illustrated in figure 10.3. During the step 5 demonstration, the stakeholder might also stand up a red team (meaning an independent team that assesses the robustness of the AI system compliant with the RAI guardrails discussed in chapter 8), which is responsible for performing an independent evaluation of whether the AI system is ready for deployment in operations.

There are several subtleties that must be included in the execution of steps 1–5. For example, the stakeholders should be brought into the I-V&V step as part of the MVP single-thread demonstration. Successes or failures should be celebrated as part of the Agile development process described in chapter 10. Failures would provide feedback to the AI team as they progress through the scrum stages of the Agile development methodology.

Another important subtlety that must also be considered is the use of a gold standard, as described in chapter 10. This is a very crucial component of the overall AI ecosystem depicted in figure 10.4. The data used in the demonstration of steps 1 through 5 must be a fairly accurate representation of the data collected and operated on in the inline operational configuration.

Recall that the AISDM strategic development plan not only is formulated with technologies—as part of the AI value proposition—but must also pay careful attention to the "people" part of the people-process-technology triad. We will present in chapters 12 and 13 a more detailed description of the key "people" characteristics outlined in table 11.1, including leadership, staff performance, and organizational topologies, to foster an innovative team environment and effective communication.

At the completion of steps 1–5, we will have traversed all the stages of MLOps:

Plan (i.e., business understanding), followed by Create (i.e., AI system architecture), Condition (i.e., data conditioning), Model (i.e., ML Modeling), Test V&V, Package, Release, Configure, Deploy (i.e., CI/CD), and Monitor.

Furthermore, these stages would have gone through a rigorous evaluation as described in the SRLs, improving the likelihood of success in transitioning from AI development to deployment.

In the next section, we describe techniques and contemporary tools that can be used within the MLOps implementation approach to make the execution of steps 1–5 easier (as part of a rigorous methodology).

11.3   MLOps Enabling Techniques and Contemporary Tools

To recap, the AI system architecture implementation framework benefits from MLOps, as discussed in the previous section, but it also introduces additional techniques to improve the likelihood of success. These techniques include:

- A set of attributes from the AI system architecture implementation framework for assessing and evaluating the I-V&V steps
- Five I-V&V steps: unit-test, subsystem, single-thread, online end-to-end AI system, and inline end-to-end AI system
- SRLs

In this section, we elaborate further on MLOps enabling techniques and contemporary tools that AI practitioners can use when executing the AI system architecture implementation framework. We begin with MLOps techniques by providing an introduction to DevSecOps. As we mentioned earlier, MLOps evolved from the well-established discipline of DevOps, and security was integrated and encapsulated later under DevSecOps.

From DevOps to MLOps

The DevOps movement began by formulating a more structured approach to delivering applications and services from the development stage to operations [16]. DevOps became a logical progression from the early days of needing agility and speed incorporating CI/CD and monitoring. The Agile methodology was accepted as the guiding principles, as stated in the Agile Manifesto in table 10.5 in chapter 10. Kim et al. described a use case, when in 2011, LinkedIn needed a more structured approach (as offered by the DevOps approach) to methodically move from development to operations and to scale with the goal of "paying down nearly a decade of technical debt" during their experienced rapid growth [16].

The five I-V&V steps, delineated in the previous section, are well aligned with the DevOps and Agile movements, breaking complex systems into manageable subsystems and employing a rapid prototyping methodology [17]. An important component of the Agile methodology, as a technique for rapid prototyping, is to adopt a product mindset, not a project mindset. A product mindset means dealing with time lines shorter than one month, with constant stakeholder feedback on the value and gaps of instances of the AI product. The team must be able to change requirements based on the user feedback.

Another very important requirement, to make the Agile approach work, is to avoid a monolithic and inflexible architecture where data is held in silos [18]. The Agile team must also be stakeholder centric, as espoused in SRLs 1 and 2, and emphasize the importance of people (e.g., customers), process, and technology [19]. The Agile technique will help AI organizations reduce what Cunningham coined as "technical debt" [20] by implementing rapid deployments and demonstrations through flexible, extensible, and viable systems [21].

SRL 4 calls for having a development and operational AI ecosystem, depicted in figure 10.4 of chapter 10. As part of the rapid prototyping, the AI team must execute measurements to evaluate the performance of the AI systems using a set of gold standards: test harness, benchmarks, and metrics. We also want to stress that benchmarks, in the context of the AI ecosystem shown in figure 10.4, are system-level. Often, AI implementers perform measurements on just the algorithms. We emphasize the importance of evaluating the full end-to-end AI system using system-level benchmarks.

As discussed earlier concerning SRL 5, each of the five I-V&V steps must be evaluated on the basis of the set of attributes, and performance metrics, such as correctness, scalability, and behavior within and outside the operating envelope, and, of course, these results must be stored in a well-defined configuration management control repository [22].

Now, let us turn our attention to the "security" part of DevSecOps. Security vulnerabilities can threaten the existence of AI systems. There are techniques, discussed earlier in the book, to encrypt data at rest, and data in motion. Techniques are in development to fortify approaches for protecting data in use. However, there are many threat surfaces accessible to malicious attackers, as discussed in chapter 7.

A technique with great potential for AI applications is processing on encrypted data. Researchers have been developing secure multiparty computation (SMPC) and homomorphic encryption. The latter is still too computationally intensive to be practical. Shen et al. have shown the value of SMPC for obfuscating the content of data and enabling information sharing and analysis [23]. As shown by other researchers, SMPC allows computation on only a subset of the data by individual parties, making access to sensitive and private data very difficult, based on cryptographic protocols [24, 25]. Also, Raizi and colleagues have established a company called CipherMode Labs that focuses on the application of SMPC to ML applications, preserving the confidentiality of the data.

Security, in the DevSecOps cycle, implies that security safeguards are in place from the beginning of the AI product or service development. Both physical security and cybersecurity are part of the AI system architecture implementation framework. The assessment of these key attributes is required as part of completing SRL 7.

Rajapakse et al. pointed out that "the adoption of DevSecOps in practice is proving to be a challenge" [26]. The authors performed a comprehensive and systematic literature review of fifty-four peer-reviewed studies. They identified twenty-one challenges and thirty-one specific solutions across four main themes: people, practices, tools, and infrastructure. For example, they identified security vulnerabilities affecting containerized approaches—we discuss containers later in this section. They also identified vulnerabilities affecting the CI/CD cycle due to the need for fast responses. Among the specific solutions recommended, they suggested implementing capabilities in a cloud-centric environment (see the discussions of cloud-native computing in chapter 5 and in the next subsection), using orchestrators to reduce the containers attack surfaces. We discuss Kubernetes (K8s) as a type of orchestrator later in this section.

Secured AI systems require more than just technical solutions. AI-driven organizations must address people (and the required training), as well as maintaining tight security practices within the organization, including role-based access control (RBAC) measures. Recently, Uber Technologies suffered a major cybersecurity attack by a hacker who gained significant access to the company's internal system [27]. Security analysts believed that the attacker was able to steal customers' credentials, providing access to the company's private network.

The US National Academies Organization completed a detailed report on the future of encryption [28]. Encryption solutions—such as the application of cryptography—will need to evolve rapidly to fend off attacks on AI systems from malicious actors. Furthermore, AI organizations must prepare themselves for when quantum computers might break encrypted solutions by making AI systems quantum-safe cryptosystems [29].

We now focus our attention on a more in-depth discussion on how MLOps can be put into practice. We structure this discussion in terms of building blocks and contemporary tools. It is important to stress that tools come and go. So we are not advocating the use of any one tool or development environment. Instead, we just introduce tools (including representative platforms) that are suitable to bringing MLOps into practice.

Raj provided a clear explanation and treatment of managing production-ready ML life cycle in his book *Engineering MLOps* [30].

MLOps Building Blocks and Contemporary Tools

Let us set the stage by reviewing what makes MLOps distinct in the implementation of AI systems. As outlined by Visengeriyeva et al. in reference to the Continuous Delivery Foundation—Special Interest Group MLOps, there is a set of capabilities that are different from traditional software engineering practices [31]:

- MLOps aims to unify the release cycle for ML and software application release.
- MLOps enables automated testing of ML artifacts (e.g., data validation, ML model testing, and ML model integration testing).
- MLOps enables the application of agile principles to ML projects.
- MLOps enables supporting ML models and data sets to build these models as first-class citizens within CI/CD systems.
- MLOps reduces technical debt across ML models.
- MLOps must be a language-, framework-, platform-, and infrastructure-agnostic practice.

In bringing CI/CD as a top priority throughout the ML life cycle, the I-V&V process must occur across the end-to-end AI pipeline illustrated in the inner circle of the AI system architecture implementation framework (see figure 10.3), including data conditioning, ML models, human-machine teaming (HMT), modern computing, and robust and responsible AI. Monitoring must include an assessment of data drifts, algorithm specification changes, and performance drifts, likely resulting in a need for refreshing the data, retraining the ML models, or both.

Furthermore, within the MLOps CI/CD steps, a set of ML artifacts must be tracked. These artifacts, as outlined by Treveil et al. are the following [5]:

- Model code and its preprocessing
- Model hyperparameters and associated model configurations
- Data for training and model validation
- The model after completing its testing phase
- Code and data for testing scenarios
- Libraries with specific versions and environment variables used in the creating of the model
- Documentation

Ormenisan et al. developed a more automated approach to logging and tracking ML artifacts at each of the stages of the ML pipeline [32]. The authors' implicit provenance avoids omitting important ML artifacts while still preserving ML automation.

These artifacts must be tracked at each of the five I-V&V steps described earlier (i.e., unit-test, subsystem, single-thread, online end-to-end AI system, and inline end-to-end AI system). We reinforce the importance of this methodical approach since rushing to deploy AI capabilities will lead to mistrust, not only of the AI organization developing the AI product or service, but also among the culture within the AI team.

The DevOps movement, as well as the need for an Agile methodology to enable rapid (i.e., short turnaround time) demonstrations, led to making applications more modular. Microservices were a solution allowing a single application to be broken into small components that could be changed, administered, and deployed as a collection of smaller services instead of a large, complex, monolithic application [33]. Microservices conform to modern service-oriented architecture, where a publish-subscribe architecture permits interfacing with databases across the computational network.

For many years, with the increasing availability of large computing resources (i.e., processing nodes, memory, and high-speed interconnects), virtual machines (VMs) became in vogue. However, VMs had the disadvantage of encapsulating not only the application and associated libraries, but also the operating system. VMs were difficult to migrate from one server to another. A more suitable approach for developing and deploying microservices was using containers. In the context of MLOps, this containerized approach became a way to deploy applications more rapidly within the CI/CD cycle.

Red Hat (a subsidiary of IBM) describes clearly, in a series of blogs, the differences between VMs and containers [34]:

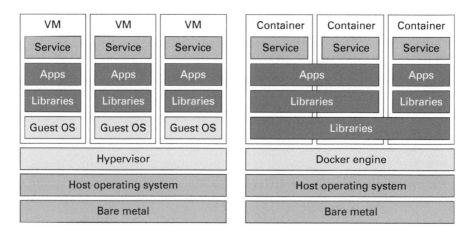

Figure 11.1   Comparison between VMs and containers. OS = operating system. Source: Andrew Morgan from MongoDB [35].

Containers: A tool to power microservices by running securely isolated applications. The Docker technology is useful for creating containers, which provided a way to encapsulate executable code, libraries, third-party apps via APIs, and other resources to achieve agile DevOps developments.

Andrew Morgan presented an illustration at the MongoDB 2016 conference [35] comparing VMs and containers, as depicted in figure 11.1.

In contrast to VMs, where the guest operating system is virtualized (making it harder to migrate across servers), containers are less coupled to the operating system. Containers source the operating system kernel from the host machine that it is running on, making it faster to spin up and migrate from one server to another. Application code and libraries can be shared, if necessary, by each container conforming to well-defined APIs. Containers also enable deploying capabilities into production by automatically rebuilding the image on the target machine since all the dependencies are containerized, meaning that it enables more rapid transitioning from an online configuration to an inline operational environment.

The Docker, as a tool, became the de facto containerization service, providing complete isolation from other applications and the host operating system. The technology was announced in 2013 and was released as open source in 2014. Docker, as a containerization technology, is used to create containers and configuration files identifying how the application and libraries and any of its dependencies must run. One can communicate and run containers in any computing environment—for example, on-premise or in a commercial cloud.

While Docker allows you to create containers, another technology that has gained popularity in the ML community is Kubernetes (or K8s). K8s is an orchestrator technology to automate the deployment, management, and scaling of applications encapsulated in containers. So microservices allow you to decompose applications into smaller services. Containers allow you to instantiate microservices as self-contained images. Docker is a well-accepted technology to create containers. K8s facilitates the orchestration of containers on-premise or in commercial clouds.

K8s orchestration is supported by many of the MLOps platforms and cloud computing providers. K8s is well matched to ML implementations because of its scalability, ability to extend across many types of hardware, and portability (so long as the tools and hardware support the K8s API).

K8s technology (initially introduced in 2014) was conceptualized at Google, making it easier to deploy containers during the CD phase of the MLOps CI/CD cycle. Google began to develop its own container-management systems under the names of Borg and Omega [36]. Later, the company, working with the Cloud Native Computing Foundation (CNCF), made the K8s tool open-source to make it easier to deploy and manage complex distributed systems by application developers.

K8s employs a declarative configuration object [37]. A declarative configuration represents a desired state of a system. K8s orchestration is responsible for making sure that the actual state matches the desired state. The desired state can be expressed using, for example, JavaScript Object Notation (JSON) files—a human-readable format for storing data (in key-value pairs), manipulating data, and defining configuration files (e.g., resources and objects).

As described by Raj, Kubeflow is a tool based on K8 enabling the orchestration of Docker-based microservices [30]. Kubeflow is a way of deploying an ML model. The container orchestration, implemented using Kubeflow, can be done by interfacing across services using the REST API (REST stands for "REpresentation State Transfer" as a publish-subscribe API). Zhou et al. published a more in-depth look at Kubeflow in the context of a K8s-based tool, for implementing a number of recognized deep learning algorithms within an ML pipeline, evaluating computation time (e.g., GPU usage), and memory performance [38].

Before we conclude this section, we strongly recommend that developers and integrators ensure strict security procedures. K8s can orchestrate distributed clusters. Pods are the smallest set of relevant containers in a K8s cluster. A pod, with several containers, uses the same underlying machine. In addition to reliable RBAC controls, there are tools, as described by Burns et al., that safeguard against malicious attacks attempting to break into K8s clusters, such as the PodSecurityPolicy API and RuntimeClass [39]. We advise AI engineers who are responsible for the infrastructure to incorporate security guardrails from the start of the AI development, as described earlier in this section and in prior chapters. Commercial cloud computing providers offer strong security controls and policies.

In the next section, we highlight some of the available MLOps platforms, their support for AutoML, and the ongoing trend toward LCNC application developments.

11.4   MLOps Platforms, AutoML, and LCNC Application Development

In the previous section, we discussed the use of MLOps techniques and explained some of the contemporary tools found within MLOps platforms. In this section, we look at MLOps platforms that are commercially available. Again, since there is great interest among the AI community in employing MLOps platforms to use available tools and techniques within the infrastructure, a very large number of companies exist. So we do not review the extensive MLOps landscape in this discussion. Instead, we present a small sample of these commercial platforms, but we refer to a number of publications that have put out more in-depth landscape reviews on what is available from small companies to large corporations.

In the literature, the term "platforms" is used, spanning all the way from implementation languages (e.g., Python, R) to multiple-billion-dollar market capitalization companies developing a full end-to-end infrastructure. In this section, we use the term "MLOps platform" as follows:

> *MLOps platform:* An environment with the ability to implement the full life cycle of an AI product or service in support of the AI system architecture implementation framework shown in figure 10.3, and to meet the requirements of an AI platform with characteristics described in table 10.6, as part of the AI ecosystem illustrated in figure 10.4.

We opt for this definition since the emphasis in our discussion is on an end-to-end holistic implementation, not just the ML algorithm implementation. To date, existing commercial MLOps platforms do not provide the full scope described in our definition. However, even though the MLOps discipline is nascent, several companies offer capable and comprehensive infrastructure environments.

Some commercial solution providers also refer to their offerings as "AI cloud platforms." In addition to the characteristics that we look for in the selection of an MLOps platform (see table 10.6), Neptune.ai writes a series of relevant blogs describing other desired MLOps platform characteristics [40].

A small set of examples of well-designed and cloud-agnostic platforms—include the following (listed in alphabetical order):

- C3.ai; https://c3.ai/
- Databricks; https://www.databricks.com/
- Dataiku; https://www.dataiku.com/
- DataRobot (together with Algorithmia); https://www.datarobot.com/
- Domino Data Lab; https://www.dominodatalab.com/
- Snowflake; https://www.snowflake.com/en/

Some of the big players in the AI and data management domains (e.g., Snowflake, Databricks, and c3.ai) are also able to interface with the other platforms mentioned in this chapter. That interplay is of great benefit to AI practitioners since the ultimate goal is to offer alternatives in the MLOps implementation. The commercial providers outlined here also have a mature ecosystem with established communities of practice, plus excellent training and customer support.

Since our goal is to highlight representative platforms, it is imperative that the AI team ultimately be responsible for doing an assessment based on the guidelines provided in chapter 10 and this chapter. The AI team must identify the best platform for their unique use case application (including their specific constraints, like the cost of acquiring a platform and the long-term sustainability and maintenance costs).

We also like to identify the comprehensive AI reports that CBInsights publishes on industries, emerging technologies, and the AI landscape [41]. Several of the major cloud computing providers, including Amazon Web Services (AWS), Microsoft Azure, and Google Cloud Platform, are also able to meet many of the AI platform characteristics described in table 10.6. The nice thing about several of these players is that they are cloud agnostic and able to provide services on top of these major cloud computing providers.

The development of AI capabilities can be time consuming and costly. Therefore, AI organizations are increasingly interested in AutoML tools. There is also strong interest in democratizing the creation of ML models that would be available to nonexperts. AutoML facilitates the rapid development and deployment of AI capabilities, beginning at the data ingestion stage and continuing through ML model development and deployment. Many of the platform providers mentioned in this discussion also offer AutoML tools.

AutoML, despite what the term might imply, cannot be fully automated. The error and implication with respect to the confidence in the machine making a decision versus the consequence of action, as discussed in chapter 6, can be detrimental if a fully automated ML system is deployed with the potential to cause grave damage. We as an AI community have not reached that point yet.

Karmaker et al. provided an excellent review article offering a seven-tiered approach to determining the achievable level of AutoML [42]. For example, they pointed out that task formulation (also referred to as the "planning stage" in the MLOps cycle) is done by data scientists and cannot be automated today. Also, they identified prediction engineering—the task of constructing and assigning labels to data and creating meaningful training and test data—as a step requiring manual intervention by data scientists, ML engineers, and data labelers. Several researchers have also published detailed content about the state of AutoML, including existing methods, tools, and challenges [43–45].

Gift and Deza, in their book *Practical MLOps*, addressed a number of use cases employing AutoML, including tools developed at Apple, Google, Microsoft Azure, and AWS [4]. An ML developer can also migrate from one environment, like Apple's CoreML, into another, like Microsoft Azure, by implementing the Open Neural Network Exchange standard (codeveloped by Facebook and Microsoft) [46], avoiding being locked in to one specific environment.

Another area receiving enormous interest lately is the use of Low-Code/No-Code (LCNC) in the development of ML models. The trend by some of the AI platform providers is to enable easy drag-and-drop capabilities for what is now called a "citizen developer" [47]. A citizen developer is someone within an organization who can assemble and compose capabilities without requiring deep knowledge of the underlying technology, with minimum or no programming skills. Imagine the ability to employ AutoML complemented with LCNC tools. This combination can allow users of such technology as robotic process automation to develop business capabilities that deliver value to an organization.

Although AutoML and LCNC are recent trends in enabling ease of use of AI tools by nonexperts, AI organizations must be very vigilant in their proper use. One concern is with cybersecurity vulnerabilities. Information technology (IT) professionals must play a role in making sure that these tools and infrastructure are compliant with security practices. Benac and Mohd performed an assessment of current low-code leading platforms and application providers [48]. Cabot made excellent points when comparing low-code approaches to model-driven engineering [49]. The ability to compose an end-to-end capability by dragging and dropping modules that encapsulate a function is not new—for example, the MATLAB Simulink environment, from MathWorks, has had such a capability for decades. What is relatively new is using such an approach to compose AI processing flows.

The interesting aspect of the LCNC trend is that it brings such functionality by AI platform providers to make assembling of capabilities easier for the nonspecialist (i.e., a citizen developer). However, the organizational and security infrastructure must be robust to support it [50]. The AI organization must have a well-established and operational digital transformation capability, as discussed in chapter 3. Today, a simple process, within a business unit, can be composed using LCNC, like forecasting sales, performing customer sentiment analysis, and looking at digital traffic for ad marketing. However, there is a need to be very cautious when trying to scale these limited AI capabilities across business units without following the rigorous approach that we have described in this book concerning the AI system architecture implementation framework. Nevertheless, these advances are very exciting, with significant potential payoffs.

In this section, we addressed some of the practices that AI practitioners can follow while developing and deploying the subsystems in the AI system architecture implementation framework. In the next section, we take more of a synoptic view of avoiding AI development and deployment pitfalls.

11.5   AI Development and Deployment: Common Pitfalls

We began this chapter by stressing the importance of following a methodical approach in transitioning from a development stage to a deployment phase, following a specific set of principles. Many AI organizations fail in this transition because they focus primarily on ML algorithm technologies and do not take a holistic approach.

In this section, we address additional techniques and processes to help AI practitioners avoid common pitfalls. As part of doing so, we also bring together some of the topics discussed in earlier sections, including SRLs:

Common Pitfall 1: Business goals and stakeholder requirements are not well defined.

Recommendation: Develop a strategic plan based on the process described under the AISDM model (see chapter 9 and the discussion of SRLs 1 and 2 in chapter 10).

Common Pitfall 2: There is a lack of well-structured development and deployment environments.

Recommendation: Define the AI system architecture and instantiate an AI ecosystem for the development and deployment environments, as illustrated in figures 1.3 and 10.4, respectively (see the discussion of SRLs 3 and 4 in chapter 10).

Common Pitfall 3: AI products and services are developed as monolithic applications, causing great difficulty in integrating, verifying, and validating key subsystem components.

Recommendation: Break AI applications into small services based on microservices and containerization technologies (see the discussion of SRL 5 in chapter 10 and of the I-V&V of subsystem components).

As we described earlier in this chapter, microservices are a technique for partitioning a single application into small components that could be changed, administered, and deployed as a collection of smaller services instead of a large and complex monolithic application. Each microservice can be developed, tested, and deployed independently [51].

Containers are an approach to powering microservices by running securely isolated applications. They provide a way of encapsulating executable code, libraries, third-party apps via APIs, and other elements to achieve agile DevOps developments. The Docker technology is one type of technique for creating containers and implementing containerized microservices.

Microservices and containerization are approaches that follow the guidance under the rubric of a cloud-native reference architecture, as espoused by the CNCF [52–54]. A cloud-native reference architecture incorporates orchestration and management as part of its required architectural layers. K8s is an orchestrator technology to automate deployment, management, and scale up or down by adding or removing containers across the underlying infrastructure. Reznik et al. provided an eloquent description of what is needed for organizations to migrate to a cloud-native architecture [52].

Fortunately for AI practitioners, CNCF was established in the summer of 2015 by the Linux Foundation to facilitate collaboration, common tools, trust, education, and best practices [53]. The CNCF organization tracks, in detail, the landscape of commercial providers for each of the elements of the cloud-native reference architecture (e.g., orchestration and management, runtime, and provisioning) [54].

Davis described Netflix's use case, where by employing cloud-native architectural patterns and best practices, it maintained a high level of availability despite the downtime experienced by the underlying cloud computing infrastructure [55]. Netflix demonstrated that by using isolated services, failure in one part of the end-to-end service did not translate to downtime of the entire system.

Common Pitfall 4: There is a lack of early demonstrations of an MVP, as espoused in an Agile development process, and compliance across all attributes (i.e., operational excellence, cost optimization, reliability and security, performance metrics, risk management, and quality assurance) shown on the outer perimeter of the AI system architecture implementation framework (see figure 10.3 and table 11.1).

Recommendation: Perform regular prototype demonstrations (i.e., AI system pilots; see the discussion of SRLs 6 and 7 in chapter 10) with stakeholders in the loop, employing an online operational configuration and cloud-native best practices [56].

Common Pitfall 5: There is an inability to operate in a highly constrained environment, such as at the edge (i.e., edge computing), when all the development was demonstrated in an enterprise-level scenario.

Recommendation: Identify early, as part of the AISDM, the implementation constraints with the goal of transitioning from an enterprise-level development to an edge environment for deployment, and from an online configuration to an end-to-end inline operational environment, in the MLOps process. (See the discussion of SRLs 8 and 9 in chapter 10.)

Much of the discussion in this chapter has focused on batch processing on-premise, private cloud, public cloud, or hybrid configurations [57]. This pitfall also implies the need to perform demonstrations that are representative of the ultimate operational environment at the enterprise and at the edge. As discussed in chapter 5, because of limitations with computing resources, ML training might be done in a cloud environment at the enterprise level, but the inference might be deployed at the edge. The latter will be more and more the case with the advent of Internet of Things (IoT) devices and the need to deploy AI capabilities on embedded devices [58, 59].

Common Pitfall 6: There is a disregard for the environmental-social-governance implications during the development and deployment of AI systems, causing businesses to incur costs and loss of revenues when their AI products or services are not accepted in the marketplace.

Recommendation: Put in place the proper guardrails, within the AI organization, to address the environmental-social-governance of AI capabilities. Environment guardrails means attention to the impact on such issues as climate change. Social guardrails imply a careful assessment of societal implications to avoid improper use of AI. Governance guardrails can facilitate compliance with the FASTEPS principles without atrophying innovation.

There are several initiatives addressing the carbon footprint and impact of AI systems on the climate and environmental changes [60–62]. Some of these efforts are also helping to provide the proper information to policy makers. Furthermore, there is a benefit in using ML-based resource management techniques to identify the optimum usage of cloud computing resources [63].

Common Pitfall 7: The level of leadership and technical talent required to successfully design, implement, integrate, and transition AI capabilities has been neglected.

Recommendation: As part of the development of a strategic road map (based on the AISDM process), identify the leadership and technical talent required for the development and deployment of an AI system. Also, the AI team must put effort into communicating effectively to their upper management and stakeholders. We will discuss these topics in more detail in the subsequent chapters of this book.

In the next sections, we conclude this chapter by summarizing the main takeaways and providing a set of exercises.

11.6   Main Takeaways

In chapter 10, we addressed a set of AI deployment guidelines that are referenced throughout chapter 11, useful in transitioning from the AI development phase to the deployment phase. These two chapters complement each other. In chapter 11, our discussion centers on MLOps as the overarching rubric encompassing all aspects of AI development and deployment.

In this chapter, we addressed MLOPs as an approach to more effectively implement elements of our AI system architecture implementation framework shown in figure 10.3 of chaper 10. The MLOps definition used throughout the chapter was:

> *MLOps:* A discipline that evolved from the practices employed in DevSecOps. It is a combination of philosophies and practices designed to enable AI teams to rapidly develop, deploy, maintain, and scale out ML models. MLOps incorporates data and models into DevOps cycles to improve the lead time and frequency of deliveries and to enable optimization of the entire ML project life cycle. Security is also a crucial component of MLOps.

The MLOps discipline, within the context of data-centric AI, promotes a set of principles to deploy AI capabilities successfully. A successful orchestration of MLOps requires following a process to instantiate, via an innovative process, philosophies and practices that have evolved from the well-established field of DevOps.

We highlighted the steps involved in DevOps, and how they extend to MLOps, as follows:

> *DevOps:* Plan, Code, Build, Test, Release, Deploy, Operate, Monitor

Each of these steps or tasks must be implemented with security support (commonly referred to as DevSecOps).

For MLOps, we modified the steps to include the full AI development and deployment life cycle.

> *MLOps:* Plan (i.e., business understanding), followed by Create
> (i.e., AI system architecture), Condition (i.e., data conditioning), Model
> (i.e., ML Modeling), Test V&V, Package, Release, Configure, Deploy
> (i.e., CI/CD), and Monitor

All these tasks must be implemented with a governance infrastructure to ascertain compliance with the FASTEPS principles discussed in chapter 8. The MLOps steps are necessary, but more need to be incorporated, such as the attributes on the outer perimeter of the framework shown in figure 10.3 (i.e., operational excellence, cost optimization, reliability and security, performance metrics, risk management, and quality assurance) and discussed as part of achieving SRL 7.

The effective implementation of MLOps requires different AI team skills, including the following disciplines: AI systems engineer/architects; data scientists; data engineers; ML scientists; and ML engineers. These roles and desired competences should not be interpreted as mutually exclusive, or that all organizations must have each of these disciplines. The talent needs of an AI organization will depend on the complexity of the AI capabilities developed and deployed. However, MLOps demands a multidisciplinary AI team.

A main takeaway from this chapter is that our AI systems engineering approach, complemented by the nascent field of MLOps, improves the likelihood of successfully transitioning from AI development into operations. We delineated, as described in chapter 2, the AI systems engineering key characteristics (see also table 11.1) that needed to be addressed in the successful implementation of an AI product or service using MLOps. Although MLOps implied a sequential process, we must not think of each step as complete until we undertake rigorous I-V&V, as shown in the right side of the Vee-model, of each of the AI building blocks and complete a performance assessment of the end-to-end AI system capability. Furthermore, as outlined in table 11.1, the successful I-V&V must incorporate a deployment readiness review.

Initial I-V&V, of each of the AI architecture subcomponents, should be evaluated starting with a simpler hardware and software prototype that evolves to a much complete system divided into the following five levels of complexity: Unit-Test; Subsystem; Single-Thread; Online End-to-End AI system; and Inline End-to-End AI system. It is important to include, in the I-V&V evaluation, the stakeholders to receive user feedback at the single-thread stage and as more complex assemblies are put together.

These five I-V&V AI architecture assemblies are well aligned with the DevOps and the Agile movements, breaking complex systems into manageable subsystems and employing a rapid prototyping methodology. These steps, as part of an Agile rapid prototyping approach, will help to reduce technical debt.

A technique with great potential for securing AI applications is processing encrypted data. Researchers have been developing SMPC and homomorphic encryption; however, the latter is still too computationally intensive to be practical. SMPC allows computation by individual parties on a subset of the data and makes access to sensitive and private data very difficult, based on cryptographic protocols.

During the implementation of MLOps, one distinction to standard software engineering practices is the need to enable automated testing of ML artifacts. At the stage of implementing the MLOps CI/CD steps, these artifacts must be tracked. Examples of ML artifacts are [5]:

- Model code and its preprocessing
- Model hyperparameters and associated model configurations
- Data for training and model validation
- The model after completing its testing phase
- Code and data for testing scenarios
- Libraries with specific versions and environment variables used in the creating of the model
- Documentation

These artifacts are a critical component of successfully implementing the MLOps process, as is the need to make applications modular. Microservices are a solution allowing a single application to be broken up into small components that could be changed, administered, and deployed as a collection of smaller services instead of as a large, complex, monolithic application. Microservices are powered by containers providing a way to encapsulate executable code, libraries, third-party apps via APIs, and other technologies to achieve agile DevOps developments. Microservices allow you to decompose applications into smaller services. Containers allow you to instantiate the microservices as self-contained images. Docker is a well-accepted technology to create containers, and K8s is one popular type of orchestration of containers, appropriate for on-premise or commercial clouds.

The techniques and approaches discussed in this chapter must be facilitated by commercially available MLOps platforms, and at times supplemented by the in-house infrastructure (e.g., proprietary data). Throughout this chapter, we used the following definition of an MLOps platform, consistent with our AI system architecture implementation framework and an AI ecosystem:

MLOps platform: An environment with the ability to implement the full life cycle of an AI product or service in support of the AI system architecture implementation framework shown in figure 10.3 in chapter 10, and able to meet the requirements of an AI platform with characteristics described in table 10.6, as part of the AI ecosystem illustrated in figure 10.4.

The development of AI capabilities can be time consuming and costly. Therefore, AI organizations are very interested in MLOps platforms with AutoML ability. There is also strong interest in democratizing the creation of ML models that are available to nonexperts. AutoML facilitates the rapid development and deployment of AI capabilities, beginning at the data ingestion stage and continuing through ML model development and deployment.

Another approach receiving enormous interest in the AI community is the use of LCNC in the development of ML models. LCNC enables AI developers to easily drag and drop composable blocks to build AI capability.

We concluded the chapter by providing the reader with seven common pitfalls often experienced during the AI development and deployment phases:

Common Pitfall 1: Business goals and stakeholder requirements are not well defined.

Common Pitfall 2: There is a lack of well-structured development and deployment environments.

Common Pitfall 3: AI products and services are developed as monolithic applications, causing great difficulty in integrating, verifying, and validating key subsystem components.

Common Pitfall 4: There is a lack of early demonstrations of an MVP, as espoused in an Agile development process, and compliance across all attributes (i.e., operational excellence, cost optimization, reliability and security, performance metrics, risk management, and quality assurance) shown on the outer perimeter of the AI system architecture implementation framework (see figure 10.3 and table 11.1).

Common Pitfall 5: There is an inability to operate in a highly constrained environment, such as at the edge, when all the development was demonstrated in an enterprise-level scenario.

Common Pitfall 6: There is disregard for the environmental-social-governance implications of developing and deploying AI systems, causing businesses to incurred costs and loss of revenues when their AI products or services are not accepted in the market place.

Common Pitfall 7: The level of leadership and technical talent required to successfully design, implement, integrate, and transition AI capabilities has been neglected.

We also discussed a set of recommendations on how to overcome these pitfalls.

11.7   Exercises

1. Elaborate on how our MLOps definition fits in the context of the DevOps process.

2. What is the difference between invention and innovation?

3. Two of the steps in the DevOps and MLOps are deploy and monitor.
   a. True
   b. False

4. Security is of no concern in the MLOps steps.
   a. True
   b. False

5. Identify and provide a short description of the attributes at the outer perimeter of the AI system architecture implementation framework illustrated in figure 10.3.

6. What is the role of the AI systems engineer/architect?

7. Elaborate on what is meant by system analysis trade-offs in the context of the AI systems engineering characteristics.

8. A successful MLOps implementation demands that all steps be done sequentially and completed before IV&V is undertaken.
   a. True
   b. False

9. What level of AI subsystem or system complexity must I-V&V testing be performed at? Pick the answer that fits best.
   a. Unit-Test
   b. Subsystem
   c. Online end-to-end AI system
   d. Inline end-to-end AI system
   e. All of the above

10. What is meant by technical debt?

11. Explain the benefit of implementing a secure multiparty computation scheme during data in use.

12. It is only necessary to track the ML model after completing its testing phase, and not the data for training and validation, as one of the ML artifacts.
    a. True
    b. False

13. What are microservices?

14. What is the value of using containers? Describe one type of technology used in the creation of containers.

15. What is the main difference between VMs and containers?

16. Describe the benefit of using Kubernetes (K8s) as a type of orchestration.

17. AutoML is a technique that automates task formulation and prediction engineering without AI team intervention.

    a. True

    b. False

18. What is a citizen developer in the implementation of the LCNC approach?

19. Choose one of the seven common pitfalls and elaborate on its significance and recommendations of how to avoid it.

20. Imagine that you are assigned the job of AI architect, based on the tools, techniques, and approaches described in this chapter. How would you mitigate the common pitfalls discussed in this chapter, working in a small entrepreneurial company with limited resources?

11.8   References

1. Khalajzadeh, H., M. Abdelrazek, J. Grundy, et al., Survey and analysis of current end-user data analytics tool support, in *IEEE Transactions on Big Data*, 2019.

2. Ng, A., *Data-centric AI*. 2022. https://datacentricai.org/.

3. Strickland, E. and Andrew Ng, AI minimalist: The machine-learning pioneer says small is the new big. *IEEE Spectrum*, 2022. 59(4): 22–50.

4. Gift, N. and A. Deza, *Practical MLOps*. 2021, O'Reilly Media

5. Treveil, M., N. Omont, C. Sternac, et al., *Introducing MLOps*. 2020, O'Reilly Media.

6. Walker, B., *Innovation vs. invention: Make the leap and reap the rewards*. 2015, WIRE.

7. Ng, A., *MIT technology review: AI-focused EmTech digital*. 2021.

8. Kelleher, J. D., B. Mac Namee, and A. D'arcy, *Fundamentals of machine learning for predictive data analytics: Algorithms, worked examples, and case studies*. 2020, MIT Press.

9. Baier, L., F. Jöhren, and S. Seebacher. Challenges in the deployment and operation of machine learning in practice, in *Proceedings of the 27th European Conference on Information Systems (ECIS)*, Stockholm and Uppsala, Sweden, June 8–14, 2019. https://aisel.aisnet.org/ecis2019_rp/163.

10. Shneiderman, B. Human-centered AI: A new synthesis, in *IFIP Conference on Human-Computer Interaction*. 2021, Springer.

11. Mäkinen, S., H. Skogström, E. Laaksonen, and T. Mikkonen, Who needs MLOps: What data scientists seek to accomplish and how can MLOps help? in *2021 IEEE/ACM 1st Workshop on AI Engineering-Software Engineering for AI (WAIN)*. 2021, IEEE.

12. Wang, D., J. D. Weisz, M. Muller, et al., Human-AI collaboration in data science: Exploring data scientists' perceptions of automated AI. *Proceedings of the ACM on Human-Computer Interaction*, 2019. 3(CSCW): 1–24.

13. Molino, P. and C. Ré, Declarative machine learning systems. *Communications of the ACM*, 2021. 65(1): 42–49.

14. Baculard, L.-P., L. Colombani, V. Flam, et al., *Orchestrating a successful digital transformation*. 2017, Bain & Company.

15. Martinez, D., R. Bond, and M. Vai, *Embedded digital signal processing for radar applications*, in *2008 IEEE Radar Conference*. 2008, IEEE.

16. Kim, G., J. Humble, P. Debois, and J. Willis, *The DevOps handbook: How to create world-class agility, reliability, & security in technology organizations*. 2021, IT Revolution.

17. Kleppmann, M., *Designing data-intensive applications: The big ideas behind reliable, scalable, and maintainable systems*. 2017, O'Reilly Media.

18. Poindexter, W. and S. Berez, Agile is not enough. *MIT Sloan Management Review*. 2017. https://sloanreview.mit.edu/article/agile-is-not-enough.

19. Rigby, D., S. Elk, and S. Berez, *Doing agile right: Transformation without chaos*. 2020, Harvard Business Press.

20. Cunningham, W., The WyCash portfolio management system. *ACM SIGPLAN OOPS Messenger*, 1992. 4(2): 29–30.

21. Forsgren, N., *2019 accelerate state of DevOps report*. 2019.

22. Forsgren, N. and M. Kersten, DevOps metrics. *Communications of the ACM*, 2018. 61(4): 44–48.

23. Shen, E., M. Varia, R. K. Cunningham, Cryptographically secure computation. *Computer*, 2015. 48(4): 78–81.

24. Cammarota, R., M. Schunter, A. Rajan, et al., *Trustworthy AI inference systems: An industry research view*, 2020. arXiv preprint arXiv:2008.04449.

25. Imani, M., Y. Kim, S. Riazi, et al., A framework for collaborative learning in secure high-dimensional space, in *2019 IEEE 12th International Conference on Cloud Computing (CLOUD)*. 2019, IEEE.

26. Rajapakse, R. N., M. Zahedi, M. A. Babar, H. Shen, Challenges and solutions when adopting DevSecOps: A systematic review. *Information and Software Technology*, 2022. 141: 106700.

27. Uber hack shows tech industry's Achilles' heel. *Wall Street Journal.* 2022.

28. National Academies of Sciences, Engineering, and Medicine, *Cryptography and the intelligence community: The future of encryption.* 2022.

29. Abuarqoub, A., S. Abuarqoub, A. Azu'bi, and A. Muhanna, The impact of quantum computing on security in emerging technologies, in *5th International Conference on Future Networks & Distributed Systems,* Dubai. 2021. https://dl.acm.org/doi/proceedings/10.1145/3508072.

30. Raj, E., *Engineering MLOps.* 2021, Packt Publishing.

31. Visengeriyeva, L., et al., *MLOps.* 2022. https://ml-ops.org/.

32. Ormenisan, A. A., M. Ismail, S. Haridi, and J. Dowling, Implicit provenance for machine learning artifacts, in *Proceedings of MLSys.* 2020. 20.

33. Newman, S., *Monolith to microservices: Evolutionary patterns to transform your monolith.* 2019: O'Reilly Media.

34. *Containers vs. VMs.* Red Hat. 2020. https://www.redhat.com/en/topics/containers/containers-vs-vms.

35. Morgan, A., Powering microservices with Docker, Kubernetes and Kafka, in *MongoDB.* EUROPE16.

36. Burns, B., B. Grant, D. Oppenheimer, et al., Borg, Omega, and Kubernetes. *Communications of the ACM*, 2016. 59(5): 50–57.

37. Burns, B., J. Beda, K. Hightower, and L. Evanson, *Kubernetes: Up and running.* 2022, O'Reilly Media

38. Zhou, Y., Y. Yu, and B. Ding. Towards MLOps: A case study of ML pipeline platform, in *2020 International Conference on Artificial Intelligence and Computer Engineering (ICAICE).* 2020, IEEE.

39. Burns, B., E. Villaba, D. Strebel, and L. Evanson, *Kubernetes best practices.* 2020, O'Reilly Media

40. Neptune.ai, *MLOps platforms blogs.*

41. CBInsights, *State of AI Q2'22 report.* 2022. https://www.cbinsights.com/research/report/ai-trends-q2-2022/.

42. Karmaker Santu, S. K., M. M. Hassan, M. J. Smith, et al., AutoML to date and beyond: Challenges and opportunities. *ACM Computing Surveys (CSUR)*, 2021. 54(8): 1–36.

43. Hutter, F., L. Kotthoff, and J. Vanschoren, *Automated machine learning: Methods, systems, challenges.* 2019, Springer Nature.

44. Ruf, P., M. Madan, C. Reich, and D. Ould-Abdeslam, Demystifying MLOps and presenting a recipe for the selection of open-source tools. *Applied Sciences*, 2021. 11(19): 8861.

45. Zöller, M.-A. and M. F. Huber, Benchmark and survey of automated machine learning frameworks. *Journal of Artificial Intelligence Research*, 2021. 70: 409–472.

46. Candela, J. Q. *Facebook and Microsoft introduce new open ecosystem for interchangeable AI frameworks.* 2017. https://research.facebook.com/.

47. Johannessen, C. and T. Davenport, When low-code/no-code development works-and when it doesn't. *Harvard Business Review*, 2021.

48. Benac, R. and T. K. Mohd. Recent trends in software development: Low-code solutions, in *Proceedings of the Future Technologies Conference.* 2021, Springer.

49. Cabot, J., Low-code vs. model-driven: Are they the same? *Modeling Languages*, 2020.

50. Yan, Z., The impacts of low/no-code development on digital transformation and software development. arXiv preprint arXiv:2112.14073, 2021.

51. Bozan, K., K. Lyytinen, and G. M. Rose, How to transition incrementally to microservice architecture. *Communications of the ACM*, 2020. 64(1): 79–85.

52. Reznik, P., J. Dobson, and M. Gienow, *Cloud native transformation: Practical patterns for innovation.* 2019, O'Reilly Media.

53. Gannon, D., R. Barga, and N. Sundaresan, *Cloud-native applications.* IEEE Cloud Computing, 2017. 4(5): 16–21.

54. *CNCF cloud native interactive landscape.* Cloud Native Landscape. https://landscape.cncf.io/.

55. Davis, C., Realizing software reliability in the face of infrastructure instability. *IEEE Cloud Computing*, 2017. 4(5): 34–40.

56. Torkura, K. A., M. I. H. Sukmana, F. Cheng, et al., Leveraging cloud native design patterns for security-as-a-service applications, in *2017 IEEE International Conference on Smart Cloud (SmartCloud).* 2017, IEEE.

57. Bieswanger, A., A. Maier, C. Mayer, et al., IBM Z in a secured hybrid cloud. *IBM Journal of Research and Development*, 2020. 64(5/6): 1.1–1.10.

58. Jacobides, M. G., S. Brusoni, and F. Candelon, The evolutionary dynamics of the artificial intelligence ecosystem. *Strategy Science*, 2021. 6(4): 412–435.

59. Prado, M. D., J. Su, R. Saeed, et al., Bonseyes AI pipeline—bringing AI to you: End-to-end integration of data, algorithms, and deployment tools. *ACM Transactions on Internet of Things*, 2020. 1(4): 1–25.

60. Dodge, J., T. Prewitt, R. T. Des Combes, et al., Measuring the carbon intensity of AI in cloud instances, in *2022 ACM Conference on Fairness, Accountability, and Transparency.* 2022.

61. Gupta, U., Y. G. Kim, S. Lee, et al., Chasing carbon: The elusive environmental footprint of computing. *IEEE Micro*, 2022. 42(4): 37–47.

62. Kaack, L. H., P. L. Donti, E. Strubell, et al., Aligning artificial intelligence with climate change mitigation. *Nature Climate Change*, 12(2022): 518–527.

63. Bianchini, R., M. Fontoura, E. Cortez, et al., Toward ML-centric cloud platforms. *Communications of the ACM*, 2020. 63(2): 50–59.

12

Fostering an Innovative Team Environment

If you think something is impossible, don't disturb who is doing it.
—Amar Bose, Bose Corporation founder

Part I and part II of the book thus far have focused on the artificial intelligence (AI) technology and process elements of the people-process-technology triad (illustrated in figure 10.2 in chapter 10). In this chapter and chapter 13, we discuss, in more detail, the "people" aspects for successfully developing and deploying AI systems.

There is a significant AI talent shortage both within the US and globally [1, 2]. There is technical talent in areas like data science and machine learning (ML) algorithm developers, but many of the premier experts in these areas are employed by large AI commercial companies (i.e., Google, Microsoft, Apple, and others). As discussed in chapters 10 and 11, AI development through deployment requires many additional skills. For example, experts on implementing machine learning operations (MLOps) have only limited availability, partly because of its nascent nature. Lack of appropriate talent is one of the biggest obstacles that organizations face in delivering AI capabilities.

Furthermore, AI organizations are not able to retain the scarce resources that they have. In this chapter, we provide practical tools and techniques to address these challenges, including best approaches to measuring progress and motivating AI talent to contribute to and achieve impactful results. We also address how to organize for using internal and external AI talent.

In chapter 9, when we introduced the AI strategic development model (AISDM) framework for developing an AI strategy and road map, we identified AI leadership and talent as paramount to creating an AI value proposition. AI organizations can have the best and most up-to-date AI technologies, complemented by well-defined processes, but unless the AI leadership and technical talent are aligned with the strategy, the AI development and deployment will likely fail to deliver the promised AI value proposition. Therefore, in this chapter, we address both AI leadership and technical talent.

In their book *How the Future Works*, Elliott et al. pointed out seven steps that organizations must take to adapt to the present demands of the workforce [3]. They emphasized, as part of these steps, the need to "focus on outcomes" and "train your leaders to make it work: soft skills matter more than ever." These are very important topics that we discuss later in the chapter. Also, after the onset of the COVID-19 pandemic, the approach to being productive and contributing to the mission and vision of an organization has changed significantly. Technical talent can come from all walks of life. Therefore, it is imperative that organizations adapt to employing resources from anywhere in the world while still maintaining the organization's core values and culture [4].

Amar Bose, a professor at the Massachusetts Institute of Technology (MIT), inventor, and cofounder of Bose Corporation, often repeated the quote: "If you think something is impossible, don't disturb who is doing it." His quote, which starts this chapter, brings to the forefront the importance of experimenting, building prototypes, and demonstrating new concepts and ideas, even if initially they might be considered impossible to do. This emphasis is very relevant to AI, since as we discussed in earlier chapters and as quoted by the leadership expert John C. Maxwell, "Fail early, fail often, but always fail forward," AI practitioners must learn from initial demonstrations. AI organizations must also understand and adapt to foster a culture of invention and innovation to balance the short-term versus the long-term impact.

In chapter 10, we outlined ten guidelines for successfully deploying AI capabilities. We repeat these ten guidelines and discussion pointers in table 12.1 for completeness. In this chapter, we focus on guidelines 3 through 6.

We begin in section 12.1 by addressing organizational culture in the context of leadership and technical talent. Without a supporting culture, AI developers, implementers, and product or service providers will have a very difficult time maintaining a competitive position in the fast-evolving field of AI. We then transition to addressing effective teams and close coupling to fostering an innovative team environment.

AI is having a profound impact on the workplace, and more broadly on economic value as well. Thus, we also discuss the global impact of AI through the lens of an economic perspective. This broad view allows us to structure the discussion around an AI organization's leadership and technical talent, taking both the short-term and long-term view.

We also take a deeper dive into tools and techniques available to AI management and staff to measure progress and take a more objective approach to assessing performance. In addition, we identify current trends in leadership as it relates to technical organizations

Table 12.1 Ten guidelines for successfully deploying AI capabilities and discussion pointers

Guideline	Description	Discussion Pointers
1	Develop a clear strategic vision and project road map for the AI system.	AI teams must understand and implement their AI strategic blueprint (see chapter 9).
2	Understand the stakeholders' AI needs.	This is a critical element of the strategic road map discussed in section 9.6 of chapter 9.
3	Strengthen the AI team by fostering internal and external relationships.	AI offerings require leveraging both in-house and subcontract work (discussed further in this chapter).
4	Build a multidisciplinary and diverse team with complementary skills.	An end-to-end AI system architecture demands a broad range of skills (discussed further in this chapter).
5	Provide measurable objectives while mentoring AI talent.	Peter Drucker, a renowned modern management scholar, had a famous quote: "What gets measured gets managed." This point is discussed further in this chapter.
6	Continue to expand AI team skills as the future of work evolves	AI as a technical field is evolving very rapidly. Talent must stay up to date. This point is discussed further in this chapter.
7	Demonstrate an initial AI capability via a prototype, then iterate.	Gain momentum by demonstrating an initial MVP (see chapter 11).
8	Verify individual subcomponents and validate the end-to-end AI system.	These are fundamental steps to a systems engineering approach (see section 10.6 in chapter 10).
9	Secure AI systems both physically and against cyberthreats.	AI systems must be robust to adversarial attacks (see chapters 7 and 11).
10	Attend to RAI principles.	AI organizations must comply with the fairness, accountability, safety, transparency, ethics, privacy, and security (FASTEPS) principles discussed in chapter 8.

and discuss the typical signs of dysfunctional teams and approaches to mitigate team discord based on the seminal work of Lencioni [5].

We conclude this chapter with a section about sustaining high-performing teams. In the last two sections, we leave the reader with the chapter's main takeaways and a set of exercises.

12.1 Organizational Culture

Organizational culture drives the ability of leaders and technical AI practitioners to obtain alignment between the overall organization's strategy and the AI strategic road map, as described in our AISDM framework. The AI strategy must be executed, and in order to do that, the organizational culture must support the strategic road map's goals and actions.

Let us look now at what is meant by "organizational culture." In simple terms, an organization's culture defines how it enacts its organizational core values. Culture is about actions, while core values are the organization's beliefs.

A more encompassing definition is described by Tharp as follows [6]:

> The culture of an organization eminently influences its myriad decisions and actions. A company's prevailing ideas, values, attitudes, and beliefs guide the way in which its employees think, feel, and act—quite often unconsciously. Therefore, understanding culture is fundamental to the description and analysis of organizational phenomena. For some, culture is considered the "glue" that holds an organization together and for others, the "compass" that provides direction.

Ed Schein, an MIT professor emeritus, is considered a prominent and renowned scholar in organizational culture and leadership. In this section, we use his organizational culture definition to frame our discussion [7]:

> The culture of a group can be defined as a pattern of shared basic assumptions learned by a group as it solved its problems of external adaptation and internal integration, which has worked well enough to be considered valid and, therefore, to be taught to new members as the correct way to perceive, think, and feel in relation to those problems.

In this definition, adapted to our AI discussion, the term "group" can refer to any AI organization, such as small entrepreneurial companies, groups in academia, business units, or a large corporation. Schein also provides further insights into culture to minimize what are often considered abstract concepts. He categorizes culture into the following three levels:

1. Artifacts
   a. Structures and processes that you can see and feel
   b. Observed behaviors
2. Espoused beliefs and values
   a. Ideals, goals, values, and aspirations
   b. Ideologies
   c. Rationalizations
3. Basic underlying assumptions
   a. Unconscious, taken-for-granted beliefs and values

Schein emphasizes: "Understanding culture at any level now requires some understanding of all of the levels. National, ethnic, occupational, organizational, and microsystem issues are all interconnected." Furthermore, an organizational culture must also incorporate the cross-cultural component of AI teams, which includes different ethnicities, upbringings, and experiences.

Both the authors of this book grew up in cultures different to what one would consider a majority in corporate America. D. Martinez had a South American upbringing (in Bolivia and Venezuela), and B. Kifle is from Ethiopia. In both of our cases, we experienced broad exposure to many different cross-cultures (as examples of ethnic differences). Understanding and putting into practice diversity, equity, and inclusion are very important to successfully develop AI products and services.

We opt to anchor our discussion on Schein's definition and his three categorization levels to reinforce that without a supporting culture, AI organizations will fail to bring value to stakeholders. Organizational culture forms the operating system of any organization.

Let us look at an example through this definition lens. As discussed in chapters 10 and 11, it is imperative that AI organizations create a minimum viable product (MVP), expected for meeting System Readiness Level (SRL) 6. AI teams will inevitably confront "problems of external adaptation and internal integration." How the AI team addresses those problems and resolves them will be guided by Schein's categorizations of culture (i.e., artifacts, espoused beliefs and values, and basic underlying assumptions).

In practice, organizations have to pay close attention to their culture, consistent with Schein's definition and categorization levels. As Daimler addressed in her article

"Why Great Employees Leave 'Great Cultures'" [8], a well-functioning culture must attend to its behaviors, systems (e.g., supporting infrastructure to hire, develop, and reward their staff), and practices (e.g., how decisions are made)—all guided by an overarching set of core values. As discussed by Daimler, when gaps start to appear in alignment, that is when you start to see problems with employee retention. In a weak organizational culture, employees either leave or become disconnected (i.e., so-called quiet quitting—namely, doing the minimum that a job requires).

Another important and tangible piece of evidence of a strong culture is when leaders demonstrate a high level of character. Crossan et al. [9] clearly explained that leaders with demonstrable character foster more effective organizational development. They defined character to include integrity, responsibility, forgiveness, and compassion. When employees rated their management as keeping to these principles, employees were more effective at their jobs, resulting in lower attrition. When competence and character were practiced hand in hand, organizational excellence and well-being were consistently achieved.

So far, we have been discussing organizational culture as it applies to any organization, regardless of its mission focus or industry sector. This culture definition and its characterization levels are broadly applicable to any industry. As we evolve into the AI era, there are benefits in embedding AI into the fabric of the organizational culture. Ransbotham et al. performed a global survey of 2,197 managers [10] and concluded that "business culture affects AI deployments, and AI deployments affect business culture." More specifically, AI-related cultural benefits, at the team and organization levels, included improvements in enterprise efficiencies, decision-making processes, personnel morale, collaboration, clarity of roles and responsibilities, and collective learning. The global survey by Ransbotham et al. reinforces how Schein's culture characterization levels manifest themselves in practice.

Traditional organizations do not need to start with a clean slate to evolve and conform to a digital-ready culture. Westerman et al. performed a comprehensive survey of digital and traditional companies [11]. They formulated four critical values of a digital culture:

1. Impact (i.e., contribute with a strong sense of purpose)
2. Speed (i.e., move fast and iterate, as emphasized in chapters 10 and 11)
3. Openness (i.e., institute an organizational atmosphere of reaching out broadly and openly)
4. Autonomy (i.e., allow people high levels of discretion in meeting the strategic goals and actions)

As Westerman and colleagues pointed out, digital-based organizations are typically stronger at rapidly experimenting (which is critically important to stay competitive in the age of AI), self-organizing, and driving decisions with data. However, traditional organizations can undertake a digital transformation, without losing their identity and heritage (i.e., their so-called birthright), to adapt their culture while still preserving what made them strong. As we emphasized earlier, leadership competence and character go hand in hand with creating stakeholder value and fostering a strong culture— these are executive-level responsibilities, resulting in a well-functioning organizational environment.

In the next section, we elaborate on creating an organization's internal structure to enable implementation of the AISDM strategic road map. This topic has direct relevance to fostering an innovative team environment.

12.2   Organizational Structure and Innovation

In the previous section, we discussed the importance of establishing an organizational culture to enable AI leadership and staff to effectively develop and deploy AI products and services. In addition to organizational culture, an AI organization must be internally structured to support the development and execution of a strategic development road map, including delivering AI capabilities that meet the stakeholders' requirements. In this section, we elaborate on options for organizing internally and the relevance of organizational structure to foster innovation.

In chapter 2, we referenced Jay Forrester, considered the father of system dynamics and a key scholar and contributor to the systems engineering history. Forrester was also instrumental in focusing organizations on the importance of the organizational structure to achieve innovative results. During the MIT Lincoln Laboratory Heritage Lecture, he expressed this belief (i.e., innovative results) in the following quote: "The social structure and management of a technical enterprise are far more important than the underlining science. If one has the right environment, it will produce the required science. But the best science will fail in an unfavorable social and managerial setting." From this sentiment, we can generalize the science to encompass all aspects of technical innovation and development.

Kotter has studied effective organizational structures for accelerating innovation in traditional organizations [12]. Figure 12.1 illustrates a traditional organizational structure based on a hierarchical topology. In a fast-moving world, such as what we are experiencing today in the age of AI, this hierarchical configuration must allow for a networked topology shown on the right of figure 12.1. As emphasized by Kotter, the organization must create a sense of urgency, build coalitions among business units, form a strategic vision and initiatives, and sustain acceleration.

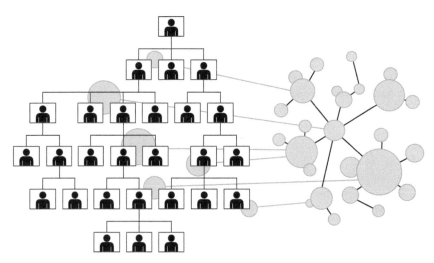

A hierarchical structure refers to a company's chain of command, typically from senior management and executives to general employees.

Accelerate innovation: build distributed but networked innovative clusters.

Figure 12.1   Building strategic agility for a faster-moving world [12].

In this hierarchical structure complemented with a networked topology, AI groups can reside in each of the business units while remaining networked to a central group. The centralized group can then be responsible for helping to facilitate the people-process-technology triad across the business units to achieve high levels of AI team effectiveness.

How an AI-based organization implements this hierarchical structure has a direct implication on how the organization is able to innovate. For example, Podolny and Hansen analyzed how Apple is able to innovate so effectively year after year [13]. Apple focuses on delivering the best products and services by dividing its business units into functions (i.e., the business units follow a functional structure). In a functional structure, business units represent domains of expertise such as design, hardware engineering, software, ML and AI, marketing, operations, law, and other functions. Those business units with the most expertise and experience in a functional area would have decision rights for that domain. In figure 12.1, each of the branches from the top for Apple would be a functional domain, and each functional domain can have AI/ML cells that connect in a networked topology to a central node with ML and AI domain knowledge. This approach also helps with collaboration and communication among the business units' subject matter experts (SMEs) and AI technologists.

The organizational structure discussed here is an internal micro-view of how companies should organize to foster effective and innovative teams. At the macro-level, several industrialized and developing countries have tried to replicate the innovative atmosphere experienced in Silicon Valley companies in California. Alec Ross, in his book *The Industries of the Future*, cites Marc Andreessen's six rules for an innovative ecosystem (with the ability to incubate ideas plus having access to labor and capital) [14]:

1. Build a big, beautiful, and fully equipped technology park.
2. Mix in research and development (R&D) labs and university centers.
3. Provide incentives to attract scientists, firms, and users.
4. Interconnect the industry through consortia and specialized suppliers.
5. Protect intellectual property and tech transfer.
6. Establish a favorable business environment and regulations.

These rules by Andreessen, an American entrepreneur, venture capital investor, and software engineer, have worked well for companies (e.g., Google, Facebook, Microsoft, and Akamai) close to universities like Stanford, the Massachusetts Institute of Technology (MIT), Carnegie Mellon University, and several others. Thus, AI organizations must assess their internal organizational structure and the positive impact on the external environment—as prescribed in the six rules—that can have in their ability to develop and deploy AI capabilities successfully. Access to talented labor can be easier if the environment permits close association with the fast pace of innovation incubated in academia—even when the AI staff are physically located far apart from each other.

Kanter addressed classic traps that organizations fall into when looking for innovative concepts with the potential to help with increasing AI uniqueness [15]. Several of the classic traps that he pointed out included counting on a few "big bets," operating in an environment where innovation is limited to siloed groups and imposing the same leadership and staff performance metrics as used in the regular business operations. Kanter offered recommendations to address these classic traps by augmenting the big bets with incremental innovations. Siloed groups can be avoided by distributing innovative cells across the organizations and embedded into the business units, as illustrated in figure 12.1. Personnel performance metrics for those exploring new ideas should be based on achieving near-term and long-term impacts.

As we described in chapter 9 and depicted in figure 9.10, innovation should be in the quadrant of low-to-mid mission-critical to the business, but offering mid-to-high company differentiators. When an innovative concept is demonstrated, either as a big bet or an incremental innovation, the AI capability can be transitioned to the exploitation quadrant (with mid-to-high mission-critical and mid-to-high company differentiators).

This approach to innovation, as part of the AISDM framework and strategic road map, will enhance a company's competitive advantage and AI differentiation. The ability to communicate effectively (see chapter 13), inside and outside the organization, is also key to avoiding several of the classic traps enumerated by Kanter.

Christensen provided groundbreaking literature on the vulnerabilities that an established corporation can face if it does not recognize how disruptive innovation from competitors can lead even great, strong businesses to fail [16]. In the age of AI, Christensen's topic of disruptive innovation is very apropos to a rapidly evolving AI field. AI organizations must be keenly focused not just on disruptive technologies, but also on the people and processes needed to maintain a competitive position in their respective markets.

We conclude this section by emphasizing the importance of the AISDM framework as a tool to formulate the AI strategic blueprint and clearly identify the strategic direction, including the organizational structure and opportunities for innovation. Peter Drucker, who is considered the father of modern management, reinforces the early discussion in his seminal work *The Discipline of Innovation* by pointing out that innovation should be managed like any other corporate function [17].

In the next section, we highlight recent research on how the future of work is evolving from the viewpoint of several economists, and how AI talent should adapt to these changes.

12.3   AI Talent and the Future of Work

In the previous two sections, our discussion has centered on organizational topics (i.e., culture and structure) as prerequisites for fostering an innovative team environment. AI practitioners should also assess the economic factors that contribute to adding AI value and growth to the economy. This economic view is important since teams are most energized when they understand that their work has merit and contributes to the organization's overall mission. In this section, we discuss an economic perspective and impact on the workforce from recent AI advancements, with direct relevance to the main theme of this chapter: fostering an innovative team environment.

In chapter 1, we described the working AI definition that we have used throughout this book within the rubric of narrow AI. There are many applications of narrow AI that are making a direct contribution to worldwide economic growth (e.g., medical applications, transportation, robotics, precision agriculture, and the supply chain, just to name a few). As Agarwal et al, discussed, it is imperative to separate the value that the AI technology, writ large, brings to prediction from the ability to augment human capabilities (e.g., in decision making or judgment) [18].

Agarwal et al. examined four direct effects with the potential to affect human labor using AI capital, as follows:

- Substituting AI capital for labor (i.e., delegating routine tasks to machines)
- Automating decision tasks (i.e., reducing the use of human labor when the confidence in the machine prediction is high and the consequence of action is low)
- Increasing labor productivity (e.g., augmenting human intelligence)
- Reducing uncertainty through AI predictions leading to new decision tasks (i.e., dedicating human labor to higher cognitive tasks)

These four effects are relevant to how AI talent can be best employed through machine intelligence. Bresnahan and Trajtenberg (from the Department of Economics at Stanford and Tel-Aviv University, respectively) coined the term "general-purpose technology," which historically has acted as an engine of growth [19]. (Note that GPT has been used with this term, but in this context, that is not the same as the Generative Pretrained Transformer (GPT) discussed earlier in this book. To make this distinction clearer, in this chapter we use the acronym GPTe to refer to this term.)

The reason for assessing the economic value of AI as a contemporary example of a GPTe is that it takes many years to see the full contribution of any GPTe as a pervasive technology. Just as it happened with other GPTe advances, such as the steam engine, electricity, microprocessors, and the internet, the rate of growth in the whole economy is slow as downstream sectors emerge in support of GPTe. Once the downstream ecosystem matures (e.g., laptop computers or smart phones), the rate of economic growth is exponential. We have not gotten there yet.

AI talent will be a critical part of this upstream and downstream evolution. Two prominent economics professors, Daron Acemoglu of MIT and Pascual Restrepo of Boston University, have argued, in their seminal paper "Artificial Intelligence, Automation, and Work," that despite the phenomenal improvements demonstrated by narrow AI, we are far from understanding how a pervasive GPTe such as AI will affect the workforce and productivity [20]. For example, they discussed how AI and robotics could create new human tasks leading to an increase in employment instead of displacing workers. However, an increase in higher cognitive tasks will likely lead to having to retrain some of the workforce [21]. Again, these macroeconomic factors will take time to fully understand and measure the economic impact due to AI being a nascent GPTe.

Throughout this book, we have focused primarily on the value provided by AI in augmenting human intelligence. Erik Brynjolfsson of Stanford makes a very powerful argument that excessive efforts in trying to achieve humanlike AI—meaning artificial general intelligence (AGI)—can lead to the so-called Turing Trap [22]. It is best to put more effort into achieving human task augmentation.

Brynjolfsson also points out that in a digital economy, a historical measure of economic growth like the gross domestic product growth rate is no longer an accurate measure since, for example, significant growth can be accrued by employing open-source software. Instead, Brynjolfsson et al. recommends GDP-B as a measure of economic benefit instead of the cost of goods and services [23]. Albert Einstein's famous quote applies well here: "Not everything that counts can be counted, and not everything that can be counted counts."

AI practitioners, business leaders, and policy makers should spend time tracking the work of these renowned economists, and others, to best formulate their companies' AI strategic road maps and assess their direct impact on AI talent (meaning human labor). AI will continue to be transformational, but today's businesses will have to go through a digital transformation to keep pace with a digital economy.

As discussed in earlier chapters, AI adoption, at the enterprise level, is still slow. These economic gains, albeit slow at the start, will accelerate, especially if leadership makes an effort to introduce AI in a phased approach, without causing concerns that employees' jobs are at stake or that staff performance will suffer. Babic et al. offered a four-phase approach for onboarding AI at the enterprise level [24]:

1. As an assistant (e.g., helping in data sorting)
2. In a monitor function to warn the user, in real time and during high-stakes decision making, to avoid the high cost of an inappropriate action
3. As a coach (i.e., personnel often desire more frequent performance feedback)
4. As a teammate (i.e., using collective intelligence acquired from what machines do well, complemented by what humans do well as per our discussion in chapter 6)

Ransbotham et al. performed a global survey of 1,741 participants representing 20 industries and 100 countries. They identified several areas where AI provides individual value in increasing staff competency, more individual autonomy, and trust in the machines, while contributing to organizational value [25]. Their findings showed that AI was most valuable when it helped individuals learn from past actions and in offering feedback on the consequence of past actions. These findings reinforce the four phases to onboard AI successfully.

In the next sections, we transition to a more specific discussion of how managers can more effectively lead AI teams, and provide techniques and approaches for motivating staff to be more effective at their jobs.

12.4   Preparing You for a Successful Career

We started this chapter by looking at the organizational culture, structure, and the economic impact that AI is likely to have on the workforce. These topics help an AI organization in understanding the importance of fostering a well-functioning environment to enable talented staff to succeed. In this section, we elaborate further on tools and techniques that can facilitate a successful career. We also address, in more detail, approaches relevant to guidelines 3–6 in table 12.1.

Over a decade ago, we developed a useful construct for anyone in a technology-based organization to assess his or her day-to-day activities in both the near term and long term. We refer to this construct as the "Circle of Balance (CoB)," depicted in figure 12.2. One of this book's authors (David Martinez), after transitioning from leading a commercial company into an R&D environment, recognized that both work environments share many similarities in staff priorities and emphasis. We will now take a deeper dive into each of the six slices of the CoB construct.

Before proceeding further in our discussion, note that the operative word in the CoB construct is "balance." No one is expected to perform and adhere to each of the slices shown in figure 12.2 perfectly. It is a balance that we all need to pay attention to; it is not

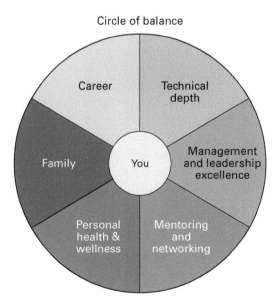

Figure 12.2   The CoB construct.

about doing one or a few of these efforts well at the expense of others. Time and time again, it has been shown that devoting exclusive effort to just one of the slices is likely to jeopardize the effectiveness of the other slices in the CoB construct, ultimately affecting anyone's job performance, family, and personal satisfaction.

Although the CoB is applicable more broadly than just in technology-based organizations, our discussion emphasizes organizations where technology prowess is of paramount importance to maintain a competitive edge, as demanded by AI-based organizations. Starting with the top-right slice, technical depth, it is imperative that staff put into practice the technical knowledge that they acquired in higher education (e.g., from community colleges, universities, and other forums). In the next section, we discuss evolution from roles requiring technical depth, to roles requiring technical breadth.

"Technical depth" refers to putting into practice the acquired educational knowledge (e.g., in data science/engineering, ML, modern computing, and other aspects of the end-to-end AI system architecture shown in figure 1.3 in chapter 1). We refer to technical depth as disciplines within the rubric of science, technology, engineering, and math (STEM).

Furthermore, in our AI system architecture, we pointed out the need for SMEs in the area of social sciences and political sciences. An example of the social sciences or social engineering is in knowing how to implement the human-machine teaming (HMT) building block. An example of someone with a background in public policy or political science is an SME with knowledge of the implication that AI has on policies driven by technology, as well as environmental-social-governance (discussed in previous chapters). Further, AI technologists must work closely with policy makers and vice versa; and policy makers must understand the AI technology, the value proposition, and implications.

As technical personnel evolve in their careers, they might want to take on more management and leadership roles. We elaborate further on this topic in the next several sections. For now, though, it is sufficient to point out that AI workers are most valuable to organizations when they can not only apply their technical knowledge as they gain more practical experience, but also in leading teams. As we will discuss later in this chapter, we do not imply that staff should have either technical skills or managerial/leadership skills. In fact, we mean the opposite. Technical talent can evolve their careers into technical leadership, where they would still need and must stay abreast of advances in their respective technical areas of expertise, but they also need management skills.

Furthermore, as staff progress in their careers, they can become most valuable to their respective organizations when they also serve as mentors to others who are less senior in the workforce. Thus, the third slice of the CoB construct, mentoring and networking, implies helping others succeed in their careers by taking time from their daily tasks to mentor and effectively network across the organization. We will discuss this topic in more detail later in this chapter.

As we traverse along the CoB construct clockwise, "personal health and wellness," the next slice, has a domino effect. The most effective performers are able to balance the demands of their technical jobs with their personal priorities. Bill George, a professor of management practice and the former chief executive officer (CEO) of Medtronic, in his Harvard Business School classes, points out: "To be equipped for the rapid-fire intensity of executive life, they [executives] cultivate daily practices that allow them to regularly renew their minds, bodies, and spirits" [26]. In addition, Khazan has developed models to use individuals' biofeedback and practice mindfulness [27].

Of course, the CoB construct is also most important when personnel are trying to balance family with the "rapid-fire" environment of a job, as per George's comment. The fifth slice of the CoB construct, "Family," stresses the need to attend to family time along with business. Progressive organizations place an emphasis and institute programs to help employees maintain a work-life balance (sometimes referred to as "work-life blending"). The term "work-life blending" is perhaps most appropriate now since personnel must continue to attend to their jobs while also being cognizant of family responsibilities—it is a daily blending of the two. We include under family parents, children, siblings, other relatives, or even very close friends who are considered family by an individual.

The last CoB slice is "Career." All the previous slices are specific areas to pay attention to while balancing the near-term activities. Career must also be part of this balance, but it represents more of a long-term view. Personnel must be supported by the organization in achieving their long-term career goals, consistent with the prior five slices of the CoB construct. Every individual should set targets based on directional and aspirational career goals. Such an approach helps an individual to set a "North Star" in his or her career.

As technical staff navigate their respective career challenges, there is always an element of fear of belonging, or even failing to do so. Staff can fall into the trap of trying too hard to prove their competences, at the expense of misbalancing (i.e., neglecting) other aspects of the CoB construct. We want to put at ease that such feelings are quite normal, especially early in someone's career. This phenomenon has been studied under the label "impostor syndrome." Tewfik has performed an extensive study on this topic, rebalancing the negative connotation associated with "impostor syndrome" against some of the benefits [28]. An example of a benefit is early-career staff doing an assessment of their tangible contributions to assigned tasks in order to make sure that they are delivering what is expected of them, while working closely with upper management to get feedback.

We conclude this section by highlighting an excellent paper from Matsudaira titled "The Evolution of Management" [29]. Matsudaira, who has had broad experience including as an individual contributor, a technology leader at Microsoft and Amazon, and an entrepreneur, discusses a career evolution that is very consistent with the insights addressed in this chapter.

In the next section, we look at career progression as technical staff transition from individual contributors (i.e., practicing their technical depth) to roles requiring more focus on technical breadth.

12.5   AI Technical Depth and Breadth

In technology-based organizations, the competence growth from deep technical expertise (technical depth) to technical breadth is very valuable to both individuals and the organization. In the rapidly evolving field of AI, technical breadth can span knowledge—referring back to our AI system architecture shown in figure 1.3—across several of the architecture building blocks.

For example, a recent graduate in a STEM field might start gaining in-depth work experience in data science or data engineering, but as the years progress, this same staff should expand their knowledge into understanding the technical underpinning of the full end-to-end architecture. This is particularly important as staff grow in their careers from individual contributors to technical team leaders. This career progression does not mean to just superficially understand other aspects of the AI architecture. Gaining in technical breadth means acquiring the functional understanding needed to ascertain that the AI products or services meet the technical needs of the customers. It also means that AI technical leaders can effectively communicate complex technical content to their upper management, other business partners, and stakeholders (see chapter 13).

Camille Fournier, an experienced leader with the unique combination of deep technical expertise, executive leadership, and engineering management, highlights an important insight in her book *The Manager's Path*: "Managing people is difficult wherever you work. But in the tech industry, where management is also a technical discipline, the learning curve can be brutal—especially when there are few tools, texts, and frameworks to help you" [30]. Fournier also explains why it is important to gain technical breadth, as captured in the following quote:

> If you go into a tech lead role and you don't feel that you fully understand the architecture you are supporting, take the time to understand it. Learn it. Get a sense for it. Visualize it. Understand its connections, where the data lives, how it flows between systems. Understand how it reflects the products it is

supporting. It's almost impossible to lead projects well when you don't understand the architecture.

We want to restate that the career progression from having deep technical expertise to gaining technical breadth does not need to happen early in someone's career. The emphasis that we espouse is to get deep expertise over the first few years as an individual contributor. Then the staff should continue to gain technical breadth by understanding the business mission and vision, envisioned future, strategic direction, AI value proposition, and stakeholder needs and requirements, as per the discussion in chapter 9 regarding the AISDM strategic road map.

In figure 12.3, we illustrate a career progression starting with a strong foundation in a STEM field. The target years are approximate time lines—there are no rigid years, and the progression must be iterative (not sequential). During the first three to five years, technical staff should focus on applying their knowledge and gaining deep expertise. Upon demonstrating uniqueness in a technical area, a staff member is better positioned to become the go-to-person in his or her area of expertise. During these few years, the staff should also begin to acquire technical breadth, especially in technical leadership roles.

The organization should foster a cultural environment that rewards external participation in conferences, panels, professional forums, and other meetings. When technical staff are recognized both internally to the organization, as well as externally by professional peers, these endorsements become a strong testament to having mastered a technical area.

In part I of this book, we explicitly discussed the technical foundation for each of the building blocks of the AI system architecture. A staff career can progress from being a recognized expert, internally and externally to the organization, in one of those system architecture building blocks, while growing in technical breadth over the first ten years of work experience.

Since there are no guarantees in anyone's career, the structure provided in figure 12.3 should be used as a guideline for the reader based on over forty years of technical experience by one of this book's authors (David Martinez), starting as an individual contributor and progressing to leading large technical divisions.

Technical leadership roles are very valuable to organizations, but mastering such discipline comes after having opportunities to learn by doing and being fully grounded in the people-process-technology triad. One can have a mastery of a technical area, but there are always challenges in managing people and applying processes to the development and deployment of AI capabilities.

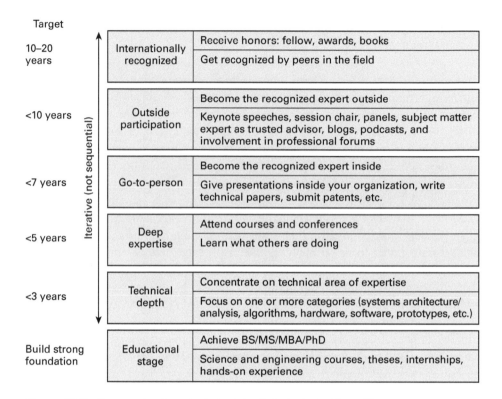

Figure 12.3   Progression from technical depth to technical breadth.

In chapter 9, we described the AISDM as a strategic framework. It is imperative to start with such a blueprint. However, as discussed in chapters 10 and 11, a strategic road map must be executed. Thus, technical expertise also includes a mastery of technical project management.

We do not want to imply that technical experts in one or more of the building blocks of the AI architecture must also need to be experts in project management. However, we do want to elevate the importance of project management to the same level as any other technical area of expertise. Technical leaders must recognize and reward those in their staff with technical project management excellence. Members of the AI team with project management expertise typically progress from a deep technical knowledge of a particular area of system implementation to a broad knowledge of the full end-to-end system.

The Harvard Business Review Press offers a number of excellent guides on managing people and project management [31, 32]. It is erroneous to think that mastering a technical area is all that is necessary to successfully develop and deploy AI capabilities. Technical expertise in managing people and associated projects is also necessary to meeting the requirements of a full end-to-end system implementation.

In the next section, we provide additional tools for managing people, including metrics for assessing progress.

12.6   Metrics for Measuring Progress and Results

For many organizations, adjusting to the digital era has required a reassessment of their performance management practices. As described by Kiron and Spindel, in 2015 IBM needed to adapt to a more current business model by putting more emphasis on AI and cloud services while preserving its high-performance culture [33]. IBM instituted a performance management system better aligned with the need for innovation and speed. The old system of yearlong cycles by management performing an annual assessment via ratings and providing feedback was outdated. IBM transitioned to a performance management system compatible with an agile methodology and an MVP performance management system, meaning assessing staff based on their performance on short and iterative deliverables.

IBM's approach of giving continuous feedback to staff is very consistent with the findings by Clear as documented in his book *Atomic Habits*, where small changes in personal habits can lead to significant gains in performance [34]. A short and constant feedback cycle, instead of the outdated yearlong assessment, lends itself well to changing or adjusting work-related performance.

Let us now look at metrics for assessing progress and results. John Doerr, the chair of Kleiner Perkins and venture capitalist, in his now-famous approach to setting up goals named "objective and key results (OKRs)," employed the OKRs from Andrew Grove, the former CEO of Intel, who originally introduced them within his company. Doerr identified the importance of goal setting that are significant, concrete, action-oriented, and inspirational [35]. He explained that OKRs are good at multiple levels, including individuals, teams, and overall organizations.

There are benefits to setting broad-level goals with the OKRs—however, these are necessary but not sufficient. Ordóñez et al. argued that goal setting without a careful assessment of side effects can actually be counterproductive [36]. Some examples of side effects can be found in the gaming of the performance system to show progress and in reducing intrinsic motivation caused by unrealistic time lines.

Priorities	Examples of skills and responsibilities
Technical excellence	• Innovation • Analytics • Hardware/software • Prototyping • Test and evaluation • Combinations of above
Team leadership	• Project leadership • Managing small teams • Multidisciplinary • Broad understanding with well-defined expertise
Moving the organization forward	• Participant in internal activities • Contributing to infrastructure advancements • Responsibility for a field site and/or testing area • Broad interaction across the organization
External involvement	• Technical/professional societies • Publications • Panel leader or member • SMEs in field of expertise

Figure 12.4   MBRs for technical staff.

We have developed and have used in practice what we called "managing-by-results (MBRs)." As shown in the table in figure 12.4, MBRs are more focused on results based on a set of priorities. The itemized list in the table includes examples of skills and responsibilities.

We explicitly emphasize four priorities: technical excellence, team leadership, moving the organization forward, and external involvement. These priorities are not prescriptive—technical staff have a wide spectrum of opportunities to exercise different sets of skills and responsibilities, with inputs from their management consistent with the overall direction of the AI project. The MBRs construct is also well matched to an agile methodology with short development and deployment cycles, meaning that staff can be assessed on how well their project performance is progressing relative to a fast-paced and iterative development approach (such as the agile methodology).

Another benefit of the MBRs for technical staff is in being able to identify how the staff are learning and advancing in their careers. For example, staff with a clear tendency toward technical depth might follow a technical senior staff path, and those with a strong interest in team leadership might follow a career path into technical leadership and management.

Moving the organization forward encourages the staff to participate in activities and areas of responsibility across the organization. Broad involvement helps junior staff to be known across the organization, not just within their working units. The last priority—external involvement—stresses the importance of getting involved outside the organization and gaining exposure to professional peers and recognition in professional societies.

A very rewarding path for technical staff is to progress to a more experienced role (sometimes referred to as "senior staff" in some organizations, while other organizations use the terminology of "principal scientist" or "principal engineer"). Leadership and management roles are also very important and highly praised in technology-based organizations. We discuss the latter in the next section. The former, technical senior staff, are often considered privileged positions in any organization, since they reflect achieving a high degree of competence in a technical area.

In figure 12.5, we showed the MBRs for senior staff. The four priorities are recognized technical excellence, technical vision and leadership, technical mentor, and renowned scholar. Note that the four priorities for senior staff evolve from the MBR priorities for technical staff, shown in figure 12.4, but they demand a much higher level of responsibility and expected set of results.

The set of examples of skills and responsibilities shown next to each priority are also very consistent with our earlier discussion on career progression illustrated in figure 12.3. For example, senior staff are expected to be able to demonstrate creativity and accomplishments in one or more fields. Another example of senior staff skill and responsibility is in leading new technical areas. Both of these representative skills come after years of experience gaining technical breadth, starting with demonstrated technical depth.

AI-driven organizations will continue to depend more and more on technical staff and senior staff performance. The field has evolved so rapidly that the skills and competences of these talented staff are and will continue to be in very high demand. However, these MBRs are guidelines. They should not be interpreted as a fixed set of prescribed areas of responsibility, but instead should be used as a tool for staff to set their priorities, and for management to provide regular and continuous feedback, ultimately helping staff succeed in their careers. There have been numerous studies showing that the inability to retain talented staff and dissatisfaction with management are most commonly caused by lack of staff feedback—management often suffers from not devoting some of their work time to providing regular feedback to subordinates.

We have found that performing an MBR review every three months (i.e., four times a year) is about the right level of occurrence (i.e., cadence). In the quarter of July through September, staff would be asked to formulate and update their MBRs according to initial

Priorities	Examples of skills and responsibilities
Recognized technical excellence	• National/international recognition • Deep subject matter expert • Leading panels, studies, task forces • History of innovation • A"go-to-person"
Technical vision and leadership	• Clear technical and system thinker • Sets technical vision • Adaptable to leading new technical areas • Demonstrated creativity and accomplishments in one or more fields
Technical mentor	• Technical impact to important national systems • Respected by peers • Able to articulate complex problems in simple terms • Track record with successful mentees
Renowned scholar	• Widely cited seminal articles • Books and/or chapters • Journal articles • Tutorials • Patents • Keynote speaker • Created or advanced a technical discipline

Figure 12.5 MBRs for senior staff.

management feedback. At the end of that quarter, in September, the direct supervisor would perform a deep MBR review with each of the staff. This review would involve going into many details and typically would last about an hour. In the subsequent quarter, October through December, management would undertake a light MBR review consisting of a course correction, if needed, and an assessment of progress and demonstrated results. In the subsequent quarter, January through March, another deep MBR review would be undertaken, again assessing progress and measuring demonstrated results. In April through June, another light MBR review would be performed, and additional discussion would take place in preparation for the start of a new cycle. These MBR review time frame would repeat every year.

When the MBRs were used as a tool to define near-term and long-term directional vectors, staff felt that their respective management were more involved and committed to achieving success. When performance problems were identified, there was enough time to take corrective action instead of waiting a full yearlong cycle.

The MBRs are also very useful in helping employees in identifying and formulating a way forward to align with their career aspirations and long-term goals via these multiple MBR reviews. Clark offered three steps for identifying staff career paths [37]:

1. Help the employee analyze patterns: Identify the employee's special aptitude and determine potential opportunities.
2. Expand their worldview: Probe into their interests and long-term career aspirations to help them set future goals.
3. Don't steer too hard: As a leader and mentor, support employees in achieving their career ambitions—not dictate them.

In this section, we have addressed technical staff and senior staff, representing two positions commonly found in many technology-based organizations. There also variants of these technical positions, such as principal scientist, principal engineer, principal senior researcher, senior research engineer, and principal staff.

In addition, Industry, in the technical sector, and R&D Laboratories, that depend on advances in technology, have established career recognition for a small percentage of their technical workforce as industry or laboratory fellows. A description of a fellow might read: "Someone who has attained the highest level of technical expertise and have achieved national or international recognition." A fellow's demonstrated set of accomplishments might include:

- World expertise in one or more important technical areas
- Demonstrated ability to develop a vision for game-changing technologies
- Track record of mentees achieving the highest levels of success
- Achieving "fellow" status in one or more recognized technical professional society

One important message that we want to leave the reader with is that all the information in this section is drawn from our experience in both the commercial sector and an R&D laboratory. There are no fixed or prescribed skills or areas of responsibilities that lead to a successful professional career. However, it is imperative that organizations make clear their set of guiding posts or directional vectors. As we move into a hybrid work environment, performance assessments with a relatively short cycles (e.g., once a quarter) will be crucial for motivating teams and retaining AI talent.

In the next section, we address the tools and techniques useful for those aspiring to be in or already are in technical leadership and management roles.

12.7   AI Leadership and Resilience

In figure 12.2, we illustrated the CoB construct. One of the important components of the construct is "management and leadership excellence." As AI staff progress in their careers, some might prefer to get into management and leadership positions. Management and leadership responsibilities are very demanding. Therefore, there is a need to be cognizant of balancing all the elements shown in the CoB construct.

We do not think of management and leadership as mutually exclusive. Peter Drucker pointed out: "Management is about doing things right; and leadership is about doing the right things." That distinction reinforces management putting more emphasis on execution and leadership putting more emphasis on strategy. However, in our experience, leaders must also attend to management roles. Technical leaders must set the AI strategy with participation and contribution from their AI teams (as discussed in chapter 9). But they must also take on management responsibilities to ascertain that the strategy is executed as laid out in the strategic development road map.

For example, in management roles, the leaders of a business unit or the total enterprise must keep close tabs on profit and loss (with input from the chief financial officer) in profit-making organizations, overall operations (with inputs from the chief operating officer), revenues backlog, costs of goods and services, personnel, stockholder value, and many other day-to-day business functions and metrics. However, transitioning from individual contributor roles to leadership roles brings a new set of required skills. Watkins highlighted some of these skills in what he refers to as "seismic shifts," given the importance and significant differences when compared to individual contributor roles [38]. Watkins discusses seven seismic shifts:

1. From specialist to generalist
2. From analyst to integrator
3. From tactician to strategist
4. From bricklayer to architect
5. From problem solver to agenda setter
6. From warrior to diplomat
7. From supporting cast member to a lead role

This set of shifts is very consistent with Jack Welch's quote: "Before you are a leader, success is all about growing yourself. When you become a leader, success is all about growing others." In management and leadership roles, you fully depend on your direct reports for your and the organization's success. Therefore, it is imperative that those in management and leadership roles spend valuable time mentoring their direct reports and serving as role models to others in the organization. Great leaders also serve the role of on-the-job teacher to their direct reports by showing and providing learning-by-doing opportunities, as discussed by Finkelstein [39].

Leadership roles and responsibilities that are not well balanced can put a lot of stress on the leader and the team, potentially leading to poor results. It is a complete myth that those at the top got it figured out—in fact, one of the common weaknesses of early career leaders is to behave as if they have all the answers. Ancona et al. [40] stressed the importance of accepting weaknesses as leaders to avoid burnout and of relying on their direct reports to amplify the collective strengths of the leadership team.

As discussed earlier, MBRs is a metric that is also relevant to leaders, but with a different set of priorities. In figure 12.6, we identify four priorities and provide examples of skills and responsibilities next to each of these pillars:

1. Vision: Define the overall AI strategic direction for the business or organization (i.e., plan the work).

Priorities	Examples of skills and responsibilities
Vision	• Deep familiarity with national and international AI strategic roadmaps • Define organization AI strategic roadmap • Update strategy based on changes in AI national and international priorities
External relationships	• Strong involvement with users and customers • Formulate/build new initiatives/programs • Coalitions across the organization • Management and technical recognition outside the organization • Table manners
Internal execution	• Empower staff to achieve timely execution • Manage contract deliverables • Prioritize • Avoid micromanagement
Mentoring and recruiting	• Recruit talent well matched to organization environment • Provide constant feedback • Mentor staff *(very important early in the staff career)*

Figure 12.6   MBRs for technical leaders.

2. External relationships: Build coalitions external to the business unit (e.g., internal to the organization), or outside the organization at large.

3. Internal execution: Execute the AI strategic road map (i.e., work the plan).

4. Mentoring and recruiting: Assemble a multidisciplinary and cross-functional team, and retain the AI talent.

Vision is an all-encompassing priority focused on defining the strategic plan consistent with the guidance under the AISDM. Again, the examples of skills and responsibilities are not prescriptive but are representative of a small sample set of what is expected of AI leaders. One example of skills, under external relationships, that often gets ignored is "table manners." Inexperienced leaders fail to pay attention to table manners, especially in technology-based organizations (e.g., by overwhelming customers with technical details instead of focusing on the AI business value offered to stakeholders). Effective leaders are able to communicate with stakeholders in terms that they can understand about the AI value proposition (see chapter 13).

Regarding internal execution, it is of paramount importance that leaders execute the plan, working closely with the stakeholders. As David Packard, a cofounder of Hewlett-Packard, famously said, "More organizations die of indigestion than starvation." Packard brought up a pitfall of business leaders who put all their attention on winning contracts and neglect execution by losing focus, trying to do too much. Eventually, the organization becomes irrelevant.

In earlier sections, we discussed the importance of culture and its impact on fostering an innovative team environment. Mentoring and recruiting are the responsibilities of leaders supported by others in the organization. Equal attention must be paid to this priority compared to other leaders' priorities, especially in a highly sought-after AI talent environment.

This set of MBRs for leaders also helps in serving as guideposts for reaching outside any leader's comfort zone. As illustrated in figure 12.7, under the so-called donut paradigm, leaders grow by taking initiatives and navigating from the well-defined to the more ambiguous scenarios. This paradigm was adapted from the teachings of David Briggs, director emeritus of the MIT Lincoln Laboratory. The right side of figure 12.7 provides more opportunities for a leader to define his or her own destiny.

After its establishment as a dominant technology-based organization, and in response to its employee inquiries, Larry Page and Sergey Brin considered running Google as a very flat organization. However, they commissioned a project called Project Oxygen to determine the value provided by having engineering managers, adding layers between the staff and the upper management. Garvin published Google's findings resulting

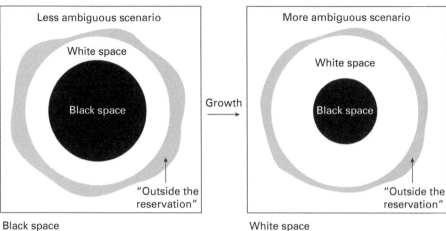

Black space
• Well-defined
• Clearly articulated by the C-suite or
  business unit leaders
• Measurable; can tell if you have done it

White space
• More ambiguous but higher paid grade
• Can't tell you what to do but expect
  you do it as best it can be done
• Important to learn how to navigate
• Opportunity to define your own destiny

Figure 12.7   Leadership growth from the well-defined to the more ambiguous scenario.

from extensive analysis of data from exit interviews, employee surveys, and performance reviews [41]. The company identified eight behaviors that a top-performing manager exhibited:

• Is a good coach
• Empowers the team and does not micromanage
• Expresses interest in and concern for team members' success and personal well-being
• Is productive and results oriented
• Is a good communicator; listens and shares information
• Helps with career development
• Has a clear vision and strategy for the team
• Has key technical skills that help him or her advise the team

These behaviors are very consistent with and complementary to the example skills and responsibilities shown in figure 12.6. We also want to expand on the last behavior (having key technical skills that help him or her advise the team), since technical leaders should not let their technical knowledge atrophy. As we emphasized in section 12.5,

Table 12.2   The five components of emotional intelligence (EQ) at work

	Definition	Hallmarks
Self-awareness	• The ability to recognize and understand your moods, emotions, and drives, as well as their effects on others	• Self-confidence • Realistic self-assessment • Self-deprecating sense of humor
Self-regulation	• The ability to control or redirect disruptive impulses and moods • The propensity to suspend judgment, to think before acting	• Trustworthiness and integrity • Comfort with ambiguity • Openness to change
Motivation	• A passion to work for reasons that go beyond money or status • A propensity to pursue goals with energy and persistence	• Strong drive to achieve • Optimism, even in the face of failure • Organizational commitment
Empathy	• The ability to understand the emotional makeup of other people • Skill in treating people according to their emotional reactions	• Expertise in building and retaining talent • Cross-cultural sensitivity • Service to client and customers
Social skills	• Proficiency in managing relationships and building networks • An ability to find common ground and build rapport	• Effectiveness in leading change • Persuasiveness • Expertise in building and leading teams

technical leaders must progress in their careers by evolving from mastering a technical area into augmenting their AI breadth of knowledge.

Another set of skills, demonstrated by effective leaders, is in their paying attention to and putting into practice emotional intelligence (EQ). Daniel Goleman, renowned psychologist, author, and lecturer, has written extensively on the subject of EQ, and how it factors into being an effective leader [42]. In table 12.2, we summarize Goleman's five components of EQ.

Goleman's five components of EQ, the specific definitions, and the hallmarks of effective leaders are so important that we strongly recommend that all leaders look back regularly at their effectiveness in these areas at a frequency consistent with the previous discussion of MBR quarterly reviews. In today's hybrid work environment and short AI product or service deliverable cycles, it is easy for leaders to minimize how critical these soft skills are. In addition to strategic and tactical skills, EQ is a force multiplier as a hallmark of leadership.

So far, we have discussed traits of leadership excellence. Now, we want to turn our attention to leadership resilience. As we pointed out earlier, AI leaders are pulled in many directions while at the same time trying to stay abreast of the fast-evolving AI field. Leadership resilience is a trait that helps in navigating these multiple day-to-day pressures.

There is a large body of excellent work dating back to Bennis and Thomas [43]. Their premise is that all leaders learn from and put personal value in major crucibles that they have experienced in their lives, careers, and upbringing. Leaders that recognize the influence that crucibles have in their leadership are better prepared to deal with unexpected pressures. Former US president Franklin D. Roosevelt had a quote that applies well in this context: "A smooth sea never made a skillful sailor."

Personal crucibles form the bedrock of another scholarly work by George et al. [44]. It is well documented that the best leaders behave and put into practice, in their day-to-day roles, attitudes, behaviors, empathy, and respect to others in a real, authentic way. Direct reports and staff can easily tell if a leader is not genuine and lacks authenticity. Leadership authenticity is becoming a high-priority training topic at many companies [45], given the many opportunities these days to cause embarrassment to organizations and make them legally responsible for any issues.

We find the teachings by John C. Maxwell, in his book *The 21 Indispensable Qualities of a Leader*, to be an excellent guide to a set of very clear leadership principles [46]. We believe, as Maxwell discusses, that leadership will evolve more and more toward a servant leadership style, meaning always putting others first. Finally, no one is born a leader, so as Maxwell also highlights, "to keep leading, keep learning."

In the next section, we address one of the most crucial skills that any leader must have—the ability to mentor, network inside and outside the organization where they work, and recruit effectively—to stay competitive in a highly demanding AI field.

12.8   Mentoring, Networking, and Recruiting AI Talent

In the CoB construct, we brought up mentoring and networking as areas that both mentors and mentees must balance among other important priorities. Mentorship can be informal or formal, but in either case, it forms part of the fabric of any organization. There are many definitions and understandings of what mentoring is or is not. We anchor our discussion in this section on the definition by Irby et al. [47]:

> *Mentoring:* A mutual learning partnership in which individuals assist each other with personal and career development through coaching, role modeling, counseling, sharing knowledge, and providing emotional support.

Kathy Kram, a professor emeritus from Boston University, is one of the best-known scholars in management and organizations, and she has published extensively on the topic of mentorship. In her teachings, she clarifies that the role of a traditional mentor as someone who "is an individual higher up in the organization or profession who provides developmental support including guidance, coaching, counseling, and friendship to a protégé" is no longer well matched to today's needs. Instead, Kram and her colleagues speak to the need for a developmental network, more of a constellation of mentors, coaches, role models and other professionals, which takes the form of a personal board of advisors [48, 49]. In their definition, a developmental network consists of "groups of people who take an active interest in and action toward advancing a protégé's career."

As discussed by Shen et al., a personal board of advisors consists of a network of individuals assisting a mentee in personal and career development [49]. The board of advisors include the following categories:

- Personal support:
  - Personal guide: Someone with whom a mentee has had a supportive relationship in the past (presently likely to have less frequent interactions)
  - Personal advisor: A person offering frequent interactions that provide a strong friendship
- Career support:
  - Full-service mentor: Also overlaps with personal support and is considered a true mentor, providing a range of career and personal support
  - Career advisor: Someone offering high-level, frequent, and active interactions providing support regarding job and professional needs
  - Career guide: Someone offering limited interactions due to a career change, for example, but still influential on the mentee's board of advisors
- Role model: Someone whom the mentee sees as a source of inspiration (interaction is typically passive and/or one-way)

This board of advisors concept is very powerful since it offers the mentee an opportunity to take advantage of and gain from other points of view inside and outside his or her organization. Dominguez and Kochan did an in-depth review of 588 articles on the subject of developmental relationships [50]. Under the rubric of mentorship, there are multiple contributors to the protégé's personal and career development, including coaches, advisors, counselors, allies, and sponsors. For example, a mentee's sponsor might be someone with influence, identifying career advancement opportunities for the mentee, but not necessarily a direct supervisor.

Cross et al. have studied factors that influence staff retention [51]. Based on data from the consulting company Booz-Allen-Hamilton, large networks did not predict longer tenure; the extent to which people built ties outside their operating units and in the broader organization mattered most.

Factors influencing retention have many dimensions. One of these was addressed by Slepian in terms of the need to build an environment of belonging [52]. The existence of a diverse, equitable, and inclusive environment is critical for any organization. However, Slepian found that in an organization with employees from multiple racial groups, genders, and sexual orientations, as well as people with different ideologies, cultural backgrounds, education, and other characteristics, retention is best achieved when organizations foster an environment with a sense of belonging (i.e., where employees can be themselves). Management should be measured, as part of their MBRs, on how well they put an identity-safe environment into practice.

Organizations must adapt and realize that talent comes from all walks of life. AI organizations will be best suited to advanced AI capabilities when they are able to introduce diversity of thought, garnered from a broad spectrum of talented personnel, not just with technical expertise but also with diverse ranges of interests. One of the best quotes applicable to this topic is by Alvin Toffler (the American writer and futurist who wrote the landmark 1970 book *Future Shock*): "The illiterate of the 21st century will not be those who cannot read or write, but those who cannot learn, unlearn, and relearn." D. Martinez first learned of Toffler's famous quote from Alex Lupafya, the deputy chief diversity and inclusion officer at MIT Lincoln Laboratory.

Fournier provides very practical and sound advice to mentors and mentees. The tips and techniques outlined here are adapted from her book *The Manager's Path* [30].

The mentor relationship has the following characteristics:

- The mentor provides guidance on the spoken and unspoken rules of the road in the organization.
- The best mentoring relationships evolve naturally and in the context of larger work.
- It is important for the mentor to understand the mentee's expectations.
- It is OK to say no if the mentor is not a good match to the mentee's expectations.
- Mentors should be recognized first-class citizens in the organization.
- Avoid mentor-mentee echo chambers with similar backgrounds (e.g., gender, education, ethnicity, beliefs).

- The best mentors are people who are further along in their mastery of the particular job skills that the mentee is trying to develop.

The mentee relationship has the following characteristics:

- It is important for the mentee to understand the mentor's expectations.
- The mentee needs to clearly identify what he or she wants to get out of the developmental relationship.
- The mentee should come to the relationship prepared to take full advantage of it.
- For both the mentor and mentee, avoid the tendency to lecture and debate (i.e., be open minded).

Many colleges and universities are making a concerted effort to reach parity among the student body and administrative staff. Fortunately, these efforts are paying off since in many educational institutions, there are as many women as men. Unfortunately, that is not the case in the commercial and nonprofit sectors. Women currently hold only 10 to 15 percent of senior leadership (C-suite) positions in corporate America [53]. Men mentoring women and women mentoring men will help each to recognize and treat the other as equal, not just in pay but also in positions of authority.

Up to this point in this section, we have been discussing primarily mentoring and networking in an organization. We conclude the section by focusing on recruiting. We left recruiting as the third topic because recruiting talented staff is an initial necessary step, but the organization must also put equal effort into building and maintaining a culture of mentorship and in giving staff opportunities to network inside and outside the organization.

In the experience of one of the book authors (David Martinez), one of the best pieces of advice he received was from a superb mentor who said, "I cannot guarantee you employment, but I can guarantee you employability." What this sage advisor meant is that leaders must make it a priority to make the organizational culture, technical projects, and work environment outstanding so employees would strive to do their best, resulting in retention. It is an absolute mistake to fear losing people because of their likelihood of being marketable, and therefore giving them only limited opportunities for growth within the organization and exposure outside the organization.

We conclude this section by leaving the reader with a set of tips for recruiting AI talent:

- Assess technical depth and breadth. For breadth, ask questions about outside interests.

- Make sure that the candidate's core values match the corporate culture.
- Look at the individual as a potential long-term addition to the organization.
- Employ the AI strategic road map, discussed in chapter 9, to clearly describe the vision and strategic direction of the organization.
- Emphasize the organization's support infrastructure.
- Clearly articulate how employees' performance will be assessed via the MBRs.

In the next section, we provide additional tools and techniques for sustaining high-performance teams.

12.9 Sustaining High-Performance Teams

In chapter 2, we presented the definition of "AI systems engineering" that we have adopted throughout this book. It is worth repeating that definition to stress that the successful development and deployment of AI capabilities, from the perspective of a systems engineering approach, requires full attention to all three elements of the people-process-technology triad:

> *AI systems engineering:* "An integrated set of AI architecture elements, subsystems, or assemblies that accomplish a defined objective. These elements include *technologies* as enablers of AI products (hardware, software, firmware) or AI services, adhering to a set of *processes*, and using *people*, information, techniques, infrastructure, and other support elements. The integrated AI system must meet a minimum set of threshold requirements, while attending to FASTEPS issues."

Thus, the prior sections of this chapter focused on tools and techniques to help AI practitioners to foster an innovative team environment and to motivate their colleagues (i.e., leaders and technical staff) to perform at their best. We conclude this discussion by exploring how to sustain these innovative and high-performance teams.

Iansiti and Nadella identified three areas that technology-driven organizations must enable in democratizing transformation: capabilities, technology, and architecture. The first of these is very relevant to the discussion in this chapter [54]. Iansiti and Nadella enumerated the capabilities needed in a tech-intense environment:

- Organizational culture
- Training and development
- Low-code/no-code (LCNC) tools
- Agile teams

Table 12.3 Lencioni's five dysfunctions of a team

Dysfunction	Typical Characteristics
Inattention to results	- Status and ego - Putting individual needs above the collective goals of the team
Avoidance of accountability	- Low standards - Lacking commitment, teams hesitating to call peers on lack of action
Lack of commitment	- Ambiguity - Lack of openness, leading to lack of buy-in
Fear of conflict	- Artificial harmony - Leads to veiled discussions and guarded comments
Absence of trust	- Invulnerability - Inability to recognize mistakes and weaknesses

- Organizational architecture
- Citizen developers using LCNC, as discussed in chapter 11
- Product management

Many of these capabilities here have been discussed in this chapter and in earlier chapters. A critical foundation of organizational culture, necessary to sustain high-performance teams, is trust. Trust is often assumed to exist, but most teams fail to deliver because of a lack of trust. Patrick Lencioni, one of the most respected authorities on team management, is best known for his work on factors leading to dysfunctional teams. He offers approaches to improve team dynamics [5] and identifies five dysfunctions in typical team dynamics (shown in table 12.3).

All five dysfunctions are commonly found in poorly operating teams. However, absence of trust is typically where everything starts to unravel in a team dynamic. It has a domino effect that engenders subsequent dysfunctions. Leaders must stay focused and pay attention to the presence of one or more of these dysfunctions.

Covey and Merrill have codified why trust is so important in teams—at any level—from small technical teams to the core leadership team in a business unit, all the way up to the C-suite. They argue that there is a crisis of trust and make the following compelling argument: "Trust is the most fundamental component to all relationships, including those within and between businesses. This includes relationships between coworkers in a company, between a leader and his/her subordinates, and between stakeholders"

Table 12.4 Behaviors highly valued by staff members

Role	Values
Self	- I have all the freedom I need to decide how to get the work done. - No matter what else is going on around me, I can stay focused on getting my work done. - I always believe that things are going to work out for the best.
Team leader	- My team leader tells me what I need to know before I need to know it. - I trust my team leader. - I am encouraged to take risks.
Senior leader	- Senior leaders are one step ahead of events. - Senior leaders always do what they say they are going to do. - I completely trust my company's senior leaders.

[55]. Trust is a force multiplier in achieving speed and reducing the costs in the development and deployment of AI capabilities.

Buckingham succinctly summarized the behaviors of individuals, team leaders, and senior leaders that staff value, contributing to sustaining high-performance teams [56]. Table 12.4 shows what staff value the most from their individual selves, their team leaders, and senior leaders [56].

The areas given in this table are most valued by staff and are very consistent with our discussion in this chapter. We want to amplify that by having an AI strategic road map (i.e., a blueprint), team leaders are better prepared to tell the staff what they need to know before they need to know it. Similarly, senior leaders are also able to be one step ahead of events and always do what they say they are going to do when guided by the AISDM framework.

In the next two sections, we summarize the main takeaways from the chapter and provide a set of exercises.

12.10 Main Takeaways

Attention to people cannot be underestimated as a way to improve the chances of successfully developing and deploying AI capabilities. Four of our ten guidelines for developing and deploying AI capabilities emphasize people:

- Strengthen the AI team by fostering internal and external relationships.
- Build a multidisciplinary and diverse team with complementary skills.

- Provide measurable objectives while mentoring AI talent (i.e., use MBRs as the metric).
- Continue to expand AI team skills as the future of work evolves.

A strong organizational culture is important for following these guidelines. An organizational culture drives the ability for leaders and technical AI practitioners to achieve alignment between the overall organization's strategy and the AI strategic road map, as described in our AISDM framework.

In addition to a strong organizational culture, AI-based organization must be internally organized to facilitate innovation, including having talented staff with a sense of urgency, while cementing coalitions inside and outside the organization. Such efforts must be consistent with the AI strategic road map.

At the macro-level, we also reviewed the economic impact that AI can have on the labor force, as seen through the perspective of several well-recognized economists. This worldwide economic perspective, as well as the impact that narrow AI is having and will continue to have on our lives, motivate talented staff to work on advancing AI products and services. Historically, GPTe can serve as the engine of growth. However, getting to a point when GPTe is used continually takes many years, since it depends on the existence of a downstream ecosystem.

After our discussions, at a macro-level, we presented a useful construct called the CoB, which has stood the test of time as a way to assess near-term and long-term priorities. The six slices of the CoB are technical depth, management and leadership excellence, mentoring and networking, personal health and wellness, family, and career. We emphasized that no one—regardless of where they are in their careers and their years of experience—can perform all these elements perfectly. The CoB is most useful when leadership and staff recognize that it is important to balance these priorities, ultimately affecting job performance, family, and personnel satisfaction.

As staff gain in-depth knowledge in a technical area of the AI work, it is advisable to continue to broaden their understanding of the full end-to-end AI architecture. By doing so, the staff are much better prepared to understand the technical underpinnings of each of the architecture building blocks. This transition from technical depth to technical breath does not imply losing technical prowess. In fact, it means the opposite. Employees must maintain their technical acumen while learning about other aspects of the AI architecture in order to be most useful to the organization as technical leaders.

Because the field of AI is evolving so rapidly, it is not advisable to adhere to a yearly staff performance evaluation. Instead, organizations should institute an evaluation cycle consistent with short and iterative deliverables of AI capabilities—either at the R&D

stage or in production. We discussed the value of setting high-level goals called OKRs, as formulated by Doerr.

Next, we went a step further by emphasizing the use of MBRs, which can be used to set up the necessary skills and responsibilities. MBRs are applicable to staff, senior members of the staff, and leadership. Of course, the specifics of each of the respective MBRs are different depending on the respective roles and responsibilities.

When the MBRs were used as a tool to define near-term and long-term directional vectors, staff felt that their respective management were more involved and committed to achieving success. When performance problems were identified, there was enough time to take corrective action instead of waiting through the full yearlong cycle.

MBRs for staff were divided into four categories: technical excellence, team leadership, moving the organization forward, and external involvement. For senior staff, and evolving from the staff MBRs, we identified four categories: recognized technical excellence, technical vision and leadership, technical mentor, and renowned scholar.

Leadership and management roles are very demanding. However, just as it is critical to perform regular assessments of the progress and results of staff and senior staff, it is equally important to assess the performance and progress in delivering results of leaders and managers via MBRs on a regular basis. Again, we treat leadership as also requiring management in the successful execution of the AI strategic plan.

The leadership MBRs also consist of four categories: vision, external relationships, internal execution, and mentoring/recruiting. The first two—vision and external relationships—are more outward-facing skills and responsibilities. The latter two—internal execution and mentoring/recruiting—are more inward focused.

Another hallmark of effective leaders is in putting into practice EQ. Goleman identified five components of EQ: self-awareness, self-regulation, motivation, empathy, and social skills. These soft skills also should be assessed as part of the MBR reviews. For example, empathy is a very important skill to have as a leader, especially when managing multicultural teams. Leaders should learn about their direct reports' backgrounds, places of origin, interests, and motivations in the near and long term.

In the CoB construct, we brought up mentoring and networking as areas that both mentors and mentees must balance among other important priorities. Mentorship can be informal or formal, but in either case, it forms part of the fabric of any organization. A very useful definition of mentoring was published by Irby et al. [47]:

Mentoring: "A mutual learning partnership in which individuals assist each other with personal and career development through coaching, role modeling, counseling, sharing knowledge, and providing emotional support."

We like this definition because it brings out the need for a board of advisors for more of a developmental network to learn from others' experiences, knowledge, backgrounds, and counseling, as discussed by Kram and colleagues. Such a board might consist of personal support, career support, and role models, assisting in mentees' personal and career development.

In addition to the many tips and techniques offered in the prior sections, we wanted to address the importance of avoiding dysfunctional teams. Lencioni identified five main reasons behind dysfunctional teams:

- Inattention to results
- Avoidance of accountability
- Lack of commitment
- Fear of conflict
- Absence of trust

Leaders must observe the team dynamics for the presence of one or more of these factors. The earlier these are addressed, the more likely that solutions can be found and implemented. Absence of trust is last on the list because—if seeing as a pyramid, from the bottom to the top of the list—that is likely to be the cause of the unraveling of the other four factors.

In this chapter, we have provided a set of guidelines, tools, and techniques that can be useful when addressing people, a very important element of achieving success in the development and deployment of AI products and services. AI leadership and talented staff must be fundamental contributors to the successful execution of the AISDM framework.

12.11 Exercises

1. Ed Schein categorizes culture into three levels: artifacts, espoused beliefs and values, and basic underlying assumptions. Elaborate on what each of these factors means and provide examples.

2. Organizational values are actions taken based on a set of organizational cultural beliefs.
   a. True
   b. False

3. A functional organizational structure consists of business units organized according to domain of expertise—such as design, hardware engineering, software, marketing, operations, and the law.
   a. True
   b. False

4. Provide examples of general purpose technologies (GPTes) and their impact on worldwide economic value.

5. List the six areas within the CoB construct and provide a short explanation of each.

6. Explain why it is important to maintain technical depth in one or more areas of the AI system architecture, while gaining expertise in understanding the technical underpinnings of other building blocks of the AI architecture.

7. OKRs are useful because they do not need to be significant, concrete, action-oriented, or inspirational.
   a. True
   b. False

8. The four categories of a staff MBRs are technical breadth, team leadership, mentorship, and (when applicable) technical depth.
   a. True
   b. False

9. Explain the meaning of the four senior staff MBRs: recognized technical excellence, technical vision and leadership, technical mentor, and renowned scholar.

10. The four categories of MBRs as they apply to AI leaders are vision, external relationships, internal execution, and mentoring/recruiting.
    a. True
    b. False

11. List and provide a short description of each of Goleman's EQ components.

12. What is the benefit of a developmental network, in contrast to a traditional mentor?

13. Explain how the AI strategic road map can be used when recruiting AI staff.

14. Describe briefly the five factors commonly found in dysfunctional teams according to Lencioni.

15. Elaborate on why trust is key to sustaining high-performance teams.

16. Can you envision how AI technologies can help a human resources unit in tracking and promoting career growth of staff and leaders? Explain your approach.

17. How would you modify the MBRs for staff and leaders to include the FASTEPS principles?

18. The CoB is a construct to help personnel balance the demands of their jobs and personal lives. How do you envision future AI advances are likely to affect these categories?

12.12   References

1. Hupfer, S., *Talent and workforce effects in the age of AI*. 2020, Deloitte Insights.
2. Zwetsloot, R., R. Heston, and Z. Arnold, *Strengthening the US AI workforce*. 2019, Center for Security and Emerging Technology.
3. Elliott, B., S. Subramanian, and H. Kupp, *How the future works: Leading flexible teams to do the best work of their lives*. 2022: John Wiley & Sons.
4. Choudhury, P. R., Our work-from-anywhere future. *Defense AR Journal*, 2021. 28(3): 350–350.
5. Lencioni, P., *The five dysfunctions of a team: A leadership fable*. 2002, Jossey-Bass: A Wiley Company.
6. Tharp, B. M., Defining "culture" and "organizational culture": From anthropology to the office. *Interpretation: A Journal of Bible and Theology*, 2009. 2(3): 1–5.
7. Schein, E. H., *Organizational culture and leadership*. Vol. 2. 2010, John Wiley & Sons.
8. Daimler, M., Why great employees leave "great cultures." *Harvard Business Review*, 2018. 11.
9. Crossan, M., W. Furlong, and R. Austin, Make leader character your competitive edge. 2022, *MIT Sloan Management Review*.
10. Ransbotham, S., F. Candelon, D. Kiron, et al., *The cultural benefits of artificial intelligence in the enterprise*. 2021, MIT Sloan Management Review and Boston Consulting Group.
11. Westerman, G., D. L. Soule, and A. Eswaran, Building digital-ready culture in traditional organizations. *MIT Sloan Management Review*, 2019. 60(4): 59–68.
12. Kotter, J. P., Accelerate! *Harvard Business Review*, 2012. 90(11): 44–52, 54.
13. Podolny, J. M. and M. T. Hansen, How Apple is organized for innovation. *Harvard Business Review*, 2020. 98(6): 86–95.
14. Ross, A., *The industries of the future*. 2016, Simon & Schuster. 320.
15. Kanter, R. M., Innovation: The classic traps. *Harvard Business Review*, 2006. 84(11): 72–83, 154.
16. Christensen, C. M., *The innovator's dilemma: When new technologies cause great firms to fail*. 2013, Harvard Business Review Press.

17. Drucker, P. F., The discipline of innovation. *Harvard Business Review*, 2002. 80(8): 95–102.

18. Agrawal, A., J. S. Gans, and A. Goldfarb, Artificial intelligence: The ambiguous labor market impact of automating prediction. *Journal of Economic Perspectives*, 2019. 33(2): 31–50.

19. Bresnahan, T. F. and M. Trajtenberg, General purpose technologies "Engines of growth"? *Journal of Econometrics*, 1995. 65(1): 83–108.

20. Acemoglu, D. and P. Restrepo, Artificial intelligence, automation, and work, in *The economics of artificial intelligence: An agenda*, 197–236. 2018, University of Chicago Press.

21. Autor, D. H., D. A. Mindell, and E. Reynolds, *The work of the future: Building better Jobs in an age of intelligent machines*. 2022, MIT Press.

22. Brynjolfsson, E., *The Turing Trap: The promise & peril of human-like artificial intelligence*. Daedalus, 2022. 151(2): 272–287.

23. Brynjolfsson, E., A Collis, W. E. Diewert, et al., *GDP-B: Accounting for the value of new and free goods in the digital economy*. 2019, National Bureau of Economic Research.

24. Babic, B., D. L. Chen, T. Evgeniou, and A.-L. Fayard, A better way to onboard AI. *Harvard Business Review*. 2021.

25. Ransbotham, S., D. Kiron, F. Candelon, et al., Achieving individual—and organizational—value with AI. 2022, *MIT Sloan Management Review*.

26. George, B., Developing mindful leaders for the C-suite. *Harvard Business Review*, 2014: 21–38.

27. Khazan, I. Z., *The clinical handbook of biofeedback: A step-by-step guide for training and practice with mindfulness*. 2013, John Wiley & Sons.

28. Tewfik, B. A., The impostor phenomenon revisited: Examining the relationship between workplace impostor thoughts and interpersonal effectiveness at work. *Academy of Management Journal*, 2022. 65(3): 988–1018.

29. Matsudaira, K., The evolution of management. *Communications of the ACM*, 2019. 62(10): 42–47.

30. Fournier, C., *The manager's path: A guide for tech leaders navigating growth and change*. 2017, O'Reilly Media.

31. Goleman, D., J. R. Kazenbach, W. C. Kim, and R. Maurborgne, *HBR's 10 must reads on managing people (with featured article "leadership that gets results," by Daniel Goleman)*. 2011, Harvard Business Press.

32. Review, H. B., *HBR guide to project management*. 2012, Harvard Business Press.

33. Kiron, D. and B. Spindel, Rebooting work for a digital era. *MIT Sloan Management Review*, 2019: 0_1–10.

34. Clear, J., *Atomic habits: An easy & proven way to build good habits & break bad ones*. 2018, Penguin.

35. Doerr, J., *Measure what matters: How Google, Bono, and the Gates Foundation rock the world with OKRs*. 2018, Penguin.

36. Ordóñez, L. D., M. E. Schweitzer, A. D. Galinsky, and M. H. Bazerman, Goals gone wild: The systematic side effects of overprescribing goal setting. *Academy of Management Perspectives*, 2009. 23(1): 6–16.

37. Clark, D., How to help an employee figure out their career goals, *Harvard Business Review* 2022. https://hbr.org/2022/10/how-to-help-an-employee-figure-out -their-career-goals.

38. Watkins, M. D., How managers become leaders. *Harvard Business Review*, 2012. 4.

39. Finkelstein, S., The best leaders are great teachers. *Harvard Business Review*, 2018. 96(1): 142–145.

40. Ancona, D., T. W. Malone, W. J. Orlikowski, and P. M. Sengé, In praise of the incomplete leader. *Harvard Business Review*, 2007. 85(2).

41. Garvin, D. A., How Google sold its engineers on management. *Harvard Business Review*, 2013. 91(12): 74–82.

42. Goleman, D., *What makes a leader?* Harvard Business Review Classics. 2017, Harvard Business Press.

43. Bennis, W. G. and R. J. Thomas, Crucibles of leadership. *Harvard Business Review*, 2020. 80.

44. George, B., P. Sims, A. N. McLean, and D. Mayer, Discovering your authentic leadership. *Harvard Business Review*, 2007. 85(2): 129.

45. Ibarra, H., The authenticity paradox. *Harvard Business Review*, 2015. 93(1/2): 53–59.

46. Maxwell, J. C., *The 21 indispensable qualities of a leader: Becoming the person others will want to follow*. 2007, HarperCollins Leadership.

47. Irby, B. J., J. N. Boswell, L. J. Searby, et al. (eds.), *Wiley international handbook of mentoring*. 2020, John Wiley & Sons.

48. Murphy, W. and K. Kram, *Strategic relationships at work: Creating your circle of mentors, sponsors, and peers for success in business and life*. 2014, McGraw Hill Professional.

49. Shen, Y., R. D. Cotton, and K. E. Kram, Assembling your personal board of advisors. 2015, *MIT Sloan Management Review*.

50. Dominguez, N. and F. Kochan, Defining mentoring: An elusive search for meaning and a path for the future, in *Wiley international handbook of mentoring: Paradigms, practices, programs, and possibilities*, 2020, John Wiley & Sons, 1–18.

51. Cross, R., T. H. Davenport, and P. Gray, Collaborate smarter, not harder. *MIT Sloan Management Review*, 2019. 61(1).

52. Slepian, M., Are your D&I efforts helping employees feel like they belong? *Harvard Business Review*. https://hbr.org/2020/08/are-your-di-efforts-helping -employees-feel-like-they-belong.

53. Johnson, W. B. and D. Smith, *Athena rising: How and why men should mentor women*. 2016, Routledge.

54. Iansiti, M. and S. Nadella, Democratizing transformation. *Harvard Business Review*, 2022. 100(5–6): 42–49.

55. Covey, S. R. and R. R. Merrill, *The speed of trust: The one thing that changes everything*. 2006, Simon and Schuster.

56. Buckingham, M., *Leaders who make a difference*. 2021, HBRLive.

13

Communicating Effectively

Your success in life will be determined largely by your ability to speak, your ability to write, and the quality of your ideas, in that order.
—Patrick Winston, Ford Professor of Artificial Intelligence and Computer Science at the Massachusetts Institute of Technology (MIT)

The ability to communicate well is a fundamental skill that is essential for human beings to express and understand their thoughts, ideas, and feelings. Good communication is equally important for the success of any organization, especially technology-based organizations where clarity, conciseness, and articulation of complex subjects in simple, understandable terms are crucial to the success of the business.

Effective communication within an organization helps align individuals around a shared mission and vision, prevents and resolves misunderstandings and conflicts, and builds trust and relationships between employees and management. In turn, this can lead to increased productivity, better decision-making, and improved teamwork and morale. In an artificial intelligence (AI) organization, effective communication can also lead to effective delivery of AI products and services consistent with the promised deliverables to the stakeholders.

The rapid adoption of technology across industries is transforming the world of work. The development of automation, enabled by robotics and AI, offers the potential for higher productivity; however, it also raises questions about the impact on jobs, the need for workforce upskilling and reskilling, and the nature of work itself [1]. While the demand for technical skills, essential to supporting organizations in their digital transformation, will continue to grow, employers are also putting greater emphasis on so-called soft skills such as social, emotional, and higher cognitive abilities. A 2021 study by McKinsey & Company identified a set of fifty-six foundational skills essential for adapting to the rapidly evolving future of work [2]. These skills were grouped into four broad categories: cognitive, digital, interpersonal, and self-leadership. Communication was a key skill group within the cognitive category.

Communication can be split into two major categories—persuasive and instructive. Communication for persuasion involves the use of language and other communication tools to convince the audience to take a specific action, adopt a certain viewpoint, or simply think favorably of you, your idea, or your product [3]. This type of communication often includes elements such as storytelling, emotional appeals, and logical arguments to persuade the audience.

On the other hand, communication for instruction involves the use of language and other communication tools to impart knowledge or information to the audience. The goal of this type of communication is to help the audience understand a concept, process, or idea, and such communication may include elements such as clear explanations, examples, storytelling, and visuals to support understanding. While both forms of communication may overlap in certain situations, the primary difference is in the goal of the communication: persuasion aims to influence the audience to accept a particular belief or argument, while instruction aims to help the audience understand a topic [4]. Both types of communication can be important in different contexts as a practitioner. In this chapter, we focus more on the elements of persuasive communication, but also highlight some heuristic rules for instructive communication. For a comprehensive review of tools and techniques for effective instructive communication, refer to Winston's book *Make It Clear: Speak and Write to Persuade and Inform* [3].

Whether you are trying to persuade a group of colleagues to support your idea or are delivering a technical talk to share the results of your experiments, effective communication can improve your relationships—for example, in building a good rapport with your AI stakeholders and advancing your career. There are various instances in which effective communication will be crucial to your professional and personal success. Some examples of technical communication in various settings across industry are highlighted in table 13.1. From effectively communicating the results of your scientific findings and progress to the company leadership, to selling a product or service to a customer or exciting an investor about your business idea, communication is key to moving ideas from you to your audience to persuade or inform them. Ultimately, effective communication allows us to better understand and be better understood by others. In this chapter, you will learn the essentials to do so effectively.

In this chapter, you will learn the frameworks, tools, and techniques of how to communicate effectively with both spoken and written content. This will include learning how to structure content, include key elements for ensuring that the spoken

Table 13.1 Examples of technical communication in industry

Setting	Example of Communication
R&D	Deliver a technical presentation or talk.
Corporate/business	Brief senior executives who are unfamiliar with your field.
	Propose a development plan to a manager/chief technology officer.
Sales and marketing	Sell a product or service to a customer.
Entrepreneurship	Pitch a business idea to an investor.
Consulting	Present the findings and recommendations from a consulting engagement.

or written content you deliver is remembered, and use tools and techniques for effective presentations and written work. As the MIT Ford Professor of AI and Computer Science Patrick Winston captured so well in the quote at the start of this chapter, your success in life will be determined largely by your ability to communicate. Thus, it follows that the investment that you make in acquiring knowledge about how to communicate effectively will be bigger than the return on any other personal development investment you make [3].

Winston believed strongly in the importance of effective communication to one's personal and professional success. As a renowned AI researcher, professor, and director of the MIT Artificial Intelligence Laboratory from 1972 to 1997, he dedicated himself equally to understanding the art of effective communication and sharing that understanding with others.

Winston's annual lecture on effective communication during the MIT Independent Activities Period, "How to Speak," has been an MIT tradition for over forty years. In this talk, he provides a set of tools, techniques, and heuristic rules for improving communication abilities, both spoken and written [5]. He compiled his decades of learning on the topic of effective communication into his book [3]. The authors of this book have personally benefited from Winston's heuristics for effective communication. One of the authors (D. Martinez) also had the great honor to serve as one of the reviewers of his book *Make It Clear: Speak and Write to Persuade and Inform*.

In this chapter, we work from substantial content from *Make It Clear: Speak and Write to Persuade and Inform* [3] to present various frameworks and principles for effective communication, with a specific focus on technical communication as demanded during AI development and deployment. In section 13.1, we discuss the Vision-Steps-News . . . Contributions (VSN-C) framework for structuring the content of your technical communication. In section 13.2, we discuss the Winston Star, a five-point graphic illustrating key elements to delivering content and being remembered as a speaker or writer. In the following sections, we discuss the fundamentals for delivering spoken and written content with clarity, as well as ways to ensure preparedness and avoid the technical deep trap. We conclude with main takeaways and a set of exercises for the reader.

It is important to note that the frameworks, principles, and tools described in this chapter are not meant to be the definitive guide to effective communication. You may not agree with all the offered guidance or even find effective communicators who violate some of these principles at times. However, no effective communicator consistently violates *all* the offered tips. As such, this information can still be useful as a guide for improving your communication skills.

13.1   VSN-C for Structuring Communications

The structure of your communication, whether written or spoken, is critical for clearly guiding your audience through the *why*, *what*, and *how* of your ideas and contributions. When delivering any spoken or written work, it is important to "show your hand" right away. Studies have shown that the average audience's attention span is only five to ten minutes, so you often have only a few minutes to convince your audience to continue listening to or reading your ideas before they lose interest [6]. Therefore, it is critical that you quickly explain the problem that you are solving, why it matters to your audience, the steps that you have taken to solve it, and any results you have achieved. There are many situations where this holds especially true, such as when briefing senior leadership on your progress or presenting your invention or business idea to an investor. By showing your hand immediately and not holding back your big idea or results until the end, you can effectively engage your audience from the start.

For written and spoken communication, the VSN-C framework (illustrated in figure 13.1) is an effective structure for organizing your content and attracting the attention and interest of your audience early. The Vision-Steps-News (VSN) part of the structure defines how your communications should start—by defining the problem, outlining the steps you have taken, and announcing the results up front. Following the news is the exposition, which is the body of your talk [3] and can elaborate on what you have done and how you achieved your results. The Contributions (C) step defines how your communications should end—namely, concluding with an outline of the contributions of your work. We'll discuss each of these elements in greater detail next.

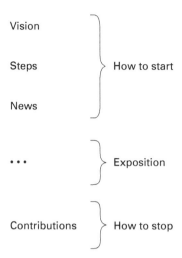

Figure 13.1   Structuring content using the VSN-C framework [3].

Vision: The first and most important part of any communication is conveying your vision to your audience. It is important to clarify for the reader that "vision" in this context is about the subject or topic that you are trying to convey—it is obviously *not* the same as the "vision" used in the AI strategic development model (AISDM) model discussed in chapter 9.

The vision consists of two key elements—the problem that you are solving and an approach to the solution. It is crucial that the problem and solution presented in the vision have relevance and interest for your audience. For example, an executive at a large corporation is likely to pay more attention to a speaker who immediately discusses how new investment areas in improving the company's product will increase market share and drive revenue gains, rather than leading with the engineering details of the feature's implementation. On the other hand, a technology investor is likely to pay more attention to a speaker who discusses starting a company to build and sell a new AI-powered solution that will disrupt an existing industry, rather than focusing on the scientific research enabling the AI solution.

Steps: After presenting your vision and capturing your audience's attention, it's important to clearly outline the steps you've taken and the progress you've made in implementing your plan. Your "Steps" section should impart confidence to your audience that you have thought through your approach and are making progress toward a solution. Instead of enumerating all the minor things that you did, focus on highlighting the key milestones and achievements in your plan. This will help keep your audience interested in what you have to say.

News: After outlining the steps you have taken and the approach you have used to solve the problem of interest, it is important to highlight your results. These results, or "News," should be presented up front and explicitly call out what you have achieved and the impact of your work (scientific results, business impact, or other effects). This will help build excitement and interest in your audience around what you have accomplished, and the rest of your communication can now focus on explaining how you achieved those results.

Contributions: The final part of your communication should emphasize the meaningful contributions that you or your organization have made. To do so, you should always conclude with an outline of the contributions, which could include advancements in engineering, scientific research, business impact, customer value, or other areas. By clearly outlining your contributions, you can leave your audience with a lasting impression of the key takeaways and impact of your work.

The VSN-C framework is a structure for written and spoken communication that emphasizes the importance of highlighting your vision, the steps you have taken, and the news as the start of your communication. After providing an overview (i.e., the VSN) of your work, the framework uses the body of the talk—the exposition—to delve into the details of what you have done in the steps and news section. Finally, the C, or "Contributions," section is a key element to highlighting the meaningful results of your work and how you have ultimately contributed to engineering, research, or business impact. By using the VSN-C framework, you can effectively persuade your audience to continue listening or reading about your work and convince them of the value and meaning of your contributions.

13.2 Winston Star: Essentials for Being Remembered

The VSN-C framework, described in the previous section, offers a structure for effective written and spoken communications. However, providing structure to your communications does not guarantee that your ideas will be remembered. To make an impact and be remembered, your work should also incorporate key elements of the Winston Star [3]. This five-point star, illustrated in figure 13.2, highlights the most

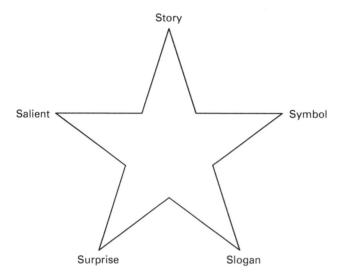

Figure 13.2 Winston Star [3].

important elements of any work that are essential for making a lasting impression and being remembered.

The elements are as follows:

Slogan: A slogan is a short and memorable word or phrase used in marketing to differentiate a brand and make it more recognizable to consumers [7]. In the context of technical communication, it serves as a handle or identifier of your work or idea. The slogan highlights the key concept that you want your audience to associate with you and your work. Some examples of slogans in the field of AI include "stable diffusion," "self-attention," and "backpropagation." These may describe key techniques or algorithms used in the research, or possibly the names of the systems, such as IBM's Watson, the *Jeopardy!*-playing program [8]. Overall, a good slogan should be catchy, memorable, and relevant to your work or idea.

Symbol: A symbol serves as an effective way to visually brand and communicate the key idea of your work. It can be a picture, chart, drawing, or graphic. Much like a slogan, a symbol also allows one to get a handle on the work. In some cases, a symbol may be used in conjunction with a slogan to create a cohesive and memorable idea. In this book, the AI system architecture shown in figure 1.3 in chapter 1 comprises the slogan and symbol that form the bases of part I of the book. Similarly, the AISDM shown in figure 9.4 in chapter 9 comprises the slogan and symbol that form the bases of part II. Overall, a good symbol should be simple to understand, recognizable, and relevant to your work or idea.

Story: Humans are inherently storytelling animals, and our ability to create, tell, and understand stories is what distinguishes us from all other animals [3]. Thus, when presenting technical ideas, it can be effective to tell stories to engage your audience. Instead of just focusing on the technical ideas and results, it is important to discuss who developed them, why they were developed, how they were developed, any fun or interesting stories about the process, and what their impact was. The ability to tell stories when presenting not only engages the audience but also makes your ideas more memorable.

Salient: A common pitfall of many technical presentations is having too many good ideas. An audience can remember only one, two, or at most three salient ideas in your technical presentation. Thus, to avoid overwhelming the listeners with too many ideas, it is important to focus on just one, two, or at most three. The term "salient" simply refers to an idea that sticks out and that you want your audience to remember as yours. A slogan can be thought of as a label for the salient idea. Ultimately, by including only a few salient ideas, you can help ensure that your audience will remember the most important points of your work.

Surprise: Humans also love surprises. They help to engage your audience and make your work memorable, so it is important to highlight any surprises in your presentation. A surprise can be any unique results achieved, substantial technical/business contributions, or a novel technique. When including surprises, it's equally important to highlight them and explain why they are significant. They also can help make your presentation more interesting and engaging.

In *Make It Clear: Speak and Write to Persuade and Inform*, Patrick Winston shares the story of how the book came to be written. He recalls that it started with an accidental conversation decades ago with one of his graduate students, Bob, to whom he was complaining about a terrible lecture that he had just attended. In response, Bob urged Winston to give a talk during the Independent Activities Period on how to speak effectively. While initially reluctant, Winston started thinking about all he had learned from various experts about how to communicate effectively, including what to do and what to avoid. His lecture drew about 100 students, and so he continued to give it in subsequent years [5]. The lecture would go on to become an MIT tradition for over forty years, drawing hundreds of listeners every year. The ideas from that lecture, along with decades of things learned from various experts, led to the publication of Winston's book.

The VSN-C framework is an effective way to structure your communications. However, to make your ideas memorable for your audience, it is important to incorporate the elements of the Winston Star into your communications. Notice how this section contains most of the elements of the Winston Star: "Winston Star" itself is the slogan; the image of the star in figure 13.2 is the symbol; the surprise is that you can make your work memorable at all; the salient idea is that all you need to do is incorporate the five elements of the star; and the story of the genesis of Winston's "How to Speak" lecture provides human interest. By incorporating these elements in your communications, you can help make a lasting impression on your audience and ensure that your ideas are remembered.

13.3   Essentials of Outlining

Outlining is a crucial step for organizing the main components of your content. Whether you are preparing a report or a presentation, an effective outline can save time and provide additional clarity on the main elements of your content and how they connect. In this section, you will learn how to deploy a "broken-glass outline," an effective method of outlining that is easier to work with in the early stages of planning your content.

There are many strategies and frameworks for outlining. For example, a formal outline, sometimes called a "Harvard outline," specifies a hierarchical structure consisting of Roman numerals at the highest level, capitalized letters in the following level, and then Arabic numerals and lowercase letters in the following two levels of the hierarchical structure [9, 10]. As illustrated in figure 13.3, the formal outline is a mechanism for visualizing the main concepts for your paper or presentation, broken into main topics and supporting topics. Such an outline can be written in fragments or complete sentences. However, the rigidity of this method, enumerating a hierarchical structure with main ideas and supporting topics, can stifle creativity and flexibility in the early stages of brainstorming and outlining content.

```
I.  A major division
        A. A major subdivision
                1. A minor subdivision
                2. Another minor subdivision
        B. Another major subdivision
II. Another major division
        A. A major subdivision in the second major division
        B. Another major subdivision in the second major division
```

Figure 13.3   A formal outline.

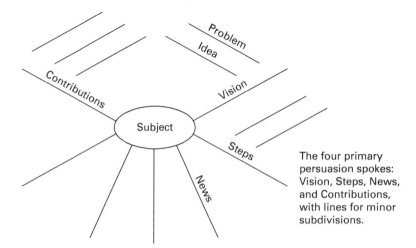

The four primary persuasion spokes: Vision, Steps, News, and Contributions, with lines for minor subdivisions.

Figure 13.4 A broken-glass outline.

Instead, you can use a more flexible approach, of which the broken-glass outline is an effective one. Named for its visual appearance, a broken-glass outline is a visual outlining process that resembles broken glass or a mind map [11]. A broken-glass outline starts with a title in the center, with spokes stemming out of it for the major subdivisions or concepts. Each of these spokes then has even more spokes for the minor subdivisions or concepts. This visual outlining process allows more creativity and flexibility in the early stages of brainstorming and outlining content.

The first key element of a broken-glass outline is the placement of standard spokes. For example, the standard spokes in your broken-glass outline for persuasive communication generally include four—one each for "Vision," "Steps," "News," and "Contribution." This is illustrated in figure 13.4. Once an overall structure is in place, begin to add detail to the outline. For example, branching off from the Vision spoke, you could add spokes outlining the problem, why it's important for your audience, and the proposed solution. For the Steps spoke, you might add spokes for each of the steps you have taken and plan to take. For the News spoke, you might include spokes that highlight some impressive recent results.

When adding detail, you may choose to start with breadth, adding spokes at the next level for all existing spokes before moving on to the next level of detail. Alternatively, you may choose to work with depth first, adding spokes down several levels as far as possible.

A broken-glass outline is flexible and easily revisable, which makes it a valuable tool for brainstorming and outlining. This flexibility helps overcome the rigidity of a formal outline and minimizes barriers to change, such that you can easily add, remove, comment, reorder, or move elements around within your outline. When using a broken-glass outline, you are encouraged to adjust, cross out spokes, move elements around with arrows, and add annotations. In the early stages of brainstorming and outlining, a broken-glass outline allows you to explore your ideas freely, discover new ones by channeling your creative and explorative thinking, and organize your thoughts and see how your ideas fit together. Once the broken-glass outline contains all the key ideas and has allowed you to freely organize your thoughts and ideas, you can move on to creating a formal outline.

With an outline in place, you have completed most of the work. What remains is building on the content within your outline to formalize your presentation or written work. In the following section, we discuss writing and presentation fundamentals.

13.4   Writing and Presentation Fundamentals

In this section, you will learn tips and techniques for delivering effective spoken and written communication. We will start by discussing writing fundamentals, including how to organize your writing and general techniques for drawing readers into reading what you have written and helping them understand what you are trying to communicate. Next, we will discuss presentation fundamentals, including best practices for how to prepare and deliver an effective presentation. Both build on the foundation of the outline for your content.

Writing Fundamentals

Seasoned readers use certain techniques to quickly understand the key elements of a piece of scientific writing. As an author, it is important to understand these techniques and use them to compose your writing in a clear and understandable way. This will help ensure that your ideas are communicated clearly and the important elements of your work are highlighted. Authors, who are deeply familiar with the subject matter that they are writing about, often are not aware of when their writing is not clear or when important elements are missing. When writing, it is important to keep in mind that seasoned readers use surface clues to extract the essence of a piece of writing. Table 13.2 highlights these key elements used for rapid decryption and their purpose.

Table 13.2   Key techniques for seasoned readers of scientific writing

Key Techniques	Purpose
Read the abstract.	Provides an overview of what you will learn from the paper
Read the introduction.	Expands on the abstract to help you decide whether to read further
Examine the conclusion.	Outlines the key points of the paper and its contributions
Note the section titles.	Enumerates the key elements of the paper
Look at the illustrations and read the captions.	Reveals essential content (i.e., results, charts, and diagrams)
Look at the lists.	Enumerates essential elements (i.e., an articulation of steps and a list of contributions)
Check the citations.	Indicates the author's research community
Read summary paragraphs and skim the topic sentences.	Provides an overview of each section and its key points

To craft your writing in a way that provides seasoned readers with the information they are looking for, you should do the following:

1. Write an abstract that summarizes your Vision, Steps, News, and Contributions.
2. Start with an introduction section that describes your Vision, Steps, and News in more detail than the abstract does.
3. Use full-sentence section titles that summarize the content of the section.
4. Use topic sentences and include summaries at the end of each section.
5. Capture key ideas in illustrations (i.e., charts, graphs, images) with descriptive captions.
6. Use lists to call attention to essential elements such as steps and contributions.
7. Conclude by describing your contributions.

By incorporating the key elements most important to your readers, you can ensure that your writing is clear, understandable, and easy to consume. We now describe an overall strategy for writing and what to include at the beginning and the end of your written work.

Organizing Your Writing

In the previous sections, we presented the VSN-C framework and broken-glass outline. We now describe an overall strategy for writing that helps focus on your contributions. This strategy is aimed at ensuring that all important elements are included and appropriately prioritized. This includes the following actions:

1. Writing the "Contributions" section first
2. Preparing any illustrations
3. Adding sections that fulfill community expectations.
4. Including the Vision, Steps, and News
5. Writing the abstract
6. Including sections on prior and future work and acknowledgments.

The "Contributions" section of your written work should be the first section you write. This section should clearly describe the contributions of your work. If you are unsure of what your contributions are, it may be a sign that you are not yet ready to write. In this case, take time to clarify the contributions of your work before proceeding. It is important to avoid using "Conclusion" as a title for this section, but instead use "Contributions," which more accurately reflects the content and purpose of the section. The proper title enforces you to focus your attention on the most important aspect of this section, and it also communicates to the reader that you have made meaningful contributions, not just presented observations or summarized existing work.

After writing the "Contributions" section, you may want to consider preparing any illustrations that will be included in your written work. These illustrations, such as graphs, charts, and other visual elements, can help to provide a clearer and more engaging presentation of your data and ideas. Once these illustrations are prepared, the task that remains is simply including writing that explains the content of the illustrations to provide additional context for the reader.

In the writing of this book, we used our MIT course lectures from the courses we taught. The charts from the course lectures helped us in formulating the VSN-C. The broken-glass structure helped us in planning the content before starting the writing of the chapters. We also used the broken-glass structure to put a time line on how long each section might take to complete. This approach helped us in estimating the required time to complete each chapter. We also gathered, read, and decided on what references to include prior to the start of writing each chapter, based on the sections shown in the broken-glass outline.

Table 13.3   Expected sections for various audiences

Field/Audience	Possible Sections
Paper for researchers and scientists	• Past Work/Related Work • Methods • Results • Discussion
Business plan for investors	• Opportunity • Why Now/Why Us? • Competitive Landscape • Team Members
Business report for senior leadership	• Executive Summary • Recommendations

After preparing any illustrations, the next step is to include the sections that comprise the main body of your written work. The specific sections that you include may vary depending on your audience and their expectations. Table 13.3 provides some examples of sections that may be expected for different types of written works and their respective audiences [3]. It is important to consider the expectations of your audience and include the appropriate sections to ensure that your work is complete and well received.

After organizing the sections for the body of your paper, the next step is to write the Vision, Steps, and News that form the basis of your introduction. Once these sections are ready, you can move on to writing the abstract. The abstract is the most important element of your paper, as it determines whether readers choose to continue reading. As such, it is essential that your abstract clearly and concisely articulates your vision, outlines the steps that you have taken, highlights the news, and enumerates your key contributions. It's important that the abstract contains all the key details of your results, such as experimental results, that make your contributions tangible.

In some cases, it may be beneficial to include a section on related or past work relating to the problem that you are exploring before delving into the details of your own contributions. This section can be titled "Past Work" or "Related Work" and should provide an overview of the relevant literature and research in the field and explain how your work builds on or extends this body of work. Similarly, it may be beneficial to include a section on future work, discussing potential areas for future research and development (R&D) that could extend the contributions of your work. This section can be titled "Future Work" and should be placed just before the "Contributions" section, providing a brief overview of future directions for R&D.

In the following section, we will describe in more detail how to compose the beginning and end sections of your written work.

How to Start

The beginning of your written work is an important place to motivate your readers and get them excited about what you are saying. Earlier in this chapter, we described the VSN-C framework, used for structuring your communications, and emphasized the importance of the VSN elements of this framework in the beginning of any written or spoken communication. Vision, which includes both the problem statement and your approach to solving it, can be expressed in different ways depending on the audience and the specific goals of your communication. Some common techniques for communicating Vision include using an "if-then" statement, asking big questions, presenting mission blockers, exploring new opportunities, or telling an interesting story [3]. Whatever approach you choose, it's important that your work is motivated with a clear vision. Depending on the audience of your communication (refer to table 13.1), the vision section may be given different headings, such as "Hypothesis," "Opportunity," "Goal," or even "Vision."

Following Vision, the Steps section should clearly enumerate the most important steps that you have taken and the steps that you plan to take to solve the problem. The Steps section is key to demonstrating that you are guided by a clear plan. Often, it can easily be developed around a clear list of items.

Finally, the News section is key to making your work current and demonstrating recent results, breakthroughs, or achievements. It's important that dates be included in the items discussed in this section to clearly communicate the recency of your developments and contributions.

Overall, the Vision, Steps, and News elements of the VSN-C framework should guide the beginning of your written work.

How to End

To effectively conclude your writing, your final section should leave your readers with a clear outline of your contributions to your field or line of work. This section, titled "Contributions," should enumerate your contributions using active verbs such as "demonstrated," "argued," "exhibited," and "identified." You may also consider other ways of

Table 13.4   Recommendations for overcoming writer's block

• Get started by working up an outline.

• Force progress by scheduling a presentation.

• Explain your examples and describe your illustrations.

• Write with abandon—you can fix everything later, once you have a draft.

• Keep a notebook; record ideas when you think of them.

• Reserve a special time for writing.

• Take a walk, go jogging, or promise yourself some kind of treat after you write for a while.

• Sleep on it, working your problems as you drift off.

• Drive out fear by focusing on a small initial step.

---

expressing contributions, such as a bold statement or as good news [3]. It is important to avoid using the title "Conclusion" or "Conclusions" for this section, as it sets the wrong expectations for your audience. The right section title is key to setting the right expectations for your audience—that you have made contributions, and they will be clearly outlined.

Overall, the "Contributions" section does serve as a conclusion, but it emphasizes the most important part of your work—your contributions.

Tips and Techniques for Writing

In this section, we highlight a set of tips and techniques for effective writing. This includes strategies for overcoming writer's block, tips for giving and receiving critiques for your speaking and writing, and guidelines for avoiding style blunders. By following these recommendations, you can improve the clarity and coherence of your writing.

Overcoming writer's block: Writer's block is a kind of inhibition that prevents you from writing. It is important to remember that many writers experience writer's block, so you are not alone. It has many potential causes, such as lack of clarity in your ideas, lack of dedicated time or space in your routine for writing, or feeling overwhelmed by the task at hand. There are many ways to overcome these various forms of writer's block. We summarize some of Winston's recommendations in table 13.4—you may find that one or some combination of them may work for you [3].

Soliciting feedback: To become a better speaker and writer, it is important to solicit feedback from others. When giving or receiving critiques, there are a few recommendations

Table 13.5   Recommendations to avoid common style blunders

- Use "I" not "we" (if you are the sole author).
- Do not write as if your paper were its own author/speaker.
- Do not use "former" and "latter."
- Do not use "above" and "below."
- Do not use "since" when you mean "because."
- Do not use "last" when you mean "past."
- Do not use "utilize" when you mean "use."
- Do not use "thing" when you can use a more specific word.
- Do not use pronouns.
- Do not switch words gratuitously.
- Do not use "which" when you mean "that."
- Do not "try and do" anything; "try to do" it.
- Do not use quotes to convey your idea; use quotes to support your idea.
- Do not use quotes unless you are quoting someone.
- Do not use apostrophes to make acronyms or time periods plural.
- Eliminate inessential words.
- Do a final scan for misused words.

---

to keep in mind [3]. First, when asking for feedback from a reviewer, it's best to present work that is complete or near complete, and never work that is incomplete. It can also be helpful to ask reviewers where they think your time and energy will be most effectively spent to achieve the greatest improvement. As the Pareto principle states, 80 percent of output comes from 20 percent of effort, so it's worth prioritizing the areas of improvement that will have the greatest impact. As a reviewer, it is important to exercise the Golden Rule: provide constructive feedback that is actionable, principled, and positive. This will help the writer or speaker make meaningful improvements to their work.

Avoiding style blunders: Writing that is full of errors can irritate readers and make it difficult for them to understand your ideas. While professional editors can help identify and correct common grammar and style mistakes, you may not always have access to an editor to review your work. Winston highlights typical grammar and style blunders and how to avoid them, as captured in table 13.5 [3].

Winston's book *Make It Clear* provides guidance on how to write various types of documents such as recommendation letters, blogs, and press releases. Readers interested in learning more about tips and techniques for preparing these and other types of documents are encouraged to refer to Winston's book [3].

Tips and Techniques for Presenting

Most of the fundamentals described earlier in this section on organizing your content, how to start, and how to stop apply to both writing and presenting. In this section, we will focus solely on the key tips and techniques for delivering effective presentations. These techniques can help you engage your audience, express your ideas clearly, and deliver a successful presentation.

How to compose slides: Slides are a powerful aid that speakers can use to enhance their message and improve its clarity and impact. The composition of your slides can vary depending on your audience and the type of presentation that you are giving. For example, a technical presentation may use slides to provide detailed information, including formulas and graphs. In contrast, a business-leadership meeting may use slides to summarize key concepts with few words and lots of illustrations. In either case, slides can be a valuable tool if used correctly, but they can be a distraction if not used properly.

When composing slides for a presentation, it is important to keep them simple and avoid overwhelming the audience with too much text. Evidence suggests that humans have only one language processor that can easily be overloaded when presented with too much information at once [3, 12]. As such, too many words on a slide are bound to jam a listener's language processor, leaving no device to listen with. To avoid this, you can reduce the amount of text on each slide and leave only the key ideas. One effective way to reduce the amount of text on your slides is to use bullet lists to enumerate essential elements. However, it is important to use bullet lists sparingly. For example, lists should be used to enumerate essential elements such as steps or contributions. By using bullet lists sparingly, you can draw more attention to your content. You can also use illustrations, such as pictures, graphs, and charts, to communicate ideas and data. When doing so, it's important that you clearly label the legends or axes. You can also simplify your slides by avoiding the use of background patterns, eliminating unnecessary clip art, removing the title, and leaving out animated transitions details.

In addition to simplifying your slides, it is important to organize them in a way that shows how your ideas fit together. As part of your presentation, you may also choose to include blackout slides. These slides, which are completely black, are there to place the emphasis on you as you speak and to transition between slides. Furthermore, to make your slides easy to read, use large and easy-to-read fonts.

Table 13.6 Preparations to make prior to a presentation

- Look over the place where you will speak far in advance.
- Make sure you have water handy.
- Make sure that all the audiovisual equipment works.
- Turn the lights on full and make sure that they can be turned on and off at will.
- Know that being nervous is good. Take a walk, breathing deeply, to keep your adrenalin under control.
- Psych up with power music (possibly played only in your head).
- Exercise your voice.
- Chat with the early arrivals.
- Establish eye contact with happy, receptive listeners.

Finally, it's important to budget enough time for each slide. In general, you should aim for at least one to three minutes per slide and be prepared to designate some slides as optional and leave them out if you run out of time. You may also consider including a progress bar to show your audience where you are in the presentation. By following these tips, you can create slides that enhance your presentation without overwhelming or distracting your audience.

Prior to and during delivery: Before delivering your presentation, it is important to work your way through a checklist to ensure that you are prepared and ready to go. This can help you address avoidable problems and focus on delivering a successful presentation. Some key elements to include in your prepresentation ritual are listed in table 13.6 [3].

The beginning and end of a talk are crucial for setting the tone and establishing your confidence and comfort. In addition to the preparations enumerated already, you may choose to compose and memorize your opening and closing sentences. This can help you start your presentation off on the right foot and end it with a polished and professional conclusion. However, it's important that you do not write out and memorize your entire presentation. Memorized lines can be useful for getting started and finishing strong, but you should allow yourself the flexibility to adapt and respond to your audience during the rest of your presentation.

In addition to prepresentation rituals, it is important to incorporate good habits and characteristics of great speakers into your own speaking style. One key aspect of effective presentation is engaging and maintaining the attention of your audience. Many people have short attention spans and can easily lose interest in your presentation. To keep your audience engaged, Medina suggests using "hooks" at regular intervals, typically every ten minutes, to regain your audiences' attention [6]. Dividing your talk into enumerated parts with a kind of oral punctuation can act as a cue to signal a change in topic and reengage those who may have fogged out.

When presenting, it is important to maintain eye contact with your audience. A good strategy is to pick out a few people scattered throughout the audience and cycle your eyes from one to the next. Another technique is to use the "W" method, moving your eyes from the left-rear part of the audience, then to the front a quarter of the way to the right, and eventually tracing out a "W." These techniques will help you engage with people all around the room and connect you to your audience. To avoid breaking eye contact with your audience, you should consider using a remote control to streamline the process of transitioning between your slides without having to walk back or be tethered to your computer. However, avoid using a laser pointer, as it requires you to look back at the screen to aim, breaking eye contact with your audience. If you need to point out something on your slide, you can approach the screen and point using your hands, or even include an arrow on a slide to highlight the element of interest. In addition, you should use or set up a display between you and the audience, as this will allow you to face your audience while presenting and avoid the need to turn back to look at your slides.

When presenting, it is also important to be mindful of your body language and appearance. Avoid playing with your hair or putting your hands in your pockets, and wear clothes that are comfortable and appropriate for the occasion. It's also important to eliminate filler words and grunts, such as "uh," "ah," and "um," which can distract from your message. Similarly, avoid uptalking, or ending declarative sentences with a rising pitch, as it can make your speech sound uncertain.

When delivering your presentation, pay attention to the time and finish when you are supposed to. It is also important to be positive and enthusiastic in your delivery, as it demonstrates to your audience your excitement and passion for the topic that you are speaking about. However, you should be prepared for a flat audience that doesn't respond with laughter or energy. After your presentation, be prepared for questions, including hostile questions or questions that you are unable to answer, or even no questions at all. If you get a question, repeat it before answering and be prepared with some standard responses for difficult or unexpected questions. If there are no questions at the end, it is useful to prepare a set to bring up and address, especially if these are questions that you have gotten before—you might have an audience that is shy but might have some of these same questions.

By incorporating and being mindful of these tips and techniques, you can improve the effectiveness of your presentations and engage your audience. Improving as a speaker is an ongoing journey. In addition to incorporating the provided tips and techniques, it is important to identify a writer or speaker you admire, ask why you like what you like, and then learn to incorporate those techniques. Continuing to observe and learn from others will be key to continually improving your skills as a writer and presenter.

13.5   Main Takeaways

In this chapter, we presented frameworks and techniques for effective communication based on the teaching of Patrick Winston. We began by outlining two major categories of communication: instructive, which aims to help the audience understand a concept; and persuasive, which aims to influence the audience to accept a particular belief or argument.

We provided these tips and techniques because we have found them useful in our own work as AI practitioners. We also want to emphasize the importance of communicating effectively to your AI team, subordinates, peers, bosses, investors, and stakeholders to develop and deploy AI capabilities successfully. You can have the most impressive technology, but if you cannot communicate something (such as the AI value proposition), your project might never get funded or off the ground, let alone completed.

To structure your communication effectively, we presented the VSN-C framework, which consists of four elements: Vision, Steps, News, and Contributions. The Vision element includes the problem that you are solving and your approach to solving it. The Steps element outlines the progress you've made and the future steps you plan to take. The News element highlights important results that you have achieved. Finally, the Contributions element emphasizes the meaningful contributions that you, your AI team, or your organization have made. In addition to presenting the VSN-C framework, we described the importance of showing your hand immediately to grab the audience's attention and interest early on.

We also presented the Winston Star, a five-point star that highlights the most important elements of any work that are essential for being remembered. This includes a slogan and an iconic symbol that serve as handles on the work, a surprising result, a salient idea, and a story. Ultimately, incorporating these elements and following the VSN-C framework can help ensure that your ideas are structured and remembered.

We also introduced the broken-glass outline, a more flexible and creative method of outlining than a traditional formal outline. A broken-glass outline starts with a title in the center with spokes stemming out for the major concepts or subdivisions. Each of these spokes can have more spokes coming off them for minor subdivisions or concepts. The visual nature of a broken glass, similar to a mind map, allows more creativity and flexibility and overcomes the rigidity of a formal outline.

Finally, we provided guidelines on writing and presentation fundamentals. We described techniques that seasoned readers use for rapid decryption of a written work and provided recommendations for how to write in a way that provides readers with the information they are looking for. In addition, we outlined an overall strategy for organizing your writing that helps you focus on your contributions. This strategy focuses on writing the "Contributions" section first; then preparing any illustrations; adding sections that fulfill community expectations, including Vision, Steps, and News; writing the abstract; and finally including sections on prior and future work and acknowledgments. We described in greater detail how to start and end your written communication, emphasizing the VSN elements in your introduction and contributions in your conclusion. We also presented general tips and techniques for writing, which included ways to overcome writer's block, critique writing and presentations, and avoid style blunders. Similarly, we presented general tips and techniques for presentation, which included composing effective slides and following general best practices for before and during your presentation.

13.6 Exercises

1. Describe each element of the VSN-C framework and why it is important.
2. The VSN-C elements should all be included in the abstract:
   a. True
   b. False
3. The types of communication can be split into two major categories. Identify these categories and provide an example of each.
4. What does it mean to "show your hand," and why is it important?
5. What are the five elements of the Winston Star?
6. Use a broken-glass outline to organize your content for an upcoming technical paper or presentation. Reflect on the value of this type of outline for freely brainstorming and outlining your content.
7. The first section of your written work that you write is:
   a. References
   b. Abstract

     c. Contributions

     d. Introduction

8. The start of your written work should contain which elements of the VSN-C framework?

9. Name three things that you should prepare or confirm before your presentation to ensure that you are ready to go.

10. Identify three things to be mindful of when delivering your presentation.

11. Why is storytelling so valuable?

12. A useful technique for effective presentations is to start by laying out the architecture of the talk. What would you incorporate into such an architecture?

13. Another very useful technique is to perform dry runs (meaning presenting to an audience to get feedback) before the actual presentation. Describe what a sequence of dry runs might incorporate.

14. For a complex study report, it is also useful to start by laying out a skeleton of the final report and then proceed to fill in pieces as the study progresses. Explain why this approach would be useful.

13.7 References

1. Manyika, J., *Technology, jobs, and the future of work*. McKinsey Global Institute, 2017. https://www.mckinsey.com/featured-insights/employment-and-growth/technology-jobs-and-the-future-of-work.

2. Dondi, M., J. Klier, F. Panier, and J. Schubert, *Defining the skills citizens will need in the future world of work*. McKinsey Global Institute, 2021. https://www.mckinsey.com/industries/public-and-social-sector/our-insights/defining-the-skills-citizens-will-need-in-the-future-world-of-work.

3. Winston, P. H., *Make it clear: Speak and write to persuade and inform*. 2020, MIT Press.

4. Hamm, P. H. and N. Dunbar, *Teaching and persuasive communication: Class presentation skills*. Harriet W. Sheridan Center for Teaching and Learning at Brown University, 2006. 10–12.

5. Winston, P., How to speak. 2018. https://ocw.mit.edu/courses/res-tll-005-how-to-speak-january-iap-2018/.

6. Medina, J., *Brain rules: 12 principles for surviving and thriving at work, home, and school*. 2011, ReadHowYouWant.com.

7.   Kohli, C., L. Leuthesser, and R. Suri, Got slogan? Guidelines for creating effective slogans. *Business Horizons*, 2007. 50(5): 415–422.

8.   High, R., The era of cognitive systems: An inside look at IBM Watson and how it works. *IBM Corporation, Redbooks*, 2012. 1: 16.

9.   Brizee, A. and E. Tardiff, *Developing an outline: "Four main components for developing an outline," "why and how to create a useful outline," and "types of outlines and samples"*. 2013.

10. Purdue, *Types of outlines and samples*. n.d., Purdue Online Writing Lab, https://owl.purdue.edu/owl/general_writing/the_writing_process/developing_an_outline/types_of_outlines.html.

11. Buzan, T. and B. Buzan, *The mind map book: How to use radiant thinking to maximize your brain's untapped potential*. 1994, Penguin Group.

12. Hermer-Vazquez, L., E. S. Spelke, and A. S. Katsnelson, Sources of flexibility in human cognition: Dual-task studies of space and language. *Cognitive Psychology*, 1999. 39(1): 3–36.

III

Human-Machine Augmentation: Use Cases

Overview

The material presented in parts I and II of this book is based largely on the material from our courses at the Massachusetts Institute of Technology (MIT) that have been delivered to advanced undergraduate students, graduate students, and practitioners ranging from early-in-career professionals to managers and C-suite executives. *Mens et manus* (minds and hands) is the MIT motto, which alludes to the belief that effective technical education requires a substantial practical component. In line with this philosophy of "learning by doing," a key deliverable for the courses was the formulation and presentation of a strategic plan for an artificial intelligence (AI) product or service, following the AI strategic development model (AISDM) framework outlined in chapter 9.

In the coming chapters, we walk the reader through strategic plans and applications of the AISDM framework in the context of different AI capabilities applied to use cases in the form of an AI product or service. The first two chapters are based on use cases formulated for two different challenges that were presented to our MIT graduate engineering class. The challenge problems were presented by the head of the Misty Robotics company, as well as AI group leaders from the Bose Health/Sleep Division. The subsequent chapters include a set of strategic plans following the AISDM framework for an AI product or service conceived by participants of a similar course delivered to industry professionals. By walking the reader through the following use cases, we hope to crystallize the key concepts presented in both parts I and II of the book.

14

Use-Case Example 1: Misty Companion Robot as Alzheimer's Application

> Compassionate artificial intelligence systems are required for looking after those unable to care for themselves, especially sick, physically challenged persons, children, or elderly people.
> —Amit Ray, the author of *Compassionate Artificial Intelligence*

In this chapter, we present the strategic road map developed by a graduate engineering team at the Massachusetts Institute of Technology (MIT) for an artificial intelligence (AI)–powered Misty companion robot designed to help assist the elderly who suffer from Alzheimer's disease. The MIT team consisted of graduate students Kristen Railey, Mesert Kebede, Rebecca Zubajlo, and Udgam Goyal.

Alzheimer's disease is a progressive neurologic disorder that gradually destroys memory, thinking, and behavior. Its symptoms eventually grow severe enough to interfere with one's ability to carry out the simplest daily tasks [1]. It is among the top ten leading causes of death in the US, and the fifth leading cause of death among adults aged 65 years or older. As of 2020, experts estimated that nearly 5.8 million Americans aged 65 years or older had Alzheimer's disease. That number is projected to nearly triple by 2060 [2].

The various symptoms of Alzheimer's disease, such as decline in memory and cognitive ability, can often increase an individual's risk of injury or even death. For example, because of the disease's impact on memory and thinking, it is common for a person with Alzheimer's to wander or become lost or confused about their location. Studies suggest that six in ten people living with dementia will wander at least once [3]. Wandering can be dangerous, and even life threatening in many cases. In addition to physical risks, there are many consequences for one's mental and emotional health and well-being. For instance, the realization of one's own declining memory and growing inability to function causes those with Alzheimer dementia to become depressed. The compounded mental, emotional, and physical effects of Alzheimer's disease, coupled with the subsequent stresses placed on family members and caregivers, can be overwhelming. Recognizing the unmet market need and the need to develop compassionate AI systems to look after those who may be unable to care for themselves, the MIT graduate engineering team proposed an AI-powered Misty companion robot equipped with various sensors and designed to reduce the stress burden on caregivers by taking on the responsibility of monitoring, mentally supporting, conversing with, and serving as a companion to elderly individuals with Alzheimer's disease.

Mission and Vision

The MIT engineering team formulated the following succinct mission and vision statements for their AI-powered Misty companion robot:

> **Mission:** Create a foundational robot that users can employ to develop personalized solutions. The Misty robot provides four core capabilities:
>
> - **Monitoring:** Misty can **track** the movement of individuals within a living space and provide data insights.
> - **Mental support:** Misty can serve as a **mental aid** by maintaining reminders, calendars, and other pertinent information.
> - **Conversation:** Misty can **inspire** people by using emotional intelligence.
> - **Companionship:** Misty can be a **friend** providing entertainment, conversation, and information.
>
> **Vision:** Become the market-leading hardware solution for in-home assistants.

The team's mission statement concisely describes the *why*—to create a platform for Misty robot that users can employ to develop personalized solutions. The subtext following the mission statement (i.e., monitoring, mental support, conversation, and companionship) provides additional context on the key core capabilities of the Misty robot. The team's vision statement complements its mission statement and provides a clear description of what the organization *aspires* to be.

Envisioned Future

The envisioned future is a key aspect of the AI strategic development model (AISDM) framework, using the horizon structure discussed in part I of the book and helping formalize the near-term, midterm, and far-term prospects for the proposed solution. The

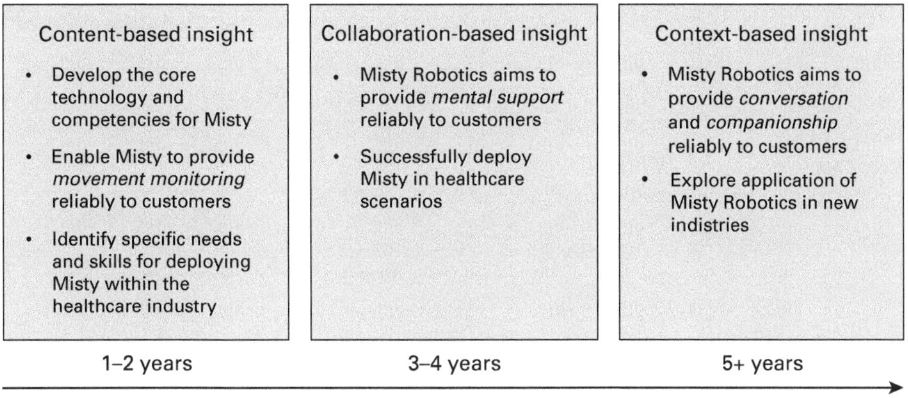

Content-based insight	Collaboration-based insight	Context-based insight
• Develop the core technology and competencies for Misty • Enable Misty to provide *movement monitoring* reliably to customers • Identify specific needs and skills for deploying Misty within the healthcare industry	• Misty Robotics aims to provide *mental support* reliably to customers • Successfully deploy Misty in healthcare scenarios	• Misty Robotics aims to provide *conversation* and *companionship* reliably to customers • Explore application of Misty Robotics in new indistries
1–2 years	3–4 years	5+ years

Figure 14.1   Envisioned future for the AI-powered Misty robot companion for elderly Alzheimer's patients.

AI-powered Misty companion robot team formulated three horizons, illustrated in figure 14.1.

The outlined envisioned future consists of three horizons. Horizon 1 (over 1–2 years) is focused on developing the core technologies and competencies for Misty. That horizon also seeks to develop reliable movement monitoring and to identify the specific requirements for developing and deploying Misty within the healthcare industry, which is identified as the first target industry given the high demand and value add, as well as low competition and barrier to entry. Horizon 2 (3–4 years) seeks to successfully deploy capabilities that would enable Misty to provide users with reliable mental support and eventually be successfully deployed in healthcare scenarios. Finally, horizon 3 (5+ years) seeks to integrate context to reliably deliver companionship and conversation to users. Such capabilities are key to mentally aiding and emotionally supporting those with Alzheimer's disease while easing the burden shouldered by family members and caregivers. The last horizon also seeks to take existing capabilities and apply them across new industries such as productivity, entertainment, and retail.

Strategic Direction

With a clearly defined mission, vision, and envisioned future, the MIT graduate engineering team developed a strategic direction for the AI-powered Misty companion robot. This strategic direction identifies the necessary competencies and technologies needed to enable the desired core capabilities, as well as the downstream system applications of those

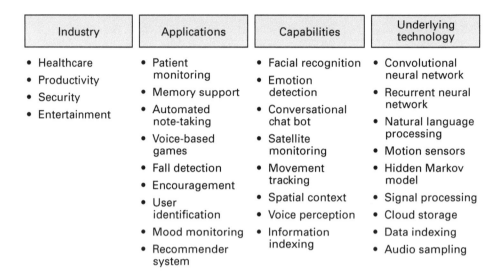

Industry	Applications	Capabilities	Underlying technology
• Healthcare • Productivity • Security • Entertainment	• Patient monitoring • Memory support • Automated note-taking • Voice-based games • Fall detection • Encouragement • User identification • Mood monitoring • Recommender system	• Facial recognition • Emotion detection • Conversational chat bot • Satellite monitoring • Movement tracking • Spatial context • Voice perception • Information indexing	• Convolutional neural network • Recurrent neural network • Natural language processing • Motion sensors • Hidden Markov model • Signal processing • Cloud storage • Data indexing • Audio sampling

Figure 14.2   Strategic direction for the AI-powered Misty companion robot for Alzheimer's patients.

core capabilities. In figure 14.2, we present the three important elements (i.e., underlying technology, capabilities, and applications) of the strategic direction for the AI-powered Misty companion robot.

The primary system application of an AI-powered companion robot for Alzheimer's patients is captured by the various applications that fall under the healthcare industry, such as patient monitoring, fall detection, memory supporting, encouragement, and mood monitoring. However, the team also defined other industries and potential applications, all of which are enabled by the same set of capabilities and underlying technology and competencies.

AI Value Proposition

With a well-defined strategic direction for the AI-powered Misty companion robot, we further expand on the elements of the AI value proposition. This serves as the foundation of the AISDM framework, bringing together the mission and vision statements, envisioned future, and strategic direction to lead to the AI strategic road map.

The value proposition for the AI-powered Misty companion robot, designed for elderly Alzheimer's patients, is illustrated in figure 14.3. The AI system architecture shows sensor-enabled data being conditioned and fed as input to various deep learning

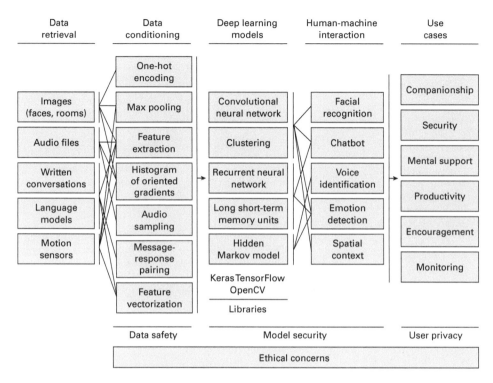

Figure 14.3   AI value proposition for AI-powered Misty companion robot for Alzheimer's patients.

techniques that power the competencies key to the AI-powered companion robot for Alzheimer's patients. The architecture employs various open-source ML libraries and is guided by the key fairness, accountability, safety, transparency, ethics, privacy, and security (FASTEPS) principles.

AI Strategic Road Map

With the AI value proposition in place, we now integrate the mission, vision and envisioned future with the strategic direction and AI value proposition to formulate the AI strategic road map. We begin by crystallizing the business need, which can be best communicated through answers to a set of key questions, an adaptation of the well-known "Heilmeier Catechism," as discussed in chapter 9.

In figure 14.4, the team presents a comprehensive overview of the industry landscape, Misty's current capabilities, and potential future developments. This bird's-eye view

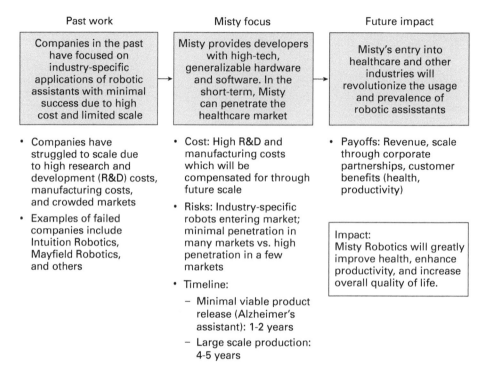

Figure 14.4   Business needs for the AI-powered Misty companion robot.

provides a better understanding of the existing solutions, their limitations and challenges, and the subsequent unmet customer needs. To further define the business needs for their solution, the team outlines the unique value proposition of their approach, evaluates the costs, risks, and payoffs, and identifies future milestones, including user and market impact.

An analysis of strengths, weaknesses, opportunities and threats (SWOT) is another key element of an AI strategic road map. In figure 14.5, we present the SWOT analysis for the AI-powered Misty companion robot.

The SWOT analysis highlights the many strengths that the Misty product (hardware plus software) and team possess, as well as the great opportunities that lie ahead for taking advantage of advances in AI and robotics to improve the status quo of how things are done across a range of industries. However, the analysis also calls out the general social apprehension toward robotics technology, as well as the fierce competition in the personal robot space as a threat to consumer adoption.

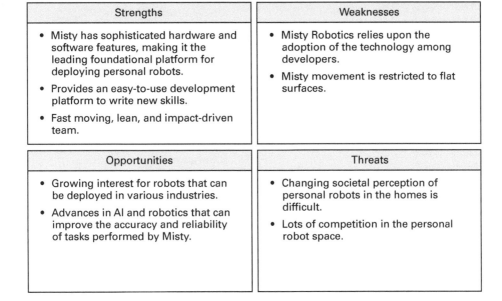

Strengths	Weaknesses
• Misty has sophisticated hardware and software features, making it the leading foundational platform for deploying personal robots. • Provides an easy-to-use development platform to write new skills. • Fast moving, lean, and impact-driven team.	• Misty Robotics relies upon the adoption of the technology among developers. • Misty movement is restricted to flat surfaces.
Opportunities	**Threats**
• Growing interest for robots that can be deployed in various industries. • Advances in AI and robotics that can improve the accuracy and reliability of tasks performed by Misty.	• Changing societal perception of personal robots in the homes is difficult. • Lots of competition in the personal robot space.

Figure 14.5   The Misty robot team SWOT analysis.

As the final step of the AI strategic road map, the MIT team delivered a demonstration of a minimum viable product (MVP) to a panel of AI academics and practitioners. This initial prototype used the Raspberry Pi, described in the appendix. A holistic overview of the capabilities demonstrated as part of the initial prototype are presented in figure 14.6. The demonstration employed deep learning for face recognition through image data. Equipped with a Raspberry Pi and camera, the team was able to successfully demonstrate the AI-powered Misty companion robot's ability to identify a person and their movements in the context of the time of day to detect abnormal behavior (i.e., potential wandering) and alert family members and caregivers.

The strategic plan presented in this chapter used key elements of the AISDM framework presented in chapter 9 to showcase the blueprint for an AI-powered companion robot for Alzheimer's patients.

14.1 Exercises

1. How would you design the Misty companion robot to adapt its offerings to unique needs, such as dealing with short-term memory loss, keeping routines, and

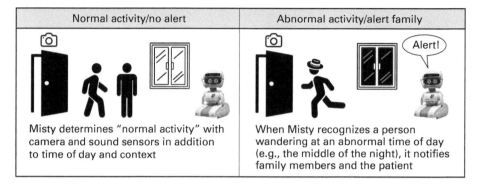

Normal activity/no alert	Abnormal activity/alert family
Misty determines "normal activity" with camera and sound sensors in addition to time of day and context	When Misty recognizes a person wandering at an abnormal time of day (e.g., the middle of the night), it notifies family members and the patient

Figure 14.6   Demo solution for the AI-powered Misty companion robot for Alzheimer's patients.

providing emotional support, as well as the preferences of individual Alzheimer's patients (communication styles, daily routines, and other elements)?

2. How would you address concerns around data privacy and security in the development and implementation of the Misty companion robot?

3. How could the Misty companion robot be designed to respond to unexpected situations, such as sudden changes in the patient's condition or emergency situations?

14.2   References

1. National Institute on Aging, National Institutes of Health. What is Alzheimer's disease? n.d. https://www.nia.nih.gov/health/what-alzheimers-disease.

2. Alzheimer's Association. Alzheimer's disease and related dementias. n.d. https://www.cdc.gov/aging/aginginfo/alzheimers.htm#:~:text=In%202020%2C%20an%20estimated%205.8,or%20older%20had%20Alzheimer's%20disease.&text=1-,This%20number%20is%20projected%20to%20nearly,14%20million%20people%20by%202060.

3. Alzheimier's Association, Wandering. 2022. https://www.alz.org/help-support/caregiving/stages-behaviors/wandering#:~:text=Six%20in%2010%20people%20living,heavily%20on%20caregivers%20and%20family.

15

Use-Case Example 2: Bose AI-Powered Cycling Coach and Warning System

> Safety should never be a priority; it should be a precondition.
> —Paul O'Neill, former CEO of Alcoa and US secretary of the treasury

In this chapter, we present the strategic road map developed by a Massachusetts Institute of Technology (MIT) graduate engineering team for a Bose artificial intelligence (AI)–powered cycling coach designed to enhance cyclists' performance and ensure their safety. The MIT team consisted of graduate students John Bailey, Bianca Lepe, John Montgomery, Tomohisa Okamoto, and Wonyoung So.

Cycling culture is rapidly growing across the US, with cycling emerging as not just a form of transportation, but recreation and a way to improve health. As of 2021, there were 51.4 million cyclists in the US [1]. As public interests for cycling continues to mature, governments have continued to invest in cycling infrastructure ranging from parking lots to dedicated cycling lanes in an effort to motivate greater adoption of this pollution-free mode of transportation. However, the uptick in cycling has come with a fair share of challenges and risks. Namely, the large increase in fatal accidents remains a major concern for lawmakers, first responders, and cyclists on the road. Between 2013 and 2017, there were nearly 4,000 reported cycling deaths due to accidents involving cyclists and others. Every year, close to 50,000 cycling accidents are reported, often caused by inattention by cyclists or vehicle drivers (or both).

Given the increasing popularity of cycling, the need for prioritizing cyclist safety, and the potential for using data and AI to enhance cyclist performance, the MIT team developed a strategic road map for an AI-powered cycling coach, with the value proposition centered on ensuring cyclists' safety and enhancing their performance. The proposed solution enabled three key functionalities:

- **Intelligent route generation:** The system generates safe and optimized routes for riders based on their objectives (i.e., sports, commuting and other travel, recreational). Workout information and goals can be integrated with existing coaching platforms to inform ride objectives.
- **Real-time accident avoidance:** The system employs sensor data to detect potential hazards and proactively alert the cyclist to avoid accidents. Sensors supplement humans' awareness and ability to perceive obstacles and predict oncoming threats.
- **Bose safe sound technology:** The system integrates audio hardware within the bicycle helmet to provide seamless, nonintrusive coaching and safety alerts to the rider through Bose sound technology.

Mission and Vision

The MIT engineering team defined clear and concise mission and vision statements for the AI-powered Bose cycling coach:

Mission: Use data to accelerate cyclists' performance.

Vision: Every cyclist has a personal coach available 24/7 that inspires performance to higher levels.

The team's mission statement clearly defines the *why* behind the product—*to accelerate cyclists' performance*. And the vision statement complements the mission statement by defining a clear vision for what the team *aspires* to achieve—*to equip every cyclist with a personal coach available 24/7 to inspire higher performance*.

Envisioned Future

The envisioned future is key to defining the near-term, midterm, and far-term milestones and prospects for the proposed solution. The AI-powered Bose cycling coach team formulated three horizons, illustrated in figure 15.1.

The proposed envisioned future, consisting of three horizons, outlines the future milestones for the team and the product. Horizon 1 (over 1–2 years) is focused on core capabilities such as object/obstacle detection, safety classification, and logging of physical ride data. Horizon 2 (3–4 years) aims to enhance the offering by enabling reliable route generation based on crowdsourced data from a network of multiple riders, as well as the ability to sync with existing gadgets to consolidate rider data. Finally, horizon 3 (5+ years) seeks to integrate the context to successfully predict and proactively identify

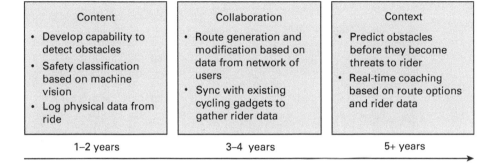

Content	Collaboration	Context
• Develop capability to detect obstacles • Safety classification based on machine vision • Log physical data from ride	• Route generation and modification based on data from network of users • Sync with existing cycling gadgets to gather rider data	• Predict obstacles before they become threats to rider • Real-time coaching based on route options and rider data
1–2 years	3–4 years	5+ years

Figure 15.1   Envisioned future for the Bose AI-powered cycling coach.

threats to ensure rider safety, while also enhancing rider performance through real-time coaching based on rider preferences, data, and progress.

Strategic Direction

We now turn our attention to the team's strategic direction, which maps out the competencies required to enable the core capabilities and subsequent system applications. Figure 15.2 presents these three key components of the strategic direction for the Bose AI-powered cycling coach. The three key functionalities previously described (intelligent route generation, real-time accident avoidance, and Bose safe sound technology) represent the system applications. These applications are enabled by various core capabilities such as routing algorithms, object detection, and classification models, integration of systems and external devices/applications, and an audiovisual interface. Finally, these capabilities require technical competencies such as map integration, data conditioning, feature extraction and detection, machine learning (ML) and neural networks, and secure data storage.

AI Value Proposition

With a clearly defined strategic direction that outlines the desired system applications, enabling capabilities, and corresponding competencies, we present the AI system architecture in figure 15.3, which illustrates the AI value proposition for the proposed AI-powered cycling coach.

The AI system architecture depicts various data sources, including labeled image data sets as well as numerous real-time sensor-enabled data, being conditioned and fed as

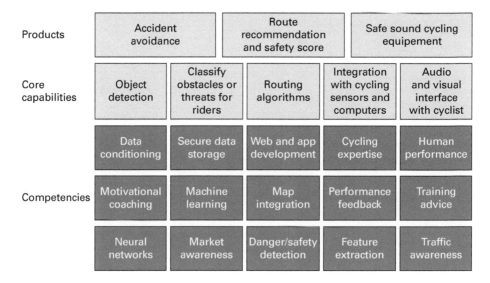

Figure 15.2   Strategic direction for the Bose AI-powered cycling coach.

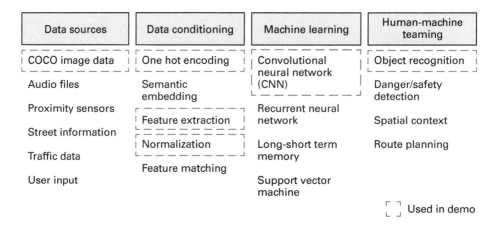

Figure 15.3   AI value proposition for Bose AI-powered cycling coach.

input to various ML algorithms and deep learning architectures such as convolutional neural networks (CNNs) and recurrent neural networks (RNNs). These data and algorithms work together to enable the various competencies required to power the three core functionalities of the AI-powered cycling coach. The architecture employs publicly available data sets such as the COCO large-scale object detection, segmentation, and captioning data set, as well as various sensors and open-source ML libraries. Such data and algorithms are paired to deliver insights that can ultimately enable successful human-machine teaming (HMT) in efficient route planning and reliable danger detection.

AI Strategic Road Map

With a clearly defined AI value proposition, we now present the AI strategic road map, which brings together the key elements of the AI strategic development model (AISDM) that have been presented thus far. The team proposed a solution motivated by the increasing adoption of cycling and the large number of accidents and fatalities associated with this growth. Furthermore, the lack of existing solutions focusing on ensuring rider safety further motivated the need for the proposed solution.

Given the importance of rider safety, there is a significant payoff if the solution is successful, as cyclists will be able to perform better and more safely. However, there are also risks involved with the proposed solution. Developing a product that consumers will rely on for rider safety requires a high level of confidence in the system's recommendations and decisions. If such a system exhibits low confidence, there is a high risk to users given the potential consequence of action. Therefore, the proposed solution must undergo extensive verification and testing to ensure a very high degree of reliability.

In addition to formalizing the business needs and proposed solution payoffs and risks, performing an analysis of strengths, weaknesses, opportunities, and threats (SWOT) is another key element of the AI strategic road map. We present the SWOT analysis for the Bose AI-powered cycling coach in figure 15.4.

The SWOT analysis presented in figure 15.4 highlights the team's strengths, especially with Bose being the market leader in audio technology and wearables. These strengths, combined with an unmet market need for innovative solutions addressing cyclists' safety needs, present great opportunities for innovation and market disruption. However, the team's lack of experience and expertise in developing cycling-related technology makes it challenging to establish credibility and name recognition in this space. This is particularly relevant considering the threats posed by current market competitors, which have a long track record of developing hardware and software products in this field.

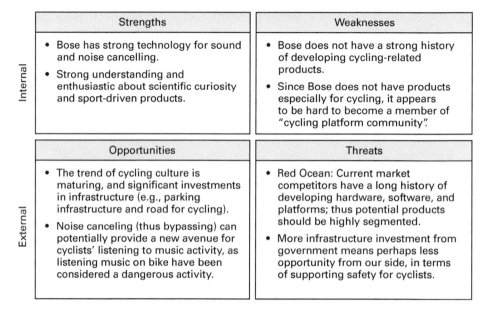

Figure 15.4   Bose cycling coach team SWOT analysis.

In line with the final element of the AI strategic road map, the MIT Bose AI-powered cycling coach team successfully demonstrated a minimum viable product (MVP) to a panel of AI academics and practitioners (see figure 15.5). The key building blocks of this demo are outlined in the AI system architecture in figure 15.3. Using the COCO data set, the team trained a CNN to track and detect objects, such as deer, cars, and trucks, to determine the safety of the surroundings. The demonstration involved loading a classifier model onto the Raspberry Pi to perform object tracking and detection from a PiCamera feed. Ultimately, this capability and system application helped ensure cyclists' safety by monitoring the cyclists' surroundings as they ride. The robust and reliable object tracking and detection algorithms can alert cyclists to impending obstacles before an accident occurs, even from their "blind spots" or when there are objects that are out of their vision (e.g., far ahead). Furthermore, in addition to object tracking and detection, AI capabilities can parse the likelihood of danger posed by an object by integrating identification and distance monitoring.

The strategic plan presented in this chapter used key elements of the AISDM framework presented in chapter 9 to showcase the blueprint for an AI-powered cycling coach seeking to ensure rider safety and enhance performance.

Our approach focuses on object detection (identification, distance) where camera sensors determine the safety of surroundings.

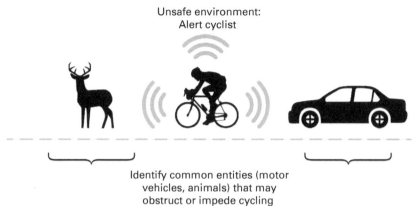

Figure 15.5   Demo solution for the Bose AI-powered cycling coach.

15.1   Exercises

1. How will you educate cyclists on the proper use and limitations of the proposed system, to prevent rider overreliance and ensure that they still remain alert and aware of their surroundings at all times?

2. How will you ensure that the system is reliable and robust enough to detect obstacles and hazards, even in low light or adverse weather conditions?

3. How would you address the potential issue of false alarms (i.e., false positives) or false negatives (i.e., missed dangers) in the recommendation and alerting system?

15.2   Reference

1. Statista, Cycling—Statistics & facts. 2022. https://www.statista.com/topics /1686/cycling/#topicHeader__wrapper.

16

Use-Case Example 3: Meal Evaluation and Attainment Logistics System (MEALS)

> To eat is a necessity, but to eat intelligently is an art.
> —Francois de La Rochefoucauld, French writer and moralist (1613–1680)

In this chapter, we present the strategic road map developed by a team of industry practitioners at the Massachusetts Institute of Technology (MIT) for an artificial intelligence (AI)–powered meal evaluation and attainment logistics system known as MEALS. The MIT team consisted of Liliana Horne, Sharna Sattiraju, Lida Dominguez, Ryan Daley, and Ramesh Selladurai.

A healthy and balanced diet is key to maintaining good health and nutrition. However, most individuals in today's fast-paced world, faced with time constraints from their school, work, and personal lives, turn to unhealthy convenience foods that are cheap, tasty, quick, and convenient as opposed to nutritious but often complex and costly home-prepared meals. This trend highlights the clear need for improving the current limitations around meal planning, ingredient shopping, and preparation of nutritious meals while reducing food waste. These limitations include the significant amount of time required to plan menus, identify and purchase ingredients, and finally prepare meals. In addition, the challenge of planning healthy, personalized meals that adhere to one's dietary or allergy restrictions and personal preferences, without making meal planning and preparation too complicated, is also difficult. Finally, this includes the task of optimally crafting meals comprised of ingredients that will fit one's budget.

MEALS is an innovative AI-powered system that aims to simplify and optimize the meal planning and preparation process. It employs sensor technologies and vital signs to intelligently recommend personalized meals based on an individual's preferences, nutritional goals, time and budget constraints, and micronutrient requirements to ensure the most nutritious meals. MEALS also enhances the end-to-end experience of meal planning, shopping, and preparation by providing seamless integration with shopping and food delivery platforms. Its ultimate objective is to develop an integrated kitchen robot capable of autonomous preparation of recommended meals.

Mission and Vision

The MIT team articulated clear mission and vision statements for the AI-powered meal evaluation and attainment logistics system:

Mission: Use data to make personalized meal planning, ingredient shopping, and preparation easy and affordable.

Vision: Create a world where every individual uses MEALS to improve their health, productivity and finances while reducing food waste.

The team's mission statement clearly lays out the *why* behind the proposed solution—*to make personalized meal planning, ingredient shopping, and preparation easy and affordable.* The team's vision statement complements its mission statement by defining a clear vision that articulates the future aspirations of the team—a world where every individual uses this solution to *improve their personal well-being* while also *reducing food waste.*

Envisioned Future

With a clearly defined mission and vision, the team formulated its envisioned future, consisting of near-term, midterm, and far-term milestones for the development of a meal evaluation and attainment logistics system. This envisioned future, consisting of three horizons, is illustrated in figure 16.1.

Horizon 1 (over 1–2 years) of the envisioned future is focused on the foundational capability of incorporating user health data to deliver personalized meal plans that meet health and budget requirements. Furthermore, this horizon also aims to provide users with intelligently curated shopping lists based on meal plans and market prices, all while continuing to track user feedback and engagement with MEALS recommendations to further improve the user experience. Horizon 2 (2–4 years) aims to incorporate health providers' inputs into meal planning and integrate with other apps and wearable devices to gather and input additional user health data to improve the health and nutrition of MEALS-curated meal plans. Finally, horizon 3 (5+ years) aims to integrate with a broader set of personal devices to track user activity and vitals, while proactively suggesting meals based on perceived patterns or health risks. This horizon also seeks to provide

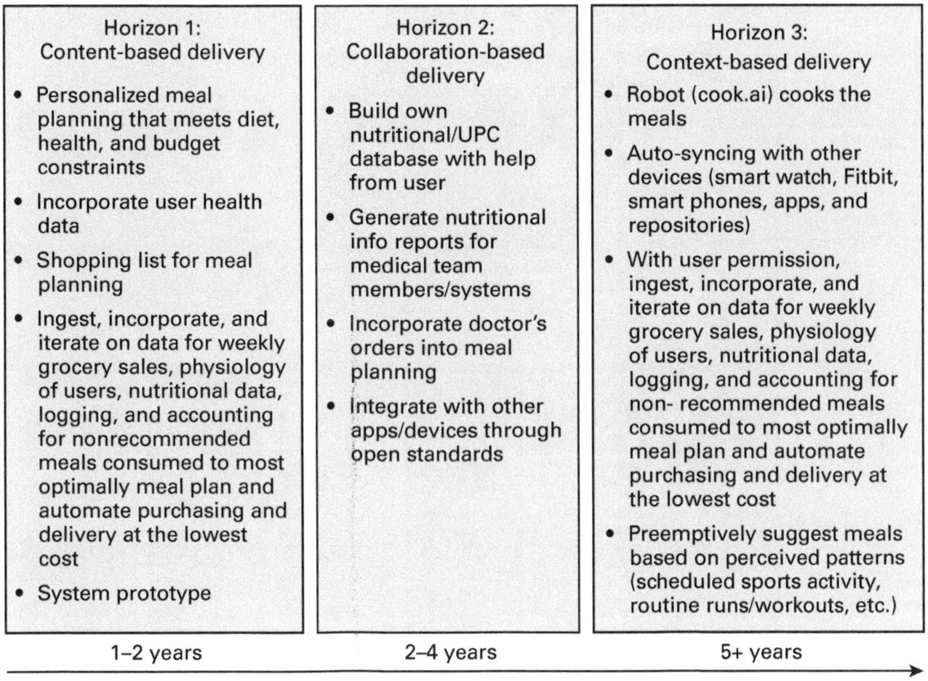

Horizon 1: Content-based delivery	Horizon 2: Collaboration-based delivery	Horizon 3: Context-based delivery
• Personalized meal planning that meets diet, health, and budget constraints • Incorporate user health data • Shopping list for meal planning • Ingest, incorporate, and iterate on data for weekly grocery sales, physiology of users, nutritional data, logging, and accounting for nonrecommended meals consumed to most optimally meal plan and automate purchasing and delivery at the lowest cost • System prototype	• Build own nutritional/UPC database with help from user • Generate nutritional info reports for medical team members/systems • Incorporate doctor's orders into meal planning • Integrate with other apps/devices through open standards	• Robot (cook.ai) cooks the meals • Auto-syncing with other devices (smart watch, Fitbit, smart phones, apps, and repositories) • With user permission, ingest, incorporate, and iterate on data for weekly grocery sales, physiology of users, nutritional data, logging, and accounting for non- recommended meals consumed to most optimally meal plan and automate purchasing and delivery at the lowest cost • Preemptively suggest meals based on perceived patterns (scheduled sports activity, routine runs/workouts, etc.)
1–2 years	2–4 years	5+ years

Figure 16.1   Envisioned future for AI-powered MEALS.

meal preparation capabilities by developing and delivering a kitchen robot capable of integrating with MEALS recommendations to prepare meals.

Strategic Direction

We now turn our focus to the team's strategic direction. The MEALS strategic direction is captured in figure 16.2, and presents the competencies and core capabilities needed to enable the three desired system applications—personalized meal planning, automated grocery shopping and inventory planning, and integrated meal preparation by a robot. These system applications require core capabilities such as the ability to provide meal recommendation based on user and sensor data, and the ability to integrate with shopping and food delivery platforms to optimize for time and budget constraints. These capabilities require competencies including the curation and preparation of data sets to

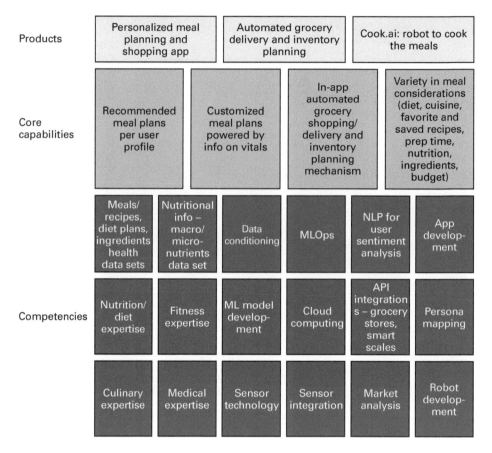

Figure 16.2   Strategic direction for AI-powered MEALS.

be used for designing a meal recommendation model, integration of sensor technologies, application programming interface (API) integration with grocery delivery platforms, robot development, and subject matter expertise in nutrition.

AI Value Proposition

We now present the team's AI system architecture as shown in figure 16.3, which demonstrates the AI value proposition for horizon 1 of the proposed AI-powered meal recommendation system application.

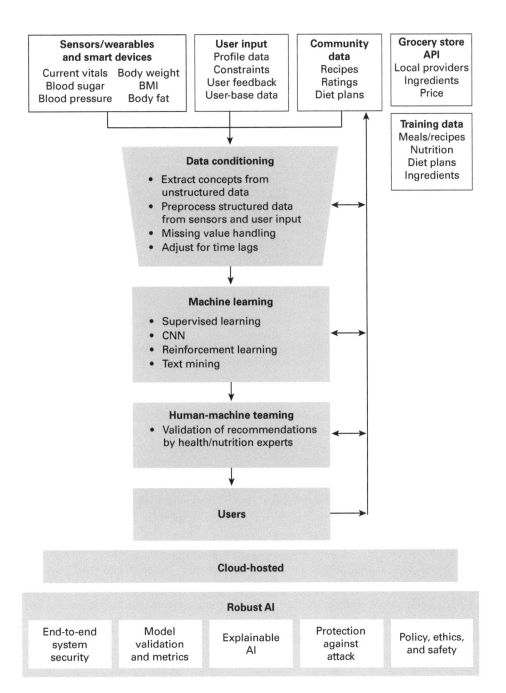

Figure 16.3 AI system architecture for MEALS.

The presented AI system architecture depicts various data sources such as real-time sensor-enabled data from wearable devices that capture important vital signs and user activity data, user-inputted profile data, external training data, and real-time grocery/market data, which are conditioned and fed as inputs to various machine learning (ML) models. The ML models provide personalized meal recommendations, which are reviewed and validated by nutrition experts. Feedback from users and nutrition experts is used as additional training data to improve the precision of future model recommendations. The system also considers user engagement with meal recommendations as additional training data for refining future recommendations. The proposed architecture aims to build, train, and deploy the ML models in a cloud-based environment, and it adheres to responsible AI (RAI) and robust AI principles, such as explainability of meal recommendations and end-to-end security of sensitive user data on the platform.

AI Strategic Road Map

We now present the team's AI strategic road map, which brings together the key elements of the AI strategic development model (AISDM) presented so far. The AI strategic road map is formalized by first articulating the business need for the proposed solution. As discussed in chapter 9, business needs can be best articulated through answers to a set of key questions, which are an adaptation of the well-known "Heilmeier Catechism." Answers to these set of questions for MEALS are presented in table 16.1.

The team identified the business needs for the proposed solution by addressing important questions, such as the limitations of existing solutions, the unmet customer requirements, the key differentiators of the proposed solution, and the risks and payoffs.

An analysis of the proposed solution's strengths, weaknesses, opportunities and threats (SWOT) is another key element of the AI strategic road map. The team's SWOT analysis highlights the integration of sensor-based data, real-time meal prediction based on user and sensor data, and end-to-end integration with grocery delivery platforms as some of the key strengths. The global interest in health, nutrition, and environmentally sustainable improvements to food systems presents a great opportunity for a solution like MEALS, which seeks to improve user well-being while reducing food waste. Furthermore, the potential for future integrations with personal digital assistants (e.g., Alexa, Google Home, and Siri) presents opportunities for a more seamless user experience.

The SWOT analysis also calls out some challenges in the form of internal weaknesses and external threats. The team calls out the high upfront cost for the computing technology needed to enable the desired solution, as well as the limited subject matter expertise in nutrition and the limited AI talent, as some of the biggest internal weaknesses. In

Table 16.1   Business needs for AI-powered MEALS

Question	Answer
What are you trying to do?	Use data to personalize meal planning, ingredient shopping, and preparation easy and affordable.
How is it done today, and what are the limits of current practices?	Planning, shopping, inventorying, and prep are too time consuming. It is very challenging to meet health goals, preferences, and budget manually.
What's new, and why do we know it will be successful?	Integrate health, dietary, and shopping data for prediction and recommendations on meal prep and purchasing. Evolve to incorporate autonomous meal preparation as well.
Who cares?	Health- and budget-conscious individuals who focus on preferences, manufacturers, grocers, and advertisers.
When successful, what difference will it make?	Users will save time and money, improve their health, and eliminate food waste.
What are the risks and the payoffs?	Risks: Low AI experience. No competing comprehensive system. Payoffs: High profits if you are first to market, with possible revenue streams from users, providers, manufacturers, advertisers, and others.
How much will it cost? How long will it take?	High initial research and development (R&D), capital expenditure, and operation expenses. Proof-of-concept platform development is estimated to take 2 mo. time, at a cost of $500,000. System will scale much better than in-store AI approach (e.g., smart carts, radio frequency identification). Revenue streams (manufacturers, grocers, advertisers, etc.) offsetting some costs early for services and capital.
What are the midterm and final "exams" to check for success?	Horizon 1 evaluated with a proof-of-concept prototype. Training to optimize meal recommendations and shopping through user profile, health data, preferences, and user feedback has been proven effective.

addition, the advantage that other competitors have with richer data that can enable stronger recommendation algorithms, as well as the need for undergoing extensive testing and regulation to measure the efficacy and long-term health effects of the provided meal recommendations, are external threats that the team would need to address.

The strategic plan presented in this chapter used key elements of the AISDM framework presented in chapter 9 to showcase the blueprint for an AI-powered meal evaluation and attainment logistics system.

16.1   Exercises

1. Imagine that you are the leader of the AI team responsible for deploying the AI-powered MEALS solution. How would you ensure user acceptance? Hint: Refer back to system readiness levels (SRLs), discussed in chapter 10.

2. What do you foresee to be the challenges in deploying the MEALS capability at the edge (i.e., edge computing)?

3. There are apps that encourage individuals to exercise via daily challenges (e.g., Apple Health). How would you formulate health challenges that can be integrated within the MEALS capability?

4. How can the MEALS team ensure that the system is designed to avoid bias in recommendations and to accommodate the diverse nutritional needs and cultural and dietary preferences of its users, while also optimizing for taste and cost?

5. What measures can be put in place to protect the privacy and security of user data in the cloud-based environment where ML models are built and deployed? What are the advantages and disadvantages of deploying models to the edge instead?

6. What ethical considerations should MEALS take into account when developing an integration kitchen robot capable of autonomous meal preparation?

17

Use-Case Example 4: Managing Energy for Smart Homes (MESH)

> The environment is where we all meet, where we all have a mutual interest; it is the one thing all of us share.
> —Lady Bird Johnson, former First Lady of the US

In this chapter, we present the strategic road map developed by a team of industry practitioners at the Massachusetts Institute of Technology (MIT) for an artificial intelligence (AI)–powered smart home energy management system known as Managing Energy for Smart Homes (MESH). The MIT team consisted of Tom Swanson, Simon Delavalle, Enrico Gebauer, Demilade Adedinsewo, and Akshit Bhatnagar.

The burning of fossil fuels such as coal, natural gas, and oil to generate electricity and heat is the largest contributor to the greenhouse gases that trap the Sun's heat in Earth's atmosphere [1]. Scientists predict that the increase of global temperatures due to human-made greenhouses gases will continue, and the effects on society will be grave if we don't collectively address the climate crisis [2]. Reducing greenhouse gas emissions by minimizing our society's reliance on fossil fuels is essential to avoid the adverse impacts of climate change [3]. Renewable energy sources such as solar energy, wind energy, and hydropower are alternative sources of energy that are key to addressing the climate crisis. We are already beginning to see citizens and businesses contributing to the solution as consumers and providers of distributed sources of renewable energy. Similarly, energy providers and governments continue to find ways to incentivize and promote the use and self-generation of renewable energy, and there continue to be valid concerns around the grid stability of renewable energy sources.

MESH aims to enable the transition to renewable energy by optimizing energy consumption and distribution through a decentralized smart energy ecosystem incorporating all participants, including businesses and individuals. This system's solution is centered around three key offerings:

- **Energy optimizer:** Analyze consumption to identify usage patterns and optimization opportunities, and ultimately provide individuals and businesses with AI-powered recommendations on how to reduce their energy usage.
- **Forecasting:** Use customer data to provide grid operators with consumption forecasts to be used for energy distribution and revenue optimization.
- **Grid stabilizer:** Aggregate data from all renewable energy generation and consumption (supply and demand) sources and employ AI to optimize the distribution of energy and ultimately ensure grid stability.

Mission and Vision

MESH seeks to enable the transition to renewable energy by optimizing the consumption and distribution of renewable energy. The team clearly defines the mission and vision statements for their energy management system as follows:

Mission: To meet emission reduction goals, mitigate climate change, and accelerate decarbonization by ensuring grid stability and optimization as the energy mix becomes more distributed and renewable.

Vision: To provide consumers and businesses with active participation in the energy system, gaining financial and sustainability benefits. Also, to permit energy grid operators to have access to a rich collection of data and dynamic predictive models, enabling a shift to distributed renewable sources as a competitive advantage.

The mission statement succinctly outlines the team's why—to accelerate decarbonization and the transition to renewable energy. The vision statement captures the future aspirations that the team holds for its two key stakeholders—consumers and businesses—as well as energy grid operators. It clearly articulates the vision and the future aspirations for these key beneficiaries of the team's proposed solution.

Envisioned Future

The team's envisioned future, consisting of three horizons, is illustrated in figure 17.1. It lays out the key near-term, midterm, and far-term milestones for the team in the development of its energy management system.

Horizon 1 (1–2 years) is focused on content-based delivery, and it seeks to use smart home hubs connecting Internet of Things (IoT) devices to enable comprehensive energy

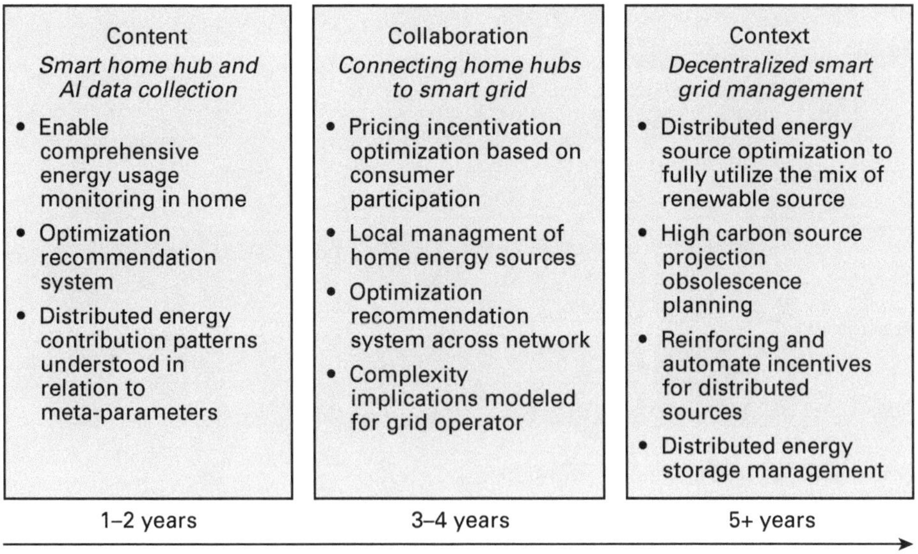

Content *Smart home hub and AI data collection*	Collaboration *Connecting home hubs to smart grid*	Context *Decentralized smart grid management*
• Enable comprehensive energy usage monitoring in home • Optimization recommendation system • Distributed energy contribution patterns understood in relation to meta-parameters	• Pricing incentivation optimization based on consumer participation • Local managment of home energy sources • Optimization recommendation system across network • Complexity implications modeled for grid operator	• Distributed energy source optimization to fully utilize the mix of renewable source • High carbon source projection obsolescence planning • Reinforcing and automate incentives for distributed sources • Distributed energy storage management
1–2 years	3–4 years	5+ years

Figure 17.1   Envisioned future for MESH.

usage monitoring. This data will be used to understand energy consumption patterns and provide users with AI-powered recommendations for optimizing energy consumption. Consumer data will also be used by energy grid operators to better understand distributed energy contribution patterns. These data sources and foundational AI capabilities will serve as the basis for horizon 2 (3–4 years), which seeks to connect home hubs to a smart grid. Doing so will enable not just consumption energy forecasting and optimization, but also pricing optimization and local management of home energy sources. An interconnected smart grid will allow MESH to provide grid operators with better supply-and-demand optimization recommendations in relation to individual hub energy generation and consumption patterns. Finally, horizon 3 (5+ years) seeks to enable a decentralized smart grid management system where a majority of energy source optimization and the obsolescence of high carbon sources are AI-powered.

Strategic Direction

MESH's strategic direction is presented in figure 17.2. It presents the core competencies necessary to enable the core capabilities that serve as the key building blocks for developing the desired system applications and products. MESH's offering is centered around three key system applications—energy optimization for homes and businesses,

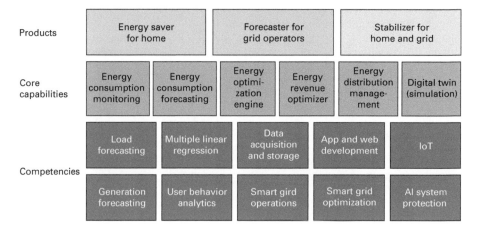

Figure 17.2   Strategic direction for MESH.

forecasting for grid operators, and smart energy grid stabilization. To enable these system-level applications, there are key core capabilities, such as energy consumption monitoring and forecasting, energy consumption and revenue optimization, and distribution management. To enable these capabilities, MESH relies on key technological competencies such as IoT, data acquisition and storage, energy consumption and generation forecasting, user behavior analytics, and smart grid optimization.

AI Value Proposition

We now turn our attention to the team's AI system architecture, presented in figure 17.3. The AI system architecture showcases the key role of various data sources, data conditioning, and machine learning (ML) techniques in delivering the desired human-machine teaming (HMT) capabilities. Namely, the presented AI system architecture showcases various data sources, such as IoT sensors, user-input data, energy consumption and supply trends, and external data such as weather data. Properly stored and processed data from these sources is employed to enable various ML techniques such as clustering, regression, and forecasting. The resulting ML outputs are delivered as recommendations to consumers and grid operators for optimizing renewable energy consumption and generation to ensure grid stability.

Data sources	Data conditioning	Machine learning	Human-machine teaming
Sensors (smart plugs)	Feature extraction	Recommender system	Energy use planning and modification
User input through smart home device	Normalization	Clustering	Energy optimization
	Clustering	Forecasting	Innovation in renewable clean energy sources
Energy supply	Data imputation	Regression	
Energy consumption			Balance between grid supply and energy produced in home
Weather data			
Calendar events			
			Asset life-cycle planning

Figure 17.3   MESH AI system architecture.

AI Strategic Road Map

We now present the team's AI strategic road map, which brings together various elements of the AI strategic development model (AISDM) presented thus far. The first step is to clearly define the business need for the proposed solution. In the case of MESH, the need for the proposed solution is clearly defined by identifying limitations in the existing solutions. Namely, energy providers today base their consumption and generation predictions on simplistic models with limited data. Similarly, consumers are only now starting to use smart technologies such as Google Nest to manage their energy. Emerging solutions only focus on pricing optimization based on individual consumption trends. However, MESH seeks to enable a decentralized energy ecosystem including all participants and their data to create a larger, more comprehensive data source for more accurate modeling and forecasting. This smart grid solution presents the potential for cost savings and grid stability, all the while helping to address the climate crisis. Ultimately, by identifying gaps in the existing solutions, the unmet customer needs, and key differentiators and business impact of the proposed solution, MESH is able to clearly define the business need for its proposed solution.

In addition to defining the business need, it is important to analyze and identify the internal and external considerations in the form of a strengths, weaknesses, opportunities, and threats (SWOT) analysis. The team's SWOT analysis, presented in figure 17.4, highlights the internal strengths of the proposed solution in using existing smart home

Strengths	Weaknesses
• Uses existing technologies • Addresses the fluctuation of renewable energies • Expands available data for energy use predictions • Positive impact for individuals and energy operators	• First step (horizon 1) requires penetration of an existing market • Requires data science experts with advanced AI skills • Benefit to grid operators is more viable if smart home devices are widely adopted by home and business owners
Opportunities	**Threats**
• Strategic platform for effective management of growing renewable energy sources • Can aid planning and transition from non-renewable to renewable energy sources • Potential market expansion with shift to renewable energy • Drive innovative energy storage solutions	• Smart home energy competitor (Google Nest + Leap) with edge computing capabilities • Challenges with user acceptance of "another" device • Limited widespread adoption if AI model does not offer significant incremental benefit compared to existing prediction models

Figure 17.4   SWOT analysis for MESH.

technologies, addressing the instability of renewable energy sources, expanding the availability of data, and the positive impact on individuals, energy operators and society at large. The analysis also identifies external opportunities. Namely, the proposed solution also demonstrates great potential for serving as a strategic platform for the transition to and management of renewable energy sources, given the growing interest and shift toward renewable energy.

The SWOT analysis also calls out some challenges in the form of internal weaknesses and external threats. The weaknesses are primarily due to the large effort needed to enter and try to transform an existing market, as well as the challenge of acquiring the right talent to help the team deliver its solution. The team also identifies external threats, including smart home energy competitors such as Google Nest and Leap, which provide edge computing capabilities. In addition, there is concern that users may resist yet another device in their home. This is especially a threat to the success of the smart grid, as the efficacy of the AI-powered energy management system relies on the widespread adoption and collection of data from a large base of users.

The strategic plan presented in this chapter included key elements of the AISDM framework presented in chapter 9 to showcase a blueprint for an AI-powered energy management system (MESH).

17.1   Exercises

1. How can MESH ensure that its forecasting algorithms accurately predict energy consumption and are able to adapt to changing consumer behaviors and energy use patterns?
2. What are some potential challenges that the MESH team may face as it scales its AI-powered energy management system to a larger user base, and how can it address these challenges?
3. What are the potential advantages and disadvantages of using federated learning over a centralized learning approach (where data is uploaded from each connected device to the cloud) to training an AI solution for optimizing energy distribution and ensuring grid stability?

17.2   References

1. US Environmental Protection Agency (EPA), *Overview of greenhouse gases*. 2022. https://www.epa.gov/ghgemissions/overview-greenhouse-gases.
2. National Aeronautics and Space Administration (NASA), *The effects of climate change*. 2022. https://climate.nasa.gov/effects/#:~:text=We%20already%20see%20effects%20scientists,will%20also%20increase%20and%20intensify.
3. United Nations, Renewable energy—Powering a safer future. n.d. https://www.un.org/en/climatechange/raising-ambition/renewable-energy.

18

Use-Case Example 5: AquaAI, an AI-Powered Modernized Marine Maintenance System

> For most of history, man has had to fight nature to survive; in this century he is beginning to realize that, in order to survive, he must protect it.
> —Jacques Cousteau, French naval officer, oceanographer, filmmaker, and author

In this chapter, we present the strategic road map developed by a team of industry practitioners at the Massachusetts Institute of Technology (MIT) for AquaAI, an artificial intelligence (AI)–powered modernized marine maintenance system. The MIT team consisted of Keithe Bennett, Luke Brownlow, Corey Mullane, Pritesh Patel, and Ian Stevenson.

Oceans cover more than 70 percent of the planet's surface, and they also hold great potential as a reliable and renewable source of energy [1]. Renewable sources of energy can be key to reducing global carbon emissions and addressing the climate crisis. However, a challenge facing the marine energy industry is the constant decay of the environmental marine infrastructure, which puts it at risk of breaking down and resulting in large-scale power losses, environmental damage, and a damaged reputation for equipment providers and the clean energy industry in general. As a preventative measure, underwater marine infrastructure requires constant, risky, and high-cost inspection, monitoring, and maintenance by a limited number of specialists.

The current process of monitoring and maintaining the underwater infrastructure can be dull, dirty, and dangerous. This includes routine tasks such as inspecting underwater environments to identify and intervene in structural defects, as well as dangerous and specialized manual work. AquaAI's proposed solution is designed to modernize marine maintenance operations by equipping offshore energy producers with an AI engine for early detection of underwater structural risks.

AquaAI aims to do this by using machine learning (ML) for rapid and more accurate reviews of underwater imagery, drawing on diverse data sources to correlate information with external data and identify and predict risks to minimize the impact of infrastructure defects. AquaAI's offering for modernizing marine maintenance is centered around three key services: anomaly detection, recommendation system, and postmaintenance auditing. Through these services, AquaAI seeks to bring value to governments, energy producers and contractors, and consumers by employing AI to facilitate the dull, dirty, and dangerous task of marine maintenance. The AI-powered AquaAI solution contributes to better environmental policies and legislation and provides safer, cheaper, and more reliable energy sources for consumers.

Mission and Vision

AquaAI seeks to modernize marine maintenance by equipping offshore energy producers with an AI engine for early detection of underwater structural risks. The team clearly articulates the mission and vision statements for their system:

Mission: Enabling a catastrophe-free marine energy sector through data-driven inspections

Vision: Equipping offshore energy producers with an AI engine for early detection of underwater structural risks

The team's mission statement describes its *why*—to enable a catastrophe-free energy sector. Similarly, the vision statement describes the aspirations of the team in equipping offshore energy producers with an AI engine for early detection of underwater structural risks.

Envisioned Future

The envisioned future is used to outline the key near-term, midterm, and far-term milestones. AquaAI's envisioned future, consisting of three horizons, is presented in figure 18.1.

Horizon 1 (1–2 years) is focused on developing a proof of concept with existing data and new technologies in order to validate the concept in the real world. This is achieved by equipping current underwater robots and divers with cameras and sensors and using data from energy producers and operators to train a structural defect classification system. In horizon 2 (3–4 years), we seek to put together the best of humans and robots by combining underwater divers and robot teams as part of human-machine teams (HMTs) for infrastructure inspection (including validating the model outputs), and enabling defect classification and maintenance decision-making. Finally, in horizon 3 (5+ years),

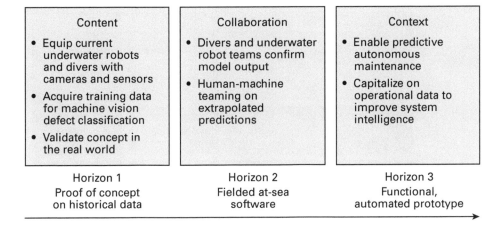

Content	Collaboration	Context
• Equip current underwater robots and divers with cameras and sensors • Acquire training data for machine vision defect classification • Validate concept in the real world	• Divers and underwater robot teams confirm model output • Human-machine teaming on extrapolated predictions	• Enable predictive autonomous maintenance • Capitalize on operational data to improve system intelligence

Horizon 1
Proof of concept
on historical data

Horizon 2
Fielded at-sea
software

Horizon 3
Functional,
automated prototype

Figure 18.1   AquaAI envisioned future.

we achieve the North Star goal of allowing fully autonomous inspection, analysis, and predictive maintenance.

Strategic Direction

A strategic direction lays out the implementation approach by presenting the core competencies that enable the core capabilities that will serve as the key building blocks for the desired system applications. AquaAI's strategic direction is presented in figure 18.2.

AquaAI's offers three key services—anomaly detection software, a recommendation system, and postmaintenance auditing. These services are enabled by computer vision capabilities, including structural defect classification, and visual identification of movement, as well as autonomous underwater vehicles and sensor fusion, which are critical to enabling the predictive autonomous maintenance described in horizon 3. Developing these core capabilities requires various data conditioning and AI development competencies, as well as subject matter expertise in marine energy and underwater infrastructure that the AquaAI team must develop.

AI Value Proposition

An AI system architecture provides a systems overview of the key building blocks such as data sources, data conditioning, and ML techniques required to deliver the desired AI capabilities. AquaAI's AI system architecture is presented in figure 18.3.

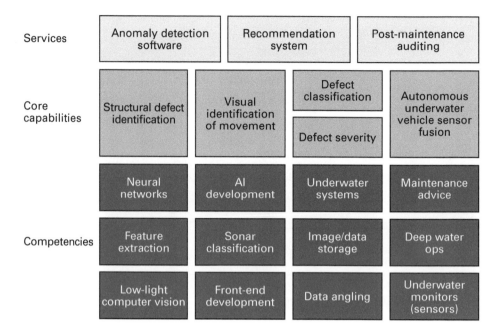

Figure 18.2 AquaAI strategic direction.

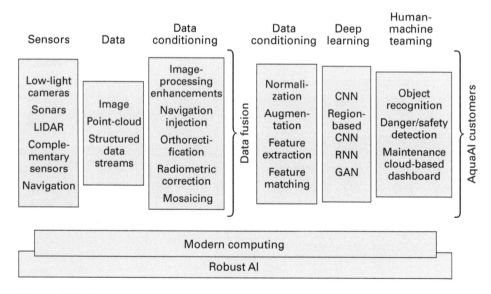

Figure 18.3 AquaAI AI system architecture.

The AI system architecture presents a diverse, multisensor data input, which includes data from sonar and LIDAR sensors, as well as navigational data and low-light cameras providing image data. These data inputs are processed using a set of preliminary yet specialized data conditioning techniques before different data sources are fused into a centralized data store. This data is conditioned using traditional data conditioning techniques. Deep learning techniques of various complexity are used to deliver three key functionalities: object recognition, danger/safety detection, and a maintenance dashboard to communicate key findings and recommendations to operators in a timely manner.

AI Strategic Road Map

We now turn our attention to AquaAI's AI strategic road map. The goal of an AI strategic road map is to bring together key elements of the AI strategic development model (AISDM) for the development of an AI product or service. One of the key steps of the AI strategic road map is clearly articulating the business needs for the proposed product or service. The business needs can be determined by providing answers to a set of eight key questions, known as the "Heilmeier Catechism." Answers to these questions provide additional clarity around the proposed solution and its novelty, the key stakeholders, and the long-term vision and associated costs and risks. The answers to these questions for the AquaAI system are presented in figure 18.4.

In addition to formalizing the business need for a proposed solution by articulating answers to the key Heilmeier questions, it is equally important to identify internal and external considerations in the form of a strengths, weaknesses, opportunities, and threats (SWOT) analysis. AquaAI's SWOT analysis, presented in figure 18.5, highlights the key internal strengths, which includes the expertise within the team in deep water, machine learning operations (MLOps), and AI, as well as the team's strong collaborations with research groups and institutions. Externally, AquaAI's pioneering work in modernizing marine maintenance presents a great opportunity for being the first in the market, using augmented reality for enhanced HMT as well as extending the applications of the core technology to other industries, such as defense.

The SWOT analysis also identifies some challenges in the form of internal weaknesses and external threats. One of the major weaknesses is the lack of ownership of historical data needed for training the desired ML capabilities and the physical assets required for further data collection. AquaAI relies on collaborating with energy producers and contractors to access the physical assets for data collection and evaluation, and the data needed to enable the desired ML capabilities. In addition, the input data sources, including low-light cameras and sensors, are constrained by the existing capabilities of those

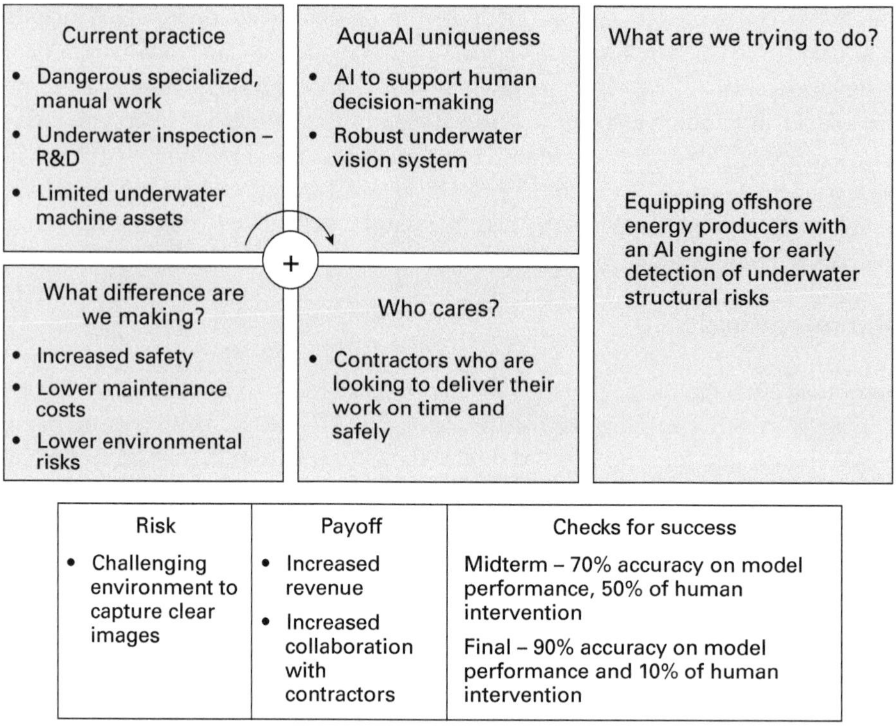

Current practice	AquaAI uniqueness	What are we trying to do?
• Dangerous specialized, manual work • Underwater inspection – R&D • Limited underwater machine assets	• AI to support human decision-making • Robust underwater vision system	Equipping offshore energy producers with an AI engine for early detection of underwater structural risks

What difference are we making?	Who cares?	
• Increased safety • Lower maintenance costs • Lower environmental risks	• Contractors who are looking to deliver their work on time and safely	

Risk	Payoff	Checks for success
• Challenging environment to capture clear images	• Increased revenue • Increased collaboration with contractors	Midterm – 70% accuracy on model performance, 50% of human intervention Final – 90% accuracy on model performance and 10% of human intervention

Figure 18.4 Business needs (Heilmeier catechism).

data sources. The team also calls out external threats, which include the inherent limitations in the data sources, emerging competition from fast-followers, and potential changes in legislation that could affect AquaAI's role and capability in modernizing marine maintenance.

The strategic plan presented in this chapter employed key elements of the AISDM framework presented in chapter 9 to showcase the blueprint for AquaAI—an AI-powered marine maintenance system.

18.1 Exercises

1. What role will HMT play in AquaAI's solution, and how can the team ensure that human operators are equipped to make informed decisions based on the system's output?

Strengths	Weaknesses
• Deep water operational experience • MLOps deployment • Proprietary AI algorithms • Collaborations with research institutes • Underwater computer vision experts • Diverse team of experts	• No owned physical assets for data collection and evaluation • Data required is owned by contractors • Funding and managerial processes that match scale of staffing needs • Dependency on existing camera/sensor technology
Opportunities	Threats
• Integration with augmented reality for enhanced human-machine interaction • No incumbents • Industry leading intellectual property • Parallel emerging industries • Counter sabotage/terrorism (defense contracts)	• Underwater image issues • Sabotage/terrorism • International waters • Changes in legislation • Competition: growing market & fast followers • Data & IP security from hardware loss • Injection into MLOps processes

Figure 18.5   AquaAI SWOT analysis.

2. What steps can be taken to ensure the accuracy and reliability of the data used to train the ML models that underpin the proposed solution (consider the impact of low-lighting, deep ocean debris, and other factors that may affect the quality of the data)?

3. How can AquaAI use other emerging technologies, such as augmented reality, to enhance the capabilities of the proposed solution?

18.2   Reference

1. Office of Energy Efficiency & Renewable Energy. n.d. https://www.energy.gov /eere/office-energy-efficiency-renewable-energy.

Appendix

A.1 Representative AI Industries and Sample Applications

The list given here itemizes artificial intelligence (AI) industries, for which narrow AI can be of great value, and sample applications within the respective industries. When this book is used as a textbook for a course, students can select an industry and sample application from this list for their team class project:

- Aerospace (e.g., drones, airplane design, satellites, air traffic flow prediction)
- Agriculture (e.g., precision farming, perishable food)
- Automation (e.g., driverless transportation)
- Chemicals (e.g., cleaning, hygiene)
- Climate change (e.g., water sustainability and purification, greenhouse effect reduction, floods, erosion, energy distribution, forest fires)
- Consumer goods and services (e.g., marketing, advertisement, manufacturing, e-commerce, lifelong learning)
- Cybersecurity (e.g., malware, adversarial AI, ransomware, deepfakes)
- Digital enterprise transformation (e.g., robotics process automation)
- Durable goods (e.g., automobiles, home appliances, construction tools, Internet of Things (IoT))
- Electronics and enterprise computing (e.g., data centers, cloud computing, microprocessor fabrication, testing)
- Engineering and construction (e.g., robotics for site planning, drones for site monitoring)
- Finance (e.g., banking, fraud detection, forecasting)
- Fitness (e.g., wearable devices, clothing, nutrition)
- Healthcare (e.g., medical diagnosis, medical equipment, anxiety/depression, surgical robots)

- Oil and gas (e.g., exploration, natural resources delivery)
- Tactile and haptic sensing (e.g., prosthetics, exoskeleton)
- Telecommunications (e.g., 5G/6G, cellular phones, wireless connectivity)
- Utilities (e.g., water, gas, electricity, power generation via solar or wind)
- X-reality (e.g., augmented, mixed, or virtual reality)

A.2   Setting up Your Interactive Development Environment (for Either PC or Mac OS Operating Systems)

In this appendix, we will walk you through the process of setting up your PC or Mac environment. Namely, we will walk you through the steps to:

- Install Anaconda Navigator.
- Install TensorFlow and set up a virtual environment to run the most recent version of Python (Python 3 at this time).
- Run a Jupyter notebook using Keras and TensorFlow.

Important note: Since the required packages are changed often by application providers, it is highly recommended to follow the latest instructions, available at the following sites:

- https://www.anaconda.com/download
- https://colab.research.google.com/?utm_source=scs-index
- https://keras.io/getting_started/
- https://www.tensorflow.org/install

Download and Install Anaconda Navigator

We assume that you have an internet browser installed (e.g., Chrome, Edge, Firefox). If not, please download your browser of choice. In the browser, navigate to https://www.anaconda.com/download to download "Anaconda Individual Edition" from Anaconda. Anaconda is a free distribution of the Python and R programming languages that aims to simplify package management and deployment. The current Python version is Python 3.11. To learn more about Anaconda, check out https://www.anaconda.com/products/distribution.

When downloading Anaconda, choose "64-Bit Graphical Installer," which will begin downloading an .exe file called Anaconda3-2020.11-Windows-x86_64 (PC) or

Anaconda3-2020.11-MacOSX-x86_64(MacOS). Follow the instructions displayed on the page and install using the default installation wizard.

Install TensorFlow and Set up a Virtual Environment

It is highly advisable that the Jupyter notebooks discussed in the rest of these appendixes be executed using Google Colab (https://www.youtube.com/watch?v=inN8seMm7UI). The benefit of running the Jupyter notebook in the Google Colab environment is that system updates, configuration files, and changes in operating systems are avoided by running the Colaboratory environment. Google Colab allows you to run Python code directly from your browser, in the cloud, with the following benefits:

- No configuration is required.
- You can get free access to graphics processing units (GPUs) computing.
- Packages like Keras and TensorFlow require no installation. They are already integrated into the Google Colab environment.
- It is easy to share the content generated among other colleagues and team members.

If the reader has to run the Jupyter notebook within a local central processing unit (CPU) environment instead of Google Colab, follow the latest instructions available in the TensorFlow installation site (https://www.tensorflow.org/install).

Launch and Run a Jupyter Notebook

Finally, we are ready to launch a Jupyter notebook. A Jupyter notebook is a web-based interactive development environment for creating and sharing computational documents. It is highly recommended to run the Jupyter notebook in the Google Colab environment mentioned earlier. An option for Google Colab appears in the top of the Jupyter notebook, which is available once you select the notebook that you want to run.

Once you have confirmed that you can launch a Jupyter notebook and import TensorFlow and Keras, you can run individual notebook cells, change cell parameters, or run all the cells from start to finish. After running the notebook, it is time to shut down the notebook and terminal. It is important to deactivate the runtime environment after running any Jupyter notebook. To do so, you must do the following:

- At the top of the Jupyter notebook, go to the Google Colab run-time option and shut down the cloud computing environment.

- In the MacOS terminal command, type CTRL+C at the same time, and type y. This will shut down the terminal

Many useful resources for learning how to navigate Jupyter notebooks are available at https://jupyter-notebook.readthedocs.io/en/stable/.

Congratulations! You have successfully installed Anaconda, set up a virtual environment to run the most recent version of Python, and launched a Jupyter notebook.

A.3   ML Performance Metrics

In this appendix, we walk through a hands-on exercise using a Jupyter notebook to run a machine learning (ML) algorithm called the stochastic gradient descent (SGD) classifier using Modified National Institute of Standards and Technology (MNIST) data. This hands-on exercise will provide an opportunity to run Python code implementing an ML algorithm and compute some of the key performance metrics discussed in chapter 4.

The material and Jupyter notebook complementing exercises A.3–A.5 are sourced from Aurelien Geron's *Hands-on Machine Learning with SciKit-Learn, Keras, and TensorFlow*, second edition (O'Reilly, 2019). You can find how to access the latest edition of the O'Reilly book, example code, and quick start instructions at https://github.com/ageron/handson-ml2.

MNIST Data Set and SGD Review

In this exercise, we will implement an SGD classifier using the MNIST data set.

The MNIST data set is a large collection of handwritten digits used for training and testing various image processing systems in the field of ML. The data set, created in 1998, consists of 70,000 images of handwritten digits. Each image is 28×28 pixels, containing a total of 784 features, where each feature simply represents a pixel's intensity from 0 (white) to 255 (black). Each image is labeled with the ground truth of the digit that it represents.

SGD is a simple and yet very efficient optimization technique used to find the values of parameters and coefficients of functions that minimize a defined cost function. SGD has been around in the ML community for a long time, but it has received considerable attention in recent years in the context of large-scale learning. In this exercise, we implement a linear classifier optimized by SGD, which implements an SGD learning routine for a defined loss function. We use the SciKit-Learn SGDClassifier module to implement the SGD classifier.

In this exercise, we will do the following:

- Train an SGD classifier.
- Evaluate the SGD classifier using cross-validation after the training is complete.
- Iterate on the hyperparameters of our model, such as the learning rate (which is a measure of how fast we want the SGD learning routine to arrive to a minimum).
- Compute Precision, Recall, and the F1-score
- Generate the receiver operating characteristic (ROC) and compute the area under the curve (AUC) to explore trade-offs and identify the best model with high recall and low false positive rates.

Recall from chapter 4 that one of the core requirements of ML is generalizability. As such, we use a hold-out test set for both cross-validation and testing to measure the performance of the model in a way that best captures how the model will perform when it is deployed in the production environment. For this exercise, we use 60,000 of the images from the data set for training and set aside 10,000 images for testing.

ML Metrics Review

In this exercise, we will compute several metrics to quantify the performance of our model. We encourage you to refer to table 4-1 in chapter 4 for a refresher on key performance metrics such as accuracy, precision, recall, and the F-1 score. We also encourage you to refer to table 4.2 for a refresher on the confusion matrix, a tabular visualization describing the performance of a classification model's predictions for a set of data for which the true values are known.

Implementing an SGD Classifier

We will begin by uploading and launching the Jupyter notebook for this exercise. The Jupyter notebook consists of a series of cells. There are two types of cells. A markdown cell is used for documentation using a lightweight markup language, while a code cell, denoted by square brackets on the left side of the cell, represents running programming code (Python, in this case). A code cell can be run by hitting Run in the top ribbon or hitting SHIFT+ENTER. When the cell is running, the square bracket on the side of the cell shows an asterisk as follows: [*]. When the cell finishes executing, a number appears in the square bracket as follows: [1], and the expected output appears below the cell.

In the first segment of the notebook, we begin by importing a few common modules (sklearn, numpy, and matplotlib), and ensuring that we are running the most recent version of Python.

Next, we fetch the MNIST data set. This consists of X, which is 70,000 images, each of which are represented as a feature vector of 784 features (i.e., pixels from the 28×28 image). We also have y, which is 70,000 1-dimensional vectors that are encoding the target labels corresponding to each feature vector in X:

```
X, y=mnist["data"], mnist["target"]
```

After gathering our data, we proceed to split it into training and test sets. As stated earlier, we will use 60,000 images (about 85 percent of the data) for training and use the remaining 10,000 (about 15 percent) as a hold-out set for testing:

```
X_train, X_test, y_train, y_test=X[:60000],
      X[60000:], y[:60000], y[60000:]
```

After splitting the data, we import the SGDClassifier from the sklearn library. We then proceed to fit the classifier to our training data, defining the maximum number of passes over the training data (i.e., epochs) as 1,000, the stopping criterion which is default set to 1e-3, and a random state that is used for shuffling the data set to 42. You can explore tuning these hyperparameters to see the resulting impact on the performance of your final model. You can read more about these hyperparameters in the official sklearn documentation for the SGDClassifier at https://scikit-learn.org/stable/modules/generated/sklearn.linear_model.SGDClassifier.html. Note that in this exercise, we are training the SGDClassifier to be able to classify an input handwritten image as the number 5 as follows:

```
from sklearn.linear_model import SGDClassifier
      sgd_clf=SGDClassifier(max_iter=1000,
            tol=1e-3, random_state=42)
      sgd_clf.fit(X_train, y_train_5)
```

Once we have fit the SGDClassifier to our training data, we proceed to compute some key performance metrics for our resulting model. We start by computing the accuracy for a three-fold cross-validation run. Recall that accuracy is a measure of the rate of predictions that our model got right from the total number of predictions. The metric can be computed as $Accuracy = \dfrac{TP+TN}{TP+FP+TN+FN}$.

The output is a three-dimensional array representing the accuracy for each run of the cross-validation.

```
from sklearn.model_selection import cross_val_score
cross_val_score(sgd_clf, X_train, y_train_5, cv=3,
                scoring="accuracy")
```

We also compute the accuracy through stratified K-folds cross-validation using the sklearn StratifiedKFold function. Our model's performance is assessed using cross-validation data. This process involves initiating cross-validation, using the model to generate predictions, and comparing these predictions against the actual ground truth data to compute our key metrics.

```
from sklearn.model_selection import cross_val_predict
     y_train_pred=cross_val_predict(sgd_clf,
          X_train, y_train_5, cv=3)
```

We start by computing our 2×2 confusion matrix using the sklearn function confusion_matrix(). Recall that a confusion matrix is a tabular visualization comparing predicted positive and negative data to true positives and true negatives in a test sample. In the case of a perfect classifier, 100 percent of true positives would be predicted as positives and 100 percent of true negatives would be predicted as negatives.

```
from sklearn.metrics import confusion_matrix
    confusion_matrix(y_train_5, y_train_pred)
```

We also use the predictions from our cross-validation to compute the precision of our model. Precision seeks to determine the proportion of identified positives (TP + FP) that were true positives (TP). It is also known as the Positive Predicted Value. The metric can be computed as $Precision = \dfrac{TP}{TP+FP}$.

```
from sklearn.metrics import
     precision_score, recall_score
precision_score(y_train_5, y_train_pred)
```

Similarly, we compute the recall, which is a measure of how confident we can be that the model found instances of a positive target level. The metric can be computed as $Recall = \dfrac{TP}{TP+FN}$. Both precision and recall can be computed directly using elements from the confusion matrix (TP, TN, FP, FN).

```
recall_score(y_train_5, y_train_pred)
```

We also compute the F1-score, which is a harmonic mean of precision and recall. The F1-score can be computed as follows: $\text{F1-Score} = 2 \times \dfrac{Precision \times Recall}{Precision + Recall}$.

```
from sklearn.metrics import f1_score
    f1_score(y_train_5, y_train_pred)
```

Now that we have computed the precision and recall of our model, we explore trade-offs between the two metrics. As precision increases, recall decreases, and vice versa. Depending on the application, you may have a greater need for operating at a higher recall or higher precision. This is known as the "precision/recall trade-off." We define a function—plot_precision_recall_vs_threshold()—that allows us to plot precision versus recall for different thresholds. We also define a function—plot_precision_vs_recall()—that allows us to plot the precision versus recall trade-off for a given threshold.

Finally, we generate our ROC, which is a visual representation of the trade-off between true positive rate (TPR) and false positive rate (FPR). The curve is plotted by computing the FPR and TPR at different decision thresholds.

```
from sklearn.metrics import roc_curve
fpr, tpr, thresholds = roc_curve(y_train_5, y_scores)
```

The ROC ultimately allows us to compute the AUC for the SGDClassifier, which is threshold invariant. Ideally, we want the AUC to be as close to 1 as possible (1 represents being able to perfectly distinguish all positive and negative classes correctly). The closer the AUC is to 1, the better the performance of the model. Calculating the AUC from the ROC is a very good way to compare ML model performance.

Congratulations! With that, we have completed our hands-on exercise implementing an SGD classifier using the MNIST data set. You have successfully completed all of the following actions:

- Trained an SGD classifier
- Evaluated the classifier using cross-validation
- Tuned different hyperparameters
- Computed key metrics such as precision, recall, and F1-score
- Generated the ROC and AUC to identify the best model with high recall and low FPR

In the following appendix, we will implement a multilayer perceptron (MLP).

A.4   Multilayer Perceptron Algorithm

In this appendix, we walk through a hands-on exercise using a Jupyter notebook to implement an MLP using the MNIST Fashion data set.

MNIST Fashion Data Set and MLP Review

The MNIST Fashion data set used in this exercise consists of 70,000 images. Each image is a 28×28 grayscale image associated with a label from one of the following ten classes:

- 0—T-shirt/top
- 1—Trouser
- 2—Pullover
- 3—Dress
- 4—Coat
- 5—Sandal
- 6—Shirt
- 7—Sneaker
- 8—Bag
- 9—Ankle boot

Each image is represented as a 784-dimension feature vector, where each feature represents a pixel's intensity from 0 (white) to 255 (black). Each image is labeled with the ground truth of the digit that it represents.

An MLP is a fully connected, feedforward artificial neural network (ANN). It consists of three types of layers—the input layer, the output layer, and hidden layers. The input layer receives the input signal to be processed. The processing is done at the hidden layers, which serves as the computational engine of the network. A final prediction/classification is generated at the output layer.

We begin by reviewing some useful definitions discussed throughout this section. We also encourage you to refer to chapter 4 for a deep dive into some of these concepts if you want more information:

- **Hyperparameter:** Tunable "knobs" available to the designer of a neural network. This includes things such as number of layers, number of nodes, type of activation function, learning rate, epochs, iterations, and batch size. These values are not learned by the algorithm itself from the data (unlike parameters), and thus are defined by the ML algorithm designer

- **Learning rate:** The rate at which the gradient descent searches for a global minimum.
- **Epoch:** A pass through the full training/validation data forward and backward.
- **Batch size:** A subset of the training/validation data set passed through the network in one iteration. The larger the batch size, the more the processing can be done in parallel using GPUs.
- **Iteration:** The number of batches needed to complete one epoch.

As discussed in section 4.4 of chapter 4, backpropagation is the engine behind the training of every neural network. Through the process of backpropagation, the weights and biases defining the connections between neurons get adjusted to minimize the loss function, which captures the difference between the output of the neural network and the desired target value during training.

In this exercise, we will accomplish the following:

- Implement a neural network based on the MLP architecture using the MNIST Fashion data set.
- Tune several model hyperparameters to achieve optimal model performance.
- Use a cross-entropy metric to measure loss and use accuracy to measure model performance.
- Train a model using backpropagation to learn 266,610 parameters and achieve close to 90 percent performance.

Implementing an MLP Classifier

We will begin by uploading and launching the Jupyter notebook for this exercise.

In the first segment of the notebook, we begin by importing a few common modules (sklearn, keras, numpy, tensorflow, matplotlib), and ensuring that we are running the most recent version of Python.

Next, we define utility functions for the various types of activation functions and their derivatives, which we will refer to later when designing our neural network. We then fetch the MNIST Fashion data set. The data set is already split for us between a training set (60,000) and a test set (10,000 images).

```
fashion_mnist=keras.datasets.fashion_mnist
(X_train_full, y_train_full), (X_test, y_test)
        =fashion_mnist.load_data()
```

We further split our training set to get a validation set of 5,000 samples. We also proceed to scale down the pixel intensities from 0–255 down to the 0–1 range and convert them to floats by dividing by 255:

```
X_valid, X_train=X_train_full[:5000] / 255.,
        X_train_full[5000:] / 255.
    y_valid, y_train=y_train_full[:5000],
            y_train_full[5000:]
        X_test=X_test / 255.
```

Let us now build our MLP model. Our MLP consists of an input layer, taking in the 784 flattened vector of the 28×28 image. The network will also consist of two hidden layers and the output layer, which will use a softmax activation to convert the output numbers to probabilities. The first hidden layer of our network has 300 neurons. The second hidden layer has 100 neurons. Both hidden layers use the rectified linear unit (ReLU) activation function. The output layer consists of ten neurons mapping to the ten classes in the MNIST Fashion data set. We will use some handy functions from keras to easily build our network:

```
model=keras.models.Sequential([
keras.layers.Flatten(input_shape=[28, 28]),
keras.layers.Dense(300, activation="relu"),
keras.layers.Dense(100, activation="relu"),
keras.layers.Dense(10, activation="softmax")
        ])
```

The total number of trainable parameters for this network can be computed as follows:

- A fully connected first hidden layer with 235,500 parameters (784 input $\times$ 300 neurons $+$ 300 bias terms)
- A fully connected second hidden layer with 30,100 parameters (300 neurons from the first hidden layer $\times$ 100 neurons for the second hidden layer $+$ 100 bias terms)
- A fully connected output layer with 1,010 parameters (100 inputs from the second hidden layer $\times$ 10 neurons $+$ 10 bias terms)

In total, our network has 266,610 total trainable parameters.

Our network is initialized with a set of random weights and biases set to 0. Before we begin the process of training our model, we must compile the model, defining our loss function and performance metric. Here, we use accuracy as our performance metric, as well as a sparse, categorical cross-entropy loss function, a common choice for tasks involving class prediction. It is formulated using probabilities as shown: $H(p, q) = -\sum p(x)$ $*\log(q(x))$, where p(x) is the true class probability and $q(x)$ is the predicted probability from the softmax activation at the output layer. The summation is applied over all ten classes. The logarithmic transformation within the formula penalizes larger discrepancies between predicted and true probabilities, which is crucial for guiding the learning process. The negative sign is needed to make the cross-entropy loss positive since we are taking the log of numbers smaller than 1.

Now we proceed with training and evaluating our model. It's important to note that training our model (with epochs set to 25) may take a couple of minutes to complete:

```
history=model.fit(X_train, y_train, epochs=25,
        validation_data=(X_valid, y_valid))
```

Each line in the output shows the results from 1 epoch (a pass through the full training data forward and backward). The cross-entropy training loss and accuracy, as well as the validation loss and accuracy, are logged. The metrics in the final epoch (25/25) denote the cross-entropy loss and accuracy of the final model. We can plot the performance of training and validation (loss and accuracy) as a function of epoch runs. We can generally see that the model converges in both accuracy and loss at around 25 epochs. Thus, we would see diminishing returns if we continued to train by increasing the number of epochs. With our final model, we can now evaluate the performance on the test set:

```
model.evaluate(X_test, y_test)
```

We can now use our MLP model to predict the performance on the first three instances of the test data set and compare it against the graphical display of the samples themselves.

Congratulations! With that, you have implemented a neural network based on the MLP architecture using the MNIST Fashion data set. In this exercise, you have successfully done the following:

- Implemented a neural network based on the MLP architecture using the MNIST Fashion data set
- Tuned several model hyperparameters to achieve optimal model performance
- Trained a model using the cross-entropy loss function and backpropagation to learn 266,610 parameters and achieve close to 90 percent performance

In the following appendix, we will implement a convolutional neural network (CNN) using the MNIST Fashion data set.

A.5   CNN with MNIST Fashion Data Set

In this appendix, we walk through a hands-on exercise using a Jupyter notebook to implement a CNN using the MNIST Fashion data set.

MNIST Fashion Data Set and CNN Review

In this exercise, we use the same MNIST Fashion data set described in the previous appendix exercise. Using this data set, we design and train a CNN, also known as ConvNet. CNNs are a type of neural network most commonly applied to analyzing visual imagery. By design, the connectivity pattern between neurons in CNNs resembles the organization of the animal visual cortex. CNNs consist of a series of layers that include an input layer and one or more convolutional layers.

The convolutional layers are critical components of CNNs. A convolution is a mathematical operation on two functions, in which two signals are combined, or "convolved," to form a third signal. In our network, a convolutional layer applies a moving filter, also known as a "convolution kernel," to the input feature map over a neighborhood of pixels defined by the filter size.

The pooling layer is another fundamental building block of a CNN. The pooling layer down-samples data by reducing the spatial size of the feature map. This is done through various techniques, such as max pooling and average pooling. The pooling layer is commonly added after the convolutional layer, and it has no trainable parameters.

The dropout layer is another key part of a CNN. The dropout layer removes randomly selected units within the network, both hidden and visible, and is used as a regularization technique for preventing overfitting.

At a high level, these pieces come together to enable CNNs to "see." This is achieved by identifying patterns at different levels of abstraction. In early layers, the network is able to detect small and lower-level features such as edges and corners. In later layers, the network can detect more complex features like texture and can assemble low-level features into higher-level features such as object parts and object classes.

In this exercise, we will accomplish the following:

- Implement a CNN using the MNIST Fashion data set.
- Tune several model hyperparameters to achieve optimal model performance.
- Train a model using a cross-entropy loss function and backpropagation to achieve close to 90 percent performance.

Implementing a CNN

We will begin by uploading and launching the Jupyter notebook for this exercise. In the first segment of the notebook, we begin by importing a few common modules (sklearn, keras, numpy, tensorflow, and matplotlib), and ensuring that we are running the most recent version of Python.

Next, we fetch the MNIST Fashion data set. As before, we further split our training set to create a validation set of 5,000 samples.

```
fashion_mnist = keras.datasets.fashion_mnist
(X_train_full, y_train_full), (X_test, y_test)
        = fashion_mnist.load_data()
```

Let us now build our CNN. We first start by defining our default two-dimensional convolutional layer, with a kernel size of 3 (i.e., 3×3) and a ReLU activation function. Our CNN consists of an input convolution layer, which takes in the input 28×28 image feature map and applies 64 filters with a convolutional kernel size of 7. This is followed by a two-dimensional max pooling layer with the pool size set to 2 to reduce the spatial size of our feature map. We repeat these layers again, applying two convolutional layers with 128 filters and another max pooling layer with the pool size set to 2. One will notice the increase in the number of filters in the convolution layer; this is because filters extract different levels of features from the input image. As we propagate through our network, we want later layers to learn more higher-level features, which are better learned with more filters. Furthermore, it is a common practice to double the number of filters after each pooling layer since the pooling layer reduces the spatial dimension of our feature map by a factor of 2. After another series of convolutional and max pooling layers, we apply a set of dense (also known as "fully connected") and dropout layers:

```
DefaultConv2D = partial(keras.layers.Conv2D,
  kernel_size = 3, activation='relu', padding = "SAME")
          model = keras.models.Sequential([
      DefaultConv2D(filters = 64, kernel_size = 7,
              input_shape = [28, 28, 1]),
      keras.layers.MaxPooling2D(pool_size = 2),
              DefaultConv2D(filters = 128),
              DefaultConv2D(filters = 128),
      keras.layers.MaxPooling2D(pool_size = 2),
              DefaultConv2D(filters = 256),
              DefaultConv2D(filters = 256),
```

```
       keras.layers.MaxPooling2D(pool_size=2),
              keras.layers.Flatten(),
    keras.layers.Dense(units=128, activation='relu'),
             keras.layers.Dropout(0.5),
    keras.layers.Dense(units=64, activation='relu'),
             keras.layers.Dropout(0.5),
   keras.layers.Dense(units=10, activation='softmax'),
                        ])
```

Before we begin the process of training our model, we must compile it by defining our loss function and performance metric. We once again use a sparse, categorical cross-entropy loss function and accuracy for the performance metric. We fit this model to our training data, setting the number of epochs to 10:

```
model.compile(loss="sparse_categorical_crossentropy",
       optimizer="nadam", metrics=["accuracy"])
    history=model.fit(X_train, y_train, epochs=10,
          validation_data=(X_valid, y_valid))
```

Each line in the output shows the results from 1 epoch (a pass through the full training data forward and backward). The cross-entropy training loss and accuracy, as well as the validation loss and accuracy, are logged. The metrics in the final epoch (10/10) denote the cross-entropy loss and accuracy of the final model. We can generally see that the model converges in both accuracy and loss at around 10 epochs. With our final model, we can now proceed to evaluate the performance on the test set:

```
      score=model.evaluate(X_test, y_test)
   X_new=X_test[:10] # pretend we have new images
           y_pred=model.predict(X_new)
```

Congratulations! In this exercise, you have implemented a CNN using the MNIST Fashion data set. You have accomplished the following:

- Implemented a CNN using the MNIST Fashion data set
- Tuned several model hyperparameters to achieve optimal model performance
- Trained a model using a cross-entropy loss function and backpropagation to achieve close to 90 percent performance

In the following appendix, we provide a high-level overview of the Raspberry Pi single-board computer. During our in-person MIT classes, we opted to have students implement a simple neural network using a Raspberry Pi as an example of a single-board computer to stress the size, weight, and power dimensions.

A.6   Raspberry Pi: Introduction and Setup

Raspberry Pi is a credit-card-sized single-board computer developed by the Raspberry Pi Foundation. There are four main generations of Raspberry Pi. To date, the most recent generation—the Raspberry Pi 4 Model B—boasts impressive specs, including a 1.5-gigahertz 64-bit quad core processor, onboard Wi-Fi, Bluetooth 5, full gigabit Ethernet, two universal serial bus (USB) 2.0 ports, two USB 3.0 ports, 1–8 gigabytes of random access memory (RAM) and dual-monitor support via a pair of micro high-definition multimedia interface (HDMI) ports for up to 4K resolution. The Raspberry Pi runs on several operating systems, including Raspbian, Windows 10 IoT Core, and Linux. The device can also be connected to a variety of sensors to interact with the physical world, making it a powerful tool for prototyping decision-based algorithms based on processing of gathered sensor data. Depending on the generation and series, Raspberry Pis can range from $4 all the way up to $75. Ultimately, the success of the Raspberry Pi, as a low-cost modular and open design system, has encouraged large adoption not only by computer and electronic experimenters, but also by the IoT industry more broadly.

There are many useful resources for getting started and setting up your Raspberry Pi. The Raspberry Pi foundation provides an excellent walk-through for beginners on setting up Raspberry Pi at https://projects.raspberrypi.org/en/projects/raspberry-pi-setting -up. Plugged into a computer monitor or television as well as a standard keyboard and mouse, Raspberry Pis are capable of doing almost everything that you would expect from a desktop computer. This includes browsing the internet, playing games, and using applications. However, the greatest use of Raspberry Pi is not as a desktop computer, but as a foundational tool for digital maker projects. With the ability to interact with the outside world by connecting to a variety of sensors (i.e., for temperature, motion, or air pressure) providing almost-human sensing capabilities, Raspberry Pis can be used to develop many powerful applications, such as an object recognition system. For a few of the use cases in part III of this book, we discussed using the Raspberry Pi as a foundational building block of an AI system.

The MIT students implemented the AI system architecture, which they formulated as part of their AI strategic development road maps, using the Raspberry Pi single-board computer. For the students' final grades, these demonstrations were presented to a panel of AI practitioners from industry and academia. Through these experiential learning demonstrations, the students were able to put into practice the teachings from part I in the book, the AI strategic development model (AISDM) discussed in chapter 9, and the effective communication discussed in chapter 13.

Abbreviations

AAAI	Association for the Advancement of AI
ACID	Atomicity, consistency, isolation, and durability
AGI	Artificial general intelligence
AISDM	AI strategic development model
ANNs	Artificial neural networks
API	Application programming interface
AUC	area under the curve
BEFAST	Balance, eyes, face, arms, speech, and time
CI/CD	Continuous integration/continuous delivery
CNCF	Cloud Native Computing Foundation
CNN	Convolutional neural network
CoB	Circle of balance
CRISP-DM	Cross-industry standard process for data mining
CPU	Central processing unit
CUDA	Compute Unified Device Architecture
DevSecOps	Development, security, and operations
DSA	Domain-specific architecture
DRAM	Dynamic random access memory
EDVAC	Electronic Discrete Variable Automatic Computer
ENIAC	Electronic Numerical Integrator and Computer
EQ	Emotional intelligence
FASTEPS	Fairness, accountability, safety, transparency, ethics, privacy, and security
FDA	US Food and Drug Administration
FLOPS	Floating-point operations per second
FPR	False positive rate
GAN	Generative adversarial network
GDP-B	Gross domestic product, but inclusive of other economic benefits
GDPR	General Data Protection Regulation

GPT	Generative pretrained transformer
GPTe	General-purpose technology (from the economic perspective)
GPU	Graphics processing unit
HBR	*Harvard Business Review*
IEC	International Electrotechnical Commission
IEEE	Institute of Electrical and Electronics Engineers
IJCAI	International Joint Conference on AI
ILSVRC	ImageNet Large-Scale Visual Recognition Challenge
INCOSE	International Council on Systems Engineering
ISO	International Organization for Standardization
ITIL	Information Technology Infrastructure Library
ITSM	Information Technology Service Management
JASON	JavaScript Object Notation
LCNC	Low-code/no-code
LIME	Local Interpretable Model-Agnostic Explanation
MAC	Multiply-accumulate arithmetic operation
MAD	Multiply-add
MAE	Mean absolute error
MBRs	Managing by results
MLOps	Machine learning operations
MLP	Multilayer perceptron
MNIST	Modified National Institute of Standards and Technology
MSE	Mean squared error
MVP	Minimum viable product
NICT	National Institute of Information and Communications Technology
NIST	National Institute for Standards and Technology
NLP	Natural-language processing
NoSQL	Not only Structured Query Language
OKRs	Objective and key results
OECD	Organisation for Economic Cooperation and Development
OSTP	Office of Science and Technology Policy
RAI	Responsible AI
RBAC	Role-based access control
REST	REpresentation State Transfer
RISC	Reduced Instruction Set Computer
RNN	Recursive neural network
ROC	Receiver operating characteristic

RPV	Resource, process, value
SGD	Stochastic gradient descent
SME	Subject matter expert
SMPC	Secure multiparty computation
SQL	Structured Query Language
SRL	System readiness level
STEAM	Science, technology, engineering, arts, and mathematics
STEM	Science, technology, engineering, and mathematics
STS	Sociotechnical system
SVM	Support vector machine
SWOT	Strengths, weaknesses, opportunities, and threats
SHAP	SHapley Additive exPlanation
TPR	True positive rate
TPU	Tensor processing unit
TX-0	Transistorized experimental computer—zero
V&V	Verification and validation
VM	Virtual machines
VSN-C	Vision, Steps, News—Contributions

Index

Note: Figures and Tables are indicated by italicized page numbers.

AAAI. *See* Association for the
   Advancement of AI
Abbeel, Pieter, 202
abbreviations, in AI, 525–527
accountability, 271
accuracy, *102*
Acemoglu, Daron, 409
ACID. *See* atomicity, consistency,
   isolation, and durability
ACM. *See* Association for Computer
   Machinery
actions, consequences and, 192–193,
   194–195, 196–197, 199–201
activation functions, *110*
adherence, 330, 354
Advanced Vector Extensions, 155
advance serverless computing, 172
adversarial AI
   attacks, 242–245, *243–244*
   brittleness and, 248
   challenges with, 245
   classes of, 229–233, *230–231*
   DNNs to, 230–231, *231*
   fundamentals of, 140
   GANs, 116, *117*, 118, 235, 240
   training, 239
Agarwal, R. S., 408–409
The Age of AI and Our Human Future
   (Kissinger), 260

AGI. *See* artificial general intelligence
Agile software, 346–347, *347*, 356, 376
AI adoption, 293, 339–341, *340*
AI algorithm era, 43–44
AI application
   in computing environments, 139–141,
      140–141
   ML algorithms and, 133
   narrow AI and, 2, *2*, *3*
AI artifacts, 403
AI blunders, 267
AI brittleness, 225
AI capabilities
   AI deployment and, 16, *16*, 31, 43, 66,
      332, *333*, 334–335, *401*
   AI development of, 9, 63, *64*
   AI implementation and, 51
   AI system architecture and, 163
   approaches to, 330
   for BEFAST, 343
   building blocks of, 20
   in business, 82–83
   core capabilities, 303
   from data conditioning, 82–83
   for delivering services, 138
   envisioned future with, 301
   human capital and, 131–132
   with IoT, 387
   with ML algorithms, 233

AI capabilities (cont.)
  ML and, 74
  monitoring of, 339
  with MVP, 288–289
  people-process-technology triad for,
    331, 415
  stakeholders and, 1, *316*
  for users, 238–239
AI community
  benchmarks in, 75
  business intelligence, 70
  open-source data for, 58–59
AI deployment
  AI, 169
  AI adoption and, 339–341, *340*
  AI capabilities and, 16, *16*, 31, 43, 66,
    332, *333*, 334–335, *401*
  in AI ecosystem, 341, *342*, 343–348, *347*
  AI implementation and, 149
  AI system architecture and, 289
  AI systems engineering for, 335,
    336–337, 337–339
  approaches to, 376–378
  benefits of, 351–352, *352–353*
  for business, 15–17, *16*, 348–351,
    353–357
  challenges in, 276, 278, 330–331,
    331–332
  common pitfalls in, 385–388, 391
  data in, 275
  deployed systems, 164, *164*
  guidelines, 327–330, *329*, 388
  to INCOSE, 35, 39
  learning exercises about, 357–358
  MLOps and, 289, 388–391
  of ML solutions, 115–116
  SWaP in, 169
AI development. *See also* machine
    learning operations
  of AI capabilities, 9, 63, *64*

AI development, security, and
    operations, 41, 289, 344–345
  of AI horizons, 17–19, *18*, 278–279
  AI operations and, *342*
  AISDM and, 45, *46*, 47, 312–313
  approaches to, 376–378
  common pitfalls in, 385–388, 391
  development operations, 15–16,
    344–345, 365–366, 375–378
  of DNNs, 242
  ethics in, 271
  history of, 4, *5*, 6–9
  life cycle, 273–275, *274*, 278
  ML and, 76–77, 350
  ML developers, 76–77
  MLOps and, 15–16, 344–345
  for users, 275
AI ecosystem
  AI deployment in, 341, *342*, 343–348,
    347
  AI governance of, 338
  AI system architecture and, 347–348,
    385
  for business, 4
  SRLs in, 343–344, 346–347, 351
  system analysis for, 232–233
  technology for, 20–21
  trustworthy AI in, *272*, 272–273
AI emphasis, *18*
AI expert system, 39
AI governance
  of AI ecosystem, 338
  data and, 74
  FASTEPS and, 305, 387
  policy and, 177
  privacy and, 262–263
AI horizons
  AI development of, 17–19, *18*, 278–279
  in AI organization, 318–319
  with AquaAI, 502–503, *503*

for business, 17–19, *18*
with HMT, *213*, 213–214
for MESH, 494–495
Misty Robotics and, 300–301
with modern computing, *171*
opportunities with, *277*
for training, *124*
AI implementation, 51, 149
AI innovation, 9, *10*, 11
AI leadership, 31, 422–427, *423*, *425–426*
AI models, *44*
AI operations. *See* machine learning
  operations
AI organization
  AI horizons in, 318–319
  at Apple, 63, 406
  communication and, 65–66
  SRLs for, 355–356
  strategy for, 291
AI outputs, 223
AI pipeline
  building blocks of, 191, 214, 287
  data sets in, 83
  HMT in, 215–216
  IoT in, 349
AI platform infrastructure
  characteristics of, 351–352, *352–353*,
    353
  cybersecurity in, 232
  digital twins in, 349, 358
  fundamentals of, 79–80
  infrastructure, 79–80
  MLOps in, 382–385
  organization infrastructure and, 63, *64*,
    382–385
AI-powered cycling, 477–479, *479–480*,
  481–483, *482–483*
AI practitioners
  benefits for, 351–352, 354
  data and, 73–74

innovative team environments for,
  408–410
R&D for, 4, *5*
SRLs for, 329–330, 335, *337*
AI products, 79
AI recruiting, 267
AISDM. *See* AI strategic AI development
  model
Asimov Institute, 116
AI staff, 291
AI strategic AI development model
  (AISDM)
  AI development and, 45, *46*, 47,
    312–313
  AI strategic road maps for, 355
  AI value proposition and, 319
  approach, 293
  for business, 65, 399–400, 402
  envisioned future with, 296, *297*, 298
  FASTEPS and, 318
  framework, 289, 294–296, *295*, *297*,
    298, 300, 303–305, 307, 317, 327,
    340–341
  Heilmeier Catechism and, 505
  with MESH, 497
  road map for, 355
  SRLs and, 335, 364
  step-by-step flow in, 343
  SWOT analysis and, 481, *482*
  symbols in, 449–450
  technology for, 375
  vision statements for, 299, 447
AI strategic blueprint
  AI strategic road maps for, 307, *308*,
    309–310, *310–311*, 312–313, *313*,
    314, 315–317, *316–317*
  for business, 318–322
  examples of, 287–288, *288*, 296, *297*,
    298
  models for, *46*, 47, 51

AI strategic road maps
  for AISDM, 355
  for AI strategic blueprint, 307, *308*,
    309–310, *310–311*, 312–313, *313*,
    314, 315–317, *316–317*
  for AI strategy, 287–289, *288*, 294–296,
    295, *297*, 298, 307, *308*, 309–310,
    310–311, 312–313, *313–314*,
    315–317, *316–317*, 337
  AI value proposition and, 473–474
  for AquaAI, 505–506, *506*
  for MEALS, 490–491, *491*
  for MESH, 497–498, *498*
  at Misty Robotics, 471–472, *472*
  at MIT, 469, 485
  for MLOps, 316–317, 363–365
  of strategic thinking, 289–293, *290*, *292*
  for SWOT analysis, 307, 309–310,
    312–313, *313*, 315
AI strategy
  with AI value proposition, 304–307
  execution of, 317–318
  formulation of, 318–321
  learning exercises for, 321–323
  road map for, 287–289, *288*, 307, *308*,
    309–310, *310–311*, 312–313,
    313–314, 315–317, *316–317*, 337
  strategic direction in, 302–303, *304*
  strategic thinking in, 289–293, *290*, *292*
  technology and, 294–296, *295*, *297*,
    298
AISysOps. *See* AI systems operations
AI system architecture
  AI capabilities and, 163
  AI deployment and, 289
  AI ecosystem and, 347–348, 385
  AI systems engineering and, 20–21
  AI vulnerabilities and, 166, *167*, 168
  at AquaAI, *504*
  building blocks of, 210

  for business, 11–15
  cybersecurity for, 177
  end-to-end, 59, 61, 337, 339, 374, 389
  functional architecture for, 40–43, *41*,
    48, 50–51
  fundamentals of, 19–20
  implementation framework, *42*, *336*
  integration with, 368
  MESH, *497*
  with MLOps, 367–368, *369–372*,
    373–375
  online end-to-end, 374, 389
  pipeline for, 9, *10*, 11
  points of vulnerability in, *227*
  simplified, *132*, *192*, *224*
  SRLs in, 350
  STEM and, 414
  subsystems, 338
  technology for, 2, *3*
  Vee-model process and, 34–35
  V&V of, 229, 343
AI systems approach, 1–2, 50
AI systems engineering
  for AI deployment, 335, *336–337*,
    337–339
  AI system architecture and, 20–21
  approach, 35, 39–40, 75
  architecture framework for, 40–43, *41*,
    42
  common definitions of, 31–33
  Conway's Law in, 65–66
  discipline in, 33–35, *34*, *36–38*, 39
  FASTEPS in, 431
  fundamentals of, 29–31, 50–51
  INCOSE on, 31–32, 312, 328, 354
  jobs, 366
  leadership, 43–45, *44*, *46*, 47, *47*
  learning exercises for, 52–53
  methodology, 19–20, 35, 39–40, 50–51
  at NASA, 29–30, 32

National Council on Systems
  Engineering, 33
people in, *38*, 49–50
process for, *36–38*, 49
risk analysis in, *36–38*, *369–372*
technology challenges in, 48
AI systems operations (AISysOps),
  327–328
AI talent
  future of work and, 408–410
  in innovative team environments,
    427–431
  monitoring, *426*, 427–431
  recruiting, 423–424, *426*, 427–431
  retaining, 419–421
  success with, 407–412, *411*
AI teams, 387, 389
AI technical breadth, *411*, 412, 414–419,
  416, 430–431
AI tools, for ML, 132–133
AI transformation playbook, *288*
AI value proposition
  for AI-powered cycling, 479, *480*, 481
  AISDM and, 319
  AI strategic road maps and, 473–474
  AI strategy with, 304–307
  at AquaAI, 503, *504*, 505
  for MEALS, 488, *489*, 490
  at MESH, 496, *497*
  with Misty Robotics, 307, *308*,
    472–473, *473*
  understanding, 292
AI vulnerabilities
  AI system architecture and, 166, *167*,
    168
  robust AI systems and, 246, *247*,
    248–249
  systems perspective on, 225–226, *227*,
    228–229
AI winters, 7, 21

AlexNet, 75–76, 120–121, 123, 151, 263
algorithms. *See also* ML algorithms
  AI algorithm era, 43–44
  architecture of, 176–177
  backpropagation, 115
  data and, 89–90
  for Deep Blue, 7
  designers of, 241–242
  DNN, 134, 143, 149–150, *150*, 153
  fixed-point arithmetic in, 348–349
  GoogLeNet, 230
  GPT, 142–143, 409–410, 434
  hardware and, 160
  HHL, 173
  high-performance, 141–142
  from ILSVRC, 75–76
  ML, 9, *10*, 11–13, 57–58
  MLP, 517–520
  predictions with, *231*
  ResNet CNN algorithm, 76
  for search-and-discovery tools, 192
  supervised learning, 73, 123–124
  for technology, 1–2
  TPUs and, 162–163
alignment, 330, 354
AlphaGo, 263
AlphaGo Zero, 8, 151
Alzheimer's disease, 202, 469–476,
  471–476
Amazon, 193, 235–236, 267
Amazon Mechanical Turk (AMT), 77
Amazon Web Services (AWS), *41*, 41–42,
  138, 307, 335
ambidextrous organization, 296
ambiguity, in business, *425*
AMD, 174
Amdahl's Law, 136–137
AMT. *See* Amazon Mechanical Turk
Anaconda Navigator, 510–511, *512*
Andreessen, Marc, 407

ANNs. *See* artificial neural networks

Apple
  AI organization at, 63, 406
  iPhones, 141
  strategic planning at, 294

applications
  LCNC application development, 365, 367, 382–385
  metaverse, 172
  system, 303
  third party, 386

application-specific domains, 154–155

application-specific integrated circuits (ASICs), 152

AquaAI, 501–503, *503–504*, 505–507, *506–507*

architecture. *See also specific topics*
  of algorithms, 176–177
  Boss, 202, *203*
  CUDA, 151
  framework, 40–43, *41, 42*
  functional, 9, *10*, 11, 40–43, *41*, 48, 50–51
  in modern computing, 158
  for neural networks, *117*, 118
  roofline metrics for, 160
  of SAGE, 32
  technology and, 50

area under the curve (AUC)
  confusion matrix and, 211
  ML and, 71
  performance of, 90
  ROCs and, 104–105, *105*, 124, 206

arithmetic precision, 145, *146, 146–147, 147*, 147–148

artificial general intelligence (AGI), 409–410

artificial intelligence (AI). *See specific topics*

Artificial Intelligence Index Report, 17

artificial neural networks (ANNs), 90, 106–107, *108*, 109

The Art of the Long View (Schwartz), 293

ASICs. *See* application-specific integrated circuits

Asimov, Isaac, 191–192

assessment, 330, 354

Association for Computer Machinery (ACM), 136, 261

Association for the Advancement of AI (AAAI), 246, 273

assumptions, 403

Atomic Habits (Clear), 417

atomicity, consistency, isolation, and durability (ACID), 67–68, *68*

attacks. *See also* adversarial AI
  cybersecurity against, 245–246
  evasion, 232
  poisoning, 231–232
  testing against, 242–245, *243–244*, 249

attributes, of data conditioning, 74–77, *76, 78, 79*

AUC. *See* area under the curve

audiences, 274, 444–452, *456*, 456–465

augmentation, with HMT, 193–194, *194–195*, 196–197

autocoding, 174

automated detection, *167*

automated machine learning (AutoML), 382–385

autonomy
  autonomous vehicles, 202–205, *203, 205*
  HMT and, 201–205, *203–205*
  after ML, 191–192
  public safety and, 264–265
  search-and-discovery tools and, 214–215

auxiliary sensors, 203–204, *204*

AWS. *See* Amazon Web Services

Babbage, Charles, 363–364
backpropagation, 109, *110–111*, 112–115, *114*
backpropagation algorithms, 115
Balance, Eyes, Face, Arms, Speech, and Time (BEFAST), 299, 307, 343
Barzilay, Regina, 211
batch sizes, 518
Beckham, David, 234
BEFAST. *See* Balance, Eyes, Face, Arms, Speech, and Time
behavior, of staff, 432–433, *433*
beliefs, 403
benchmarks, 348–351
benefits, of AI deployment, 351–352, *352–353*
Bennett, Keithe, 501
Bennis, Warren, 44–45
BERT, 123
Bhatt, U., 241
bias
  collective harms and, 269
  data, 72
  emergent, 264, 268–269
  fractions and, 147–148
  in NLP, 266
  preexisting, 267–268
  racial, 265
  to RAI, 228–229, 264–267
  societal harm from, 270
  technical, 268
big data, 58, *60*, *61*, 61–62, 89, 175
Bigtable, 66–67
biological neurons, 107, *108*
black-boxes, 231, 240, 248
Bose, 400, 477–479, *479–480*, 481–483, *482–483*
Bose, Amar, 399–400
Boss architecture, 202, *203*
Boston Consulting Group, 16

Bowman, Nina, 292
brain floating-points, 147
Bresnahan, T. F., 409
Briggs, David, 289–290, 424
Brin, Sergey, 424–425
brittleness, 248
broken-glass outlines, 451, *452*, 463–464
Brynjolfsson, Erik, 62, 131, 409–410
building blocks. *See also* AI system architecture
  of AI, 11–15
  of AI capabilities, 20
  of AI pipeline, 191, 214, 287
  of AI system architecture, 210
  contemporary tools and, 378–382, *380*
  of HMT, 196
  of measurement, 205–206, *207*, 208–210, *209*
  for ML, 123–125, *124*
  for ML algorithms, 57–58
  Moore's Law and, 175
  for people-process-technology, 35
  performance of, 192–193
  of robust AI systems, 225–226, *227*, 228–229
  from Vee-model, 51
"Building Digital-Ready Culture in Traditional Organizations" (Westerman), 306
Buolamwini, J., 264
Burns, B., 382
Bush, Vannevar, 29
business
  AI capabilities in, 82–83
  AI deployment for, 15–17, *16*, 348–351, 353–357
  AI ecosystem for, 4
  AI horizons for, 17–19, *18*
  AISDM for, 65, 399–400, 402
  AI strategic blueprint for, 318–322

business (cont.)
  AI system architecture for, 11–15
  ambiguity in, *425*
  best practices in, 262
  big data for, 58
  CEOs in, 287–288, 305
  CRISP-DM process for, 66
  cybersecurity for, 233–234
  digital, 63
  goals, 43–44, 313, *314*, 315, 348,
    403
  at Google, 424–425
  hidden technical debt in, 351, *352*
  humans in, 237
  intelligence, 70
  macro-level assessment in, *317*
  management, 414–417
  mentoring in, 427–431, 435–436
  ML models for, 237–238
  MLOps and, 321
  MVP in, 334
  needs, *310–311, 474, 506*
  open-source data in, 66–67
  operations, 65
  organizational maturity clusters in,
    339–341, *340*
  organization infrastructure of, 63, *64*
  planning, 291–292, *292*
  progress metrics in, 417–421, *418, 420*
  R&D funding for, 7
  reorientation in, 306
  responsibility in, 289–290, *290*
  sample applications in, 509–510
  SME in, 307
  social media, 235–236
  SRLs in, *337*, 337–339, 376–378
  stakeholders assessment in, 51
  stakeholders in, 71, 298
  strategic principles for, 2, *3*, 49
  SWOT analysis for, 315

technology for, 19–22
women in, 430
Byrne, R., 241

C3 AI Digital Twin, 349
Caceres, R., 225, 232–233, 343
California Privacy Rights Act, 82
Cambrian Explosions, 176
Cambridge Analytica, 265
carbon dioxide, 134, 170
careers, *411*, 411–414, 428
Carnegie Mellon University, 11, 202,
  203
central processing units (CPUs), 8
  DSA for, 168, 176
  energy for, 153–154, *154*
  FPGAs and, 158
  GPUs and, 132
  history of, 149–152
  for ML, 137
CEOs. *See* chief executive officers
CERTs. *See* Computer Emergency
  Response Teams
Chakraborty, A., 243–244
Challenger explosion, 30
challenges
  with adversarial AI, 245
  in AI deployment, 276, 278, 330–331,
    331–332
  data conditioning, 79–82, *81*
  to effectiveness, 67
  to HMT, 210–214, *213*
  macro-level, 210–211, *332*
  micro-level, 210–211
  ML, 123
  modern computing, 169–174, *171*
  people, 49–50
  of physical environments, 213
  process, 49
  RAI, 276, *277*, 278

for robust AI systems, 245–246
technology, 48
Chang, Morris, 287–288, 317–318
ChatGPT, 142–143, 276. *See also specific topics*
chief executive officers (CEOs), 40, 59, 287–288, 305
China, 229
chips. *See specific topics*
Christensen, C. M., 293, 408
Churchill, Winston, 259–260
CI/CD. *See* continuous integration/ continuous delivery
CipherMode Labs, 377
Circle of Balance (CoB), *411*, 411–413, 422, 427–428, 434–435
Clark, Wes, 135
classification, 92–93, *93*
cleaning, of data, 70–74, *72*
Clear, J., 417
climate change, 493
cloud computing
    at AWS, 138
    Cloud Native Computing Foundation, 381, 386
    commercial, 382
    MLOps platforms for, 382–383
    paradigms, 138–139
    for robotics, 307
    storage in, 164, *164*
clustering, 97–98, *98*, 155–156
CMPC. *See* secure multiparty computation
CNNs. *See* convolutional neural networks
CoB. *See* Circle of Balance
COCO dataset, 77, *78*
Codd, Edgar, 66
code/coding, 39–40, 155, 174, 356–357. *See also specific topics*
Codex, 174

collaboration-based insights, 17–19, *18*, *81*
collective harms, 269
"Coming to a New Awareness of Organizational Culture" (Schein), 305–306
commercial cloud computing, 382
commercial face recognition, 264
commitment, 330, 355
communication
    AI organization and, 65–66
    effective, 443–446, *445*, 463–464
    learning exercises about, 464–465
    outlining, *451–452*, 451–453, 463–464
    in presentations, 453–463, *454*, *456*, *458–459*, *461*
    processing and, 164, *164*
    skills, 445–446
    technical, 444, *445*
    thinking skills in, 293
    VSN-C for, 445–448, *447*
    Winston Star and, 448–451, *449*
Compassionate Artificial Intelligence (Ray), 469
COMPAS system. *See* Correctional Offender Management Profiling for Alternative Sanctions system
competencies, 303
compliance, with FASTEPS, 80, *81*, 209–210
computational complexity, 99, 112–114, 134
computer chips. *See specific topics*
Computer Emergency Response Teams (CERTs), 245–246
Computer History Museum, 136
computers. *See specific topics*
Computer Science and Artificial Intelligence Laboratory, MIT (CSAIL), 6, 77, 294–295

Compute Unified Device Architecture
  (CUDA), 151
compute usage, *44*
computing engines, 155–156, *157*,
  158–160
computing environments, 139–141,
  140–141
"Computing Machinery and Intelligence"
  (Turing), 4
confidence
  in cybersecurity, *195*
  in machine decision making, 192–193,
    194–195, 196–197, 199–201
  in machines, 2
  in ML, 2, *3*, 14–15, 17–19, *18*, *194*
  in ML algorithms, 205–206
  in performance, 215–216
confusion matrix, *104*, 206, 211
connectionists, 6
consequences, actions and, 192–193,
  194–195, 196–197, 199–201, 228
consumers, 11
containers, 380, *380*
contemporary computing engines,
  155–156, *157*, 158–160
contemporary tools, 375–382, *380*
content-based insights, 17–19, *18*, *81*
content structuring, 446–447, *447*
context-based insights, 17–19, *18*, *81*
continuous integration/continuous
  delivery (CI/CD), 327–328, 339,
  344–346, 364–366, 374–381, 390
contributions, 448, 455–458
convolutional neural networks (CNNs)
  convolutional layers for, 109, *111*,
    121
  fundamentals of, 90–91, 120–121, *122*,
    123
  GANs and, 235
  hyperparameters in, 119–120

ML algorithms and, 116
ML techniques with, 125
with MNIST, 521–523
pooling layers for, 121
RNNs and, 481
Conway, Melvin, 65–66
Conway's Law, 65–66
core capabilities, 303
Core ML modeling tools, 141
Correctional Offender Management
  Profiling for Alternative Sanctions
  (COMPAS) system, 265
cost functions, 112–113
cost optimization, *207*, 350
counterfeit parts, *167*
Cousteau, Jacques, 501
Covey, S., 432–433
COVID-19 virus, 196, 266, 400
CPUs. *See* central processing units
CRISP-DM. *See* cross-industry standard
  process for data mining
Cross, R., 429
Crossan, M., 404
cross-industry standard process for data
  mining (CRISP-DM), 66, 273, 309,
  365
cross-validation, 95–96, *97*
CrowdStrike, 166
cryptosystems, 378
CSAIL. *See* Computer Science and
  Artificial Intelligence Laboratory
CUDA. *See* Compute Unified Device
  Architecture
curated data sets, 74–77, *76*, *78*, *79*
curse of dimensionality, 73
customers, 4
Cybenko, A., 234
Cybenko, G., 234
cyberhygiene, 164–165
cyber kill chains, 165, 177

cybersecurity
  in AI platform infrastructure, 232
  for AI system architecture, 177
  against attacks, 245–246
  for business, 233–234
  community, 229
  confidence in, *195*
  domain, 215
  fundamentals of, 163–166
  R&D for, *167*, 168–169
  for users, 177
  US Senate Armed Services Committee
    on Cybersecurity, 169

Daimler, M., 403–404
DALL-E, 123
Dally, William, 137, 153–155
DARPA. *See* Defense Advanced Research
  Projects Agency
data. *See also specific subjects*
  access, 160–161
  for AI, 57
  in AI deployment, 275
  AI governance and, 74
  AI practitioners and, 73–74
  algorithms and, 89–90
  analytics, 352
  availability, 149
  bias, 72
  big data, 58, *60*, *61*, 61–62, 89, 175
  Cambrian Explosions with, 176
  cascades, 73–74
  cleaning, 70–74, *72*
  components, *146*
  cross-validation of, 95–96, *97*
  curated data sets, 74–77, *76*, *78*, *79*
  databases, 66–70, *67–69*
  data-centric AI, 70–71, 75
  data lakehouse, 70
  for DNNs, 236

drift, 73
engineers, 71, 366
exploration, *72*
exponential data growth, 59, *60*, *61*,
  61–62
force multipliers, 62
gathering, 93, *94*
GDPR, 82, 237, 242, 263
inputs, 74–75
labeled, 92, 212
marketplace, 300
metadata, 12
mining, 273, 309
ML algorithms and, 11–12
multimodal, 173
NoSQL, 66–68, *68–69*
open-source, 58–59, 66–67
organization of, 79–80
parallelism, 159
poisoning, *227*
preparation, 70–74, *72*
preprocess, 93, *94*
privacy, 82, 138–139, 265
products, 74
quality, 70–74, *72*
sanitization, 239
schema, 83
scientists, 57, 71, 366
segmented, 94, *94*
semistructured, 59
on social media, *60*, 198, 226
SQL database, 66–68, *68*
streaming, 80
structured, 12, 59, 66–67
training, 92–93, 99–100, *100*, *139*
unstructured, 12, 59
variety of, *61*, 61–62
V&V and, 239
database management, 69–70
Databricks, 70

data conditioning
  AI capabilities from, 82–83
  attributes of, 74–77, *76, 78, 79*
  challenges, 79–82, *81*
  databases for, 66–70, *67–69*
  for data-centric AI, 70–74, *72*
  digital transformation in, 63, *64,* 65–66
  exponential data growth for, 59, *60, 61,*
    61–62
  learning exercises for, 84
  ML algorithms and, *10,* 11–13
  ML and, 57–59, *227*
data drift, 73
data operations. *See* machine learning
    operations
data sets
  in AI pipeline, 83
  COCO dataset, 77, *78*
  collection for, 274–275
  curated, *78*
  ML, 82
  MNIST, 512–513, 517–518
Daugherty, P. R., 200
Davenport, T. H., 62, 196–197
Davis, E., 193–194
Dawnbench, 349
Dean, Jeff, 151–152
decision-making. *See also specific topics*
  action and, 2
  by machines, 192–193, *194–195,*
    196–197, 199–201
  tools, 265, 269
  by users, 15
deconvolutional layers, *111*
Deep Blue, *5,* 7–8
deep CNNs. *See* convolutional neural
    networks (CNNs)
deepfakes, 233–236, 245
deep learning
  DNNs and, 123

history of, 125
  neural nets and, 106–107, *107–108,* 109
DeepLearning AI, 59
"Deep Learning Hardware" (LeCun), 169
DeepMind Technologies Limited, 8, 174
deep neural networks (DNNs)
  to adversarial AI, 230–231, *231*
  AI development of, 242
  algorithms, 134, 143, 149–150, *150,* 153
  ANNs and, 106–107, 109
  to DARPA, 228
  data for, 236
  deep learning and, 123
  fundamentals of, 8
  hardware, 171
  IoT and, 171
  ML algorithms with, 112
  ML-oriented accelerators and, 170
  models, 172
  at NVIDIA, 151, 159–160
  PGD and, 240
  saliency maps of, 242–244, *243*
Defense Advanced Research Projects
    Agency (DARPA)
  DNNs to, 228
  GARD for, 232
  Grand Challenge, 7
  R&D for, 9
  regulation of, 236–237
  Urban Grand Challenge, 202, *203*
  US and, 204–205
Defense Science Board, US, 168
defense stages, 242–245, *244*
defensive distillation, 239
defensive-GANs, 240
delivering services, 138
Deloitte Global Boardroom, 306–307
demo solutions, *483*
DENDRAL project, 7, 135–136
Dennard Scaling, 136–137

dense/fully connected layers, 121
Department of Transportation, US,
    204–205
designers
    of algorithms, 241–242
    hardware, 148
    stakeholders and, 273–274
designing
    ACID and, 67–68, *68*
    DSA, 154–155
    hyperparameters for, *119*
    IEE-754 floating point standard in,
        145, 147–148
    ML algorithms, 151–152
    neural networks, 115–116, *117*,
        118–120, *119*
    parallelism in, 158–160
Deza, A., 384
Dietterich, Thomas, 212
digital business, 63
digital transformation, 63, *64*, 65–66,
    82–83
digital twins, 349, 358
dimensionality, 73, *98*, 98–99
discipline, in AI systems engineering,
    33–35, *34*, *36–38*, 39
The Discipline of Innovation (Drucker),
    408
distributed platforms, 172
DNNs. *See* deep neural networks
Docker, 380–381
Doerr, John, 417
domain-specific architecture (DSA)
    for AI, 132
    for CPUs, 168, 176
    designing, 154–155
    ML algorithms and, 148
    TPUs in, 152–153
domain-specific hardware, 152–155, *154*
domain-specific software, 152–155, *154*

Dominguez, N., 428
Domo, 59
DRAM, 160–162, *161*
dropouts, in neural network layers, *111*,
    119, 121
Drucker, Peter, 408, 422
DSA. *See* domain-specific architecture
Dwork, C., 241–242
Dynabench, 81, 349

edge computing, 138–141, *139–141*,
    171, 387
EDVAC. *See* Electronic Discrete Variable
    Automatic Computer
effective communication, 443–446, *445*,
    463–464. *See also* communication
effective implementation, 71
efficiency, 172
Eisner, Howard, 33
Electronic Discrete Variable Automatic
    Computer (EDVAC), 135
Electronic Numerical Integrator and
    Computer (ENIAC), 134–135
Elliott, B., 400
ELT. *See* Extract-Load-Transform
Embedded EthiCS, 262
emergent bias, 264, 268–269
*Emerging Areas at the Intersection of
    Artificial Intelligence and Cybersecurity*,
    165
emotional intelligence (EQ), 425–426,
    426, 435
empirical evidence, 73–74
end-to-end AI system architecture, 59,
    61, 337, 339, 374, 389
energy
    for CPUs, 153–154, *154*
    at Google, 170
    Green AI for, 177
    MESH for, 493–499, *495–498*

energy (cont.)
optimizers, 494
renewable, 493–494
usage assessment, 159
US Federal Energy Regulatory
Commission, 77
engagement, with attacks, 242
engineers. *See also specific subjects*
data, 71, 366
engineered systems, 9, *10*, 11
essential skills for, *47*
at Google, 136
IEEE, 261
ML, 367
Moore School of Electrical Engineering,
134–135
to NAE, 45, *46*, 47, 51
National Council on Systems
Engineering, 33
National Society of Professional
Engineers, 261
performance to, 175
software, 59
ENIAC. *See* Electronic Numerical
Integrator and Computer
enterprise computing, 138–141,
139–141
Environmental Protection Agency (EPA),
30
envisioned future
with AI capabilities, 301
for AI-powered cycling, 478–479,
479
with AISDM, 296, *297*, 298
at AquaAI, 502–503, *503*
for MEALS, 486–487, *487*
at MESH, 494–495, *495*
mission/vision and, 299–301, *301*
with Misty Robotics, 202, 299–301,
301, 470–471, *471*

EPA. *See* Environmental Protection Agency
epochs, 518
EQ. *See* emotional intelligence
errors, 101, *103*, 124
essential skills, *47*
ethics. *See also* fairness, accountability,
safety, transparency, ethics, privacy,
and security
in AI development, 271
of data privacy, 82
Embedded EthiCS, 262
history of, 278
ETL. *See* Extract-Transform-Load
Europe, 229, 237, 263
evaluation
hardware, 166
MEALS, 485–488, *487–489*, 490–492,
491
testing and, 96–97, *97*
TEV&V, 49, 93–97, *94*, *96–97*
validation and, *94*, 95–96, *97*
evasion attacks, 232
evolution
management and, 66–70, *67–69*
of modern computing, 138–141,
139–141
of organization infrastructure, 63, *64*
"The Evolution of Management"
(Matsudaira), 413–414
execution, of AI strategy, 317–318
experimenters, 340, *340*
explainable AI (XAI)
deepfakes and, 245
explainability, 239–240, 249, 271
mitigation techniques with, 223, 225
R&D in, 236–237
robust AI systems and, 236–238
tools for, 246
V&V with, 228
exploit model parallelism, 158

exponential data growth, 59, *60*, *61*, 61–62
Extract-Load-Transform (ELT), 70
Extract-Transform-Load (ETL), 70

Facebook, 226, 235–236, 265
fairness, accountability, safety, transparency, ethics, privacy, and security (FASTEPS)
  AI governance and, 305, 387
  AISDM and, 318
  in AI systems engineering, 431
  compliance with, 80, *81*, 209–210
  guardrails, 261
  issues, 354
  principles, 58, 211, 215, 260, 270, 273, 278, 341, 366
  RAI and, 205–206, 302–303, *333*, 334
  security with, 272
  social media and, 259
  STSs and, 260–261, 267–270, 278
false negatives (FN), *102*, 104, *104*
false positives (FP), *102*, 104, *104*
FASTEPS. *See* fairness, accountability, safety, transparency, ethics, privacy, and security
FDA. *See* Food and Drug Administration
feature selection, 73
feature squeezing, 239–240
Federal Energy Regulatory Commission, US, 77
feedback, 458–459
Feigenbaum, Ed, *5*, 7, 135–136
field-programmable gate arrays (FPGAs), 132, 158
FirstMark, 70
fixed-point arithmetic, 348–349
floating-point operations per second (FLOPS), 134, 143, *147*
FN. *See* false negatives

Food and Drug Administration, US (FDA), 315
force multipliers, 62
forecasting, 494
Forgie, James, *5*
formulation, 330, 355
Forrester, Jay, 405
Forsberg, K., 33–34
forward propagation, *146*
foundational R&D, 133–134
foundation models, 173–174
Fournier, Camille, 414–415, 429
FP. *See* false positives
FPGAs. *See* field-programmable gate arrays
fractions, 147–148
functional architecture
  for AI system architecture, 40–43, *41*, 48, 50–51
  as terminology, 9, *10*, 11
functions
  cost, 112–113
  for ReLUs, 143, *144*, 145
  sigmoid, *110*
  step, *110*
  tanh, *110*
fundamentals
  of adversarial AI, 140
  of AI platform infrastructure, 79–80
  of AI system architecture, 19–20
  of AI systems engineering, 29–31, 50–51
  of CNNs, 90–91, 120–121, *122*, 123
  of cybersecurity, 163–166
  of DNNs, 8
  of MLOps, 365–367
  of Moore's Law, 43–44
  of presentations, 460–463, *461*
  of SWOT analysis, 295, 298
  of writing, 453–460, *454*, *456*, *458–459*

funding, 7
"Future Directions in Human Machine
    Teaming" (workshop), 208
Future Forward (Rifkin), 290
Future of Life Institute, 276
future of work, 408–410

GANs. *See* generative adversarial networks
GARD. *See* Guaranteeing AI Robustness
    against Deception
Garvin, D. A., 424–425
gathering data, 93, *94*
GDPR. *See* General Data Protection
    Regulation
Gebru, T., 82, 264
GEMM operator. *See* GEneral Matrix
    Multiplication operator
General Data Protection Regulation
    (GDPR), 82, 237, 242, 263
GEneral Matrix Multiplication (GEMM)
    operator, 143, *144*
generative adversarial networks (GANs),
    116, *117*, 118, 235, 240
Generative Pretrained Transformer (GPT)
    algorithms, 142–143, 409–410, 434
George, Bill, 413, 427
Gift, N., 384
GitHub, 174
goals, business, 43–44, 313, *314*, 315,
    348, 403
Goleman, Daniel, 426
good fits, *96*
Goodwin, O., 343
Google
    brain floating-points by, 147
    business at, 424–425
    Cloud, 138, 307
    energy at, 170
    engineers at, 136
    EQ to, 425–426, *426*

GoogLeNet, 121, 123
    image verification at, 230, *230*
    MAC and, 162
    model training at, 141
    NLP at, 118
    NVIDIA and, 153
    robotics at, 193
    science at, 151–152
    TPUs, 13, 156
GPT-1, 142
GPT-2, 142
GPT-3, 123, 142–143, 173–174, 224
GPT-4, 142–143, 276
GPT algorithms. *See* Generative
    Pretrained Transformer algorithms
GPTe maturity, 131
GPUs. *See* graphics processing units
"Gradient-Based Learning Applied to
    Document Recognition" (LeCun),
    120
gradient descents, *114*, 114–115
graduate studies, 450–451
graphics processing units (GPUs)
    at AMD, 174
    CPUs and, 132
    hardware, 8, 149
    ILSVRC and, 76–77
    at NVIDIA, 150–151
    technology, 149–151
    TPUs and, 89, 134, 142, 148, 151–152,
        175–176
gray-boxes, 231, 248
Green AI, 170, 177
grid stabilizers, 494
Guaranteeing AI Robustness against
    Deception (GARD), 232

Hadoop Distributed File System
    (HDFS), 67
Hamel, G., 303

Hansen, M. T., 406
hardware
    algorithms and, 160
    designers, 148
    DNN, 171
    domain-specific, 152–155, *154*
    evaluation, 166
    GPUs, 8, 149
    malicious actors for, 226, *227*, 228
    performance, 156
    performance tools for, 172
    software and, 158–160, 175
Harrow, Aram, 173
Harvard Business Review, 292
Hassidim, Avinatan, 173
Hawking, Stephen, 89
HDFS. *See* Hadoop Distributed File
    System
Heilmeier, George H., 309
Heilmeier Catechism, 505, *506*
Heilmeier Criteria, 309, *310*
Hennessy, John, 136–137, 155, 175
Hewlett-Packard, 424
HHL algorithms, 173
hidden layers, *144*, *146*
hidden technical debt, 351, *352*
high-performance algorithms, 141–142
high-performance teams, 431–433,
    432–433
high-power devices, 156, *157*, 158
Hinton, Geoffrey, 5, 8, 13, 75–76
history
    of AI algorithm era, 43–44
    of AI development, 4, *5*, 6–9
    of AI success, 40
    of ChatGBT, 142–143
    Computer History Museum, 136
    of computing technologies, 134–137
    of CPUs, 149–152
    of deep learning, 125

of ethics, 278
of HMT, 191–193, *192–193*
of ML, 13, 89–91, *91*
of ML algorithms, 169
of modern computing, 131–134,
    132–133
of RAI, 259–261
HMT. *See* human-machine teaming
Horowitz, M., 154
Horvitz, Eric, 169
How the Future Works (Elliott), 400
human-machine augmentation, 467.
    See also use cases
human-machine teaming (HMT)
    AI and, 14, 81, 163, 199–201, *201*,
        332
    AI horizons with, *213*, 213–214
    in AI pipeline, 215–216
    augmentation with, 193–194, *194–195*,
        196–197
    autonomy and, 201–205, *203–205*
    building blocks of, 196
    challenges to, 210–214, *213*
    consequences of, 228
    history of, 191–193, *192–193*
    learning exercises for, 216–217
    ML and, 226, 378, 496
    modern computing and, *227*
    performance metrics for, 205–206, *207*,
        208–210, *209*
    planning in, 481
    search-and-discovery tools with,
        197–199, *199*
    SME and, 412
    spectrum, 192–193, *193*, 200–202
    technology and, 214–216
humans
    in business, 237
    capabilities of, 193–194, *194–195*,
        196–197

humans (cont.)
  human capital, 131–132
  Human-Centered Artificial Intelligence
    Institute, 17
  human intelligence and, 89–90
  human language technology, *195*
  human-machine augmentation, 2, *3*
  human-machine models, 211–212
  human-machine teaching, *10*
  human-machine teaming, 14, 81,
    163
  IT and, 197–198
  knowledge of, *193*
  machine intelligence and, 199–200,
    214–215
  NLP and, 90
  robotics and, 191–193
  search and discovery tools for, 196
hybrid approaches, 159
hyperparameters
  in CNNs, 119–120
  for designing, *119*
  parameters and, 118–119
  resources for, 517

Iansiti, M., 431–432
IBM
  Deep Blue, 5, 7–8
  ML at, 6
  ML models at, 240
  organizational culture at, 417
  R&D at, 66, 173
  Red Hat, 380–381
  Watson, 5, 7–8
IEE-754 floating point standard, 145,
  147–148
IEEE. *See* Institute of Electronics
  Engineers
IJCAI. *See* International Joint Conference
  on Artificial Intelligence

ImageNet Large Scale Visual Recognition
    Challenge (ILSVRC)
  algorithms from, 75–76
  competition, 75–76
  data sets from, *78*
  DNN algorithms in, 149
  GPUs and, 76–77
  images, 120
  ML and, 13
  MNIST and, 83
image verification, 230, *230*
impact, of AI emphasis, *18*
implementation framework, *42*
imputation, 72
inclusiveness, 270–271
INCOSE. *See* International Council on
  Systems Engineering
individual harms, 269
individual leaders, 291
industrial IoT, 197
The Industries of the Future (Ross), 407
inference classifiers, *139*
information technology (IT), 197–198
Information Technology Infrastructure
  Library (ITIL), 346
Information Technology Service
  Management (ITSM), 345–346
infrastructure. *See* AI platform
  infrastructure
inline end-to-end AI systems, 374, 389
innovative team environments
  for AI practitioners, 408–410
  AI talent in, 427–431
  careers in, *411*, 411–414
  fostering, 399–400, *401*, 402, 433–436
  high-performance teams for, 431–433,
    432–433
  learning exercises about, 436–438
  measurement of, 417–421, *418*, *420*
  networking for, *426*, 427–431

organizational culture in, 402–405
organizational structure in, 405–408, *406*
resilience in, 422–427, *423*, *425–426*
technology and, 414–417, *416*
Institute of Electronics Engineers (IEEE), 261
integer representations, *147*
integrated systems, 155–156, *157*, 158–160
interactive development environment, 510–512
International Council on Systems Engineering (INCOSE)
  AI deployment to, 35, 39
  on AI systems engineering, 31–32, 312, 328, 354
  NAE and, 47, 51
  risk management to, *314*, 320–321
  standards, 31–32, 50
  on subsystems, 32–33
  Vee-model and, 33–35, *34*
International Joint Conference on Artificial Intelligence (IJCAI), 246, 273
Internet of Things (IoT)
  AI capabilities with, 387
  in AI pipeline, 349
  big data and, 89
  devices, 138
  DNNs and, 171
  industrial, 197
  ML algorithms and, 301
  timeline for, 494–495
  VPAs in, 215
investigators, 340, *340*
IoT. *See* Internet of Things
IT. *See* information technology
iterations, 518
ITIL. *See* Information Technology Infrastructure Library
ITSM. *See* Information Technology Service Management

Japan, 350
JavaScript Object Notation files, 381
Jennings, Ken, *5*, 7–8
Jeopardy! (TV show), *5*, 7–8
Jetson TX1 board, 159
Jobs, Steve, 131
Johnson, Lady Bird, 493
Juniper Research, 197
Jupyter Notebook, 511–512

K8s. *See* Kubernetes
Kaggle, 77
Kahneman, Daniel, 8, 212, 226
Kanter, R. M., 407–408
Karmaker, Santu, 384
Kasparov, Garry, *5*, 7
Keras, *133*
kernels, in ML algorithms, 175–176
key computational kernels, 141–143, *144*, 145
Khazan, I. Z., 413
Kim, G., 345
Kirby, J., 196–197
Kiron, D., 417
Kissinger, H. A., 260
knowledge, 14, *193*
Kochan, F., 428
Koizel, E., 166
Kotter, J., 405–406
Kram, Kathy, 428, 436
Krizhevsky, Alex, 120, *122*
Kubernetes (K8s), 381–382

labeled data, 92, 212
labor market, 266
LADAR. *See* laser detection and ranging
Landing AI, 59
La Rochefoucauld, Francois de, 485
Larson, Robert, *5*
laser detection and ranging (LADAR), 30

latency requirements, 175
layers
  convolutional, 109, *111*, 121
  deconvolutional, *111*
  dense/fully connected, 121
  hidden, *144, 146*
  neural network, *111, 119*, 121
  pooling, 121
LCNC application development. *See*
  low-code/no-code application
  development
leadership
  AI, 31, 422–427, *423, 425–426*
  AI systems engineering, 43–45, *44, 46,
  47, 47*
  Bennis on, 44–45
  from CEOs, 40
  management and, 49, 51
  at MIT, 29
  PLD, 294
  for systems thinkers, 43–45, *44, 46*, 47,
  47
Lean Product Development, 345
learning exercises
  for AI, 21–22
  about AI deployment, 357–358
  for AI-powered cycling, 483
  for AI strategy, 321–323
  for AI systems engineering, 52–53
  about AquaAI, 506–507, *507*
  about communication, 464–465
  for data conditioning, 84
  for HMT, 216–217
  about innovative team environments,
  436–438
  for MEALS, 492
  with MESH, 499
  for ML, 125–127
  with MLOps, 392–393
  for modern computing, 178–180

  with RAI, 279–280
  with robotics, 475–476, *476*
  from robust AI systems, 249–251
learning procedures, 112, 212–213
learning rates, *119*, 518
LeCun, Yann, 8, 75, 120, 137, 169,
  212–213
Lee, W., 234–235
Lencioni, Patrick, 402, 432, *432*,
  436–437
leverage quantum computing, 173
Li, Fei-Fei, 13
LIME. *See* Local Interpretable Model-
  Agnostic Explanation
Llinas, 39–40
Lloyd, Seth, 173
Local Interpretable Model-Agnostic
  Explanation (LIME), 240, 249
local SRAM, 160–161
logic machines, 135
long short-term memory (LSTM), 118
low-code/no-code (LCNC) application
  development, 365, 367, 382–385
low-power devices, 156, *157*, 158
LSTM. *See* long short-term memory

MAC. *See* multiply-accumulate
machine decision making, 192–193,
  194–195, 196–197, 199–201
machine intelligence, 199–200, 214–215
machine learning (ML)
  accelerators, 175–176
  advancements in, 131
  AI capabilities and, 74
  AI development and, 76–77, 350
  AI tools for, 132–133
  for AquaAI, 501–502
  artifacts, 379
  AUC and, 71
  AutoML, 382–385

autonomy after, 191–192

backpropagation in, 109, *110–111*, 112–115, *114*

BEFAST and, 307

building blocks for, 123–125, *124*

carbon dioxide and, 170

challenges, 123

classes, 91–92

classifiers, 75

confidence in, 2, *3*, 14–15, 17–19, *18*, *194*

convolutional neural networks in, 120–121, *122*, 123

Core ML modeling tools, 141

CPUs for, 137

data conditioning and, 57–59, *227*

data drift in, 73

data sets, 82

deep learning and, 106–107, *107–108*, 109

demands, 152

engineers, 367

framework, *41*, 41–42

history of, 13, 89–91, *91*

HMT and, 226, 378, 496

at IBM, 6

ILSVRC and, 13

implementation, 351

LADAR and, 30

learning exercises for, 125–127

Machine Learning Enabled Security Platform, 166

Machine Learning Technology Readiness Levels, 244–245

methods, 91–92

ML-oriented accelerators, 170

models, 39, 65, 72, *72*, 81, 101, *102*, 155, 223, 237–238, 240, 388

neural network designs for, 115–116, *117*, 118–120, *119*

NLP and, 231

output, 193, 241

performance measurements with, 48, 100–101, *102–105*, 104–106

performance metrics, 512–516

pipeline, 381

privacy and, 172–173

production, 378

Project Jupyter and, 20

R&D in, 233

reinforcement, *91*, 92, 99–100, *100*

rigor, 349

robust AI systems and, 223–225, *224*

scientists, 366

SHAP for, 240–241

software, *133*

stages, 57

supervised, *91*, 91–97, *93–94*, *96–97*

systems, 165

techniques, 200–201, *201*

training, 43–44, *44*, 145, 174, 387

University of California Irvine Machine Learning Repository, 77

unsupervised, *91*, 92, 97–99, *98*

Machine Learning (Murphy), 90

machine learning (ML) algorithms

AI application and, 133

AI capabilities with, 233

big data and, 175

building blocks for, 57–58

CNN algorithms and, 116

confidence in, 205–206

data and, 11–12

data conditioning and, *10*, 11–13

designing, 151–152

with DNNs, 112

DSA and, 148

HHL algorithms and, 173

history of, 169

implementation of, 357

machine learning (ML) algorithms (cont.)
  improvements, 149–152, *150*
  inference classifiers in, *139*
  IoT and, 301
  kernels in, 175–176
  MLOps and, 399
  models for, 9
  Moore's Law and, 158
  output of, 197
  PCA and, 99
  performance of, 80–81, 106
  requirements of, 137
  SME with, 236
  training, 91–92, 141
machine learning operations (MLOps)
  AI deployment and, 289, 388–391
  AI development operations and, 15–16,
    344–345
  in AI platform infrastructure, 382–385
  AISysOps and, 327–328
  AI system architecture with, 367–368,
    369–372, 373–375
  AWS and, 41
  business and, 321
  common pitfalls in, 385–388
  enabling techniques, 375–382, *380*
  fundamentals of, 365–367
  learning exercises with, 392–393
  ML algorithms and, 399
  platforms, 382–383, 390–391
  process, 346
  road map for, 316–317, 363–365
  SWOT analysis and, 505
  technology for, 352
  well-operating, 16–17
machines. *See specific topics*
Machines We Trust (Pelillo and
  Scantamburlo), 229
macro-level
  assessment, *317*

  challenges, 210–211, *332*
  micro-level and, 74–75
MAE. *See* mean absolute error
Magic Quadrant, 352
Make It Clear (Winston), 443–446,
  450–451, 459–460
malicious actors
  for hardware, 226, *227*, 228
  malicious cyberdiscussions, *199*
  for software, 174, 226, *227*, 228
management. *See also* risk management
  business, 414–417
  database, 69–70
  evolution and, 66–70, *67–69*
  leadership and, 49, 51
  managing-by-results, *418*, 418–421,
    420, *423*, 423–424, 426, 429, 435
  of ML production, 378
  of public health, 266
  upper, 291
The Manager's Path (Fournier), 414–415,
  429
Managing Energy for Smart Homes
  (MESH), 493–499, *495–498*
"Manifesto for Agile Software
  Development," 346–347, *347*, 356
MapReduce, 67
Marcus, Gary, 193–194, 224
marine maintenance, 501–503, *503–504*,
  505–507, *506–507*
marketplace data, 300
Massachusetts Institute of Technology
  (MIT). *See also* use cases
  AI-powered cycling at, 477
  AI strategic road maps at, 469, 485
  Artificial Intelligence Laboratory, 135
  Center for Information Systems
    Research, 63
  CSAIL, 6, 294–295
  education at, 445

leadership at, 29
Media Lab, 235
ML techniques at, 200–201, *201*
Moral Machine experiment, 265
MVP at, 475, 482
Raspberry Pi at, 524
R&D at, 63, 211, 477, 493, 501
SERC at, 14–15, 262
students at, 20
mathematics, 145, *146–147*, 147–148,
153
MathWorks, 384
MATLAB, 384
Matsudaira, K., 413–414
maxpools, *111*
Maxwell, John C., 400, 427
McCarthy, John, 4, *5*, 135
McCulloch, Warren, 135
McGovern, Patrick, 290
McKinsey & Company, 16, 293,
443–444
McKinsey Global Institute, 201
Mead, Carver, 136
meal evaluation and attainment logistics
system (MEALS), 485–488, *487–489*,
490–492, *491*
mean absolute error (MAE), 101, *103*,
124
mean squared error (MSE), 101, *102*,
124
measurement
building blocks of, 205–206, *207*,
208–210, *209*
of innovative team environments,
417–421, *418*, *420*
performance, 48, 100–101, *102–105*,
104–106
medium neural networks, *107*
memory, 160–161, *161*
mentoring, 427–431, 435–436

Merrill, R. R., 432–433
MESH. *See* Managing Energy for Smart
Homes
metacognition, 212
metadata, 12
metaverse, 172
methodology
for adversarial AI attacks, 242–245,
243–244
AI systems engineering, 19–20, 35,
39–40, 75
of PCA, 73
metrics
performance, 205–206, *207*, 208–210,
209, 348–351, 512–516
progress, 417–421, *418*, *420*
review, 513
roofline, 160–163, *161*
Michelangelo, 1
micro-level, 74–75
micro-level challenges, 210–211
Microsoft
AI at, 49
Azure, 138, 307
deepfakes to, 235–236
emergent bias at, 264
R&D at, 76
minimum viable product (MVP)
with AI, *331*, 334
AI capabilities with, 288–289
at MIT, 475, 482
SRLs and, 386, 403
to stakeholders, 374
users of, 338
Minsky, Marvin, 4, *5*, 6, 135
Mirsky, Y., 234–235
misclassification rates, *102*
mission/vision
for AI-powered cycling, 478
at AquaAI, 502

mission/vision (cont.)
  envisioned future and, 299–301, *301*
  for MEALS, 486
  at MESH, 494
Misty Robotics
  AI value proposition with, 307, *308*,
    472–473, *473*
  on business needs, *311*, *474*
  envisioned future with, 202, 299–301,
    301, 470–471, *471*
  strategic development with, *304*
  SWOT analysis with, 312, *313*
  use case with, 469–476, *471–476*
MIT. *See* Massachusetts Institute of
    Technology
mitigation, for RAI, 123
mitigation techniques, 223, 225,
    238–242, 249
ML. *See* machine learning
ML algorithms. *See* machine learning
    algorithms
MLOps. *See* machine learning operations
MLP. *See* multilayer perceptron
MLPerf, 349
ML training, *44*
MLTRL. *See Technology Readiness Levels
    for Machine Learning Systems*
MNIST. *See* Modified National Institute
    of Standards and Technology
models
  for AI strategic blueprint, *46*, 47, 51
  DNN, 172
  foundation, 173–174
  human-machine, 211–212
  ML, 39, 65, 72, *72*, 81, 101, *102*, 155,
    223, 237–238, 240, 388
  model-based frameworks, *209*
  modeling, 174
  model inversion, 232
  model parallelism, 159

model performance accuracy, 142
  neural networks, 139, *139*
  Open Systems Interconnection model,
    169
  strategic development model, *295*
  training, *94*, 95, 105–106, 141
modern computing
  AI and, 13–14
  architecture in, 158
  arithmetic precision in, 145, *146*, *147*,
    147–148
  challenges, 169–174, *171*
  cybersecurity with, 163–166
  domain-specific hardware/software for,
    152–155, *154*
  evolution of, 138–141, *139–141*
  history of, 131–134, *132–133*
  HMT and, *227*
  integrated systems in, 155–156, *157*,
    158–160
  learning exercises for, 178–180
  ML algorithms improvements in,
    149–152, *150*
  neural networks in, 141–143, *144*,
    145
  roofline metrics in, 160–163, *161*
  securing, 163–164, *164*
  security in, 166, *167*, 168–169
  technology for, 134–137, 175–177
Modified National Institute of Standards
    and Technology (MNIST)
  CNNs with, 521–523
  data sets, 512–513, 517–518
  ILSVRC and, 83
  R&D at, 75–77, *76*
monitoring AI talent, *426*, 427–431
Moore, Andrew, 11, 136
Moore, Gordon, 136
Moore School of Electrical Engineering,
    134–135

Moore's Law
  for application-specific domains,
    154–155
  building blocks and, 175
  computer advancements with, 134
  Dennard Scaling and, 137
  fundamentals of, 43–44
  limitations from, 151
  ML algorithms and, 158
  VLSI and, 136
Mooz, H., 33–34
Moral Machine experiment, 265
Morgan, Andrew, 380, *380*
Mozilla, 77
MSE. *See* mean squared error
multilayer perceptron (MLP)
  algorithms, 517–520
  neural network, 145
multimodal data, 173
multiply-accumulate (MAC), 153,
    162
Murphy, Kevin, 90
MVP. *See* minimum viable product

Nadella, S., 431–432
NAE. *See* National Academy of
  Engineering
narrow AI
  AI application and, 2, *2*, *3*
  optimism with, 224
NASA. *See* National Aeronautics and
  Space Administration
National Academy of Engineering
  (NAE), 45, *46*, 47, 51
National Aeronautics and Space
  Administration (NASA), 29–30, 32,
  331
National AI Initiative Act, 276
National Council on Systems
  Engineering, 33

National Institute of Standards and
  Technology, Japan, 350
National Institute of Standards and
  Technology, US (NIST), 229, 276,
  350
National Science and Technology
  Council, US (NSTC), 166, 168–169
National Society of Professional
  Engineers, 261
natural-language processing (NLP)
  bias in, 266
  at Google, 118
  humans and, 90
  ML and, 231
  robotics and, 197–198
  for VPAs, 210–211
networking, *426*, 427–431
Networking and Information Technology
  Research and Development (NITRD),
  166
Neural Information Processing Systems
  (NeurIPS) conference, 246
neural networks. *See also specific neural
    networks*
  ANNs, 90
  architecture for, *117*, 118
  deconvolutional layers in, *111*
  designing, 115–116, *117*, 118–120, *119*
  forward propagation in, *146*
  MLP, 145
  models, 139, *139*
  in modern computing, 141–143, *144*,
    145
  neural nets, 106–107, *107–108*, 109
  ReLUs in, 109
  RNNs, 118, 235, 481
  training, 109, *110–111*, 112–115, *114*
NeurIPS conference. *See* Neural
  Information Processing Systems
  conference

neuromorphic computing, 172
neuro-symbolic AI, 8, 212
news, 448
Ng, Andrew, 15–16, 40, 59, 106, *107*,
    288, 363–365
Nilsson, Nils, 4, *5*, 134
NIST. *See* National Institute of Standards
    and Technology
NITRD. *See* Networking and
    Information Technology Research and
    Development
NLP. *See* natural-language processing
Northcutt, C., 80
Not-only Structured Query Language
    (NoSQL), 66–68, *68–69*, 80, 83
NSTC. *See* National Science and
    Technology Council, US
NVIDIA
  DNN at, 151, 159–160
  FLOPS and, 143
  Google and, 153
  GPU technology at, 150–151
  R&D at, 137

Oak Ridge National Laboratory (ORNL),
    174
object classification, *122*
objective and key results (OKRs), 417, 435
OECD. *See* Organisation for Economic
    Cooperation and Development
Office of Science and Technology Policy
    (OSTP), 82
Okravi, Hamed, 169
OKRs. *See* objective and key results
OLAP. *See* online analytical processing
OLTP. *See* online transactional processing
O'Neill, Paul, 477
online analytical processing (OLAP), 58
online end-to-end AI system architecture,
    374, 389

online transactional processing (OLTP),
    58
OpenAI, 224. *See also* ChatGPT
open-source data, 58–59, 66–67
Open Systems Interconnection model,
    169
O'Reilly, C. A., 296
Organisation for Economic Cooperation
    and Development (OECD), 263, 270
organizational culture
  AI, 443
  at IBM, 417
  in innovative team environments,
    402–405
organizational maturity clusters, 339–341,
    340
organizational structure, 405–408, *406*
organization core values, 302–303, *304*
organization infrastructure, 63, *64*,
    382–385
Ormenisan, A. A., 379
ORNL. *See* Oak Ridge National
    Laboratory
OSTP. *See* Office of Science and
    Technology Policy
outlining communication, *451–452*,
    451–453, 463–464
overfitting, *96*
Ozkaya, Ipek, 39

Packard, David, 424
Page, Larry, 424–425
Papert, Seymour, 6
parallelism, 158–160
parameters, 118–119
passives, 340, *340*
Patterson, David, 136–137, 155, 170,
    175
"Pause Giant AI Experiments" (Future of
    Life Institute), 276

PCA. *See* principal component analysis
peak performance, *157*
Pelillo, M., 229
people, *38*, 49–50. *See also* humans
people-process-technology triad, 35, 51, 248, *331*, 339, 415. *See also* AI deployment
Perceptrons (Minsky and Papert), 6
performance
    of AUC, 90
    of building blocks, 192–193
    confidence in, 215–216
    to engineers, 175
    hardware, 156
    high-performance algorithms, 141–142
    high-performance teams, 431–433, *432–433*
    measuring, 100–101, *102–105*, 104–106
    metrics, 205–206, *207*, 208–210, *209*, 348–351, 512–516
    of ML algorithms, 80–81, 106
    peak, *157*
    tools, 172
Perkins, Kleiner, 417
peta–floating-point operations (peta-FLOPS), 43
Petroski, Henry, 209–210
PGD. *See* projected gradient descent
pioneers, 340, *340*
pipeline. *See* AI pipeline
pipeline parallelism, 159
Pitts, W., 135
platform infrastructure. *See* AI platform infrastructure
PLD. *See* Program for Leadership Development
Podolny, J. M., 406
poisoning attacks, 231–232

policy
    AI governance and, 177
    OECD, 263, 270
    OSTP, 82
    in society, 260–261
pooling layers, 121
Practical MLOps (Gift and Deza), 384
Prahalad, C., 303
precision, *102*
predictions
    with algorithms, *231*
    in confusion matrix, *104*
predictive analytics, 165
preexisting bias, 267–268
preparation, of data, 70–74, *72*
preprocess data, 93, *94*
presentations
    communication in, 453–463, *454*, *456*, *458–459*, *461*
    fundamentals of, 460–463, *461*
principal component analysis (PCA), 73, 99, 165–166, *167*
privacy. *See also* fairness, accountability, safety, transparency, ethics, privacy, and security
    AI governance and, 262–263
    data, 82, 138–139, 265
    ML and, 172–173
    public health and, 266
    standards, 201
    of users, 271
process
    for AI systems engineering, *36–38*, 49
    challenges, 49
    CRISP-DM, 66
    data preprocess, 93, *94*
    MLOps, 346
    OLAP, 58
    people-process-technology triad, 35, 51, 248, *331*, 339, 415

process (cont.)
  processing, 164, *164*
  RPV theory, 293
  for trustworthy AI, 237–238
  Vee-model, 34–35
products, data, 74
Program for Leadership Development
  (PLD), 294
programming, 299
progress, 417–421, *418*, *420*
projected gradient descent (PGD), 240
Project Jupyter, 20
Project MAC, 135
pruning, 73, 176
public health, 266
public safety, 264–265
Python, 155, 510, 512
PyTorch, *133*

qualitative performance metrics, 205–206,
  207, 208–210, *209*
quality, of data, 70–74, *72*
quantitative performance metrics,
  205–206, *207*, 208–210, *209*
quantum computing, 173
The Quest for Artificial Intelligence
  (Nilsson), 134

racial bias, 265
Rackspace Technology, 62
RAI. *See* responsible AI
Raj, E., 378, 381
Rajapakse, R., 377
Ramo, Simon, 33
Ransbotham, S., 196–197, 306, 339,
  355, 404, 410
Raspberry Pi, *308*, 317, 475, 482, 524
Ray, Amit, 469
RBAC. *See* role-based access control
R&D. *See* research and AI development

Rebooting AI (Marcus and Davis),
  193–194
recall, *102*
receiver operating characteristics (ROCs),
  104–105, *105*, 124, 206
recruiting AI, 267
recruiting AI talent, 423–424, *426*,
  427–431
rectified linear units (ReLUs), 109, 143,
  144, 145
recurring neural networks (RNNs), 118,
  235, 481
Red Hat, 380–381
Redman, T., 62
Reduced Instruction Set Computer
  (RISC), 136
reduction, dimensionality, *98*, 98–99
regression, 92–93, *93*
regulation
  of DARPA, 236–237
  with FASTEPS, 80, *81*
  Federal Energy Regulatory Commission,
    77
  GDPR, 82, 237, 242, 263
  of RAI, 14–15
  SERC and, 14–15
  V&V and, 30
reinforcement machine learning, *91*, 92,
  99–100, *100*
Rejto, Stephen, 288
reliability, 239, 249, 271
ReLUs. *See* rectified linear units
renewable energy, 493–494
reorientation, in business, 306
representational learning, 125
representative AI industries, 509–510
rescaling, 73
research and AI development (R&D)
  advances in, 17
  in AI, 434–435

for AI innovation, 9, *10*, 11
for AI practitioners, 4, *5*
for auxiliary sensors, 203–204, *204*
CoB and, 411, *411*
contributions to, 456–457
at CSAIL, 6
for cybersecurity, *167*, 168–169
for DARPA, 9
foundational, 133–134
funding, 7
at IBM, 66, 173
laboratories, 49, 421
at Microsoft, 76
at MIT, 63, 211, 477, 493, 501
in ML, 233
MLTRL and, 331
at MNIST, 75–77, *76*
near-term barriers in, 201–205,
    203–205
at NVIDIA, 137
in RAI, 262
RISC and, 136
study areas, 81
in XAI, 236–237
resilience, 239, 249, 422–427, *423*,
    425–426
Resilient Mission Computer (RMC), 169
ResNet, 121, 123, 151
ResNet CNN algorithm, 76
resource, process, value (RPV) theory, 293
responsibility, in business, 289–290, *290*
responsible AI (RAI)
    in AI development life cycle, 273–275,
        274
    bias to, 228–229, 264–267
    challenges, 276, *277*, 278
    data bias in, 72
    embedding, 328
    FASTEPS and, 205–206, 302–303,
        333, 334

history of, 259–261
    learning exercises with, 279–280
    mitigation for, 123
    principles, 223–224, 270–273, *272*
    regulation of, 14–15
    robust AI systems and, 11, 228
    for society, 261–263
    sociotechnical systems for, 267–270
    technology for, 278–279
retaining AI talent, 419–421
Rifkin, Glenn, 290
RISC. *See* Reduced Instruction Set
    Computer
risk analysis, 35, *36–38*, *369–372*
risk management
    in AI, *207*, 298
    to INCOSE, *314*, 320–321
    to NIST, 276
    security and, *207*
    SWOT analysis and, 320
RMC. *See* Resilient Mission Computer
RNNs. *See* recurring neural networks
robotics. *See also specific topics*
    at Amazon, 193
    in autonomous vehicles, 202–205, *203*,
        205
    auxiliary sensors for, 203–204, *204*
    BEFAST in, 299
    cloud computing for, 307
    at Google, 193
    humans and, 191–193
    labor market, 266
    learning exercises with, 475–476,
        476
    machine intelligence, 199–200
    marketplace data in, 300
    NLP and, 197–198
    programming in, 299
    Rules of Robotics, 192
    SWOT analysis with, 474–475, *475*

robust AI systems
  adversarial AI and, 229–233, *230–231*
  against adversarial AI attacks, 242–245, *243–244*
  AI vulnerabilities and, 246, *247*, 248–249
  building blocks of, 225–226, *227*, 228–229
  challenges for, 245–246
  deepfakes for, 233–236
  knowledge of, 14
  learning exercises from, 249–251
  mitigation techniques with, 238–242
  ML and, 223–225, *224*
  RAI and, 11, 228
  XAI and, 236–238
ROCs. *See* receiver operating characteristics
role-based access control (RBAC), 377, 381–382
roofline metrics, 160–163, *161*
Roosevelt, Franklin D., 29, 427
Rosenblatt, Franklin, *5*, 6
Ross, Alec, 407
Roth, A., 241–242
RPV theory. *See* resource, process, value theory
Rules of Robotics, 192
Rus, Daniela, 294–295
Rutter, Brad, *5*, 7–8

safety, 271
SAGE. *See* Semi-Automatic Ground Environment
salience, 450
saliency maps, 242–244, *243*
Sambasivan, N., 73–74
sample applications, 509–510
Samuel, Arthur, *5*, 6
scale computing, 173

scale driving deep learning progress, 106–107, *107*
Scantamburlo, T., 229
Schein, Edgar, 305–306, 402–404
Schwartz, R., 170, 293
science
  CSAIL, 6, 77, 294–295
  data scientists, 57, 71, 366
  at Google, 151–152
  ML scientists, 366
  NSTC, 166, 168–169
  OSTP, 82
  scientific writing, 453–454, *454*
  US Defense Science Board, 168
Sculley, D., 40
search-and-discovery tools, 192, 196–199, *199*, 214–215
secure multiparty computation (CMPC), 377, 389–390
securing modern computing, 163–164, *164*
security. *See also specific topics*
  AI development, security, and operations, 41, 289, 344–345
  cybersecurity, 163–166
  with FASTEPS, 272
  Machine Learning Enabled Security Platform, 166
  in modern computing, 166, *167*, 168–169
  risk management and, *207*
Sedol, Lee, *5*, 8
segmented data, 94, *94*
Selfridge, Oliver, 4, *5*, 135
self-supervised learning, 212–213
Semi-Automatic Ground Environment (SAGE), 32
semiconductors, 137, 287–288
semistructured data, 59
sensitivity, *102*

sensors, 12
SERC. *See* social and ethical responsibilities
SGD. *See* stochastic gradient descent
shallow neural networks, *107*
Shannon, Claude, 135
SHapley Additive exPlanation (SHAP), 240–241, 249
sigmoid functions, *110*
simplified AI system architecture, *132, 192, 224*
single-threads, 373–374
size, weight, and power (SWaP), 133, 169, 171
skills, *47*, 445–446
Slagle, James, *5*
Slepian, M., 429
slides, for presentations, 460–461
slogans, 449
SME. *See* subject matter expert
Snowflake, 70
Sobol, M. G., 39
social and ethical responsibilities (SERC), 14–15, 262
social media
  big data from, *60*
  business, 235–236
  data on, *60*, 198, 226
  FASTEPS and, 259
  malicious cyberdiscussions on, *199*
society, 260–263, 270
sociotechnical systems (STSs), 260–261, 267–270, 278
software
  Agile, 346–347, *347*, 356, 376
  Anaconda Navigator, 510–511
  domain-specific, 152–155, *154*
  engineers, 59
  hardware and, 158–160, 175
  interactive development environment for, 510–512

malicious actors for, 174, 226, *227*, 228
ML, *133*
performance tools for, 172
sources, 12
spectrum, HMT, 192–193, *193*, 200–202
Spindel, B., 417
SQL database. *See* Structured Query Language database
SRLs. *See* system readiness levels
Stadler, C., 294
staff. *See* innovative team environments
stakeholders
  AI capabilities and, 1, *316*
  AI teams and, 387
  assessment, 51
  in business, 71, 298
  business goals for, 348
  designers and, 273–274
  MVP to, 374
  needs of, 368
  requirements, 43–44, 329, 337
  systems requirements, 34–35
standards, 31–32, 50
Stanford Digital Economy Lab, 131
STEM, 414–415
step-by-step flow, 343
step functions, *110*
stochastic gradient descent (SGD)
  gradient descents and, *114*, 114–115
  review, 512–516
storage, in cloud computing, 164, *164*
stories, 450, 457
strategic agility, *406*
strategic development model, *295*
strategic direction
  for AI-powered cycling, 479
  at AquaAI, 503, *503*
  for MEALS, 487–488, *488*

strategic direction (cont.)
  at MESH, 495–496, *496*
  organization core values and, 302–303,
    304
strategic principles, 2, *3*, 49
strategic thinking, 289–293, *290*, *292*
streaming data, 80
strengths, weaknesses, opportunities, and
    threats (SWOT) analysis
  AISDM and, 481, *482*
  AI strategic road maps for, 307,
    309–310, 312–313, *313*, 315
  of AquaAI, 505, *507*
  fundamentals of, 295, 298
  of MESH, 497–498, *498*
  MLOps and, 505
  risk management and, 320
  with robotics, 474–475, *475*
  threats in, 490–491, 497–498, *498*, 505
structured data, 12, 59, 66–67
Structured Query Language (SQL)
    database, 66–68, *68*, 80, 83
STSs. *See* sociotechnical systems
students
  common definitions for, 31–33
  in competition, 75–76
  at MIT, 20
style blunders, *459*, 459–460
subject matter expert (SME)
  at Apple, 406
  in business, 307
  HMT and, 412
  inclusiveness to, 270–271
  with ML algorithms, 236
subsystem integration, 226, 373
subsystems, 32–33
supervised learning algorithms, 73,
    123–124
supervised machine learning, *91*, 91–97,
    93–94, *96–97*
support, in careers, 428

support vector machines (SVMs), 97,
    117, 165–166, *167*, 240
surprises, 450
Sutskever, Ilya, 224
SVMs. *See* support vector machines
SWaP. *See* size, weight, and power
SWOT analysis. *See* strengths,
    weaknesses, opportunities, and threats
    analysis
*A Symbolic Analysis of Relay and Switching
    Circuits* (Shannon), 135
symbols, 449–450
system analysis, 232–233
system applications, 303
system readiness levels (SRLs)
  in AI ecosystem, 343–344, 346–347,
    351
  for AI organization, 355–356
  for AI practitioners, 329–330, 335,
    337
  AISDM and, 335, 364
  in AI system architecture, 350
  in business, *337*, 337–339, 376–378
  explanations for, 337–339
  MVP and, 386, 403
  V&V and, 368, 373–375
Systems Engineering (Eisner), 33
systems perspective, 225–226, *227*,
    228–229
systems thinkers, 43–45, *44*, *46*, 47, *47*
Szegedy, C., 229–230

tactical divides, *290*
tanh functions, *110*
Taiwan Semiconductor Manufacturing
    Company (TSMC), 287–288
target requirements, 35
Tay (chatbot), 264
technical communication, 444, *445*
technical depth, *411*, 412, 414–419, *416*,
    430–431

technology. *See also specific technology*
  for AI, 29
  for AI ecosystem, 20–21
  for AISDM, 375
  AI strategy and, 294–296, *295*, *297*,
    298
  for AI system architecture, 2, *3*
  algorithms for, 1–2
  architecture and, 50
  for business, 19–22
  challenges, 48
  commercial face recognition, 264
  contemporary tools, 375–382, *380*
  GAN, 235
  GPT-3, 173–174
  GPUs, 149–151
  HMT and, 214–216
  human language, *195*
  innovative team environments and,
    414–417, *416*
  for MLOps, 352
  for modern computing, 134–137,
    175–177
  OSTP, 82
  people-process-technology triad, 35,
    51
  Rackspace Technology, 62
  for RAI, 278–279
  from semiconductors, 137
  for systems analysis, *36*
  technical bias, 268
  for V&V, 238
*Technology Readiness Levels for Machine
  Learning Systems* (MLTRL), 331
TensorFlow, *133*, 511
tensor processing units (TPUs)
  algorithms and, 162–163
  DRAM and, 161–162
  in DSA, 152–153
  FPGAs and, 132
  at Google, 13, 156

GPUs and, 89, 134, 142, 148, 151–152,
    175–176
Tesla, 204–205
test, evaluation, verification, and
    validation (TEV&V), 49, 93–97, *94*,
    96–97
testing
  against attacks, 242–245, *243–244*, 249
  test harness, 348–351
  on Twitter, 264
  unit-tests, 373
TEV&V. *See* test, evaluation, verification,
    and validation
Thinking, Fast and Slow (Kahneman), 8,
    212
Thinking Strategically (Harvard Business
    Review), 292
third Industrial Revolution, 260
third-party applications, 386
Thompson-Ramo-Wooldridge, 33
threats. *See also* strengths, weaknesses,
    opportunities, and threats analysis
  external, 506
  identification of, 177
  predictive analytics for, 165
  in SWOT analysis, 490–491, 497–498,
    498, 505
  vulnerabilities and, 166, *167*, 168
threshold requirements, 35
TN. *See* true negatives
To Engineer Is Human (Petroski),
    209–210
TOPS/W power efficiency. *See* trillion
    operations per second per watt power
    efficiency
TP. *See* true positives
TPUs. *See* tensor processing units
training
  adversarial AI, 239
  AI horizons for, *124*
  computational complexity and, 112–114

training (cont.)
  data, 92–93, 99–100, *100, 139*
  generizability and, 100–101
  ML, 43–44, *44,* 145, 174, 387
  ML algorithms, 91–92, 141
  models, *94,* 95, 105–106, 141
  neural networks, 109, *110–111,*
    112–115, *114*
  pruning in, 73
  with ResNet CNN algorithm, 76
  supervised learning algorithms, 123–124
  with transformers, 123
Trajtenberg, M., 409
transformers, 123
transparency, 271
trillion operations per second per watt
    (TOPS/W) power efficiency, 145
true negatives (TN), *102,* 104, *104*
true positives (TP), *102,* 104, *104*
Truman, Harry S., 29
trustworthy AI, 229, 237–238, *272,*
    272–273
TSMC. *See* Taiwan Semiconductor
    Manufacturing Company
Turck, Matt, 70
Turing, Alan, 4, 8, 134–137, 410
Tushman, Michael, 294, 296
Twain, Mark, 57
*A 20-Year Community Roadmap for
    Artificial Intelligence Research in the
    U.S.* (report), 208
The 21 Indispensable Qualities of a Leader
    (Maxwell), 427
Twitter, 226, 264
TX-0, 135

Uber, 30, 264, 377
underfitting, *96*
United Nations, 229
United States (US)
  DARPA and, 204–205
  deepfakes to, 234–235
  Defense Science Board, 168
  Department of Transportation,
    204–205
  FDA, 315
  Federal Energy Regulatory Commission,
    77
  National AI Initiative Act, 276
  NIST, 229, 276, 350
  NSTC, 166, 168–169
  OSTP in, 82
  Senate Armed Services Committee on
    Cybersecurity, 169
  voting in, 265
unit-tests, 373
University of California Irvine Machine
    Learning Repository, 77
unstructured data, 12, 59
unsupervised machine learning, *91,* 92,
    97–99, *98*
upper management, 291
US. *See* United States
use cases
  AquaAI, 501–503, *503–504,* 505–507,
    506–507
  Bose AI, 477–479, *479–480,* 481–483,
    482–483
  for human-machine augmentation, 467
  MEALS, 485–488, *487–489,* 490–492,
    491
  MESH, 493–499, *495–498*
  Misty, 469–476, *471–476*
users
  AI capabilities for, 238–239
  AI development for, 275
  customers compared to, 4
  cybersecurity for, 177
  decisions by, 15
  deployed systems and, 164, *164*
  of MVP, 338
  privacy of, 271

user problems, 8–9
user protection, 165

validation, *94*, 95–96, *97*
values, 403
Van Maanen, J., 306
variety
  of big data, *61*, 61–62
  of data, *61*, 61–62
Vee-model
  building blocks from, 51
  INCOSE and, 33–35, *34*
  modified, *329*
  representation, 50
  SRLs in, 346–347
  V&V with, 306
velocity, *61*, 61–62
veracity, *61*, 61–62
verification and validation (V&V)
  of AI system architecture, 229, 343
  data and, 239
  development operations and, 376–378
  mitigation techniques for, 249
  of MLOps, 389
  regulation and, 30
  SRLs and, 368, 373–375
  during subsystem integration, 226
  technology for, 238
  with Vee-model, 306
  with XAI, 228
very-large-scale integration (VLSI), 136
VGGNet, 121, 123
virtual environments, 511
virtual machines (VMs), 379–380, *380*
virtual personal assistants (VPAs), 197,
  210–211, 215
Visengeriyeva, L., 378
vision/mission. *See* mission/vision
Vision-Steps-News-Contributions
  (VSN-C), 445–452, *447*, 452,
  454–457, 463–465

VLSI. *See* very-large-scale integration
VMs. *See* virtual machines
Volkova, Anastasia, 202
Volkswagen, 30
volume, *61*, 61–62
von Neumann, John, 135, 173, 175, 178
VPAs. *See* virtual personal assistants
VSN-C. *See*
  Vision-Steps-News-Contributions
V&V. *See* verification and validation

Walker, B., 364
Watkins, M. D., 422–423
Watson, *5*, 7–8
Welch, Jack, 422–423
well-operating MLOps, 16–17
Westerman, G., 306, 404–405
white-boxes, 231, 248
Wiener, Norbert, 223–224
Wilson, H. J., 200
Winston, Patrick, 443–446, 450–451,
  458–460, *459*, 463
Winston Star, 446, 448–451, *449*,
  463–464
Wissner-Gross, Alexander, 11–12
Wolfram, Stephen, 143
women, 430
word embeddings, 266
writing
  broken-glass outlines, 451, *452*
  fundamentals of, 453–460, *454*, *456*,
    458–459
  scientific, 453–454, *454*
  writer's block, 458, *458*

XAI. *See* explainable AI

Yosinski, J., 244

Zissman, Marc, 288